THE BEST OF
FLYING

THE
BEST
OF

FLYING
FLYING
FLYING
YING

Compiled by the Editors of *Flying* Magazine

VNR VAN NOSTRAND REINHOLD COMPANY
NEW YORK CINCINNATI TORONTO LONDON MELBOURNE

Photo Credits

Flying Photofile 2, 4, 13, 19, 21, 23, 24, 25, 26, 29, 39, 41, 44, 45, 48, 49, 57, 59, 61, 62, 63, 78-79, 98, 99, 100, 101, 107, 110, 115, 119, 147, 150, 155, 156, 163-167, 171, 173, 181, 183, 200, 214, 233, 235, 239, 241, 244, 255, 257, 265, 269, 277, 282, 291, 293, 296, 300-301, 304-305.

Acme Photo 112, 157, 298.

American Airlines 228.

Joyce Arons 91.

H. M. Benner 96.

British Air Ministry 175.

Mike Charles 246-247, 250.

Chicago International 274.

Consolidated Vultee Aircraft 264.

Art Davis 143.

Downie and Associates 4.

Experimental Aircraft Association 194-195.

William Flynn 223.

Peter Garrison 92, 205.

James Gilbert 5, 198, 201.

Hans Groenhoff 66, 221, 227, 272, 349.

Halsman 154.

International News Photo 169.

Japanese American News Photos 179.

Kaman Aircraft Corporation 191.

Published in 1977 by Van Nostrand Reinhold Company
A division of Litton Educational Publishing, Inc.
450 West 33rd Street, New York, NY 10001, U.S.A.

Van Nostrand Reinhold Limited
1410 Birchmount Road
Scarborough, Ontario M1P 2E7, Canada

Van Nostrand Reinhold Australia Pty. Ltd.
17 Queen Street
Mitcham, Victoria 3132, Australia

Van Nostrand Reinhold Company Limited
Molly Millars Lane
Wokingham, Berkshire, England

16 15 14 13 12 11 10 9 8 7 6 5 4 3 2 1

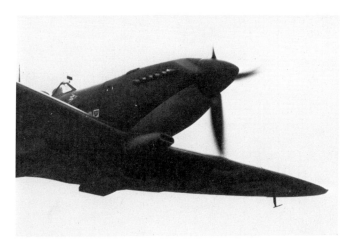

Preface

The mission: to gather together all the material that a monthly magazine has published over half a century and distill it into a single book that will convey the essence of the publication in content, thrust, and quality.

There have been easier missions. However, for all its difficulty and the agonizing that accompanied it, it has been particularly pleasant at the same time. The task has given us an opportunity to look back at where we and aviation have been and what we have done. It makes for a heady experience, which we wish to share with you.

As Bob Parke, who was Editor and Publisher of *Flying* for more than 12 years, describes in his introduction to this book, we proceeded by reviewing everything and noting those items we felt would best represent the magazine in its various periods and stages of development. We were, of course, overwhelmed by the magnitude of our selection. A basic strategic decision—a "painful reappraisal," as people in Government might call it—was made in order to render the material more workable, given the severe space limitations we faced. As a result, we have emphasized feature articles, unfortunately at the expense of the magazine's many fine ongoing columns and departments. Furthermore, we've sought to provide the best or most provocative or representative stories during each stage of *Flying*'s evolution, so as to provide the flavor of the 1920s and 1930s, the War Years, the postwar doldrums, and so on.

In providing this essence, we have in several cases photographed titles and pictures directly from the magazine. Most of the captions are those that graced the original articles. In other cases, we've updated, so to speak, the pictures and captions. In all cases we've sought to be true to the spirit and effect of each story. This has often meant retaining the type and spelling styles of the original articles. *Flying*'s style and "look" have gone through many changes over the years, and we have allowed this book to reflect that process of change.

What we've sought is good balance and mix and a hefty dose of that all-important "flavor of the times." We hope that what we've had to leave out of this book is more than made up for by what we've included and that, if the "flavor" turns out bittersweet, it will be enjoyable enough to cause our readers to call for a second helping.

The Editors, *Flying* Magazine
June 1977

CONTENTS

149 PART IV: **Air Power**

211 PART V: **Airways**

INTRODUCTION

BY ROBERT B. PARKE

Editorial Director and Publisher, *Flying* Magazine

Flying's first name, *Popular Aviation*, surely represented in the mind of the publisher, William B. Ziff, a wish more than a reality. He was a dedicated man, and although he had been involved in aviation in the years prior to Lindbergh's Paris flight, he, like a good many others, was caught up in the excitement and opportunity that the Lindbergh era seemed to present.

Aviation was not popular in 1927 in the sense that a great many people were flying. Nor did many people have confidence in the future of aviation, either as a commercial or popular activity and there was certainly very little attachment to the airplane on the part of the military. While Ziff believed fervently in the future of air transportation and exuded confidence in the proliferation of the "small airplane," it was in the latter category that he felt deeply that aviation was being short-changed. He was an ardent follower of General Billy Mitchell and an eager advocate of air power, and he never lost an opportunity to point to the woeful lack of vision on the part of the United States Air Service compared with that of Germany.

All this was reflected in the fledgling *Popular Aviation* and later in *Flying*. During the 30s, as war seemed inevitable, *Flying* veered sharply in the direction of drawing attention to military aircraft and tactics of both the United States and European countries. While giving commercial and private flying their due, *Flying* had a strong and recognizable military bent that carried through strongly in the war years and to a lesser extent in the half-dozen years that followed.

If military coverage became somewhat muted in the postwar years, this was certainly due to the apparent boom of private flying, which was becoming known as general aviation, and the surge of activity in the commercial airline field. But by far the largest number of advertisers in the postwar years were from general-aviation manufacturers, flying schools, and a variety of service organizations.

While the immediate postwar years seemed filled with promise for general aviation as thousands of veterans learned to fly using GI Bill benefits, few were in a position to buy airplanes with their own money. In the short span of three years, general aviation went effectively from a booming, confident industry that was sure its success was going to parallel that of the automobile industry to a shaken, shrunken group of three airplane manufacturers and a dozen major equipment and service groups. *Flying*'s fortunes plummeted along with those of its advertisers. By the early 1950s, the magazine was in trouble, and there was talk of folding it. It staggered on, however, largely because it had been the first magazine in what had now become a moderately successful Ziff-Davis Publishing Company, and, frankly, William Ziff hated to close it down.

In what may have been a last desperate effort to find a formula that could revive the ailing property, Ziff sought out aviation writer Gill Robb Wilson and urged him to come on as editor and publisher. Wilson, an ex-minister of the gospel, a veteran of the Lafayette Escadrille, lately State Director of Aviation in the state of New Jersey and aviation writer for the New York *Herald Tribune*, a sometime poet and a powerful speaker, was also a patsy for a challenge, and he accepted. He increased the subscription rates and thereby lowered circulation and reduced advertising rates. He cut the staff unmercifully and called on his army of friends to write for him for nothing or a modest $25 a story. He induced his industry buddies to support his effort with advertising, and in a matter of a few years, *Flying* was in the black and ready to grow. The magazine has been growing ever since.

The untimely death of Bill Ziff, Sr. left the publishing company rudderless for a short time before William Ziff, Jr. decided to join the organization. Although he was quite young and had had minimal business experience, he brought a quick and open mind to the job as well as an astonishing ability to work practically nonstop for weeks at a time as he learned not only about the magazines he published but about the publishing business itself. His success is certainly one of the strongest endorsements for on-the-job training.

While many of the magazines in the company underwent substantial redirection under the guidance of Ziff, Jr., *Flying* under Gill Robb went serenely on. Circulation grew, and advertiser confidence in the vitality of the magazine won it healthy support. The editorial product was a strong mix of adventure, history, flying technique, and reports on airplanes and equipment—military, commercial, and general aviation—all flavored with the unabashed romanticism that Gill Robb Wilson identified with aviation, or, as he loved to call it, the airman's world.

In celebration of the pilots, the airplanes, and the air ocean in which they lived, Gill composed a series of verses appropriately titled *The Airman's World*, which were published in book form in 1957. To this day, *Flying* gets requests from devoted readers for this treasured volume, which has long since been out of print. For many, the high point of *Flying* was the little more than that decade when Gill Robb Wilson was in every sense the poet laureate of aviation, Mr. Flying and Editor-in-chief of *Flying* Magazine.

With Gill's retirement, *Flying*, recognizing the increasing utility of general aviation, sharpened its focus on this aspect of the airman's world. Service features and product evaluations dominated the editorial mix, leavened by colorful photo essays of antiques, World War II aircraft or current military jets.

In many respects, *Flying* has been the father of general aviation, as it is today. Starting from the days of its origins, *Flying* has fostered the growth and encouraged the development of the "little guy." Along with other aviation publications and aviation organizations, *Flying* has done its share to upbraid, denounce, cajole, ridicule, demand of, and command the powers and bureaucracies that have from time to time attempted to slight the potential of general aviation. And, if we may say so, *Flying* has invariably done this, as well as all its distractions and duties, with class.

From the time when general aviation could be taken seriously as an avocation, a serious and practical means of transportation, a hobby or a profession, *Flying* has tried to find the most colorful and skilled writers to reflect the moods and realities of general aviation's coming of age. Time and again we have had stories and articles reprinted and we have had readers write to admire and enthuse about stories and articles and authors and layouts and style. It is gratifying.

It is also what prompts *The Best of Flying*. There have been so many stories we treasure and so many articles that have clearly moved and stimulated many readers that selecting the best for everyone is of course impossible. What we have done is what you would do: made a list of the pieces we liked best, found it ten times as much as a book could hold, and then cut it down, trying to provide a balance and a mix that at least could be representative of what *Flying* was and is and perhaps will be.

We've come to a full fifty years, middle age. For most of that time we have carried that little line under our *Flying* logo—"world's most widely read aviation magazine." It doesn't mean much when you're sitting in Dubuque or Dallas or New York until business or pleasure or chance takes you to some remote corner of the world and you wander out to the airport to look around. Like as not, you'll find to your surprise and comfort, a copy of *Flying*, dog-eared, tattered and months old, and best of all you'll find people who have read it and know you better and easier because of that. After all, he, too, is a citizen of the airman's world.

PART I
Prologue

Popular Aviation was created in an age of euphoria, as "The Outstanding Feat of the Ages," included in this section, indicates. The point of the magazine was clearly to sell aviation to the general public, and Lindbergh's successful attempt to cross the Atlantic alone was the perfect springboard, but by the time the magazine first reached the public, other transatlantic flights had been accomplished and a great barrier to aerial progress seemed to have been removed. The airplane was now more widely seen as reliable. Where the emphasis of interest had been mainly on aviation as a weapon of war, a form of show business and *perhaps* a means of transportation, the thrust now shifted to practicality.

Yet the journalistic imperatives still governed *Popular Aviation*—to inform and to entertain, in whichever order would assure success. That very first feature article, "The Outstanding Feat of the Ages," was guaranteed to do both, but to the discerning eye, there was also drama to be found in "Forecasts and Prophecies." The visionaries were destined to be right—both for good and ill.

Popularizing Aviation

*T*HIS is a changing world. The toy of yesterday becomes the necessity of today. The miracle of today is the commonplace of tomorrow.

When young Charles Lindbergh sailed over the blue Atlantic as though it were a trip across an inland lake he was more than a courageous, adventuring American boy defying fate and imagination. He became a standard bearer.

A new and more intense civilization is rushing in on the world out of the great unknown. Just as the wildest dream of Jules Verne could not encompass the achievements of the past few decades, so no living man can surely visualize the commanding future!

The entire complexion of civilization is changing.

Aviation will be one of the largest factors in this change.

Transportation of goods and men will be via the air.

It is not too much to predict that within twenty years the developments will be so great as to make the vast automobile industry look like a minor achievement in wealth and power.

Flying craft will dot the air.

The air is free—no rails—no streets.

Men flying several hundred miles per hour will work in the cities and live in the mountains and hills and pleasant places far away.

It sounds weird, but that day is here now. A few more years and it will seem commonplace. Twenty years ago a dangerous experiment—today, a flight from New York to Paris, or Berlin, in less than two days.

To countless thousands of intelligent and able young men of today the business of the air will bring wealth and achievement. The inventor, the explorer, the capitalist, the investor—all will enter into this vast program.

Popular Aviation is issued secure in the knowledge that these things are not a dream.

It means to place in the hands of the young men of the nation the facts of this new great human adventure. Its columns will be easy to read, but authoritative; instructive, but written so that all can understand.

If it will help to speed the great day that is coming and bring incentive to wealth and power to these young men whose will to achieve is ushering in a new era, we will feel that we have also contributed something to the accomplishment.

Any extended story of the record breaking flight of Charles Lindbergh comes, at this time, as anti-climax. Never before has any event been given the newspaper and magazine space accorded this deed. POPULAR AVIATION has, accordingly, merely touched the high places of this adventure so that those who wish to keep their copies of POPULAR AVIATION may have a permanent record of this accomplishment.

The Outstanding
Feat of the Ages

Lindbergh inspecting his Ryan monoplane, The Spirit of St. Louis, at Le Bourget, near Paris, the day after his flight.

A BLASE world has been thrilled. A smug 20th century has been aroused out of its phlegm to pay homage to a 25-year-old flier who spanned the 3,600 miles between New York and Paris in a single hop.

Colonel Charles A. Lindbergh (to give him his present title) took off from San Diego, California, at four o'clock on the morning of May 10, 1927. He was flying his new Ryan monoplane, "The Spirit of St. Louis," provided by the business men of St. Louis, Mo., and in which he had invested his savings. The Ryan plane was engined with a Wright air cooled motor, and was fitted with the best and latest aeronautical instruments, but stripped of all unnecessary weight to provide extra gasoline capacity.

Stopping at St. Louis a day to pay his respects to his backers and for a brief visit to his home town, he left for New York on the morning of May 12, arriving in New York the same day. It seems incidental, in view of his greater triumph, to mention the fact that his transcontinental time of 21 hours and 45 minutes was also a record, being the fastest time in which this continent has been spanned.

In New York three planes were biding their time in anticipation of flight to Paris. Weather conditions were never quite right, and the flyers waited so that as many factors of safety as possible might favor them. For days the sporting world waited for news of some action. None came. The delay may have taken the flyers off the front pages of the newspapers, but it did not decrease the interest of the great public in the proposed flight. Crowds were in constant attendance at the field.

After days of inaction the country was electrified by the news that Lindbergh, playing a lone hand, had left for Paris. No beating of drums, no crowds of spectators, no cordons of police—just a calm young man off for another flight. At 7:52 in the morning, May 20, 1927, the Spirit of St. Louis left Roosevelt Field.

In order that the layman who has never been in a plane can understand and appreciate the problems which Lindbergh had to confront on his spectacular flight, it might not be amiss to picture the dashboard of the Ryan monoplane. The reader will then realize that Lindbergh had to watch and understand each record given by the instruments. Neglect, a moment's inattention probably spelled Death.

First of all the altimeter, which gives a constant record of the plane's height above sea level. At night this must be watched every moment in case the plane falls imperceptibly and drops into the waves.

The compass, of the earth inductor type, whose needle showed every variation from the line of flight determined upon. This compass is so arranged that a course is set by means of one dial, just as a station is tuned in by the turn of a knob on a radio, while a pointer on another dial shows even the slightest turn to left or right of the set course.

The turn and bank indicator, showing at a glance the position of the plane, with reference to the earth's surface. Without it, at night particularly, or in clouds, an aviator might not be flying horizontally and not know it. The turn and bank indicator and the earth inductor compass are used together in directional flight.

The oil gauge on the dash shows the amount of oil left in the tank, the oil pressure gauge shows that the oil pump is supplying the motor with ample oil, and the oil thermometer showing the temperature of the oil in the motor. Each one of these had to be watched carefully for indications of trouble.

The tachometer records the motor revolutions, a drop indicating trouble. To obviate the most frequent cause of trouble, in the gasoline line, Lindbergh had two gas lines installed, either of which could carry the full load of the motor. In case of trouble with one it could be turned off and the other could be used.

The air speed indicator shows the velocity of the wind. A comparison of the tachometer and the air speed indicator will immediately tell the pilot how fast he is going and how much headway he is losing against the wind.

In addition to these instruments there were the usual ignition switch, carburetor choke and throttle.

Lindbergh made no preparations to take sextant observations, but relied on what navigators call "dead reckoning." This was characterized by Allen Cobham, the English aviator, as a wonderful feat in itself. Lindbergh had carefully checked the accuracy of his new instruments and had made the most careful preparations. As he said before leaving New York, "People forget that I will be able to check my course by the map all the way from New York to Newfoundland. Even at night I could do that, for unless the night is very dark it is possible to distinguish water from land. The most difficult hours of the flight, the first fourteen, will be over land most of the way, and if I can get a good sight of the coast of Newfoundland as I leave it I should be able to tell my drift and get my bearings with some degree of accuracy."

An interested nation listened to radio reports the whole day, and the flight was followed in anxious suspense by hundreds of thousands of people who had never heard this young man's name before that day.

After taking off, Lindbergh was soon out of sight of land. From Cape Cod to Nova Scotia the route lay over open sea. The motor performed to perfection, and the plane carried the load of 451 gallons of gasoline and 20 gallons of oil. This fuel gave the ship the greatest cruising radius of any plane of its type.

At St. John's our flyer went out of his way a few miles to check on his course and flew low so that his location could be definitely established. St. John's was his last sight of land before taking to the open sea.

In spite of the fact that he was enclosed in his plane and required a periscope in order to see straight ahead, Lindbergh had sufficient view through the side windows to navigate with ease.

After leaving Newfoundland he began to see icebergs through the low fog. He struck clouds and ascended to an altitude of 10,000 feet and remained at that height until early morning. In his own words, "It was like driving an auto over a smooth road, only it was easier."

As the morning light became somewhat stronger the clouds increased and he was forced to fly under some and over others. Sleet in all of these clouds began to cling to the plane. Sleet is the bane of the aviator, as it adds to weight and cuts down the efficiency of the plane's lifting surfaces. There was no remedy for the situation, so he continued on his way to Ireland.

Early in the afternoon he sighted a fishing fleet and knew that he was not far from the coast. An hour later he sighted land, shortly before 4 o'clock, a point on the Irish coast, as he afterward discovered, only three miles from the theoretic point chosen. This in itself is a remarkable proof of his navigating abilities.

Flying low over Ireland, he crossed to England, also flying low, crossed the English Channel at a greater elevation and passed somewhat west of Cherbourg.

With victory in his grasp, he followed the Seine until Paris, with the brilliantly lighted Eiffel Tower, lay under him. When only a half hour from Paris rockets and Very lights were sent up from the flying field, and he had an inkling of the attention he was attracting, he circled Paris once and flew over LeBourget flying field, but, mistaking its direction from Paris, he again flew back to the city, looking for another air field. Not finding one, he knew that he had been right the first time and returned to LeBourget, where he stepped into one of the most uproarious and crushing celebrations that the French capital had ever seen.

He was all but overwhelmed by a sea of humanity that swept into the field to greet him and see the victorious plane. Only the prompt action of the airport authorities, aided by platoons of gendarmes, prevented the Spirit of St. Louis from being torn to pieces by souvenir hunters.

His first words on landing were typical of the modesty of this youth who has captured the hearts of not only his countrymen but of all the foreign nations who have seen him. No blatant boasting of his deed, no facetious statement of triumph, nothing but a mere statement of identification—"I am Charles Lindbergh."

With a foretaste of celebrations that were to accompany him for weeks, he was brought to the United States Embassy as a guest of Ambassador Herrick, who hailed him as a flying ambassador of good will for the United States.

The cheers and adulation which he received in Paris, and subsequently in Brussels and London, are old stories by now. It is sufficient to say that the welcome given him and the encomiums bestowed upon him were induced as much by his own conduct as by his feat.

To make the New York to Paris flight alone required daring and knowledge. To win the love of a nation, depressed by the tragic failure of one of their own admired flyers to accomplish the same deed, required something more—diplomacy.

A Washington statesman said that a better emissary could not have been chosen for the purpose of sustaining friendly relations had a deliberate search been made. His modesty and simplicity, entire absence of vain-glory, and the little sentimental touches, which the French love, were all there. The visit to Nungesser's mother, visits to memorials dear to the French heart, attendance at benefits—all these were the acts of an unspoiled hero.

The heads of three governments bestowed decorations on this youth and he was showered with offers aggregating millions of dollars. These did not turn his head. Declining to consider offers unrelated to the business of aviation and wisely seeing that time would not permit him to visit every European capital, he prepared to cut his visit short.

A message from President Coolidge, through a cabinet committee, asked the flying hero to be the guest of the nation at the capital immediately on his return, and placed at his disposal the cruiser Memphis, the flagship of the American fleet then in European waters.

Leaving his European admirers behind, and disappointing thousands who had hoped to see him, Lindbergh sailed for the United States on the Memphis. He was met near Cape Henry by an escort of destroyers and planes. On landing he was first met by his mother, the quiet teacher of chemistry in a Detroit school, so envied by hundreds of thousands of mothers.

Stepping into a plane, he was piloted to Washington, where he was received by the official committee and decorated by the President. Thousands of people poured into Washington June 11 to get a sight of this man.

Here he received the honors he had been awarded and made the brief speeches which are so characteristic of him. Function after function were attended by "Lindy" and his mother, who were the guests of the President at the temporary White House.

His later reception in New York eclipsed anything the city had ever staged. Somewhat tired, but unbewildered by the constant crowds and the prearranged programs, Lindbergh is sacrificing his privacy for the benefit of those whose wish is to see him and, if possible, speak to him. For days, and probably weeks to come, he will be unable to go about the ordinary affairs of life without attracting the attention given to those in the popular eye.

Tons of newspaper clippings, a half million pieces of mail, thousands of telegrams, uncounted gifts, still await the attention of this American, who by a single act of courage and skill drew to himself the loving admiration of many nations.

Lindbergh's welcome in St. Louis and his efforts to catch up with the enormous mail awaiting him were interrupted by a call from Washington. Government officials of the Air Corps, Navy Department and the Bureau of Aeronautics asked his presence for a series of conferences.

To date no indication has come from the Colonel as to what his future plans are to be, except that he has steadily maintained that whatever he does will be connected with aviation. Offers from the vaudeville circuits and the motion pictures apparently do not interest him.

It has been announced by the Guggenheim Foundation that Lindbergh will first make an air tour of the country in the Spirit of St. Louis for the purpose of promoting interest in aviation.

Two moving picture men were talking over the New York welcome to Lindbergh and estimated that to reproduce the sight in the movies would cost somewhere in the neighborhood of $50,000,000.

"We have been putting on gigantic spectacles for years, but have never put over anything like that show of New York for Lindy." One of the men figured that scenery alone would cost about a quarter of a million per block, with the hiring of thousands of extras, planes, police, ships and the like to be added to say nothing of paper to throw out of windows.

Forecasts and Prophecies

For I dipt into the future, far as human eye could see,
Saw the vision of the world, and all the wonders that
would be;
Saw the heavens fill with commerce, argosies of
magic sails
Pilots of thy purple twilight, dropping down with costly bales;
Hear the heavens filled with shouting, and there rained a
ghastly dew,
From the nations' airy navies grappling in the central blue;
Far along the world-wide whisper of the south and
rushing warm,
With the standards of the people plunging through the
thunderstorm.

—Tennyson, Locksley Hall.

"I expect to see regular commercial flights across the Atlantic ocean made swiftly and safely and profitably."—Charles Lindbergh.

"The feat performed by Chamberlin and Levine again convinces us that transatlantic flying is getting to the practicable point. Soon there will be big planes with many engines in a regular ocean crossing service."—Guiseppe M. Bellanca.

"Transoceanic flying can not be a commercial fact until our instruments have been perfected to a degree where such navigation is shorn of its elements of uncertainty. One drifts too easily so far off the course that there is yet no comparison between safe aerial navigation and marine sailings."—Orville Wright.

"Chamberlin's flight to Germany, close on the heels of Lindbergh's air journey to France, has demonstrated that the Atlantic can be crossed by airplane without the use of astronomical observations."—Commander Richard E. Byrd.

"Within the next two years there will be more commercial flying in the United States than in all the rest of the world put together."—Charles L. Lawrence, president, Wright Aeronautical Corporation.

"It seems to me that the first development of transatlantic aviation will be the mail service. Think of the vast amount of important mail which now crosses the ocean in six, seven and eight days. Suppose it could be transported in a day and a half. What would that not be worth to business men on two sides of the Atlantic?"—Charles Lindbergh.

"We have plans for a new airplane which we believe can insure safety in intercontinental traveling."—Charles A. Levine.

"The key to intercontinental flying is simply a better plane. This means more speed and weight capacity in proportion to fuel consumption."—Clarence D. Chamberlin.

"German aviation interests are contemplating a direct transatlantic air route from Europe to the United States following practically the same route taken by the United States world flyers in 1924."—Vice Consul Ellis A. Johnson, Copenhagen, Denmark.

"The transatlantic flights may be expected to have an immediate and favorable influence on the foreign demand for American aircraft."—H. H. Kelly, automotive trade commissioner, United States Department of Com.

"The time is ripe for serious organization of trans-oceanic aviation lines on the basis of the journey being divided into several laps. It is necessary to split the trip into several laps and to await the construction of a new machine capable of greater mobility with a more limited consumption of gasoline."—Francesco de Pinedo.

"It is not too fantastic to say that in the near future regular transatlantic air service will be developed, for it is now only a problem of suitable planes, because the courage of aviators has already solved the problem of man power. Read, Alcock, Coutenho and Cabrai, Franco, De Barros, Lindbergh, Chamberlin and Levine, and De Pinedo are the pioneers who have opened up the skies to civilization."—Musolini.

PART II
High Adventure

In each stage of his aeronautical development, man has entered the skies as something of an innocent. The dangers that pilots have faced, even as their equipment has improved over the years, have invested aviation with an almost mythic quality. People interested in flight have, for the most part, embraced this romantic side of what is usually best done in a scientific, dispassionate manner. Yet how could one dispassionately think of actually flying 200 miles, when the year was 1911 and when the idea of "going up" at all was still strangely new and held an aura of improbability?

With the coming of the 1920s, the romance of flight took on other dimensions. The daring aces of World War I were replaced in the public imagination by the itinerant pilots, men and women who faced injury and even death by barnstorming, attempting previously unheard of feats or simply thrusting themselves into harm's way by routinely flying in dangerous portions of the world. By the 1930s, barnstorming had waned, to be replaced by air racing as the major aviation crowd pleaser.

In the pages of *Flying* during that era, a dual trend could be found. Again, it was a matter of informing and entertaining, of boosting aviation and making it exciting, of nourishing the romantic side—which meant, in part, the dangerous side—and also of being "responsible" about aviation, of pushing it as commercially viable, sound and oriented toward developing safer equipment and procedures. In such a cause, one would find in the pages of the magazine an article by Jimmy Doolittle about testing racing planes and not long after, articles condemning air racing as an especially lethal sport.

The true excitement, the magazine seemed to be saying, lay among the bush pilots and among pioneers such as Ira Eaker as he dared to fly across the United States, *entirely on instruments;* among pilots who voluntarily or perforce "stepped into space" for a parachute jump, as did Max Karant, who, by the way, was destined to be one of the leading writers in aviation.

As aviation developed further, as the open-cockpit environment of helmet and goggles was replaced by relatively comfortable cabins, the romantic side changed; it became more internal. The Editors of *Flying* recognized that however practical aviation might become, to the individual pilot, the act of flying, of meeting the requirements for licenses and ratings are still enormously exciting, as is building one's own airplane, as did Peter Garrison, and then trusting it to take one across an ocean. And as is flying an airplane among looming canyons to set it down into an impossibly small clearing.

Just before World War II, *Flying* came up with a "formula" feature that survives to this day as one of the most popular elements of the magazine, "I Learned About Flying From That." Every aviator has at least one experience in his background, often a dangerous one, that has taught him an important lesson about flying in general and about *his* flying in particular. The magazine has, over the years, published hundreds of unsolicited stories by its readers as part of this feature, some of them by well-known personalities. There has also been a touch of the prophetic to the series, as betokened by the appearance of a small piece by an airline captain named Ernest Gann, in 1942, and of one by an Air Force officer named Robert B. Parke, in 1953. Both of these writers were to figure prominently in the future of *Flying.*

WINNING $10,000

Leave the crowded airways and glide back to 1911 and the first cross country air race. An article written after the race by the man who won it.

BY EARLE OVINGTON

NO LESS THAN $10,000 in three hours and six minutes! Seems like easy money, doesn't it. And yet I earned every cent of that $10,000.

The flying men of today (1911) earn, in most cases, far more than they receive. An aviator, at the control of a flying machine, is an acrobat balancing his frail craft upon the invisible currents of the air, just as a tight-rope walker balances himself upon his narrow support. But there is this great difference between the two: if the aviator makes one wrong move he pays with his life for his mistake, for he has no net to catch him as he falls earthward—and usually it's a long way down.

During the aviation meet held at Squantum, Mass. in September 1911, the Boston Globe offered $10,000 to the winning aviator in the first cross-country aeroplane race ever held in America. The course was to be from Boston, Mass. to Nashua, N. H. to Worcester, Mass. to Providence, R. I. and back again to Boston.

There were only nine aviators at this meet: Captain Thomas Baldwin; Harry Atwood; Tom Sopwith, the English aviator; Lincoln Beachy, the reckless daredevil flyer; Eugene Ely, his pal; Lt. T. D. Milling; Arthur Stone; Grahame-White, another English aviator and, of course, myself.

Before entering the race, we decided to first go over the course to find out the condition of the country over which we were to fly. We would go in our automobiles, to get an idea of what we would have to land on in case our motors should conk out in mid-air.

Between Boston and Nashua we found a good many open fields, which would make fairly safe emergency landing places, but between Nashua and Worcester there was nothing but dense woods and thickly populated towns. We all decided that if our motors quit in the Nashua-Worcester leg of the race, we would probably have to land in a tree, or on somebody's housetop.

The condition of the country between Worcester and Providence was better, but still dangerous from an aviator's standpoint; whereas between Providence and Boston there were a good many landing places. We figured that if we got as far as Providence we had a fair chance of finishing the race.

After going over the course in our cars, and seeing how dangerous it was, at the final count there were only four entrants: Harry Atwood, in his Burgess-Wright biplane; Lt. T. D. Milling in a machine of the same make; Arthur Stone in a Queen monoplane; and myself in my Bleriot monoplane, which I called The Dragonfly. The others decided that it was too risky and refused to participate.

Finally the day came—the day of the great race—Labor Day, September 4th, 1911. It dawned clear and bright, the wind blowing not more than 15 miles an hour—ideal for the contest that was to take place.

Long before the announced starting time people began to gather in the grandstand and on the field. Great was the interest in this, the first cross-country aeroplane race ever held in America.

The four contestants waited for the starting bomb, two of us in monoplanes, the other two in biplanes.

I climbed into my seat, donned my safetybelt, and strapped myself into position. I got this strapping habit abroad, as all foreign aviators, in case of accident, believe in sticking with the machine.

Alfred, my chief mechanic, started the engine, and its seven cylinders roared out reassuringly, while the big propeller stirred up a hurricane in the rear of the machine. Six strong men were required to hold the plane to the ground. They dug their feet into the earth and pulled with all their might as my engine rapidly warmed up to its work.

All was in perfect running order so, at the sound of the starting bomb, I raised my hand as a signal for the men to let go.

The other three aviators took the air at the same time.

Across the field shot my plane, like an arrow from a bow,

Above: The author's Dragonfly, which flew to victory in America's first cross-country air race. Left: Ovington lifted to shoulder of aviator Grahame-White after completing successful landing to win race.

gathering speed at every instant. About 200 feet from my starting point I pulled my elevator control toward me, and rose into the air. Up, up I went, climbing the long invisible hill, and rapidly approaching the city of Boston. At 600 feet I was over the harbor, and my altimeter registered just 1,000 feet as I passed over the State House dome.

Soon I was sailing far above the Charles River and its bridges.

I continued to climb until my altimeter showed me that I was a mile above the earth. My speedometer told me how fast I was traveling, and my map was drawn to scale. I figured, therefore, that I should be in the neighborhood of Lowell and Lawrence at about this time, and sure enough, there were the two cities a mile below me. I knew Nashua was just beyond, so I had no difficulty in locating it, and landed there without any trouble.

My reception was a warm one, since I was the first aviator in the race to get there. The other monoplane, I was told, as well as one of the biplanes had already gone to earth, and their pilots had given up the contest, before even reaching the first stop. The remaining biplane, flown by Lt. Milling was following in my wake, but at a much slower speed.

After stopping for only a few minutes at Nashua, I proceeded on my way. Rising to a height of about a mile I steered directly for Worcester. At the end of three-quarters of an hour I knew that it should be in sight, and straight ahead of me I saw it. I had just passed over the reservoir at Clinton when I recognized the twin spires of the Union Station.

At Worcester I had much the same reception that I had had at Nashua, except that it was even more enthusiastic.

When the time came to leave, I found that the tiny aerodrome, surrounded by tall buildings and trees, was never meant for a flying field. How I ever got out of there with my monoplane will always be a mystery to me.

Once in the air I rose rapidly to 6,000 feet and headed for Providence, which was comparatively easy to locate from the air, because of the contour of the coast. And in 40 minutes from the time I left Worcester, I was in sight of Narragansett Pier in Providence.

Looking down from my height in the air, I knew from the size of the black ring around the grass oval where I was to land that there was an unusually large crowd waiting for me. I cut off my motor and made a straight glide to earth, landing so lightly that I didn't even feel the wheels touch the ground.

I have seen enthusiastic crowds before, but I have never seen anything to equal the crowd at Providence. I thought more than once that they would tear me to pieces in their joy at seeing a real live aviator at close range.

I had barely gotten out of my plane, leaving it in the care of Alfred, before the governor of Rhode Island, the mayor of Providence, and several other officials, came forward. They gave me a hearty welcome and we made our way toward the grandstand.

As we passed through the surging mass of people, some of them came up and touched my coat, just to say that they had done so. You may laugh, but this is actually what happened. They treated me as though I were a being from another world. I felt like the Man from Mars.

Two kitchen chairs were now brought forward, one for the governor, and one for me. We must have made a strange pair. There I sat, after being over two hours in the air dressed in flannel shirt and knickerbockers, and pretty much begrimed with oil and dust. And there beside me sat the jolly little governor, his manner and dress most formal, complete with an official smile and a high silk hat. Behind him stood the mayor and several military men of rank, in full uniform. Tired and dirty as I was, these men hung on every word I said, as I related my experiences from the time I left Boston.

But the race was still on, and already I had spent far too much time here, so I said goodbye and gave my attention to my aeroplane.

"How is the Dragonfly, Alfred?" I asked my chief mechanic, although I knew from his broad grin that everything must be in perfect condition.

He simply looked at me and smiled, twirling his big mustache as only a Frenchman can.

Then I turned to the police officer, told him I had to leave, and to please clear the way in front of me.

He started out on his magnificent black horse, as if he really expected to do so, but I soon saw that it would be impossible for him to do anything with the crowd.

I then said to one of the officials, "Let me have your megaphone for a minute, will you?" He handed it to me and I stood up in the chair.

As soon as the crowd saw that I was going to speak to them they quieted down.

Through the megaphone I told them how important it was for me to get away as soon as possible, and how necessary it was for me to have plenty of clear space from which to start.

"I've got to get back within the hour," I shouted, "in order to win the prize. And that $10,000 looks mighty good to me."

They fell back on both sides and in front of me, leaving a narrow lane about 200 feet long, and wide enough to give me about ten feet on each side of my machine.

Placing the megaphone again to my lips, I begged for more room. It still didn't work. I decided that I would have to appeal to their reason.

"If this 70-horse-power engine of mine, driving this eight-foot four-inch propeller and running at 1,300 revolutions a minute should get mixed up with you people, it would go through your ranks like a rotary snowplow when it clears the railroad tracks, and I don't want that to happen to you."

They fell back, perhaps two feet more on each side of me. My lane was still far too narrow and much too short, but I saw that it would be impossible for me to get more space.

Should I take the chance? I figured quickly. I knew what my engine would do, and decided to risk it.

Again I turned to Alfred, and asked him if everything was in perfect order, as the big propeller would have to run at full horsepower in order to raise the plane from so small a starting ground.

"Ze moteur is parfait, monsieur," he replied promptly.

I strapped myself to my seat, put my heavy helmet in place, adjusted my goggles, turned on the gasoline, and signalled for Alfred to spin the big propeller. Half a dozen good sized men were holding the tail of my machine until I raised my hand as a signal to let go.

"Good luck, monsieur," shouted Alfred, as he threw his weight upon the big propeller, while the surrounding crowd echoed the cry, "Good luck! Good luck! Good luck!"

The motor started with its reassuring roar. On each side of me I could see the densely packed throng, and my muscles stiffened as I realized how accurately I must steer my machine to void plowing into this living mass, not a dozen feet away from either side of my wing-tips, and only 200 feet from in front of the machine.

Down I bore on them, the motor roaring in my ears, at a speed of nearly a mile a minute. One hundred feet now separated me from this human wall. Fifty feet. Forty feet. Would the plane *never* rise? Thirty feet. Twenty feet. I glanced at my speedometer and saw that at last I had sufficient speed to pull my elevator quickly toward me and send the machine into the air—*not five feet above their heads.*

Five feet had separated these people and me from certain death.

I was more than a hundred feet above them before my wildly beating heart slowed down to its regular, even tempo. I breathed a deep sigh of relief as I realized that at last I was in the air: that the danger to them was over; and that I was still alive.

When I finally straightened out my course for Boston, my altimeter showed me that I had risen to over a thousand feet. I climbed until I was more than a mile above the earth, and kept on rising until I was at an altitude of 8,000 feet.

After traveling for some time, a glance at my clock told me that I was almost at Squantum. Without difficulty I found the aerodrome.

I cut off my engine and started the long, downward volplane to earth. As my landing wheels touched the ground, the crowd rushed toward me, and the band struck up The Star Spangled Banner.

Once again I found myself in the midst of a chaotic mob—newspaper men, officials, aviators, and photographers—all gone mad in the excitement of the moment. But I turned away from them all until I had greeted my wife. I reached down and swept her into my arms, and felt her tremble as I did so. Poor little girl. What a strain she must have been under that day!

As I jumped from The Dragonfly I was seized by my fellow aviators and carried triumphantly on their shoulders around the field. The wild shouts and cheers from the crowd were deafening.

I had covered 186 miles as the crow flies in three hours and six minutes, and won the Boston Globe's prize of $10,000.

The Vanishing Gypsy

Immediately after the war the aircraft industry was in a more or less chaotic state. Production had been speeded up for war needs, and the capacity of the various factories making aircraft and all pertaining to them was far in excess of a regular peace time demand.

Used planes and surplus planes were a drug on the market. The country was full of pilotless planes and planeless pilots who had been trained in the hard school of war.

Here and there a venturesome spirit turned to the air as a means of livelihood. The people as a whole were interested in flying and air circuses became an attraction at fairs and carnivals.

The Gypsy Flyer soon became a more or less familiar figure. A fair held adjoining a good road, any kind of a crate that would fly, a come-on sign and a pole with a sock on it—enough equipment for anyone. Rides at $25.00, at $10.00, at $5.00— for any sum to pay for gas, oil and the absolutely necessary repairs.

But no one can laugh at these Gypsies. They were the men who were keeping flying before the public, bridging the gap between war flight and the coming commercial and pleasure use of the plane. They initiated thousands of people into flight, and they implanted the flying germ in many of the younger pilots of today.

Trained in the rough school of war, ingenious, used to taking chances, these men barnstormed about the country. As they gradually drifted away from the precarious business of earning a living by flying in exhibitions and running up greenhorns for thrills, they were succeeded by the younger generation of airmen.

The new crop had not had the benefit of war work. Also, they were not under some of the disadvantages of that type of training. The country was full of youngsters anxious to fly for the thrill and romance of it. Here was a coming business that would be fun, fun mixed with spice and with little resemblance to work. Once in a while would be found a serious youth or an old-timer, still in the business, that had a true vision of the future of flying—of the possibilities of air transport.

There was no Safety Code in existence. Traffic rules for the air were a matter of common practice only, but no one had formulated them. There were no regulations as to use of fields, nothing to prevent a pilot from taking up an utterly unfit ship.

With the safety regulations existing today, involving rules covering the structure, design and fabrication of aircraft, specifications, design, assembly and tests for powerplants and equipment, and qualifications for aviators, it seems almost incredible that so short a time ago the business of flying was carried on in such a hodge-podge fashion.

Here are some excerpts from the story of a flyer, still a young man, who has gone through the mill from the stunt-flying, gypsy period down to the present day of regular air lines, scheduled traffic, thorough schools and rigidly enforced air regulations.

No one can read of these experiences without feeling some sort of a thrill. The old-timer's reaction will be different from that of the casual reader. He will be able to appreciate the chances taken and the "percentage."

"In February, 1922, I bought an old dilapidated Jenny at Waseca, Minn., for $600. "Speed" Holman, the winner of the Stunt Event at the St. Louis Air Meet, was to fly it to Chicago for me, and arrived at the Checkerboard Field, west of Chicago, with no trouble. In making the twelve mile hop to the Ashburn Field at Chicago, he ran into a snow storm and missed the field. The men at Ashburn heard him fly over the field, but could not see him next morning and it was assumed that he had been forced to land in the lake, and was lost.

"As a matter of fact he had descended at Dwyer, Indiana, and had neglected to telephone his whereabouts. He appeared at the field just as the newspaper reporters were talking to me about his loss in the cold waters of Lake Michigan.

"I took a couple of hours instruction at Ashburn and then had the ship brought to a field on the north side of Chicago, nearer my home. I rented it to a pilot who used it for passenger carrying at Deerfield, but one day he took the telegraph wires at the end of the field down with my Jenny and she had to be rebuilt completely.

"I put in about twelve hours solo time and then traded the Jenny for a Thomas Morse scout. The Tommy was a faster plane than the Jenny and I expected to install dual control in it as I would need more training with an instructor before I could really fly the Tommy. It took so long for them to get at the job that I finally got tired of waiting and took the Tommy out and managed to fly it without too much trouble.

"One day, in landing at Paducah, Ky., I broke the four rear longerons in trying to get into a small field. As it would take about two weeks to get the necessary supplies to Paducah so that I could make the splices on the broken longerons, I got eight pieces of scrap iron and tied them to the broken longerons with hay wire.

"The longerons are the four pieces of steel tubing, or wood, that form the main support and structure of the fuselage. Should they break the control surfaces of the plane would fall off and the plane would become unmanageable and fall to the ground.

"Beginner's luck held good and I reached St. Louis just at dark. An aircraft company at St. Louis spliced the longerons for me and I took off for Chicago. Outside of running into a bad fog and being forced down on account of a broken throttle rod I reached Chicago without a mishap.

"The Tommy had been eating money up and, bringing nothing in, which was not my idea of aviation, so I decided to trade it off for a machine that would hold more than one passenger and would not need such a large field in which to land and take off. I finally got in touch with a man in Pittsburgh and arranged to trade with him for an 0X5 Standard.

"Outside of breaking a cylinder stud the trip to Pittsburgh came off all right. When the stud broke and let the water out of the engine I was in sight of the field at Cleveland and was just able to glide into it. Glenn Martin was there and through his courtesy one of his mechanics made the repair for me. Half way to Pittsburgh I ran into a snow storm and had to wander around for two hours looking for the aviation field. The gas gave out and I had to land two miles from the field. I made the trade for the Standard and flew it back to Chicago without any trouble.

"We had tried carrying passengers for joy rides in the small towns around Chicago, but did not have much success as we had no way to attract people to the field. Finally I bought an exhibition parachute. This form of chute is packed in a big bag strapped to one of the wing struts. The jumper climbs out on the wing to the strut and then straps on the parachute harness.

When he arrives in a position from which he believes he can drift into the field he pulls the cord at the mouth of the bag, the chute opens, fills and drags him off the wing.

"We put out several thousand circulars announcing a parachute drop from 2,000 feet at Winnetka, Ill. We had only one plane, but there were enough people on the field to watch the event to net us $200 for the evening. One of the mechanics had volunteered to make the jump, so I tied the parachute bag to the wing strut and took off. When I motioned the mechanic to get up on the wing (we were up 2,000 feet) he started to do it, but the blast of the propeller struck him with such force that he got cold feet.

"I circled the field several times, trying to get him out on the wing to make the jump, but he wouldn't do it, so I flew the plane over to another field and took on a pilot. I was determined not to disappoint the crowd and decided to make the jump myself. All I remember about it is that I was so busy thinking about how the harness should be strapped on, and how I was to climb out on the wing that I didn't have time to be afraid. The parachute opened perfectly and I landed in a plowed field a short distance from the landing field. The parachute was a great success in getting people to the field and we used it many succeeding Sundays with good results.

"Then we secured a contract to make a jump to land on the roof of a rug factory on the west side of Chicago. I had a professional jumper hired for the jump, which was to net me $500, and just before the event was scheduled the jumper made a jump for fun and broke his leg. That put him out of the money and I couldn't get any one to agree to try it, so I had to plan on making the jump myself. On the day the jump was advertised we took off in a rain which became so heavy by the time we were up a thousand feet that the motor began to miss. We were just over the center of Chicago when the motor died completely. The pilot was an old-timer, and he side-slipped her almost all the way to the ground, and by ground-looping when he landed we managed to get down safely on the football field at Humboldt Park.

"Next morning a crew from one of the local factories came out to disassemble the plane and haul it to the field. Before starting their work they measured the field and found it to be 800 feet long and talked me into flying the plane out of the field. Fool's luck held and I made it, with a few inches to spare over the top of a three story apartment building.

"I began to receive commercial contracts for flying for advertising purposes. My plane was a billboard and I distributed circulars and samples from the air. One morning I had a chance to talk it over with the judge. I dropped candy samples in the crowd at Stagg Field during a football game between Chicago and Northwestern University. The whole mess brought up the question of flying over cities at a proper height and I was discharged on condition that I warn aviators about the Chicago ordinance which forbids flying over the city at a height of less than 2,000 feet.

"Really, there was actually no danger in the advertising stunt which I pulled off, no danger to the crowd, as I had a landing place within gliding distance even if the motor should cut on me.

"Not long after I got the idea for a jump from a burning

plane. I was in Cocoa, Florida, and had just lost a good contract with a real estate firm. Curly Burns told me how often the trick had been tried, always with a failure. The spectacular part of the stunt appealed to me and I could see how big a drawing card it might be. I lay awake one night figuring out how it might be done and the next morning told Curly that if he would book the stunt I would pull it off for twenty-five percent of the gate receipts.

"Two days later he had the show booked for Cocoa Beach, and besides the drop from the burning plane there was a complete flying circus. I hunted all over Florida for an old plane and finally located an old Jenny at Daytona that could be put into flying shape without too much trouble. I flew it over to Cocoa and took it apart and then set it up again on a vacant lot in the main part of town. Whenever I didn't have anything else to do I climbed into the cockpit and opened the motor wide. It stopped traffic, brought people out of stores and proved pretty good advertising.

"The day before the show I rounded up the aviation enthusiasts and set them to work taking the plane apart. One good thing about aviation, then, was that an aviator with a little diplomacy could get half the town to help when there was any work to do about a ship. A truck took both wings over to the beach and Probst, my mechanic, went over with the fuselage in a small Ford truck.

"The members of the flying circus were discussing newer and better ways to commit suicide than in doing it from a burning plane, and giving me the benefit of their cheering comment, when Probst came to tell me that a big truck had run into the tail surfaces of the Jenny and it wouldn't be in shape to fly the next day.

"There was no material in town to repair the tail surfaces, and Scotty and I began to figure that the main drawing card of the show was off. But the head of the flying circus came to our rescue and sold me the surfaces of a Jenny that he was rebuilding at Orlando. It was only sixty miles away and in ten minutes we had a mechanic on his way to get them. We worked all night on the old crate and at dawn she was ready to fly.

"People came to Cocoa from Jacksonville and Miami—as they said, 'to see a fool commit suicide.' At the last minute the small back-pack parachute that I had wired for failed to arrive and I had to plan on using a cambric exhibition chute. A back-pack chute is made of silk and folds into a small pack which fits your back. With one of them you can jump clear of the ship and pull the chute open when you are clear of the plane.

"The exhibition chute is a big affair and about the last thing to use for a jump of the kind I was to make. It was too big and bulky and there were too many chances of fouling.

"The plane had been fixed so that it could be set on fire by throwing a switch. The electricians had run four sets of wires, one to each wing tip, and each connected with a spark plug that was packed in a small cloth bag filled with powder. A switch to throw the juice of an auto battery, and a small Ford coil completed the apparatus.

"The plane was saturated with twenty-five gallons of gasoline and ten gallons of crude oil had been poured over the wings. Cocoa Beach was packed with people and I had to make the start. The parachute was tied to me with a rope and was so

bulky that I could hardly move. I was helped into the cockpit and gave the Jenny the gun.

"All went well until I had climbed about 1,200 feet. Then the old engine started missing badly. I looked at the beach to see if there was a place for me to land and remedy the trouble, but it was packed with people. There was only one thing to do so I started for the ocean. A half-mile off shore the motor cut completely.

"I had planned to climb to about 3,000 feet and when I was a mile off shore to touch off the plane. Then, when it was burning good, to stand up in the seat and shove the stick forward with my foot. This would throw the old crate into a nose dive and would toss me, with my chute, clear.

"Well figured, but it simply didn't happen. I was going to stand up in the seat when the chute got all tangled up in the controls. Here the darn thing was burning all around me and I was tied to it. I got out on the step and hung on with one hand and worked at untangling the ropes with the other. The plane stalled and then went into a dive just as I got free.

"It was a race for a minute, no, a year or two, between the blazing ship and myself. Then the chute opened and I made a nice landing in the water. There weren't any hungry sharks about at the time and a boat crew soon picked me up.

"The thrills were over for me, but not for the crowd, because the Jenny came out of her nose dive when she was about three hundred feet from the ocean and, upside down, started to glide toward the crowd on the beach. It looked for a minute as if the crowd was due for a few escapes from a burning ship, but Jenny changed her mind and went to bed in the water not far from the shore."

There is the gypsy flyer—in his own words. Ingenious; working all night on a cracked-up ship so as to be able to keep his engagement; secure in his knowledge of his own ability and resourcefulness; taking chances, some of them desperate; all in order to stay in a business that he loved.

These are the men who kept aviation before the public in the days before the air mail became well known, when regular passenger traffic in cabin planes running on schedule was unknown. They are the link between the old days and the new.

All in the day's work for a barnstormer—a broken prop and bent landing gear. Barnstorming star Clyde Pangborn made this ignominious landing amid the sand dunes of a San Francisco beach.

23

Over the ice floes on the Newfoundland sealing expedition.

Free-Lancing in the Air

Romance—adventure—thrills—hardships—all these attend the life of the roving free lance of the air. It is not often that such a clear picture of this life can be given as in this fascinating story of real life.

AS TOLD BY C. S. CALDWELL TO E. L. CHICANOT

FREE lancing, no matter what the manner of pursuit, exerts a very powerful appeal which most men by force of circumstances are compelled to resist. There are some of us so constituted, however, as to be unable to tolerate life in a groove, and our number has greatly increased since the war. The most interesting and engrossing of routines, because it is a routine, palls in time, and the urge comes to break it and seek new places, new people, and new conditions of life. It may be that a rolling stone gathers no moss, but it accumulates some wonderfully compensating experiences as long as it continues to meander. There is certainly variety to the life of the man who never permits himself to feel bound by the chain of economic circumstances, but at liberty at all times to follow his own sweet will and move hither and yon as the impulse takes him.

And of all roving careers, it seems to me that of a free-lance air pilot, willing to take on any commission, to fly anywhere, is about the most alluring, as capable of satisfying every craving for change and excitement. When in addition the sphere of operation is the broad and raw Dominion, it seems to me that the very apogee of adventure and romance must be reached. That, at least, is how I feel looking back over the last four years in which I have done some of the most out-of-the-way flying in Canada.

It has only just occurred to me to think of this. I have seen so much written of the exploits of trans-continental, trans-Atlantic and other spectacular flying that I have come to see the phase of civilian flying in Canada in which I have been engaged in a new light. When I learn of all the assistance rendered and the elaborate precaution taken to ward off possibility of mishap in these flights, I think of the hundreds of hours spent in the air when a slight accident might have resulted in a catastrophe of which the public might never have heard anything. When I consider in addition the unusual phases of Canadian life I have seen, replete with episodes of every kind, I begin to appreciate that it has been a fascinating career, and it occurs to me that a brief account might be of some interest.

It was three years after the war when, having become steadily more bored with ground existence, I seized the opportunity to get back to flying and joined the Laurentide Air Service as air-mechanic. The company was that summer engaged in aerial survey for the Ontario Government in the James Bay area, operating five planes and flying crews. Our base was at Remi Lake, about fifty miles west of Cochrane, and we spent the entire summer in the territory over more or less routine sketching. It sounds prosaic and generally was, though it might not have been.

Each day each pilot made a trip of four and a half hours, during which the mapping and sketching were done. For part of the season's work we operated from a base at Moose Factory. There the day's flight carried us over a huge marsh or floating muskeg extending for forty miles south of the Bay. It was an indescribably desolate looking region, nothing but oozy swamp, with here and there little patches of water showing. There was not a landing place in all of its extent. The four and a half hours' sketching was spent in flying from one possible landing place to another, with lots of time in between to wonder what would happen if one had to make a forced landing.

Forced landings were made in peculiar circumstances at times. Once when flying further south, returning to the base after surveying all day, we ran into a heavy rainstorm and, after buffeting it for some time, were forced to land in a small lake which came conveniently to hand. It was only after the storm had subsided that we discovered how painfully small it really was. We attempted to take off, but the machine would not climb quickly enough to clear the trees at the end of the lake. I did some rapid thinking, but there seemed nothing else to do but turn and hope for the best. As was almost inevitable, the machine struck the water very awkwardly and commenced to sink right away. It was all we could do to clear ourselves, and we had no time to recover emergency rations, matches, or anything else before it disappeared with a gurgle, leaving us to swim in our clothing to the shore.

Luckily this was not far away or we might never have made it, but we realized thirty miles of practically uninhabited wilderness separated us from our base. The only route to follow in the gathering darkness was along the bank of a river flowing out of the lake, mud bottomed and brush covered. Throughout the day we forged along in the blackness, practically feeling our way, and it was a very weary, grimy, and famished pair that next morning stumbled into a trapper's shack many miles from the flying base.

On another occasion a pilot out on the survey miscalculated the amount of gasoline he carried or tried to make it go too far. Anyhow, he suddenly found himself thirty miles from the base with a spluttering motor, and the only body of water in sight was a lake of painfully small dimensions. Gravity will take a machine a long way under such circumstances, but there are limits, and there was nothing to do but endeavor to hit that lake. Luckily the pilot accomplished this safely. Realizing that the plane would be the best means of attracting attention to their plight, the three occupants moored the boat fast in the middle of the lake and swam ashore with the emergency rations, prepared to spend the night.

Flying down from Moose Factory that same evening, I crossed directly over the lake and spied the plane there. Realizing that the pilot would never have alighted there by choice, we came down and circled the lake several times before we made out the three figures on the shore gesticulating wildly to draw our attention. It was quite impossible for us to land with that plane already occupying the lake, and in addition that machine was the only one of the fleet possessing quick enough take-off and rapid climbing powers to get out of such a restricted area. There was nothing to do with dusk closing in

but to continue on to the base and hand in the information that one of the machines was down in a duck pond thirty miles away.

Early on the following morning I set out again in a machine with a mechanic to see what I could do, but without the glimmering of an idea of what form assistance was going to take. Reaching the lake, I flew down low and began circling about. Almost immediately my attention was directed to the words "DROP GAS" printed in gigantic letters in the mud of the beach. Deciding that this plan was better than any of mine, I returned right away to the base. Here we secured five stout five-gallon cans, filled them with gasoline, and crated them with strong boards. Loading them on the plane, I tuned up and set out again for the lake where the three were marooned.

C. S. Caldwell

Arrived over it, I flew the boat as low as was prudent parallel to the shore, and on each circuit at precisely the same spot the mechanic released a case of the fuel. I could see the pilot of the stranded plane standing stark naked upon the beach, eager and expectant. As each consignment hurtled down he made a wild dash through the water after it. We had no way of determining what success we had, which to me was rather problematical, as at that height and at the comparatively low speed at which we were traveling water is practically solid. The three seemed to be signalling that all was well, however, and we returned to the base. The missing plane came in the same afternoon. Two of the five containers had survived the terrible impact, and the trio was able to fill up the tank and get out of the lake.

After an idle winter, I engaged for the following summer in what was comparatively prosaic flying work, aerial forest fire patrol for a pulp and paper company, which owned large reserves in Quebec. There was nothing especially out of the ordinary about that summer's flying except that the season was one of the worst for many years from the standpoint of fires. Most of the patrol work was carried out every day in a thick haze of smoke which made flying somewhat difficult. Fall was reached, however, without any mishap or the experience of any episode sufficiently unusual to have been registered permanently on my mind.

The only amusing incident I recall of that summer pertains to another pilot who was also doing forest patrol work. His one dread of traveling through the woods alone appeared to be bears. When one afternoon he was compelled to make a forced landing and abandon his plane far from his base his principal apprehension was that he might encounter one of these usually harmless animals. Accordingly part of his salvage from the disabled plane took the form of a tin can, and as he proceeded through the dense brush endeavoring to make his way back to the base he kept up a ceaseless tattoo, with the object of scaring off any bears in the neighborhood. The plan was successful in another direction, for the strange hammering was heard by a trapper whose place the airman was unconsciously passing and he crawled curiously through the bush to investigate. The two came upon one another to mutual surprise, with the result that the stranded pilot found food and shelter long before he had anticipated them and was put safely upon a trail which led him back to his base.

Caldwell and the plane flown to the Quebec gold fields.

During the year 1924 the Laurentide Air Service, with which I was associated, established the first passenger and express air service in Canada. Strange as it may seem, this was not between any of the main centers of population, but from the edge of civilization into the newly discovered gold field of northwestern Quebec. The plane jumped off from Haileybury on Lake Temiskaming and landed in the primordial wilderness of Rouyn. In the pioneer days of that mining field I think it must have been the most unique flying service in the world, and mining authorities have since paid it the tribute of having set the development of the Quebec gold field ahead by two years at least.

I am sure the aeroplane has never been requisitioned for such multifarious services or the proud ship of the air been subjected to such indignities in the way of loadings, as in this daily service to the gold field. I have loaded into my plane for transport to the mines every imaginable kind of commodity—fresh meat, fruit, vegetables, eggs, dynamite by the box, horseshoes, and even dogs and cats. I once flew over the sixty

miles of wilderness with fifteen-foot lengths of heavy piping strapped to the wings, and on another occasion a five hundred pound pneumatic drill was dismantled and stowed away in various parts of the plane's capacious interior and was delivered safely to a Rouyn mine. An entire drum of gasoline was taken in on one trip through dividing it into two parts and putting one in either cockpit. There was very little stuff we had to turn down. So greatly did the mining camps come to depend upon us and rely upon our service that I once came upon a teamster who was convinced that a new team of horses which had been ordered and was expected would come in by plane.

About the only thing we did not knowingly carry was liquor. We sometimes took it in when we had no control over the consignment. One day two noisily happy and quite inebriated individuals, miners or prospectors returning from celebrating, approached me as I was tuning up and expressed a hiccoughed desire to be transported by air into the Rouyn. As they had their fares I could not well refuse them, even had I felt inclined. However, suspecting the possibility of there being more liquor in their packs in addition to what they were carrying on their persons, I stowed these well away aft in the boat and placed the two side by side in the forward cockpit.

The first experience of a take-off is just a bit disturbing, and this had a decidedly sobering effect upon my two passengers. For some time they remained quiet as they glued themselves to their seats. This seemed to very rapidly wear off. I saw the two heads temporarily disappearing as they went down periodically, apparently for refreshment. I was not apprehensive at all, however, until suddenly a quart bottle, seeming to me about the size of a balloon, whizzed past the propeller. Then I observed one of my passengers was up on his feet, swaying unsteadily, swinging his arms with clenched fists and in the most bellicose manner inviting the other to combat. A second later the fight was on, the two grappling and struggling in their narrow, unstable quarters. I didn't know what to do. In fact, I knew there was nothing which could be done but fly on as straight and steadily as possible. I watched fascinated as the two men in front of me swayed from side to side, tightly locked in each other's embrace, momentarily expecting them to topple out as one gained a temporary advantage and pressed the other ominously back over the edge of the cockpit.

Fortunately nature decided to take a hand. We struck one of those sudden gusts of wind and the machine dropped several hundred feet in that sickening manner which makes the novice think his stomach has left him, ending up with an abrupt jolt as if the earth had come up to meet us. Just exactly how it registered upon the minds of my two passengers I don't know, but they subsided in the bottom of the cockpit and I neither saw nor heard any more of them. I understood why when on landing the two were discovered to be so utterly incapacitated as to have to be carried out and to shore bodily.

I have never ceased to marvel at the way in which the comparatively low hazards of flying are regarded by men who day after day work on the edge of far more imminent catastrophe. It reminds me of an occasion on which an engineer in the Rouyn engaged a crew of especially heavy and husky miners to sink a shaft, and the work being pressing arranged to have the men flown in. I think I took in not only the first mining crew to ever be transported by air but probably one of the heaviest aggregations of humanity ever flown. I contracted for this by the head, though I was sorry after seeing my prospective passengers that I had not taken the job by the pound. Not one of them was under 200 pounds and several considerably over. They were all foreigners, and it was easy to determine by the way they regarded the plane they were not anticipating the journey ahead of them with any great pleasure.

Deciding to get the worst over at first I loaded up with the four heftiest of the scared crew. The pick of the quartette, who seemed to be also the most terror-stricken, I placed on the seat beside my own, assuring him this was a position of much greater safety than that occupied by the others. As a matter of fact I knew that the balance of the machine would not be so much upset with his colossal weight nearer the centre of gravity. I left him there to tune up but could observe the blonde giant was becoming more and more nervous. He insisted on trying to talk to me and nothing I could do would discourage him. Finally he became so earnestly importunate, standing up in his place, endeavoring to shout something it was impossible to hear above the roar of the engine, that I came to the conclusion it was something important. After having had some difficulty in tuning up I shut off the engine and asked him what was wrong. In a quavering voice he said "If anything happens to me tell my mother. She lives in Poland."

I informed him somewhat curtly that if anything did happen I expected to be in on the party and tuned up again. After running an undue distance along the water the plane staggered off into the air, and an hour later I delivered a nerve-wracked mass at the mine where next day these men who had been thrown into a panic by a brief journey in the air would quite heedlessly expose themselves to the hazards of blastings, cave-ins, and other accidents, all in the routine of mine workers.

It was on a flight out of the Rouyn that I had an experience which might easily have had a fatal ending. I had no passengers from the gold field and made up a load with odds and ends of express. Included in this assortment were two personal pack sacks and a box of ore samples. Not long after leaving the Rouyn I detected a strong aroma of burning paper but at first paid no attention to it as ground odors are much in evidence several thousand feet in the air. The odor became so strong however, that I became a little nervous and this increased until I decided to make a landing and picked out a convenient lake for the purpose.

It proved a wise course as I discovered on investigating that the box of ore samples which had a high sulphur content were in a state of combustion and very shortly in the stiff breeze would have burst into flame. That would have been a pretty predicament two thousand feet off terra firma.

Aircraft managed to revolutionize gold mining without any great trouble or opposition, but the same could hardly be said of my next sphere of activity—the sealing banks of Newfoundland. There I not only undertook what I think is the most hazardous form of flying in Canada, but did so in the face of the most deep-rooted conservatism and unreasoning scepticism. I don't think there is any doubt now but that the sealing captains have been brought round to a conviction of the unique value of

the plane in their annual expedition, but the bulwarks of tradition have been broken down with difficulty and the admission will only be made with extreme reluctance.

Every spring for the past 163 years the sealing vessels have sailed or steamed out of St. Johns, Newfoundland to prosecute the search for seals in a manner that has, up to the last year or two, remained almost identically the same. The actual business of locating the seals is very much like a game of hide and seek as the expansive ice floes upon which the young seals have been born break up badly and drift with the elements in any direction. The vessels merely make the best of their way north, breasting the icy seas, battling with blizzards, forcing a slow and perilous way among the ice floes. Each captain has his own idea of where the seals may be that particular spring and heads roughly in that direction, making sometimes only a few miles a day through the ice field, while a vigilant lookout is kept for indications of the seal herds.

The principal one is the formation of the ice, which is of three kinds. Slob ice is newly formed and blackish in color owing to the water showing through. Arctic ice, formed far north and drifted down, is rough and hummocky and generally contains much snow. Sheet ice is the second stage of slob ice and is found in large flat sheets. This is the ice on which the seals are found, as the old seals must make their bobbing holes when the ice is comparatively thin. These are apertures kept open as the animals come up periodically to breathe.

In the past observation for such ice has been from barrel towers at the mast head, eighty feet above the deck, by lookouts scanning the icy prospects with powerful telescopes. The difficulties can be appreciated. Small openings are not easily seen and the young seals, snow white, are practically indistinguishable against the icy background. It is even difficult at times to see the older, darker seals where the ice is rough and hummocky, and they may be concealed in the hollows. Often vessels have steamed for weeks without encountering seal patches, passing within a few miles of thickly dotted floes.

The possibility of utilizing the aeroplane in sealing was first suggested to the sealer owners of Newfoundland when the machine especially constructed for the Shackleton expedition was left behind and remained long unutilized. Finally the owners of the sealing vessels jointly purchased it and in 1923 they brought out a pilot from England to operate it on the spring expedition. The attitude of the sealer captains towards this innovation was pronouncedly skeptical, not to say a trifle hostile, and as a result the fleet returned from the hunt with the plane still fasted aft, having never been flown. In 1924 it was sent out again with Capt. B. Grandy, a veteran airman of the Great War and himself a former sealer commander. He was scarcely more successful but at length, the seal hunt being protracted, days and weeks passing without any trace of the seals being encountered, the captain of the vessel carrying the plane consented to stop long enough to land it and give him opportunity for a flight. But the one flight was made but it was responsible for locating several large patches of seals, one containing upwards of 50,000, four miles long and eight miles wide, and putting the sealing vessels in touch with them.

In the spring of 1925 the sealer owners seemed to experience some difficulty in securing a pilot to go out with the fleet, and at length got in touch with me. As the result of successful negotiations I went down to St. Johns and joined the fleet just prior to its sailing. The plane I found was a Baby Avro, equipped with wheels, floats, or skids, so that it could be flown off the ground, water, snow, or ice. It was a light machine, easily assembled and transported, and carried sufficient fuel for two hours flight. It left with the fleet mounted upon a specially constructed aft deck, this being for transport only.

I thoroughly enjoyed the sealing expedition in its novelty and daily hazard, though I couldn't help but feel from the beginning that I was regarded as superfluous and intrusive. Men who had gone out with the sealing fleet from the time they were boys were naturally sceptical of such revolutionary innovations, and their attitude expressed resentment that anyone from the effete outside world should imagine he could improve upon the traditional methods of the hunt. These antiquated methods annoyed me and made me impatient. I knew that in half an hour I could cover as much observation as was being done in a day. But the sealer captain is absolute lord and master on the expedition and it was useless trying to press the matter.

Finally, however, I was given my chance. Time passed and no signs of the seals were encountered. The captain had an idea they were in towards the land, where the way was obstructed by heavy ice, through which progress would have been the slowest. He asked me to make a flight over it. The sealer was accordingly tied up to the floe and the plane lowered. The crew were put to work smoothing out the rough surface and soon prepared sufficient space for a take-off. I hopped off successfully.

I was away an hour and flew between thirty-five and forty miles. I encountered no seals but on the contrary ascertained by the formation of the ice that the desired animals could not be in that direction. I returned to the ship with this information which could otherwise have been secured only after the loss of days of valuable time. It decided the captain to steam in the diametrically opposite direction from that which he had intended.

Two days later I made a second flight under orders from the captain, flying between 70 and 80 miles and almost immediately spotted a small patch of seals—stragglers, but indicating the position of the main herd. Without the help of the plane they would certainly have been missed altogether. The information I brought back decided the captain to change his course with the result that at noon the next day the seal herds were encountered and the objective of the expedition attained.

Accompanying the sealing fleet I have no hesitation in saying is the most hazardous form of flying in Canada. First one is operating in an intensely cold temperature and it is difficult to get the machine started. It is not always easy to find a suitable place for a take-off. Often a long runway will be cleared on a narrow floe of ice and by the time one returns the wind has changed and it is necessary to make a landing on the narrow width. The machine is necessarily small and the equipment not very elaborate. The only instruments the pilot has are an at times inaccurate compass and a watch. A forced landing while on a flight means very great risk. The ships

would, of course, set out to look for the pilot if he failed to return in reasonable time, but ice floes are never stationary nor is it possible always to determine their trend, so that by the time they had reached the place where he had come down he might have drifted very far away.

On returning from the sealing fleet in the spring of 1925 I joined the expedition of a syndicate of United States mining capitalists, for exploration work in the unexplored regions of the Yukon and Northern British Columbia, as assistant pilot-mechanic. Incidentally this was, to the best of my knowledge, the first mineral prospecting party to plan the use of the aeroplane. Colonel Williams, who was in charge of aerial transport, and I appreciated quite keenly that the territory to be flown over was isolated, largely unexplored, and but very inaccurately mapped, and realized the hazards and difficulties of such an expedition as we were setting out upon. We dug up all the available data on the country but this was painfully fragmentary and left us little better off. The machine we were to fly, a Vickers Viking amphibian, was shipped by rail from Three Rivers, Quebec, to Prince Rupert, British Columbia, and we were there ready, with the plane assembled, to commence our air trip on June 1st. Altogether we made up a party of nine.

We flew from Prince Rupert to Wrangell Island where we refuelled, thence directly over the coast range by way of the Stikine Valley, and on to Telegraph Creek. It was a route of exquisite wild loveliness, as we sailed over glittering glaciers, high whitecapped mountain peaks, and saw the ribbons of innumerable rivers winding below. We made a landing on the Stikine River and established camp there to await the breaking up of ice on Dease Lake.

On June 13th word was received that the ice on Dease Lake had broken up and on the following morning we made an early start with the entire outfit, consisting of nine men, tents, dogs, and other equipment, to fly to our main base which was to be located at the head of Dease Lake. The distance from Telegraph Creek was more than seventy miles over a country absolutely devoid of landing places. A forced landing over this stretch of country would have spelt not only failure to the expedition but in all probability disaster to all of us. We had fortunately no such mishap and the entire miscellany was transported to the head of Dease Lake in forty-five minutes. Ours was the first machine to have ever crossed this height of land. An altitude of three-thousand feet having been maintained after leaving Telegraph Creek we found we were only five hundred feet above water over Dease Lake, as that body is located twenty-five hundred feet above sea level. After having ascended three thousand feet on the take-off we had only to descend five hundred feet to land.

Our first trip from Dease Lake was down the Dease River as far as McDames, where there is a Hudson Bay post and a couple of mining companies are engaged in dredging operations. This distance of ninety miles took up just fifty minutes. We found it quite impossible to convince the Indians of this as it took them one week to make the return trip. They plainly regarded the machine as something diabolical and could not be induced to approach closer than three hundred yards, while they seemed to believe there was something supernatural about us. "It flies like a duck and lands like a goose," one observing Indian put it. It was the sole topic of the tribes and fresh Indians came in from miles every day to view it. The younger set for long periods sedulously practiced the roar of the engine and with their arms imitated the movements of the plane in turning and banking, evidently so as to be able to take back an accurate and convincing story to their brethren in the bush.

While at McDames we chartered a large scow capable of carrying five tons of supplies and had the Indians bring it up to the head of Dease Lake, where it was loaded with gasoline, supplies, and equipment and sent down the river to Liard Post, located at the junction of the Dease and Liard rivers. The trip by scow took twenty days. We did it by plane in just under three hours.

Our arrival at Liard caused excitement enough, the Indians collecting on the river bank very much alarmed. One particularly wild-looking fellow immediately started in a panic for the brush, but as we circled preparing to land the shadow of our machine intercepted his flight and terrified he turned deciding the river was his safest retreat, and made for it. Just as he got to the bank, however, he realized we were landing there and frenziedly turned again and plunged terror-stricken into the bush. He lay hidden there for some while before slinking forth to secure his pail, rifle, and dog, and set about the construction of a rude raft. That evening he set out on a four hundred mile journey to Nelson to tell the big chief of the fearsome thing which had arrived. I still wonder what happened to him.

While the scow was en route we transported the prospectors by air to various locations. Considering all conditions and the difficulties a forced landing would have put us in it was not the safest kind of flying.

The scow eventually arrived on July 11th and the two redmen who had brought it down intended returning on foot, a matter of five days and nights travel. As we were going back to Dease Lake and had to pass the Indian camp we invited them to go along. At first they would not consider it at all and so strong was their reluctance and so amusing to us the attitude of these two Indians who had piloted the clumsy scow down all manner of rough water that we decided to persuade them to fly if we could. It was hard work. "No, me weak heart. You big man. You strong heart." We at length managed to overcome this objection, then it was "Me got dogs. No can take the air." "Sure we can take them," I said, and with excuses running out they began to look shamefaced, and we summarily bundled them into the plane.

We had seated them and got as far as putting helmets on their heads and cotton wool in their ears when the nose of one of the two began to bleed profusely. This was regarded as a very bad omen indeed and the trip was nearly off. However, more persuasive eloquence and copious cold water poured down the back of the neck stopped the bleeding, and losing no more time we took off.

These two Indians would have thought nothing of running rapids with a few poles lashed together with moose hide strings, but the plane plainly had them terror-stricken. As soon as we left the water their heads disappeared beneath the cockpit and did not show again until after we had landed fifty minutes later. It would have taken them five days to cover the

distance on foot. I will never forget their expressions as they tried to realize the distance they had come. Yet greater, if possible, was the wonder and bewilderment of their friends who had congregated upon the bank to see the big bird land as they saw two of their own people emerge from the bowels of the monster. I am sure they would be made chiefs or killed as wizards.

On the evening of July 5th we picked up the prospectors at French Creek, which runs into Dease River some forty-five miles from Liard post, and flew them to Liard. As the machine approached this post for the first time the whole population, Hudson's Bay factory and Indians gathered on the river bank trying to solve the strange noise. Finally they spotted the big bird coming straight for them a couple of thousand feet in the air. The Hudson's Bay man, so we discovered later, thinking the machine was going to pass and get lost, instructed the natives to get their guns and fire a fusillade, imagining that the reports would be heard by us. We knew nothing about this, but a few hours after landing discovered considerable water in the hull of the machine, and on investigation found a fountain shooting up at the bottom just at the rear of the gas tanks. The fact that the hole was clean, and most suggestively about the size of a 30-30 bullet, caused us to make inquiries and we were informed about the shooting. Little damage was done though it is difficult to say what might have happened had the gas tanks been pierced.

Flying boat at the Factory in James Bay.

The game of that sparsely inhabited country is unbelievable. The lakes and rivers teem with the finest fish which in addition to being an important article of our diet furnished us with superb sport. Flying over the country it was a common sight to see moose, sometimes as many as a dozen at a time, standing at the shore line of small lakes. Sailing along close to the high-rigged mountains caribou, goat, grizzly bear and other animals could be seen at almost any time. Sometime it will be a wonderful sporting country. On one occasion I saw a pure white moose. Another time when I was fishing I saw a big moose and called to Col. Williams to come and see it as he was casting further up stream. As he did not come immediately I went to see why and discovered he had been intercepted by a bear with two cubs which he was busily bluffing. Luckily his bluff wasn't called.

The last place we visited before coming out was the fabled "Tropical Valley," colored and exaggerated press accounts of which have been so sceptically received. We heard vague accounts of the valley while at Liard Post, based mainly upon the meagre information of Indians, and also about a prospector and trapper, Tom Smith, and his daughter who had wandered overland from the Yukon, had gone into that region two years before and had not been seen or heard from since. We decided to make a trip to see if the valley really existed and if possible get trace of the missing pair.

We knew that the mysterious region was approximately two hundred miles from the post and accordingly took off and followed the course of the river, which is punctuated with frequent terrifying rapids. We had flown what we judged to be approximately the distance when we reached a region of peculiar looking lakes. We thought them peculiar but there was perhaps nothing unusual about their aspect. They were striking to us since there were no other similar bodies in the country for some considerable distance in any direction. Deciding this must be the place we made a landing on the river and tied up. With the machine secure we started out to scout.

Almost immediately we struck a trail, faint and apparently not used for a long while, but unmistakably a trail. We followed this for about four hundred yards from the river and then surprisingly encountered a board nailed across a tree. Upon a piece of wood painted in rude characters were the words "A. B. C. Code" followed by a jumble of figures and signed Tom Smith. It baffled us for a little while but finally the combined wits of the party deciphered out the words "Message in bottle at foot of tree."

We dug at the bottom of the tree and just below the surface struck a bottle which upon opening was found to contain two sheets of paper covered with very legible handwriting. I cannot remember the exact wording of the message but it started off "To any white man finding this message" and went on to the effect that by following the trail back for two miles a cabin would be found near the first hot springs and a short distance beyond a garden "we have left planted to potatoes and onions." By further following the trail, the message continued, the finder would come to a large hot spring overlooking a meadow "in which moose are to be found every morning and evening during the summer months." The message, signed Tom Smith, concluded by saying that he and his daughter had been living there for two years and had seen no white man and were leaving for Fort Simpson which "barring trouble with the Indians we expect to reach in the spring."

We progressed up the trail, the atmosphere becoming more torrid and langorous as we advanced. Shortly we reached the first hot spring and just beyond that found the deserted cabin. A little further away was the garden, rank and overgrown. From here we started out to explore a limited area of the valley which appeared to be about ten miles square in extent.

Hot springs sprang from the ground all over. Some of them just bubbled up and ran straight away. Others formed the pools

of varying dimensions we had seen from the air. The rich and luxuriant growth and foliage was distinctly suggestive of a tropical region. Ferns grew to an enormous height and size, and vines spread all ways in a tangled mesh. Berries of many kinds were growing in profusion and of an extraordinary size. Flattened patches indicated where bears came in to dine. About many of the little boiling lakes were large patches of purple violets of a size and beauty I had never before known. Much of the growth was unfamiliar to me and such as I had known in other parts of Canada of an extraordinary lavishness and production. I was brought up in Alabama and nothing has ever so reminded me of that southern state.

We stayed two days and then flew back to Liard. Later when flying out we reported what we had found to the Mounted Police at Fort Simpson. They told us the rest of the story of Tom Smith and his daughter. Coming down the river in a canoe the craft had capsized in the rapids. The girl had been taken down the river and swept up unconscious on a sandbar. When she came to her senses there was no trace of her father, an old man, to be found. She had made her way to Fort Simpson and was working for the Hudson's Bay Company.

On our way out with the summer's work completed we were forced to make a landing on a lake about one hundred miles from Fort Fitzgerald, and while here waiting for the weather to clear two Indians of the "caribou eaters" paddled out to where we were camped and inquired in part English and part French whence we had come. We replied from the barrenlands, or the Land of No Sticks, as they call it, and this amazed them. They were immediately eager to know if we had seen any caribou and when we replied in the affirmative they were greatly pleased as this meant "plenty meat" for the winter.

One of them was anxious to send a note to some friends at Fort Fitzgerald and asked if we would deliver it. We replied that we would be only too glad and he produced a notebook and with much laborious effort which brought beads of perspiration to his brow he penned a queer geometric note. After its slow completion he handed it to me with a dollar bill. I declined it and asked him how long it would take him to go to Fort Fitzgerald himself, and he replied, "Long time. Lots of portages. Mebbe one month." When I told him we could get to the fort in under an hour the pair was dumbfounded and incredulous, one of them saying "Far, far, very far; no not far," evidently intending to convey that what was a long distance to them was little to us.

The next day we found the weather sufficiently clear to permit us to resume our way and reached Fort Fitzgerald. From there we flew to the government aerodrome at High River, Alberta, after passing over Edmonton. In all we covered about seven thousand miles without trouble or mishap. The staunch little Amphibian is hibernating while I am tasting the things of civilization again. I have a little time to reflect and certainly there seems to have been a lot crowded into the last four years. But each experience merely whets the appetite for more. Soon the spring will be round again and the Red Gods beckoning back to the romantic and adventurous life a free lance airman.

BY JACK R. HUNT

A man who envisioned transatlantic flights by balloon proved that a great river of air flowed eastward.

The Greatest Air Voyage Ever Made

IT WAS DUSK July 1, 1859 as four excited men in St. Louis, Mo., boarded the *Atlantic,* the largest balloon in the world. Thousands of paying spectators were clamoring for its release. Most of them had been there throughout the hot, sultry day. They were afraid the flight might be postponed.

The sight of such an anxious crowd was no new experience for the pilot of the big balloon. John Wise had been only 27 in 1835, when he made his first ascent from the corner of Ninth and Green Streets in Philadelphia, Pa. He had then been just a country boy with a yen to fly. His first balloon was homemade from plans which he drew after studying the atmosphere, hydrostatics, and pneumatics. He had planned on making a single flight to experience the thrill of being in the air; however, once he tasted success, he devoted his life to aeronautics. He had risen a mile aloft over Philadelphia! That was 34 years and 230 flights ago. This flight was for much higher stakes.

Eventually the crowd broke through the police lines and, converging upon John Wise, demanded to know whether his intentions were honorable; or was this another hoax to take their money and cancel the flight as had often been done at county fairs. The intrepid aeronaut and Director in Chief of the Transatlantic Balloon Company, paused long enough to address them.

"Ladies and Gentlemen! I have never been a party to a balloon ascension that did not come off as scheduled. On more than one occasion, I ascended in an unsafe balloon—at great personal risk, rather than disappoint my admirers. Ladies and gentlemen, the field of Aeronautics has become a science. I assure you, we will ascend in very short order."

Pointing toward a much smaller balloon a short distance away, he continued, "A well known local aeronaut has offered to act as our guide on this historic event. He will take yonder balloon and lead us upward into that great river of air that sweeps ever eastward. Ladies and gentlemen, I give you Mr. Robert Brooks."

Mr. Brooks bowed, climbed into the basket of his one-man balloon, and with a wave of his hat ascended.

Back at the large balloon, John Wise was almost ready for the launching. The *Atlantic* was a huge, beautiful machine conceived as being capable of crossing the Atlantic—hence its name. John Lamountane, one of the crew, had been in charge of its construction. It was 50 feet in diameter and 60 perpendicular, made of lacquered silk covered by a woven hemp net. At the bottom of the net was an iron load ring, below which was suspended a wicker basket. A special lifeboat, capable of carrying 1,000 pounds, was suspended 15 feet below the basket. The boat was encased in a heavy canvas jacket which acted as a sling and protective cover. A rope ladder enabled the crew to climb between boat and basket.

This flight, St. Louis to New York, was a test to prove the airworthiness of the balloon and demonstrate a theory advanced by Wise that a river of air flowed west to east and might serve as an avenue of travel around the world. Wise had often noted at various altitudes that the air flowed east and kept notes regarding his observations. In May, 1842, he wrote in a log of one of his flights, "It is now beyond a doubt in my mind established *that a current from west to east in the atmosphere is constantly in motion* within the height of 12,000 feet above the ocean. Nearly all of my trips are strong proof of this."

This particular flight would end in New York and, after preparation, the flight across the Atlantic would begin from there. St. Louis had erected an arena on the city common and had allowed the promoters of the flight to charge admission to defray expenses, a common practice of that time.

Plans for the translantic flight called for a two-man crew with enough ballast to complete the journey. Since this leg of the flight to Europe would not be as long as the North Atlantic crossing, Wise decided that a larger crew could be taken and some ballast dispensed with. The crew, therefore, was to consist of himself, John Lamountane, O. A. Gager, promoter of the venture, and a last minute addition, a reporter named Hyde from the St. Louis *Republican,* who proposed that he should go along to write a description of the flight. The American Express Company also asked Wise to carry a few sacks of mail to dignitaries on the East Coast. These, too, were accepted and a few more ballast bags were left behind.

As Brooks, in the pilot balloon, drifted out of sight eastward, Wise boarded the wicker basket where he could control the valve cord. Below, in the lifeboat, the other three made themselves comfortable. Supplies consisted of a bucket of lemonade, a lunch basket with a well cooked turkey, and a valise of champagne. The mail sacks and ballast bags were stowed in the lifeboat. At 6:45 p.m., the anchor lines were cut and, as thousands cheered, the adventurous quartet drifted

majestically skyward.

The first night of the voyage was uneventful. The crew took turns standing watch. At first it was easy to spot towns by the lights but, as the night wore on, there were fewer and fewer lights. The man on watch amused himself by shouting to the world below. The shouting would cause the dogs to bark. Whenever the balloon was over a village, hundreds of dogs could be heard barking at once. Often this commotion would cause lights to come on and angry voices could be heard cussing the dogs.

The night had been clear and cold in the upper air. As dawn broke, Lake Erie was seen to the east. Smoke rising from farm house chimneys suggested the thought of hot coffee to the fliers and it was decided to drop in on some farm family for breakfast. Gas was valved to start the descent but as the crew neared the ground they became aware that they were traveling at a surprising rate of speed. A line dropped over the side tangled with some trees and momentarily stopped the lifeboat with such force that the occupants were almost thrown out. A hurried consultation resulted and the idea of a hot breakfast was abandoned. Ballast was dumped and the balloon rose back into the calm upper air.

Exactly 12 hours after take-off, the *Atlantic* crossed the shore line of Lake Erie near Toledo. Clouds were forming to the south and the surface of the lake was very rough. The balloon was drifting at a ground speed of better than 60 mph. The remainder of the morning was spent uneventfully as the balloon passed over several boats and traveled the 250 miles to Niagara Falls. In an endeavor to shift course more to the south, ballast was jettisoned. The balloon rose to 10,000 feet. From that altitude, the Niagara river looked like a silver cord linking Lakes Erie and Ontario. Clouds seemed to be manufactured in the falls and to take up an orderly line of march toward the east, expanding as they traveled along beneath the balloonists.

By now, a feeling of apprehension gripped Wise, who had become aware of the significance of the clouds closing in on them and of the ever-increasing velocity of the wind. He was facing a storm with a crew of novices. Hyde was on his first ride. Gager had made but one previous ascension. Lamountane had completed a mere half dozen flights and was really being groomed as a crew member because of his experience in seamanship which might stand them in good stead if they were forced down during the ocean flight. As the city of Rochester came into view, Gager looked up to the wicker basket and asked, "Professor, what keeps you so quiet?"

Startled, Wise replied, "Do you not see anything extraordinary down there?"

"Why, yes," answered Gager, "the wind is very strong. I could even hear the limbs of the trees crack as if they were splitting from the trunks, but," he hesitated, "I was thinking we might make a landing near Rochester, drop off Hyde and myself to take the mail down to New York while you and Lamountane continue the journey as long as you care to."

"All very well considered, my friend, but do you realize we are traveling better than 90 miles an hour?"

Hyde and Lamountane had also become aware of the situation below, although they could not completely grasp the gravity of it. Now that everyone was concerned about their predicament, a consultation was called to decide on a course of action. In spite of Wise's warning, the first suggestion was to attempt a landing. However, before much could be said, the balloon began settling. When they had descended far enough to see the trees bending and hear the wind as they sped madly along, the problems of a landing became quite evident. The crew in the lifeboat were fairly hypnotized by the destruction below.

Lamountane broke in. "Professor, what can we do?"

"Throw everything overboard!"

The mail sacks were quickly jettisoned and shortly the balloon was rising again and the occupants could momentarily relax. As this crisis passed, Hyde, who until now had contributed very little to the discussion, worriedly asked, "Is there great danger?"

"Yes!" replied Wise. "I see no earthly chance unless we land in the water."

He called Gager up into the basket and the two debated hurriedly. "The wisest plan is to swamp the balloon in the lake, if we can intercept a boat," confided the pilot.

Gager, without a thought for the loss of the money invested in the venture, said, "Whatever you say, but do we stand a chance of being picked up in those mountainous waves down there?"

The two looked over the side at this point and saw they were again falling. Wise threw over the valise of champagne with which he had planned to celebrate their arrival in New York. The only ballast left in the basket now was a grapnel on a rope and a small hatchet.

Wise looked into the boat below and saw Hyde scribbling on a pad. He wondered if it were notes for a story, or the reporter's last will and testament. Lamountane was exploring the boat for any form of ballast. Wise told them to climb up into the basket so they could cut the boat loose and thus remain in the air a little longer. Hyde immediately climbed up but Lamountane elected to stay below and asked for the hatchet. He intended cutting out the double bottom of the boat to use for piecemeal ballast. The hatchet was passed down and Lamountane began in earnest to tear the boat apart. A few sticks had been tossed overboard when the boat smashed into the water with a huge splash. When the spray died down, Lamountane's hat could be seen floating on the crest of a wave. Wise shouted, "Lamountane's gone!"

But the words had barely left his lips when he heard a voice sing out, "No I'm not, it's only my hat!"

The sailor was lying in the bottom of the boat with his arms wound around one of the cross seats. He resumed his hacking as the lifeboat bounced along on the crest of the waves. Finally, the double bottom was gone and the airship gained a few hundred feet altitude; Lamountane joined the others in the wicker basket.

The storm clouds were now thick and black around the balloon. The lake was surging and foaming like a thing gone mad. Wise suddenly pointed. "I think I see a steamer out there! Will we swamp the balloon?"

A unanimous "No!" was the response.

Lamountane complained of being sick and unable to swim. Hyde said, "If we are to die, let's die on the land—if we can reach it."

Gager expressed the same sentiment.

It was evident they had all forgotten how rough the land had looked before and they were now hoping against hope for a miracle to happen. Minutes seemed like ages as everyone watched for something other than clouds and spray. Suddenly, something could be seen up ahead. It was a steamer, the passenger ship *Young America,* and it was crossing directly in their path.

Wise resurrected the idea of landing on the water and of being picked up by the ship. He recalled a similar incident on Lake Erie. He had become lost and was rescued by a brig after landing in the water. When he again put it to the rest of the hapless crew, they were still solidly against it. Wise now calculated the hapless group had traveled over 100 miles across the lake and should be making landfall within a half-hour. The rest of the crew seemed to gain solace from this report. At 1:35 p.m. Hyde declared he saw land, but that it looked like a million miles off.

"Too near for our comfort," was the aeronaut's retort.

Wise now reasoned that he might be able to swamp the balloon in the water just before it reached the shore. He cautioned them all to hang on tight, and busied himself with the valve cord. Wise claimed it was a turn of fate. Other aeronauts say he misjudged the rate of descent. Whatever the cause, before he could bring the gas bag down to water level, it bounced with a violent crash upon the shore. Instantly, he threw out the grapnel and with the others, held on for dear life.

The balloon rebounded and shot up over the treetops. The grapnel caught a tree but it was like a fishhook anchoring a battleship. In another moment the balloon went dragging along through the treetops like a mad elephant through a jungle. The thing that saved the occupants was the large iron ring Mr. Wise had installed above the basket to distribute the load of the car and the lifeboat along the net. The balloon went careening over the countryside at a terrific rate and it seemed unbelievable that anyone could survive this dashing and crashing process. Eventually the basket lodged in the side of a high tree, the balloon split with a loud bang, and then, like a dying whale, collapsed.

There they were, the boat still fastened by three ropes, hanging at a 60-degree angle. The men examined themselves to learn if there were serious injuries. Much to their surprise, no one was hurt. Ignominiously they climbed down the inclined deck of the boat and dropped to the ground.

Several people had watched the drama and had come running to where the flight ended. After expressing their sentiments about anyone who would venture forth in such a contraption, they told the travelers that they were near the town of Henderson, N.Y., 1,200 miles from St. Louis—as the crow flies; but they hastened to add that this was not a fit day for a crow to be flying.

John Wise walked to a small stump, wearily sat down, pulled his notebook from his pocket and dutifully wrote:

"July 2, 1859, 35 minutes past two p.m., landed in a tree on the place of Truman Whitney, in the township of Henderson, in the county of Jefferson, in the State of New York. Thus ended the *Greatest Air-Voyage Ever Made.*"

Testing Racing Planes

BY JIMMY DOOLITTLE

A most interesting account of racing plane tests by one of the most noted racing pilots in the world. An address delivered by Mr. Doolittle before the Institute of the Aeronautical Sciences.

INASMUCH as my work for the past three years has been the developing and selling of petroleum products for aviation usage, I am scarcely in a position to deliver a truly up-to-date dissertation on scientific flight testing.

With your indulgence, I will, however, detail a few personal experiences encountered while testing and grooming a modern racing airplane. Since experimental high speed airplanes differ radically in certain phases of design, admitting numerous specific problems peculiar to each of several types; for the sake of brevity, I shall confine this discussion to the recent Laird 400 Racer.

In the mid-part of 1931, the ailerons and a good sized piece of right wing fluttered off my Travel Air "Mystery S" airplane which had just been rebuilt, cleaned up and obviously speeded up. This failure was probably due to the Frieze type ailerons being overbalanced at the higher speed or to flexibility in the aileron control tubes.

The failure of my own airplane made available the new racing plane that was built by Mr. E. M. Laird and financed by the Cleveland Speed Foundation. This little job had a 21 foot span, 108 square feet of wing area, carried 112½ gallons of gasoline, 8 gallons of oil and weighed 1,580 pounds when light and about 2,500 pounds fully loaded. (With the geared engine it weighed about 75 pounds more.)

The first test flight was made from the old Aero Club Field, south of the Chicago Municipal Airport. Laird felt or hoped that the high speed of the airplane would be around 300 m.p.h. The geared Pratt and Whitney Wasp Jr., originally mounted in the airplane, developed 560 h.p. at 2,500 r.p.m. The propeller was 9 feet in diameter and set at 37½ degree pitch at the 42 inch station. Due to the high pitch angle, the ground revs and static thrust were very low.

The airplane ran about a half a mile before it could be pulled into the air and then flew for about two miles more before it picked up suffcent speed to come under complete control. In succeeding flights, the propeller setting was reduced 5 degrees and the takeoff was satisfactory though the engine over-revved somewhat (2,600 r.p.m.) in level flight at full throttle.

The tail surfaces were identical with those of the last year's racer. The airplane was stable longitudinally and laterally, but extremely unstable directionally. This directional instability increased with speed and the airplane was barely manageable at a speed of 200 m.p.h.

This was due to the increased fin area forward, resulting from the longer N. A. C. A. cowl employed with the geared engine, the large pants and the fairing which filled in the space between the front and rear landing gear struts. All of this fin area was forward of the c. g.

The culpability of pants and fairing were proved by removing them and making a flight. The airplane was directionally stable with pants and fairing off. To correct the directional instability, the fin and rudder were increased in height nine inches. Thereafter the airplane was directionally stable, but not as stable laterally as before.

There did not appear to be any appreciable torque resulting from the large propeller and geared engine except an accelerational torque when the throttle was moved quickly.

Although the pilot was sitting on 50 pounds of lead shot, the airplane was so stable longitudinally that it was difficult to get the tail down in landing and the airplane landed very fast. The fast landing tendencies of this airplane were attributable largely to the fact that the tail could not be brought all the way down and advantage taken of the maximum angle of attack when landing, and also the blanketing effect of the large propeller disc on the small wings.

This is indicated by the fact that the airplane, when mounting the direct drive engine and 8 foot propeller, was only slightly lighter but was less stable and landed considerably slower; I should say fully five miles per hour.

In order to ventilate the pilot's cockpit, fresh air was led through a flexible conduit from two holes located in the leading edge of the upper wing. These holes were about three feet from the center line of the wing, located in the slip stream, but well outside the "fume area."

The center of each of these circular holes was somewhat above the center-line of the leading edge and the tops of the

"Jimmy" Doolittle, then a U.S. Army pilot, is shown with the U.S. Army racing plane that he flew in the last Schneider Trophy race to be held in America. It was this race that brought Doolittle into public view.

holes were aft of the bottoms. In flight the starboard hole blew slightly and the port hole sucked slightly. The efficiency of the ventilating system increased somewhat as the speed increased. The trouble was corrected by putting a scoop on the top of each hole.

The geared engine ran much cooler than the direct-drive. The head temperature averaged from 50 to 75 degrees lower. This appeared to be due to two things. First: The nose of the N. A. C. A. cowl on the geared engine was longer and tended to direct the air over the cylinder heads better than the blunt nosed N. A. C. A. on the direct-drive engine. Second: The geared engine was throwing more oil than the direct-drive engine. This was indicated by greater oil consumption and higher oil temperature in the use of the geared engine.

At a later date, the direct-drive engine was returned to the factory. The bearing clearances were increased so it would "sling" more oil and overheating difficulties were eliminated. The oil consumption was increased from about one quart to three quarts per hour. This increased oil consumption, however, was not a problem as there was ample oil tankage. Using the oil for a cooling medium caused the oil temperature to run too high and it was necessary to cool the oil by directing cold air taken from in front of the engine and over the oil tank.

The fuel used was a straight-run gasoline containing 3 cc's of tetraethyl lead and having a knock rating of 87 octane. For full throttle operation, an 89 octane gasoline, containing 5 cc's of lead was used. This was a safety measure with the 6:1, 10:1 Wasp Jr. as the 87 octane fuel operated satisfactorily.

The 87 octane fuel was probably on the border line as I later flew a 6:1, 12:1 Wasp Sr. and with this engine at full throttle operation, the 87 octane fuel started detonating almost at once and the thermocouple showed a head temperature of over 600° F. With the 89 octane fuel there was no detonation and

head temperature was steady at about 520° F. The 87 octane fuel was satisfactory at cruising speed.

The 400 Racer was extremely temperamental to rig. It seemed impossible to adjust the landing and flying wires so that it would not be wing heavy on one side or the other. On one flight it balanced perfectly for a while and then gradually became left wing heavy.

Finally, in desperation, the stick was slapped over hard to lift the left wing and the wing heaviness after leveling off was greatly increased. This gave me an idea. The right wing was depressed and the stick moved sharply to the left. The airplane was then only slightly left wing heavy. The maneuver was repeated and the airplane balanced perfectly. Repeated again and actual right wing heaviness resulted. Here was an airplane that could be rigged in flight!

The difficulty was that it wouldn't hold its rig. This fault became continually and rapidly more apparent and annoying. On one flight, in very rough air, the rigging became so flabby that an actual lateral motion of the trailing edge of the upper wing could be observed when the ailerons were moved and when a bump hit one wing more severely than the other.

Before the flight was completed, it was found impossible to get the airplane out of the left bank at cruising speed without throttling back; so all turns, even around the landing field, were made to the right. In this airplane the main wing truss was incomplete. The auxiliary wing truss had depended upon a fitting around the center of the continuous rear spar in the upper wing to take unevenly distributed wing loads. A careful inspection showed that the spar had crushed at this point and the bolt holes had elongated.

As a temporary expedient, an eighth inch thick piece of sheet steel was driven between the fitting and the spar to take up the play. This corrected the trouble temporarily, but after a few hours' flying it again appeared due to further crushing of the spar.

The incomplete wing trussing, mentioned above, explains a difficulty experienced when running the 3 km. speed trials. The airplane was clocked at 255 m.p.h. with the direct drive engine. Up to 240 m.p.h. no trouble was experienced, but above this speed a tendency to turn to the left and difficulty in removing the airplane from a left bank was observed. A right bank at full throttle gave very little trouble.

The direct drive engine was removed and the 3.2 geared engine installed. The air-speed indicator showed about eight miles per hour more speed with this engine. There was a very strong tendency to roll to the left. The airplane was then rigged very right wing heavy in order to correct this tendency at full throttle, but it is doubtful if it could be handled in a race.

The course was flown in practice for a few laps at 240 m.p.h., then the throttle was gradually opened until wide. A team mate on the ground was instructed to watch until I rocked my wings and then to time the next lap as it would be the only one at full throttle.

Coming down the home stretch, I rocked the stick laterally but the rolling motion of the airplane was so slight that I was afraid my ground observer would not be able to notice it. The 10 mile course in 1931 was an irregular pentagon. The first two pylons were executed successfully but at the third, where the angle was sharper, the left wing would not come up and I was unable to recover from the bank until after overturning. Rolled in to a steep right bank to get back on the course and had difficulty getting out of the right.

Banked to the left of No. 4 pylon. The bank increased even with controls reversed and it was necessary to throttle to regain control of the airplane. Obviously the geared engine could not be used in the Thompson Trophy Race. Not because of any torque difficulties but because wing warping at high speeds induced the aileron reversal. This condition became critical at about 260 m.p.h.

In order to check the speed of the airplane with "geared-versus-direct-drive-engine," an attempt was made to fly the 3 km. course. On the first run the airplane gradually rolled to the left until out of control. Throttled and tried again. This time entered the course with the right wing down about 30°. After about 1 km., the airplane was level and at the 2 k. mark the left wing was down some 30° and depressing rapidly.

I throttled and landed, unable to make even one run across the 3 km. speed course. It would have been possible to make runs at say 220 m/h and 250m/h, note the corresponding R. P. M.'s; plot air-speed versus R. P. M. and, knowing that the maximum level flight R. P. M. was 2600, produce the curve and obtain the high speed.

However, the Thompson Trophy Race was scheduled for the next day and flights previously made indicated that the direct drive engine could surely be handled, though sloppily, on the course. The engines were changed over night and the morning of race day spent trying to rig the airplane.

The difficulty experienced in test flights and in the race itself indicated that something was loosening up and that the wing warping tendencies were rapidly becoming worse. The airplane was finally rigged very right-wing-heavy to facilitate

getting out of left banks and entered in the race. It was forced down in the seventh lap due to overheating and piston failure.

Had we been able to use the cooler running geared engine; had the wing trussing been complete, or the center rear cabane fitting more secure so the wings couldn't warp; had we known as much then as we know now, none of these difficulties would have arisen; but that is experience. The main difficulties in this airplane were:

1. Engine overheating.
2. Wing warping.
3. Poor vision (a. when flying, b. when landing).
4. Insufficient gasoline capacity.
5. Poor takeoff characteristics.
6. Excessive longitudinal stability.
7. It was too slow.
8. Poor cockpit ventilation.

In the 1932 400 Racer these inherent defects were corrected as follows:

1. A longer sharper nose was put on cowling. The engine was adjusted to throw more oil and the oil tank was air cooled.

2. Wing trussing was redesigned and made complete.

3. The pilot was raised 10 inches so he could see over the upper wing. A slide door was arranged in cockpit covering so pilot could stick his head and shoulders out and see ahead when landing.

4. To increase gas capacity, the fuselage was fattened out near the c. g. without decreasing fineness ratio or streamlining.

5. A controllable pitch propeller greatly improved take-off characteristics.

6. The c. g. was moved aft to correct the excessive longitudinal stability.

7. A retractable landing gear was designed and incorporated to increase speed.

8. The cockpit ventilation intake holes were put exactly on the center line of the leading edge.

It seemed that we had corrected all faults in the original design and the first test flight of the redesigned airplane tended to prove our belief correct until time came to land. Then a new problem arose. The landing gear, in ground tests, dropped all the way out and then was spread and locked into place.

In actual flight, the air loads and the rotation of the slip stream spread the gear before it had dropped out, locked it in an intermediate position and it was necessary to make the first landing on the bottom of the fuselage without landing gear. This was corrected by means of a rubber shock cord which held the wheels together until the telescoping struts were fully extended.

In later flights, it was found that the tail fluttered badly when gliding in slowly for a landing. The exact reason for this has not, as yet, been determined, but an effort is now being made to correct it through the use of larger rear fillet between the lower wing and the fuselage. It may or may not work.

There is no work as intensely interesting as testing and improving high speed airplanes. Not even air racing. But I have yet to hear of the first case of anyone engaged in this work dying of old age.

DEATH STALKS THE AIR RACERS

An instructive article by a well-known race-pilot who reviews the hazards and thrills experienced by the pylon polishers at the air races. It is an intensely interesting story by an authority on the subject.

BY WILLIAM A. ONG

AT SIX-THIRTY on the afternoon of September 4, 1933, the curtain came down upon a stage filled for four consecutive days with daredevil pilots, colorful personalities and hair-raising performances. The stage was Curtiss-Reynolds Airport near Chicago and the show was the International Air Races, official aviation extravaganza of Chicago's Century of Progress.

For four thrill packed days, thousands of spectators had gaped skyward at the spectacle of Lieutenant Tito Falconi, Italian ace, nonchalantly flying upside down past the stands a scant hundred feet above the earth; at Spud Manning, a bashful quiet young fellow from Los Angeles, leaping from a plane at sixteen thousand feet and falling like a black meteor a full fifteen thousand feet before releasing his parachute. They shuddered as Frank Clark of Hollywood, fresh from movieland, threw his white ship into a vicious power spin at six hundred feet, recovering just in time to nearly run his wheels on the ground as he zoomed upward.

Once or twice each year, an everchanging little group of race pilots, chute jumpers and aerial acrobats gather to put on the show that inevitably takes its toll of human life. Chicago's last major aviation spectacle was the National Air Races held on the same airport in 1930. Less than half the competing pilots of that year faced the starters' flag on Labor Day, 1933. Oddly enough, the toll has been heaviest among the most prominent and skilled flyers.

Speed Holman, beloved pilot of the Northwest, crashed upside down at over two hundred miles an hour. Dale Jackson, co-holder of a world's endurance record and premier stunt pilot, was carried to his death in the crumpled wreckage of his plane, unable to extricate himself from the tangled mass to use his parachute. Lowell Bayles hurtled into the ground at nearly three hundred miles an hour as he set a new record for landplanes. Russell Boardman, who flew from New York to

Turkey nonstop, was fatally injured taking off at Indianapolis during the Bendix Trophy Race on July 1st. So they go, cheerful iron-nerved fatalists, shooting the works to give the crowd a thrill.

Now, back to Curtiss-Reynolds Airport on the afternoon of September 4. Fifty thousand thrill seekers had exchanged their pieces of silver for a seat in the crowded grandstands. Thrills they got, and plenty of them. Major Ernst Udet put their hearts in their mouths with a mad exhibition of crazy flying, miraculously missing the ground time after time. Spud Manning brought them right out of their seats with a fifteen thousand foot delayed parachute drop. Falconi, with the zest for showmanship characteristic of his race, spared neither himself nor his ship in an effort to show up Udet. The picked pilots of the Army and the Quantico Marines did their stuff in a frenzied bid for the favor of the crowd. Then came the final event of the four day show, and spectators sat back in their seats as the announcer called the feature of the program, the 100 mile Phillips Trophy Race, ten laps of a ten mile triangular course. Then James Rowan Ewing, acclaimed the best air race announcer on the air, threw a bombshell.

"—now we have a surprise for you. Florence Klingensmith of Minneapolis has just filed her entry in the Phillips Trophy Race. She is the only woman pilot ever to compete in a race of this importance, the fastest race in the world for landplanes. Florence, step up to the microphone and say 'hello' to the crowd."

Let's pause a moment to look at Florence Klingensmith, who has so calmly walked into the very den of the hottest flying bunch of pilots kicking a rudder today. Twenty-six years old. A striking blonde, blue eyed, with a fair skin that no amount of scorching sun or biting dust seemed to mar. Unlike many woman pilots she retained all her attractive femininity. Even now, clad in light green jodphurs and brown Norfolk jacket with a saucy little beret to match her slim figure, she kept her girlish charm.

Florence could fly anything with wings on it and knew it. So did every woman pilot in the country; which may have been the reason she held a warmer spot in the hearts of men pilots, who respected her unusual ability. At a party or around a bridge table, the girls may have their catty moments. But on the flying field feminine rivalry is bitter indeed.

Florence paid little heed to comment, good or bad. In her supremely confident way she continued to fly everything she could find, smoothly, beautifully, perfectly. Now, as she stepped to the microphone, her white teeth flashed with that cocky little smile. "—so I'm going out to try to give the boys a race. I know it's a lot of competition and all that, but I'll tell you more about it after it's over!"

Seven racing airplanes, motors idling, were on the line for a race hero start. Among them Jimmie Wedell, who an hour before had set a new world's speed record of 305 miles an hour, Roscoe Turner, holder of the East-West transcontinental record, and the slim blonde from Minneapolis. Not smiling now, her goggled eyes glued to the flag in the upraised hand of the starter. The flag dropped, idling engines broke into an ear splitting roar, and the Phillips Trophy Race was on.

Florence was away to a fast start, but as the planes came down the home stretch ending the first lap, the superior speed of the Wedell and Turner entries gave evidence that, barring accident, one of the two, would win first place. However, an interesting struggle began with Miles, Minor, Gelbach, and Florence Klingensmith fighting fiercely for follow-up honors. The blonde girl pilot was giving her low-wing monoplane everything it had. A stream of black smoke trailed comet-like behind the ivory and scarlet fuselage. As the racing ships hurtled down the stretch the crowd came to its feet. Wedell in the lead, Turner right on his tail.

Both pilots took the turn nicely, allowing but little distance clearance as they rounded the steel pylon. A cheer came from the crowd,—then swelled to a roar of enthusiastic approval as Florence Klingensmith, rolling her bullet shaped racer vertically on its side, pulled the ship around the marker so closely it seemed her wing must actually brush the pylon! With that turn Florence Klingensmith stole the show from under the very eyes of some of the finest race pilots in the game. From then on, every man and woman in the stands was pulling for her to win and curt comments of admiration for this supreme exhibition of sheer nerve and flying ability came from the pilots.

Sixty miles completed, forty yet to go. Captain James Haizlip, holder of the West-East transcontinental record, was smiling as he checked the stop watch in his hand. At his side, excitedly squeezing her famous husband's arm, was Mae Haizlip, the girl pilot who now holds the world's speed record for women. Her eyes were shining as she watched Florence Klingensmith, flying a beautiful race, go into one of those breath taking turns, ending the sixth lap.

And then, suddenly as a flash of lightning, came tragedy. As Florence came out of the turn the fabric covering on the right wing failed under the strain. Pulled away from the wing as swiftly as one would tear a sheet of paper from a pad. Bits of linen floated lazily earthward and a wave of despairing helplessness swept over the watching pilots on the ground. The ship did not instantly plunge to earth. Like a flash the girl pilot, so terribly alone in the doomed plane, turned the nose from the course and flashed across the field. No one will ever know the thoughts that filled her mind. The instant of indecision—to attempt a jump at a perilously low altitude,—the realization that the ship might plunge into the massed thousands of parked automobiles bordering the airport.

Mae Haizlip buried her face on her husband's shoulder. Jim beat his clenched fist upon my back and shouted hoarsely. "Bail out—jump, Florence,—jump NOW!"

Those age long seconds; twenty, thirty,—and the ship, still travelling over two hundred miles an hour, abruptly began to lose altitude. It cleared the parked cars and then suddenly, without an instant's warning, the nose dropped like a plummet, and the plane plunged to earth. Then did a slim blonde girl join the ghostly squadron whose names and deeds have made high courage.

The crowd got its thrill, and another scar of bitterness was carved upon the hearts of those pilots who watched—and still will carry on in the most dangerous game there is. John Livingston, ace of American race pilots, crushed his cigarette savagely in his open palm and spoke abruptly. "Is there any sense to this at all? How long are we going to pay such a price,—and for what?"

The answer is simple and old as the ages. Natural progression. Glory. Publicity. The roar of the crowd. So long as we seek to outdo our competitor, the price will not come down. So long as the lust of conquest, the thirst to win, the hunger for acclaim beats in human hearts there will be no shortage of candidates for the sacrificial altars of Winged Mercury.

Member No. 1 — Caterpillar Club

A thrilling story of how the first member of the famous Caterpillar Club became eligible for admission to this select body. Few become members by choice and Mr. Wacker's initiation was sure rough.

BY HELEN S. WATERHOUSE

IT was July 21, 1919, when the first Caterpillar of the now famous Caterpillar Club came into existence.

The story of Henry Wacker's leap in an old Spencer-type parachute from a flaming airship and his crash on the seventh story cornice of a building from which he fell two stories to a fire escape is one of the high spots in the saga of lighter-than-air. But this thrilling page of airship history has been left out of most aviation books.

While the experiments of Count Zeppelin, the triumphant flights of the Graf Zeppelin, the Akron and the Macon, and even the Shenandoah tragedy are pretty well known, the name "Wingfoot Express" suggests nothing to the average person. Yet the Wingfoot Express was the first commercial passenger-carrying ship ever built by the Goodyear-Zeppelin corporation before it was called by this name, and in the days when army airships had been successfully built by the Goodyear and flown from the army field in Akron, Ohio.

It was with all confidence, therefore, that Goodyear launched its new project, a commercial passenger-ship, built to carry eight people. Since there was no hangar large enough to hold the 200,000 cubic foot ship in Akron, it was shipped to Chicago where it was assembled ready for flight at the only available hangar of that time in White City Park.

Left: The Wingfoot Express landing. Note the position of the elevators. Above: A close-up of the rotary engine suspension which may or may not have been the cause of the trouble.

The ship was a non-rigid filled with hydrogen of the regulation cigar-shape, with an open gondola suspended below the bag and two motors in the rear.

A mechanic in the back of the gondola took signals from the pilot in front. Jack Boettner, who had taken training in lighter-than-aircraft handling and who has since been in charge of the Goodyear blimp fleet for many years, was the pilot on the ill-fated trip. And Henry Wacker, also a trained lighter-than-air man, went along as observer.

"I can still picture the other poor fellows who went with us, thrilled at the prospect of being the first airship passengers in an American-built commercial ship," said Wacker. "There was Norton, a newspaper camera man from one of the big Chicago dailies. There was Davenport, a publicity man from White City Park and Buck Weaver, our motor mechanic.

"The ship had already made a successful trial flight or two out over the lake. It had made its first landing at Grant Park, where the two Goodyear ground-men in charge of operations, Herman Craft and Bill Young, who had come over from Akron with the ship, were promptly arrested for landing in a park without a permit. This little difficulty having been successfully ironed out, the ship once more took off with two army officers as passengers and returned safely to Grant Park.

"Then we five climbed in ready for a last trip back to the hangar," related Wacker. "We circled out over the lake at 2800 to 3000 feet, riding along smoothly and easily. But suddenly, as we swung in again over the city, the ship gave a lurch. It was so much of a lurch that I realized immediately that something was wrong, and being the only trained airman on board with the exception of Jack Boettner, I left my seat and looked around to see what happened.

"I checked to see if a suspension rod had let go, and looked over the rudder and elevators, but couldn't find anything wrong. Hardly had I sat down again, however, when there came a much bigger lurch or bump. This time Jack also stood up and said 'Something must be wrong.' And at that moment I discovered that the left hand side of the bag, about at the equator line of the ship in the rear, was all in flames. Ten chances to one it was static which had caused it, though I didn't stop to figure that out then.

"The passengers were all very calm and quiet, not realizing that anything serious was happening and unaware of the fire. I knew immediately that the ship was doomed and I told Buck Weaver to jump. Fortunately, we were all equipped with parachutes. Then I hollered to Davenport who was behind me to jump and Boettner told the others. It all happened in an incredibly brief space of time.

"I remember I waved goodbye to Jack and let my body fall over the edge of the ship. Being an old type parachute, it fell 150 to 200 feet before opening. I looked up but couldn't see the ship or any part of it, but I did hear a big noise. Then I saw with horror that my own parachute was on fire in two places. It had apparently become ignited from the flaming bag as I left the ship.

"I watched those two holes smouldering as I fell slowly towards the earth. As they grew larger, the air blowing through them started the chute spinning. Just as I had about given up all hopes of ever landing safely, I saw with relief that the fire in my chute was dying down, and presently it went out.

"My next concern was where was I to land, since I was coming down directly over the heart of Chicago. I remember passing one of the tallest buildings of that day, I believe it was the Gas building, and I missed it by a distance of only fifteen feet. In fact, I passed so close to that building that people leaning from the windows to watch the flaming balloon, reached out their hands and tried to grasp me as I went by.

"Then, to my horror, I saw that I was headed directly for the elevated railroad and all the stories I had ever read about the fatal third rail came into my head. I was afraid that my parachute would drag me right on that rail and I resolved that, after getting a bit closer to the earth, I would cut my parachute so that I would drop clear of it and not be dragged.

"The spinning of the chute was terrific by this time and, on that account, I was unable to see the building into which I suddenly crashed. I hit a cornice on the seventh story and the force of the blow made my parachute collapse. I then fell two stories to a fire escape on the fifth floor of the building. My chute was still attached to a vent pipe on the seventh story when they found me.

"Everything was blank for nine days after that. When I came back to consciousness, I found myself in the Chicago Presbyterian Hospital with two skull fractures, all my ribs broken, a badly injured back and my right arm completely torn from its socket."

It was then that Wacker learned the tragic story of the disaster. The bag had burned completely in mid-air. The gondola and engines went hurtling down through the skylight of the only low structure in that section of the city—the Illinois Savings and Trust Bank. Eleven people, including bank employees and bank customers, were crushed to death beneath the heavy weight of the big engines and gas tank. And of the five passengers who had set forth so gaily in the Wingfoot Express that Monday morning, only Jack Boettner, the pilot, and I had survived the crash.

Jack, it seemed, had been the last one to leave the ship. The flames were beating in his face as he jumped, and he burned one hand badly in shielding his face from them, but that was his only injury. While his chute, as well as Wacker's and all the others caught fire, he was within a few feet of the roof of a building when it finally burned through, and released him. He dropped with a hard bump, becoming Caterpillar Number Two.

When the question arose later, as to whether Wacker or he should be termed Caterpillar Number One, Jack readily conceded first place to his observer, since Wacker left the ship first and landed first.

Poor Davenport, the publicity man from White City, had had no chance to save himself, for his parachute snagged on a cable of the ship before he left it. When it was finally burned free, his body dropped straight as a plummet to the roof of the bank building, where it was found later.

The newspaper man, Norton, with his parachute burning as Wacker's had been, was blown violently against the side of a building and when his parachute collapsed at the blow, he fell to his death. Weaver, the mechanic, was burned to death in mid-air, from the flames of his own parachute.

"What to most flyers is an accident, I have done intentionally. I can crash any type of ship." This renowned stunt flyer reveals the science of his perilous profession.

Crashing Planes for the Movies

BY DICK GRACE

MINE is an unusual line of work. It involves the science of aviation. It involves personal risk, judgment and money. From the very beginning it is thoroughly destructive. In the long years I have been flying it has been my pleasure to wash out over half a million dollars worth of flying equipment.

My work is done for motion picture companies with a specific purpose in mind. No one outside of the cameramen who photograph the thrill and myself are physically involved. Absolutely everyone else is strictly barred from the danger area. There is much preparation before, and much precaution during a crash. Numerous apparently insignificant details are crowded into a moment's time. Taken compositely they are seen in a motion picture as the greatest of thrills —the actual crash of an airplane, such as you saw in "Wings," "Lilac Time," "Young Eagles" and the many other productions.

My crashes are harder than the ordinary washout with which a pilot is confronted. I am forced to hit on an exact spot, at a specified time, on certain signals and with due regard to the cameras. It is not sufficient that I am told by producers when I sign my contracts just how the ship is to lie afterward.

Some of my contracts demand that the ship shall end on its back; and others stipulate that it is *not* to end in such a position. Perhaps I am told to wash out the right wing or the fuselage or hit the nose first. I may be called upon to sideslip or to skid or dive in. It is a most exacting and strenuous bill of

instructions which I receive. Sometimes I have to fly into buildings or into cliffs or to crash on the edge of a stream. There is always one little order which makes it highly difficult to satisfy.

Consider then that these are but a few of the things required of me. Now, with the advent of the sound apparatus, it is not sufficient for me to fulfill the above mentioned details but I must crash with regard to microphones. That means that the motor must have almost full throttle when I hit. It seems that I don't dare cut the switch; the roaring of the motor must be heard on the screen just before the sound of the actual crash.

If the scene requires a dead stick landing then I am confronted with the problem of making an absolute spot crash, with no chance to gun the motor should I overshoot. I want to ask you if you think this is easy? In the washout on the edge of the narrow stream in "Young Eagles" I ended so close to one camera that he couldn't get all the ship in the picture—not by a good margin. The wreckage was less than six feet from his lens and about five from the nearest microphone.

Buddy Rogers was the star of "Young Eagles" and the picture was directed by William Wellman, of "Wings" fame.

"And don't forget," were Wellman's final instructions, "It's got to end on its back—if it doesn't, we can't use it in the picture. Come in, motor on, hit the spot, and slide over on the back—much the same as the crash you did with the Spad in

No-Man's-Land for 'Wings'."

I was seated in the cockpit when the director signaled that all was ready. My lap belt was in place. Chest belts were fastened and the shoulder straps buckled. I manipulated the gas and throttle with three fingers, as I would have to in a few moments. The motor was turned up and tested for the last time. Then I gave orders to remove the chucks.

Away went the old ship down the fairway. Soon the tail lifted out of the mud. We gathered speed. A few hundred feet down the slight incline I pulled the stick back and we were in the air.

Swinging over toward the troupe I saw that all of the cameramen were by their equipment. The red flag from the ground signaled that the microphones were ready to record the sound of the motor and of the crash. Then I banked away, gained a little altitude, made a quick turn.

My altimeter, placed conveniently on my left, indicated that I was a hundred and fifty feet. Now my eyes were on the row of trees. I could see the tall derrick through them, with cameramen and a few of the privileged technical staff on its top landing.

Throttling the motor slightly, I lost about fifty feet. From this height I decided to plunge in.

Nearer and nearer we came.

Few people realize just how fast eighty or ninety miles an hour is until the tops of trees are clipped. I don't think that anyone has the opportunity to measure sensation that I do.

Closer and closer was I coming. Now it was almost time to control into the spot. An accident! A catastrophe! For just a bare second my foot slipped from the loose rudder! To find the wobbly control again I had to take my eyes from the ground. Would there be time? A decision had to be made. I could not climb sufficiently to clear the obstacles ahead. I must crash! If it weren't in the spot it would be into that maze of automobiles and spectators or the tall tower. If the latter, what horrible results could only be conjectured.

A slight glance. My foot was there! Too late. I couldn't get in where I was supposed to. I banked quickly to the left. No one would dare continue in the direction in which I was headed.

Below me, some little distance ahead, was a lone cameraman. I was pocketed again.

No time now to pause and consider a multitude of possible alternatives. No thought of saving the ship or of getting the shot. Certainly no thought of what would happen to me. I was wrong. But once again—when in an airplane under such control—if you're wrong, go ahead. Instantaneous decision is ninety per cent of being right at all times in an emergency. It is the better part of judgment. To be wrong and follow through the mistake is better than to be right too late.

With a yank I pulled the stick toward me as far as it would go. Then the rudder was kicked to the right. My throttle-hand pushed the gun full on.

It was a tailspin position. A tailspin with power but there was a cameraman down there—a man I had worked with on many pictures—Elmer Dyer. The nerviest, most daring individual who ever turned a crank. I knew he was grinding now. I knew he'd grind until I hit or hit him.

The ship almost turned on its back. Then it plunged groundward in a spin. Reaching over I cut my switch and waited.

The ground was coming up at an amazing rate. It grew in size much like a picture does that is moved toward a magnifying glass.

Crash! The motor hit and plunged through the gas tank. The wheels and the landing carriage jutted into the fuselage. Then the entire ship swung around and flew backwards for about twenty-five feet. The tail-skid, the rear of the fuselage, the rear controls broke away. There was an angry snarling of ship and motor parts, as though the old boat was giving vent in its last moments to an ungovernable rage. It was still. It was all over.

Elmer Dyer rushed to the side of the cockpit. His eyes were bright as he looked at me seriously.

"Thanks, Dick, for saving my life," he said.

"Pretty close," I countered.

"You bet, but wait until you see the shot I got!" he said, a faint smile breaking over his tanned face. And I believed him, for he sure was close enough.

What to most flyers is an accident, I have done intentionally, and I believe I speak with authority when I say I can crash *any* type of ship. It can be an old war-time Spad or Fokker D7, or it may be a new production job. I can crash it completely and be able to walk out.

Of course I'd like the benefits of a few of the precautions I observe when crashes are done intentionally, if I were to accidentally be forced down, but I've proven to my own satisfaction that even redressing a ship for a crash is unnecessary. Five out of the last six ships I have used were current stock models of modern types. There was no rebuilding. All of these crack-ups were with power on—all were difficult. One of them was completely stripped of its wings, another hit in so much of a dive that the tires were not blown nor the landing carriage injured, though I landed upside down. On the third I crashed in a flat spin.

Taking into consideration the preparations that are necessary for a crash, we have to start with the pilot. In the first place a highly keyed, unimaginative person should have no thought of succeeding in this line of work. A fear of self produces a disquietude of mind which reacts on the nervous system. It causes a loss of sleep and produces a high muscular tension. No one in such physical shape is capable of cool calculation. When driving a ship into the dirt you cannot duck. With the speed necessary it is most important that everything be seen, or the spot may be missed and several cameramen messed.

Many times I've had people ask me, "Don't you have a few drinks before doing such a thing?" My answer to this is that if I ever had any stimulant before doing a crash I am pretty sure I wouldn't be here now. If I ever need a drink to bolster my nerve I'd better quit. Then I'd be sure that I had lost my intestinal fortitude.

Precautions for the ship vary according to the effect desired and the speed necessary. Crashing of an airplane successfully is nothing more or less than finding a way for the dissipation of forces. If I come in at 95 m. p. h. and 1400 revs. there must be *some* way of diverting that force to the motor and the ship, *before* it is transferred directly to the fuselage and to me. Perhaps a wing strikes first and absorbs a certain amount of shock. Then the propeller and radiator hit. When the motor plunges into the firewall and gas tank there is but a fraction of the initial force transferred to the pit and to me.

One of the most important items of a successful crash is the ground crew which coördinates with me. The riggers and mechanics, after the ship is fitted for the crash, are transferred to the rescue crew. This most important detail is always under the command of a man in whom I have implicit confidence—a naval engineer by the name of Harry Reynolds. Harry gives each of the men just one implement of rescue. Some have pliers and nippers. Others have hack and wood saws, or sledges, wire-cutters, axes. Only on word from their commander are they allowed to approach a wreck and then but one works at a time. It is an extremely quick and efficient system.

Hand in hand with the rescue crew is the fire detail. Two forty-gallon chemical tanks and half a dozen three-gallon fire extinguishers are adequately manned by men experienced in fire fighting. Besides chemicals, the crew is equipped with shovels. If the crash is to be in water a pulmotor is on hand.

A doctor and nurse with emergency kit are near a waiting ambulance or standby car. No detail is overlooked.

I am safe in saying that I receive less thrill than anyone of those who watch me. I am the only one who does not hear anything. The explosive noises make no impression on my auditory nerves. I see little dust, and get no sensation other than that of satisfaction or disappointment that the ship is or is not doing as I wish.

Spectators see the plane come in from afar. They hear the fateful drone of the motor. Their desire to talk aloud is supressed by stern orders of the police, who demand absolute quiet. Nothing but crash sounds must record on the film.

Gradually the ship gets larger before them. They see a slight object flutter from the ship—the pilot's goggles. Then the cameras start to grind—the sound trucks hum their monotonous song of operation.

Seconds seem to lag as the plane speeds along.

Then it seems to shoot suddenly forward. The nose is raised, the right wing drops, the tail goes into the air. With a vicious roar—a splitting, splattering sound, the prop breaks; the motor digs in, the wing tears off.

Perhaps the ship hits so hard that it bounces. It twists over on its back. Directly below the pilot's head a heap of broken struts and steel tubes gather. The spinning wheels fly into the air, the linen of the fuselage is ripped—all four wings are torn off.

With a last quiver the plane settles slowly back—not a rib, not a control useful.

I don't like to crash and I don't dislike to. It is a matter of profession with me. There is no thrill, no sensation, nothing to be proud of in my line of work. But there is the satisfaction of doing a job as specified and doing it in a scientific way.

In "Wings," crashing into a shellpitted no-man's land in a little French Spad gave me some difficult problems.

A washout with this type of ship at any time is serious. In the first place it is so small that it is impossible to get the head below the cowling. No one likes to balance an inverted ship by the muscles of the neck or by the spinal cord.

Again, with wooden longerons, there is danger of splinters and broken sections piercing the back. So these members were taped from the motor mount to the tail. The cockpit was strengthened with light shelby tubing. Special gas tanks were built to replace the pressure system used in this type of ship and the inter bay struts were weakened on the right to allow that wing, which was to hit first, a proper opportunity to fold.

When and where I hit, and my escape without a scratch, probably always will be a thing unbelievable to many of those who saw it. Suffice to say that I tore out a section of barbed wire entanglements for about seventy-five feet—and when I say I tore them out I mean it.

The cedar posts, to which the wire was wrapped, were imbedded two feet in the ground and protruded four. Thirteen of them were completely torn out. On the last roll which the ship made I ducked forward so that I would not be pinned to the ground. Just at that moment a post pierced the cockpit eleven inches back of my head, ripping part of my flying coat away.

I use no personal protection other than a padded helmet. Once a piece of propeller passed between the side of my coat and my controlling hand with sufficient force to pierce a one-eighth-inch steel plate and a steel longeron. How much armor would be needed to give protection to the body—even for such a small splinter?

The solution is not in armor of the body—but in knowledge beforehand of what will splinter and where those splinters will go.

Belts are of the utmost importance. In addition to the regular lap belt I install a set of wide webs across my chest, which are adjusted so tightly before the last takeoff, that it is impossible for me to take a full breath.

I wear the flyer's regular clothing. A helmet, goggles, Bedford cord trousers, oxfords and woolen socks. The reason for the latter in place of boots is that in case of leg injury boots are harder to remove.

Personal precautions are few. Instead of ducking I must watch the ground speeding under me. I have to see the spot approaching. Sometimes I am forced to side-slip and at others skid and fish-tail so that this is possible.

Until I hit, my head must be out on the side of the ship which first crashes. My unprotected eyes are subjected to the full propeller blast, but I know from the several times my head has been dashed into the pit or instrument board that my eyes would have been destroyed had I used goggles. Moreover any sort of optical protection creates a certain amount of distortion and also limits the angle of sight.

Watch your crashes in the movies. Nine times out of ten you'll find some part of the right side of the ship go in first—the reason being the shock will be transferred to the right side of my body. Remember that the heart is on the left. Sometimes the shock I receive is a terrific one—and shock has been known to stop the heart. Moreover the stomach is not in the center but inclined to the left.

Finally when the force is transferred to the cockpit and to me I'm probably taking but forty to fifty miles an hour shock and the scene is the more spectacular for it.

THE Alaskan aviator leads a hazardous and adventurous life.

In addition to his regular work, he often receives weird and unusual assignments. Such as carrying prospectors to unexplored regions. Hunting for lost and frozen men. Locating passes through mountain ranges. Directing schooners through the ice, and even herding reindeer.

Besides this, hardly a week passes that he is not called upon to rescue some injured prospector or trapper out in the wilds. Ever since Ben Eielson first set the example by transporting in his plane a desperately sick man, who would have died had he been carried "over the trail" by dogteam, hundreds of rescue flights have been made in Alaska.

One of the most dramatic of these emergency trips was made by A. A. Bennett into the Chandalar, where is situated the farthest north gold camp in the world.

Here two men named Shaw and Dunlap were testing fuses one night in their cabin. They noticed about eighty caps lying on a table nearby but thought they were far enough away for safety. However, a spark escaped and set the caps off. In the ensuing explosion both men received the full impact of the mass of burning metal in their faces and were at once

DEFYING DEATH IN FROZEN ALASKA

Barren Wastes . . .
Howling Blizzards . . .
Devastating Cold . . .

ALASKA!
the land of adventure.

BY E. STOY REED

Above: How would you like to fly over this country? A typical stretch near Juneau. Below, a Russian Junkers, the first Russian plane ever to reach Fairbanks.

completely blinded.

They lay helpless and groaning on the floor and would have died if a miner from some distance away had not happened to drop in on them that night. He found them so terribly injured that he did not dare leave them. So he scribbled a note and tied it to the collar of his "lead" dog.

How he ever made the animal understand that he was to go for help, remains one of the mysteries of the north. But this malemute climbed across a mountain range to Little Squaw mining camp, twenty miles away, where the note was read and a dogteam dispatched to bring in the blinded men.

Though no airplane had ever been up in the Chandalar before and there was no landing field there, an urgent appeal was telegraphed in to Fairbanks. Pilot Bennett started out in fog so thick that he found his way to Beaver, the half-way point, only by flying ten or twelve feet above the Yukon River.

Stormbound there many days, he at last set out through the mountains. It was snowing so heavily that he could not see the earth at five hundred feet. Hugging the ground, he flew up and down branching creeks, perpetually in danger of crashing into a mountain, feeling his way to the frozen lake at Little Squaw.

"If I ever saw a happy bunch of men in my life," said Pilot Bennett later, "I saw them then. The feeling there was that 'God had sent an angel' and that was the expression they used. Every mine had closed down and the men were working in crews of twos, taking care of Shaw and Dunlap. It was forty below zero that day, and as the plane was only an open two-seater, the only one I had at the time, I did not feel like exposing the injured men to such severe cold."

"Next morning it was a little warmer, so we started out. I had not flown two hours before a sixty-mile blizzard hit me, forcing me down in a hurry. I just managed to spot a small cabin before my view was cut off by a solid wall of wind and snow. How I brought my ship down beside that cabin only heaven knows."

On bursting into the hut, Bennett found that the occupant was away, leaving him without any possible help. Although spare of build himself, he lifted his two fur-clad charges from the plane, one of them weighing at least two hundred pounds, and carried them into the cabin.

For three days they were marooned in this trapper's shanty. Built of moss-chinked logs, with only a Yukon stove inside and a stove-pipe of tomato cans telescoped together, it offered very little resistance against unbelievable fury and death-dealing cold of the blizzard, which swept down upon them from across endless stretches of wilderness.

During those three days Bennett contrived somehow to feed the two helpless men, to care for their eyes, attend to their wants, and to rustle firewood as well, to keep them warm. Anyone knowing this pilot, with his half flippant, half "hard-boiled" exterior, would find it hard to imagine him in the role of an "angel of mercy." But afterwards in the hospital, the two men said: "No one could have been kinder. We owe our lives to Bennett."

The storm died down at last, but in the dead calm that followed, it was fifty degrees below zero. Bennett did not know whether the make of plane he then had would fly in such a low temperature. However, driven by desperation, he gassed the

ship, heated the oil which he had drained on landing, and under a canvas hood warmed the motor with a blow torch to about ninety degrees.

Then he packed Shaw and Dunlap down in their seat and covered their faces with fur. The snow had crusted, enabling him to take off easily with his skiis. He nosed the machine sharply upward with wide open throttle and in this way kept the engine from freezing until he succeeded in reaching an altitude where it was only thirty below.

He groped his way to Fairbanks through the perpetual fog with which interior Alaska is blanketed during severe cold. That afternoon the two suffering men reached at last what seemed like heaven to them, two warm beds in St. Joseph's Hospital at Fairbanks. Dunlap never recovered his sight. Shaw, however, regained the use of one eye.

Many instances, such as the following, are remembered in Alaska of Bennett's never-failing resourcefulness in an emergency.

One day, owing to an exhausted oil supply, he was forced to land near an Indian village. Somewhat disconsolate at his predicament he walked into the big community log cabin, where four or five families lived. He began to wonder how in the world he could ever get back home without any oil.

One of the squaws was making pastry. She was kneading into the flour that favorite Alaskan brand of lard, bear grease. The aviator, as he watched, was seized by a sudden inspiration.

He jumped up and took the can from her, exclaiming: "You sell me bear grease. I pay you heap plenty!" Then he set it on the stove. When it was melted, he added it to his depleted oil supply and continued on to Fairbanks!

At another time he made a forced landing in the interior at forty below zero. The skiis, fusilage, radiator and one wing, were broken. He mended the broken wing with willow limbs and gunnysack twine. The fusilage was made "just as good" with baling wire and willows.

He packed the engine to a roadhouse three miles away and rebuilt it with a hammer, cold chisel and monkey wrench. Then with his passengers, he walked twenty miles to their destination. Returning, he started out in his airplane for Fairbanks, accompanied by one lone, chance-taking passenger. With a leaking radiator, one cylinder out of comission and fire spouting from the exhaust, Bennett crawled back, barely clearing the lowest passes. In a popping, sputtering, fluttering, barely moving ship, he just made the home field!

Frank Dorbandt, versatile Alaskan trail-blazer, remembers as one of his most unique experiences a reindeer round-up which he conducted by airplane. Nine thousand deer were missing in northern Seward Peninsula.

Twelve men had been out looking for them for three weeks. Dorbandt flew over a high range of hills and at last located the runaway deer. He found them scattered in little bands, far and wide over the vast, treeless tundra, some up on hills, others down in the valleys.

"One mountain top, as I flew alongside of it," explained Dorbandt, "looked as if it had suddenly sprouted a forest. A herd was standing there gazing at my plane. Their enormous, branching antlers looked like trees outlined against the sky.

"When I discovered how frightened these animals were of my plane, I decided to attempt driving them with it. Swooping low over each band in turn, I started them one by one, down on their way towards a great plain, where the main herd was already gathered. Terrified by the roar of my engine, they would try to escape by galloping ahead of me.

"As reindeer generally travel in a straight line, all I had to do was to start them off in the right direction. In an hour and a half I had the whole herd of nine thousand rounded up. Then I flew over to the coast to get the owners."

One of the most spectacular flights in Alaska was made by Ralph Wien, the brother and pupil of Noel Wien.

With only eleven hours previous solo flying, he flew across Alaska to save a mail contract when his brother was missing at Point Barrow. Though he had been over the route twice before with Noel as a passenger, he had never as yet set his plane down on any other field than the one at Fairbanks. On this three-day trip he made ten successful landings on unfamiliar fields, fields which were both rough and small.

Noel Wien is one of the greatest living Alaskan aviators. Numerous tales of skill and daring center around his name. Of his many hazardous trips, one of the most exciting was his pursuit of a criminal into the wild and unknown Endicott Mountains.

In March, 1928, a trapper snowshoed many miles into the Indian trading post of Fort Yukon to swear out a warrant against another trapper who, after staying with him in his isolated cabin, had escaped into the mountains.

As it was impossible to follow him by dogteam at that time of year, the Assistant District Attorney of Fairbanks, to whom the case was referred, decided to attempt pursuit by airplane. Piloted by Noel Wien in the Hamilton monoplane which was afterwards wrecked in Siberia, he flew to Fort Yukon, where he conferred with the marshal and the plaintiff.

The trapper drew a rough map for them of the terrain as far north as his cabin on the Coleen River. Of the country farther north nothing was known. As the accused man had escaped into this unexplored region it seemed futile to ask Wien to go after him.

They were about to give up the case when they happened to think of a strange band of Eskimos who had come into town the day before. The first of their kind ever to appear in fort Yukon, they had created a great sensation, especially among the Indians. It turned out that one of these Eskimos knew a few words of English.

On being questioned, he admitted that he had seen a white man in the range. This man had sought shelter with their tribe during a blizzard and for all he knew, might still be camping with them. He said that his tribe had come from the Arctic Coast the year before, crossing the Endicott Range through a pass unknown to white men (Barter River Pass). The southern entrance to this canyon was marked by a great "Black" mountain. The wanderers had found fox and caribou plentiful here and had established a temporary settlement.

"Do you think you could locate this 'black mountain?' " asked the Assistant District Attorney, turning to Wien. "Would you care to risk such a hazardous trip?"

"I'm game if you're game," replied the aviator. "We will gas up here and if we can't find any place to land in the range, we may still have sufficient gas to return."

The local commissioner refused to fly. This necessitated bringing the prisoner and witnesses back with them. Wien pulled out at noontime, carrying the trapper as a guide, the local marshal and the Assistant District Attorney. They left behind them a duplicate of the trapper's map and the instructions: "If we are not back by tomorrow night, wire for a plane from Fairbanks to go in search of us."

Wien flew northeast, following the flat, wooded stretches of the Porcupine River almost to the Canadian border. Then he turned north of west up to Coleen. That river led them into the rugged foothills of the Endicott or Arctic Range. The only possible landing places below were occasional lakes, which were still frozen, although it was nearly May.

At last, just before they reached the timber line of northern Alaska, the trapper pointed out his cabin beside the frozen river. Then they headed for the range, which towered before them as unknown and forbidding as the mountains of another planet.

Wien at first tried following the Colleen to its source. Then, when this one landmark failed him, he struck out blindly, flying at a great height across a maze of treeless, snow-capped mountains, bound on the seemingly hopeless quest of singling out the "black" peak from among hundreds of thousands of peaks.

When they had gone about ninety miles they saw a great pass leading towards the Arctic. It was guarded on the south, not by one, but by three mountains, which appeared "black" compared to the surrounding country because windstorms had almost denuded them of snow.

Although this scene did not fit the Eskimo's description, Wien circled the nearest of the three peaks and then went on to the next one. He was about to fly over this one also and give up the search, as it was growing late, when he happened to notice some dark spots on the snow beneath.

He circled lower suspiciously. Then he was amazed to see tiny black figures running from round, snow-covered huts, which were invisible from the air. He found out afterwards that the dark spots which had attracted his attention were where staked dogs had trampled and packed down the snow. As the snow-covered village itself was indistinguishable from the surrounding country, it seemed a miracle that he had ever found the place.

About a mile away, on a plateau near the summit of the mountain, was situated a smooth, rounded knoll. Wien flew very low over this several times to see if there were any rocks jutting up from under the snow. Then he motioned to his passengers that he was going to attempt landing. While they braced themselves against the shock of possible disaster, he landed at the foot of the incline. When the giant monoplane stopped taxiing, there they were perched on the very top of the knoll!

The Eskimos began running out towards them. But halfway to the plane they stopped. Nothing could induce them to come any closer to this terrifying monster of the air.

On stepping out into the snow, everyone sank to their knees. Though the crust held up the Hamilton, which was supported on broad skiis, human beings could not walk through it. The party donned snow shoes and made their way over to the Eskimos, who, with sun-blackened faces and rough fur clothing, were a very wild and primitive looking species of man. They had children with them, and one of the squaws carried in her parka hood a baby not more than a few weeks old.

"We come for whiteman," said the marshal, trying to make the staring natives understand by sign language. At last a glimmer of understanding dawned in the slit black eyes of one of them, who knew a few words of English. He pointed beyond the village of skin tents.

"Whiteman go," and he made motions that they had driven the man out of their camps.

"How far he go?"

"Not far. Wind he blow," he held up five fingers and whirled his arms to illustrate a terrific blizzard, and then pointed to the sun, giving them to understand that today had been the first clear day.

"You help us find him? You drive us with dogteams?"

The Eskimo agreed. Under his directions the tribe hitched up two teams of wolfish dogs to toboggan sleds. While they were doing this, Noel Wien looked apprehensively at the lowering sun.

"I can't leave the ship perched on that knoll very long," he said. "The least little wind that comes up will wreck her."

The Assistant District Attorney agreed to stay with him and said to the marshal: "We can only allow you two hours. If you are not back within that time, we will have to fly back to the Coleen and spend the night on one of those lakes where there is some shelter from the wind, and come back for you in the morning."

The marshal nodded, and taking the trapper with him, motioned to the Eskimo drivers that he was ready to start. In spite of the light toboggans, the soft, drifted snow made gruelling work for both men and dogs. When they had gone about three miles, they saw ahead of them a mound under the snow with a streak of smoke coming from it. Nearby were curled up some very worn looking malemutes.

The marshal called out, but received no answer. Then he peered inside the skin hut and saw a heavily bearded man lying on the ground wrapped in caribou hides. He started up, gun in hand. The marshal covered him and he sank back.

"What do you want?" he growled. But a glimpse of the stern-faced trapper behind the officer silenced him.

"Get your things on," said the marshal. "There isn't a minute to spare."

"I'm sick."

"Here, I'll help you."

They found that the man's illness was far from being feigned. He had attempted living all winter on the straight meat diet of the primitive Eskimo, with the result that he was a mental and physical wreck. His dogs also were sick and could not have taken him much further. He would have perished half way through the pass.

It was growing dark. The marshal looked at his watch and saw that the time was already up. He hurriedly turned the prisoner's dogs loose and helped him on a sled. Any moment now he expected to see the great monoplane soaring overhead. He was not keen on being left behind in this wild land. He knew there might not be another clear day for weeks and he doubted if Wien could ever find the place again. He motioned to the drivers to hurry up their teams. They lashed the struggling dogs onward.

They were nearing the village, when they heard the ominous whine and roar of the plane. The marshal jumped off his sled and floundered through the snow, wildly brandishing his arms and shouting.

Wien saw them. He had just decided to give them twenty minutes longer, when he caught sight of the long, dark dogteams winding down over a snowfield. He kept the engine going while the natives drove their teams to within a few yards of the plane, refusing to come any closer.

The prisoner was carried the rest of the way and lifted into the cabin. Then Wien taxied through the soft, deep snow. His passengers braced themselves tensely. Would he strike a hidden rock? Could he ever attain sufficient speed to rise? If it had not been down an incline, probably he never would have made it. But as it was, the ponderous monoplane took to the air readily from half way down the knoll safely bound for Fort Yukon.

Alone Across America— BLIND!

by LIEUT.-COL. IRA C. EAKER

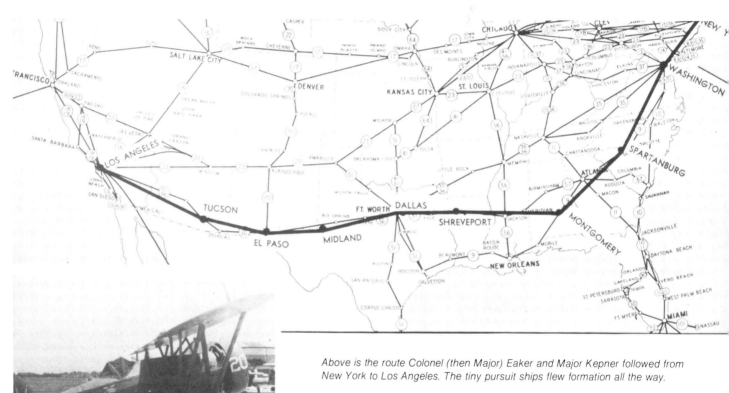

Above is the route Colonel (then Major) Eaker and Major Kepner followed from New York to Los Angeles. The tiny pursuit ships flew formation all the way.

Above is the author's tiny Boeing P-12 fighter with its special hood.

"YOU are over the ocean west of Mines Field now, Ira, you can uncover," came the calm voice of Bill Kepner. I reached up, unlatched the cover which shrouded my cockpit, rolled it back and stuck my head out into the clear, California sunshine. A deep breath of sea-washed air was a great relief after smelling burning oil and feeling the hot breath of my engine for more than 23 hours. Just as my imagination had pictured it so many times on the long trek across the continent, there below me lay red-roofed Santa Monica; on the left was the blue Pacific; to the right sprawled magic Los Angeles; in front were the brown Santa Monica Hills. It was the end of the journey. The first flight across the continent blind, entirely by instrument aid, was finished. . .

This transcontinental blind flight was not the result of a sudden impulse. A long trail of incidents and accidents spread

over more than 18 years of flying spawned the idea and prompted the deed.

The genesis was in the Philippine Islands on a Sunday afternoon. Occasional rain squalls were beating the airport on Pasay Beach, but I was unfamiliar then with tropic rains. Expected back at the Army post at Stotsenburg. I took off and turned north toward dark clouds hanging over the end of Manila Bay. In a few minutes I ploughed into the van of a typhoon which was whipping up out of the China Sea. In a moment it was dark as night, though it was mid-afternoon. The rain descended in liquid sheets. I could see nothing. The plane, my old black Dehavilland, began to do queer things. One moment the controls seemed light and of no effect. The next I was thrown up against one side of the cockpit. I knew the plane was not right side up, there was no telling which side was up. We had no parachutes in those days. It looked bad. The plane evidently was headed downward at an alarming rate, but I seemed unable to do anything about it. As I held a hand before my eyes to keep out the stinging rain, suddenly a yellow mat appeared below; it made a horizon; I jerked the plane to the horizontal and pulled up the falling nose just before it struck the water. It was a close call. But that was not what alarmed me. For the first time I found that I could not fly when I could not see! Had it not been for the fact that the waters of Manila Bay near shore are yellow, making a contrast with the dark green rain, I should have dived into that placid surface. Some say that the flat shores of Manila Bay and the muddy, passive water present an uninteresting prospect. To me they will always be a beautiful land-seascape.

That night I must have been reticent and moody at dinner, for Charles T. Phillips, now a Lieutenant-Colonel in the Air Corps, my roommate, said: "Was the señorita unresponsive this week-end, boy?" It was not the señorita; it was not the fact that I had that afternoon tweaked the long beard of the old man with the scythe and hour-glass. It was the fact that I had discovered an alarming weakness in my flying ability. I did not know that other pilots could not fly blind. I assumed that the trouble lay with me. In those days a flyer did not admit an inferiority. Flying men were known by their deeds. If in the morning a pilot landed out of one turn of a spin, that afternoon you landed out of two turns of a spin. Only the best survived. Well, perhaps that is not true. I have always said that the greatest flyers are dead. Perhaps it was the lucky who lived.

On a morning not long after, low clouds hung over our flying field. I suggested to Lieutenant Longfellow that he take up a DH and fly a while in the clouds. I rode in the back seat. A few minutes after he pulled up into the fleecy canopy, the plane fell into a tail spin. I could see that he looked worried and chagrinned, but I could have cheered for joy. Another pilot, one whose flying ability I admired, could not fly blind. We landed and put our heads together. I told him of my earlier experience. We began to practice cloud flying and when the clouds were not there we chalked our goggles to simulate blindness. We got better, but we were never able to hold a true course. The compass would eventually start rotating and the plane would take up unwanted maneuvers. We borrowed a carpenter's level and set it on the top longeron in the cockpit to tell us by its bubble when the plane was true fore and aft. This helped some, but it did not stop the turning tendency. We were licked. Our answer was to stay out of zero visibility, not to fly when we could not see the ground.

A few months later, we received from the States a little clock-like instrument. At its base it had a liquid tube with a steel ball in it, something like a carpenter's level, but at its top it had a needle which would turn right when the ship turned right and left when the plane turned left. It was called a turn and bank indicator. That was what we had been dreaming about. With it, after a little practice, we could fly in the clouds; we could hold a course when the horizon disappeared. For us the dawn of instrument flying was breaking.

The early reactions of the old flyers toward airplane instruments was a peculiar thing. Real flyers flew by "feel", they said. These fellows who had to have a lot of gadgets to fly were inferior sissies. "What," said they, "does a man need a thermometer for; can't he smell the burning oil, can't he see the radiator boiling over?" It was the same with the flight instruments. "Just a few more trinkets to hang on the old bus," they opined. "Pretty soon the old Christmas tree will be so loaded down she won't take off."

A few months later, when we returned from Air Mail and took up our regular routine of military flying at March Field, I began to worry about our situation there. We had pursuit planes, little single-seaters, equipped with but one flight instrument, the turn and bank indicator. There was not room for more on the tiny instrument board; even the compass had to be mounted on the top wing, because of the congestion in the cockpit. I had learned by experience and practice that there was a distinct difference between flying the big transport planes fully equipped with the latest instruments and flying pursuit planes under similar conditions of poor visibility. I asked General Arnold, our Wing commander, for permission to adapt a cockpit cover for our pursuit planes and give our pilots practice in flying single-seaters blind. Ready (as he always had been since being taught to fly by the Wright brothers in 1911) to undertake or support any experiment which would promote the progress or safety of flight, he said, "Go ahead, but stay up high so that you will have time to uncover if you get in trouble."

Lieutenant Frank Cook of March field and I devised a blind canopy and he built it, doing an excellent job; it served our purpose perfectly. The little cockpit was completely enclosed and the pilot was bereft of any outside vision when our cover was closed. Flight had to be by instrument aid. With that device installed on a Boeing P-12 pursuit I began the tests. Due to the greater maneuverability and instability designed into this little ship to give it suitable acrobatic qualities for air fighting, it was much more difficult to pilot blind than larger planes with more instruments and greater inherent stability. After about three hours under this hood, however, I could steer a reasonably good course and could hold the plane in level flying attitude. It was with that cockpit cover and a P-12 plane

This action shot shows Eaker's ship (No. 20) and Kepner's on the flight.

powered by a Pratt & Whitney *Wasp* engine that I made the first transcontinental blind flight. The only other change found necessary was to mount a compass inside the pilot's compartment, for, when the canopy, was closed, sealing the cockpit, vision of the compass on the top wing was necessarily cut off.

Come sit with me in the little plane on a leg of this flight. As a matter of fact, you can't do that, as the tiny ship is a single-seater, there is scarely room for me, so I'll tell you about it. Just after take-off, and when I have climbed to 2,000 feet over the field, I pull the canopy forward and latch it just under the windshield. This canopy is built on the principle of a baby buggy top. It has a series of oval ribs like a fan, covered with heavy canvas. When it is fastened forward the canvas is taut, making a tent under which the pilot sits. It is built to conform to the shape of the cockpit perimeter, excluding not only outside vision but most of the light as well. To care for that, there is a little electric light bulb set in the center of the instrument board which sheds its light on the board. The power for this light is taken off the battery. By its glow can be seen the turn and bank indicator, on which I keep a constant watch, for by it alone can I tell when the plane is flying true. Just to the left of it the compass face can be seen to tell me my course; by holding this compass needle at 270 degrees, for example, I know the plane is flying west. There are also on the instrument board before me, which I could touch with my head if I leaned forward, the engine instruments—an oil temperature gauge, an ammeter, an oil pressure gauge, and a rate of engine revolution indicator. These I glance at only occasionally to make sure the engine is running true. To my right is the dial for setting my radio at the desired frequency; to my left is the throttle by which I control the engine speed; below, between my feet, mounted on the floor is the generator for the radio and the radio installation. My feet rest on the rudder pedals, and up between my knees is the little stick control which my right hand grasps as I steer the little craft. I sit so low that my right elbow can rest on my knee and that is well, for it holds down the map. On this map I plot my course. That is not necessary, for I can tell by the radio where I am, but it helps to pass the time, as it shows me graphically the progress of the flight. Inside my helmet are earphones so that I constantly hear the radio. There is little chance for change of bodily position, as my shoulders almost touch either side of the cockpit, and if I move, it causes my feet to kick the rudder pedals, disturbing the forward movement of the plane in its straight course. The air in the little cabin is none too pure; I suspect from the headaches I get that there is carbon monoxide gas; also there is a smell of burning oil and it is very hot, as the ventilation is poor; we have sealed the cockpit well.

When an airplane is flying blind there is always a possibility that it may be a menace to other planes in flight on that route, especially those headed in an opposite direction. Generally, when we are practicing blind flight at our home airdromes, we do it in two-seaters and a safety pilot rides behind who can look out ahead. If there be an approaching plane he warns the pilot inside the covered front cockpit to turn, using the interphone for communication. Naturally I could not cruise the national airways without a safety pilot. There was no place for one in

the plane. We licked that situation by having another pursuit plane of the same type fly formation—slightly above and to the rear of my plane, so that he could serve as the lookout for other craft in flight. Each plane was equipped with a two-way radio so that the safety pilot in the plane behind could transmit any warning.

The radio was an invaluable aid to us as it has been continually to all airline flying. It not only told us where we were by the radio beam, but through it we could hear the weather broadcasts ahead, and with it we could hold conversations as though we sat at either end of a telephone; in fact, it was a radio telephone.

The business of selecting a flyer to accompany me, fly formation on my plane all the way, and serve as safety pilot, was not a chance affair. Every flying man thinks, usually quite rightly, that his safety aloft is dependent upon his flying judgment as well as his flying skill. The former is probably the more material factor. Thus the judgment of the man I selected for safety pilot became the ticket for my safe transport across the continent. I must exchange my flying judgment for his. I gave more thought to that selection than to any other one factor surrounding the expedition. I finally asked Major William E. Kepner, Air Corps, if he would accompany me on the flight. He accepted with alacrity.

Bill Kepner has not been in the military flying game as long as some. He was doughboy in 1918 at Chateau Thierry when the Germans shot a hole through his jaw and stuck a bayonet through his face. But he refused to die. When the war was over he decided that the chances for adventure were better "upstairs," so he transferred to the lighter-than-air section of the Army Air Corps. Thus he became a "balloonitic" and dirigible pilot. With Captain A. W. Stevens he teamed up to attempt the first Army-National Geographic stratosphere project. He rode that great, bursted bag down to within 500 feet of the earth before he took to his trusty parachute to save his neck.

In the meantime, he had learned to fly airplanes. I flew pursuit formation with him during the year at the Air Corps Tactical School and found that he could stick in an acrobatic formation or a "dog fight" with the best of them. Furthermore (and what really decided me), he had a true love for the air game; adventure was his next door neighbor and best friend.

The safety pilot had another function to perform. He was to keep my cockpit cover under constant observation, so that he could certify at the end of the journey that I had made the flight without outside vision.

So it was that we climbed into two Boeing P-12s 30 minutes after we graduated from the Tactical School and headed for New York for the beginning of the transcontinental blind flight. On the way we practiced our system and perfected our signals and routine.

On one point we had spent no little time in thought and discussion. We agreed that we needed some positive signal for me to uncover promptly in case the radio failed and some emergency arose demanding immediate attention. Between Wilmington and Philadelphia on the way north I was flying under the hood when Bill's voice came over the radio: "I am going to fly close above you and jazz my engine to see if you can hear it above the roar of yours." There was a spell of silence, then: "Did you hear that?" I had to admit that I had heard nothing save the thunder of my own nine cylinders. In a few minutes he came in again, "Where am I now?"

"You are in front of me close and your propeller wash is about to blow me out of the sky," I said, sketchily, as I fought to regain control of my crazily maneuvering plane. A pilot never mistakes a good dose of propeller wash. He seemed pleased as pie. His voice came in filled with satisfaction, "That will be our signal," he said, "if I can't raise you on the radio and I want you to uncover at once, I shall dive down immediately in front of you, give you a strong whiff of my propeller wash. When you fall out of control you will know that it is time to uncover." I admitted that I would.

At noon June 3, we took off from Mitchell Field and headed for Washington on the first leg of our blind flight. At five minute intervals Bill's voice came in over the radio to tell me that all was well. As we had agreed, if he did not call me at five minute intervals I was to give him a call; if unable to raise him on the radio, I was to rock my wings, which was our positive signal for him to call me. If he did not then call, I was to uncover. It would mean that our communication system, essential to safe blind flight, had broken down completely. At the end of an hour and 50 minutes of flying without incident, Bill called me and said, "You are now over Bowie racetrack. Alter your course five degrees to the right and you will come in over Bolling Field."

On the next leg, from Washington to Spartansburg, S. C., we had our first unprogrammed event. We had just passed Richmond when my radio brought me an order, "Change your course 30 degrees to the right." I knew we were on the beam and could not be that much off our course. That was more of a change than would be necessary to avoid an oncoming plane. It must mean bad weather ahead. Shortly another insistent message said: "Change your course 40 degrees to the left." It was not long until it got so dark inside that I had to turn on the light to see the instruments. Then mist began to sift into my little cubicle. We were in a squall. I knew that, too, by the way the plane was bouncing around. Thereafter, for the next few minutes, messages came in thick and fast for me to change direction and to go lower. Finally, Kepner said, "Fly as straight as you can, I am coming in to fly right on your wing." For a time the only messages which came said, "O.K." Bill told me later that we had reached a point where he could see nothing due to the heavy storm, so he had pulled into tight formation and sat there, both of us flying blind. He could not watch his instruments and see my plane at the same time, so he just kept the top wing of his plane lined up with mine. At one time he said, "Perhaps you had better uncover, Ira; we are down to a thousand feet and it looks very dark ahead, I have been guiding you around thunder storms, but now I can't see any openings. We shall have to turn back or plow through." I told him that I

preferred to plow through. "O.K.," he said. "I have just heard the Department of Commerce weather broadcast, and it is all right about fifty miles further along." It was very rough and we bounced around a lot. It was very hard to hold a course. Eventually we came through into clear skies again. We had not been afraid of colliding with other planes on the airways, because we were flying at the westward altitude. By Department of Commerce ruling, planes flying blind going West cruise at odd numbered thousands of feet, one, three, five, etc., while planes flying East cruise at even numbered thousands of feet.

Our next untoward incident occurred on the leg between Ft. Worth and Midland, Texas. Shortly after passing Ft. Worth, I began to get directions to go lower and then lower. Finally, he told me to pull up to 3,000 feet slowly and fly as straight as possible. I knew what that meant. He was blind again, and was going to fly formation on me until we came out on the other side. In about half an hour he said: "O.K., now, we are in the bright sunshine. I didn't tell you before because I didn't want to put anything else on your mind, but you pulled us through about 50 miles of clouds again. I was as blind as you were."

Over west of Deming, New Mexico, I had another uncomfortable spell. Bill had been telling me at half-hour intervals where we were so that I could plot it on my map. At one time he said: "You are now five miles south of Rodeo." I marked the position on my map and saw that directly in our path lay a peak considerably above our present altitude. If we continued on our present course we could not avoid it; in fact, it was very near at hand. To me, it meant either that Bill did not know our position or we must have more altitude to clear the peak. I did not want to disclose any lack of confidence in my director by asking questions, but I did sneak in another five hundred feet of altitude. In talking it over later, he said that I had misunderstood his statement of our position. He had said that we were five miles north of Rodeo, not south of it.

As we came up out of Imperial Valley and neared San Gorgonio Pass, I told Bill that I wanted to go through the pass at about 6,000 feet. I knew that it was so narrow that in less than five minutes I could hit either side. He said, "I am sorry, but we can't go through at that altitude on account of the low clouds, but I shall call you every minute so that you will know that everything is all right." As we approached the pass it became very rough. My little plane was tossed from side to side like a terrier shakes a rat. But the reassuring voice came over the radio, "You are 500 feet above the ground now; we are just entering the pass; hold your present course and we shall split it in the middle." Then again, "You are over Banning now," and later, "We just passed Beaumont." I knew that already, for the air had calmed down, it was cooler, and I could even smell the orange scent in the air. I knew where we were as well as he; we had broken through to the coastal plain.

I STEPPED OUT

Not many people would walk out the door of an airplane 2,000 feet in the air. So, when a POPULAR AVIATION editor volunteered to try it, we got a bang-up yard on "how it feels." ...

You'd think that, with the parachute so conclusively proved a God-send to mankind, the average person would have some respect for it. And inasmuch as a parachute has never failed to open of its own accord, you'd think the average person would look upon the device as a boon. But very few people will.

Very little has been written in defense of the parachute. Likewise, very little has been written describing the *actual* sensations and thoughts during a (and in my case, my first) 'chute jump.

The actual sensations and thoughts, however, presented a different problem. I had no files to dig into for that material and, if POPULAR AVIATION readers ever were to get any idea of what runs through a fellow's mind up there as he tumbles through space, well, I either had to find someone who could write his feelings for us—or do it myself. . . .

On July 18, a little after 5 p.m., I stepped out the cabin door of a new Stinson 2,000 feet above Curtiss-Reynolds airport at Glenview, Illinois. . . .

The jump itself took a mere two minutes from the time I reached for the rip cord until my feet touched the ground. But Lieut. E. Verne Stewart and John Wickum, veteran Chicago parachute men, had been slyly making me parachute-minded over a five-year period. During the course of my wanderings around various airports and air meets, I had seen Verne and Johnny handle their Triangle 'chutes hundreds of times. Each time built up more and more confidence in the mysterious umbrellas until, just after the recent International Aerobatic exhibition at St. Louis where I'd seen them give their 'chutes their umpteenth workout, I got a letter. Coming from Verne, it was abrupt and quite to the point:

"I think you are ready to jump. I am confident that you now look upon parachutes sensibly. Which Sunday will be the most convenient for you?"

Just like that.

By the time "the" Sunday rolled around I had been told I was crazy by every expert within 100 miles. Experts, incidentally, who never had been any closer to a parachute than the airport fence.

Those five years of gradual familiarization with Verne's and Johnny's parachutes really filled the bill. Scared? Honestly, no. Yes, I was *worried*. But not over the sort of thing the usual wisecrackers had a field day with:"We'll scoop you up," "I've got a sponge and bucket ready," etc. What bothered me was the practical worry that I might pull that rip cord too soon and wrap the 'chute around the plane's tail. Other than that, however, I had no fears.

Walking back from that airplane ride was one of the greatest experiences I have ever had. The wind had died down to six miles an hour and, as the zero hour approached, a mechanic got the Stinson ready by removing the cabin door. Then Verne and Johnny helped me into the harness that held the two 'chutes. (Bureau of Air Commerce regulations require that all premeditated jumps be made with two parachutes. On routine flights, however, the standard single 'chute can be worn.)

Wearing that harness just before my jump was the most unpleasant part of the whole experience. Two parachutes are heavy, and with the harness pulled up until it was almost impossible to walk, well, I was actually looking forward to getting out of the ship and getting that 'chute open to relieve the weight.

Within ten minutes I had practiced dragging that load out the open cabin door (on the ground), we had climbed aboard and were climbing rapidly toward my 2,000-foot ethereal diving board. As I was to be the first one out (Verne jumped right after me) I was sitting almost in the open doorway watching the earth drop farther and farther below our ship. I deliberately kept watching the earth below to see whether I would become as frightened as people believed I would. But not once, throughout the entire jump, did I feel any of that fear I'd been warned about. My warners, you see, had never had any actual experience with parachutes, a fact I should have paid far more attention to.

Friends on the ground told me later that they were breathless as our ship slowly edged up to the point in the sky I bailed out at. Up there in the ship, however, things were entirely different. At a nod from Verne, who had been watching the ship's position in relation to where I was supposed to land on the airport, I dragged those two 'chutes through the cabin door and stood with my right foot on the step outside, my left hand on the Stinson's wing strut and my right hand inside the cabin door to hold me in position.

Actually, I was hanging out the cabin door at about a 60 degree angle. The cool slipstream from the propeller was pounding on my body and I had to hang on tight to keep from being blown off the ship before I got Verne's slap on the back as my signal to go. But even while standing 2,000 feet in the air, with just that thin metal step between my feet and the ground, I did not experience any actual fear. Perhaps it was because I had flown so much before and was familiar with practically every feeling of flying except the actual jump itself.

Though I leaned out the door only a few seconds, I noticed

INTO SPACE

BY MAX KARANT

many things on the ground. Two airplanes were coming in for a landing right under me. I wondered whether they knew we were right above them and would look out for us. I saw a great many upturned white faces. I was somewhat surprised when I looked back at the tail and noticed that it rides high above the cabin level in full flight. Not so much to worry about at that. Then suddenly. . . .

Thump! I felt Verne's slap on my back. I flashed another look downward, then just let go with both hands at the same time. Immediately I left the ship face downward. My right foot hadn't left that step when the rest of my body left the ship with the result that, instead of falling head over heels as I'd originally intended, I fell off on my side in something like a continual barrel-roll. One of the first things I noticed was the streaking blur of airports, skies and airplanes that whirled past me as I seemed to lie perfectly still in mid-air.

There was not the slightest sensation of falling. Quite subconsciously, I recalled the airplanes that flashed before my face. Actually, it was the ship I'd just left that I saw several times as I rolled down the sky. Also quite subconsciously, I knew I was clear of the ship and yanked the rip cord. The big Triangle 'chute didn't open instantly. Actually, the pilot 'chute was dragging the main 28-foot canopy out into the air as I continued to fall. But before the main 'chute took hold I realized I still was falling. Then I remembered I had forgotten to count to three, then pull. Nevertheless I saw my right hand out in front of my face with the rip cord in it and the cord's steel cables flapping in the wind. Even at this point, instead of screaming in terror as several know-alls assured me I'd surely do, I was noticing the neat job that had been done in joining the rip cord cables to the ring itself.

Then suddenly the whole world stopped spinning around my body and I just swished up into a sitting position. The main canopy had filled out. There I sat, about 1,800 feet above that sea of faces, yelling happily over the fact that I'd succeeded in hanging onto that rip cord. It had been impressed on me that keeping the rip cord all the way down on your first jump is some sort of an honor. So the first thing I did was to carefully unbutton the right trouser leg pocket in my coveralls, tuck the rip cord in and rebutton the pocket. Then I noticed how silent it was and that I was sitting in a big chair way above the earth and turning gently as I drifted down. When I thought I heard someone yelling I unbuckled my helmet and looked around. It was Verne, who had jumped right after I had. He was floating off behind me and yelling instructions I couldn't understand. By this time I had found that bailing out is about the greatest feeling one can have and I was thoroughly enjoying the few moments before I had to land. First I pulled the forward risers, then the rear ones (there are four). The riser, incidentally, is that section of the harness to which the shroud lines are fastened in even groups.

I found that the steering qualities claimed for the Triangle parachute were true. The 'chute turned easily as I pulled first one riser than another. Then I heard a yell below me. Down on the airport—I was floating directly over the main section of the field—I saw an open car bouncing in my direction. I recognized a friend of mine in the car's back seat as he tried to stand up and shoot motion pictures of me. He looked funny as he jerked around with that camera pointed at me. Johnny, meanwhile, had leaped out of the car and was running across the field toward the approximate spot I would land on. He was yelling excitedly. Something about turning my 'chute. Glancing down past my feet I noticed I was drifting sideways. I reached up, pulled down the riser I'd found would turn me in the direction I wanted, and held it. The whole 'chute swung round and, when I was headed directly into the wind (just like an airplane landing) I released the riser and the turn stopped instantly. The ground started to come up a little faster. I noticed that all yelling on the ground had stopped. As I floated closer to the field I could see that everyone had stopped talking and were intent on my landing. Only sound was an excited announcer yelling a description of the "parachute landing procedure" at the quiet crowd.

Suddenly the ground appeared to be rushing at me. I relaxed my legs so that I'd collapse easily. Then, before I hardly realized it, my fall seemed to slacken and my feet touched gently in the airport's high grass. I started to fall over but found that I'd landed so lightly I could remain standing right where I had landed. It was a standup landing!

I had expected a breath-taking plunge as I dropped into space. Instead, I got the most pleasant "ride" I've ever had. In other words, nothing my—and various other—imaginations had plotted for me actually took place.

The Beginning of the Gates Circus

BY BILL RHODE

A FEW short years ago, aviation was in its most precarious position, a condition in which the now established industry hung on by the skin of its teeth.

The period I refer to is from 1919 to 1929, but the fact remains that aviation did hang on and credit for this goes only to the barnstormers, that carefree bunch of aviators ranging from unimportant gypsy flyers to the full-fledged flying circuses of four and five planes and pilots. Although this credit has never been stressed before I am going to point it out now.

In 1916, aviation was exhibition flying. Airplanes in those days were frail and apt to fall apart at any time. The world was thrilled to the exploits of Beachy, Masson and the EARLY BIRDS. Then the war broke out. The U. S. Government had no Air Force to speak of. The Allies had proved an air force necessary. Accordingly, the U. S. Air Service was formed and the Government set about buying planes and instructing the youth of the land.

At the close of the war more than 5,000 army trained flyers were turned loose. Most all of these were to return to their old jobs for they saw no future to aviation. At that time there were no flying schools, no commercial airlines, and plane manufacture took a nosedive for there was no demand for aircraft. The majority considered airplanes as instruments of war, lacking commercial use, much the same as artillery or battleships.

A few, however, were optimistic and decided there were some things to be done with a plane and they purchased a ship and set out over the country seeking fame, fortune or at least their daily bread. These were the barnstormers, who kept the fire of aviation glowing during its darkest period, until the flights of Lindbergh, Chamberlin and others focused attention on the neglected offspring of the modern era of transportation.

Among the flock of so-called flying circuses, the Gates Flying Circus was the most famous and greatest of all time, and the one upon which this story is based. But, first we are going to start with the name of Clyde E. (Upsidedown) Pangborn, who was perhaps the nucleus of the flying circus.

Clyde Edward Pangborn was born in Douglas County, Washington; a real westerner who broke broncos, rode logs down the chutes of lumber camps and pursued surveying until the time his country went to war. He then went to join up with the Air Service. He successively passed his physical and mental exams and in May, 1918, he appeared at Eberts Field, Arkansas, for dual instruction. His instructor was Max Miller and, after seven and a half hours dual time, he soloed.

One day, Pang climbed the Jenny trainer 3000 feet to practise figure-eights and fell into a tailspin. Tailspins in those days were little understood and they were fatal in nine cases out of ten. They were not attempted either by instructors or exhibition flyers.

As was the rule in a spin, the rookie usually froze on the controls not knowing what to do. Pang remembered Miller's bidding, "When in a spin reverse the controls and trust to luck." Pangborn collected his thoughts and attempted to pull out. Those on the ground cranked an ambulance and sirens shrieked. Pangborn kept his head, reversed the controls, and

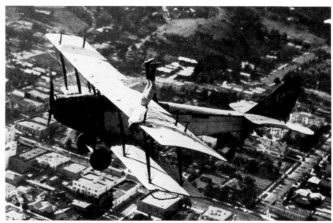

Gates stunter Ormer Locklear headstanding over Los Angeles.

Four famed members of the Gates troupe—(left to right) Lowell Yerex, Milton Girton, Clyde Pangborn and Freddy Lund. Yerex later founded TACA Airlines in Central America and became a millionaire.

pulled out at the tree tops. Just a narrow squeeze for him.

It was then that he acquired the self-confidence that later was responsible for his reputation as a competent pilot. He tried loops doing twelve in a row.

Now, it so occurred that Pangborn and Miller were both sweet on a young flapper living near the field. Pang, unknown to Miller, decided to fly out to the girl's house which was a few miles from the field. Pang picked a pasture to set down in. The field was level but not long. Pang got down, but tossed the Jenny into a fence, mangling a wing-tip. He was in a tough position. He was sure that if Miller heard about him being out with his girl, Pang might just as well start packing home for St. Marie, Idaho, his hometown.

He begged the girl to be silent, climbed into the battered Jenny and flew back to Eberts. When Miller saw the ship he was nonplussed. "Jee-whiz" he bawled, "You flew back in that? You must have had a forced landing."

Pang left Eberts for advanced training at Love Field, Dallas, Texas, in October, 1918. He was there three months. Early in 1919 he was sent to Ellington Field, Houston, Texas, as an instructor. It was here he met his wartime buddy, Ralph A. Reid.

The average army flyer of 1919 didn't care about getting his pants dirty, or getting grease on his boots.

Lt's Reid and Pangborn were different. The C. O. approached them one day. "How would you two fellows care to have your own personal ship while here at the field?" They lost no daytime in telling him they'd take his mother-in-law "for a ride," just for the chance. His agreement was that if they patched and repaired an old Jenny on the field, they could use it as their personal ship as long as they were on the field. As a result they learned to patch and repair a plane and could then fly whenever and wherever they cared to. But straightaway flying is tiresome, so they began thinking up things to do.

They would fly high until they saw a barnyard and then dive down hedgehopping. Horses leaped fences, cows kicked over milk buckets, hens stopped laying. They tried to take the knob off the community flagpole and tried to pick the hands from the town hall clock. They waved to stenographers in Houston office buildings. They would fly high, until they would sight an inter-urban trolley car. Deciding it was full of Germans, they would dive down in front of it and wag their tail in the astonished conductor's face. Complaints came pouring in. The Commandant called them on the carpet and gave 'em hell.

Had they been carrying on, endangering their own and other people's lives with wild stunts? Why no, of course not, they had been away several miles at that time, shooting landings on an emergency field. The number was 314? "Well, well, must be another ship with the same number."

They went shopping for some washable show-card paint and turned the number into a puzzle. They made the "3" look like an "8," and the "1" look like a "4," etc. Sometimes the numbers 841, 341, or 344 were reported, but never 314, and they continued to strafe trolleys, and to run their wheels on the roofs of Pullman coaches.

Just before they were let out of the Air Service, they stole a plane. It so happened that the Jennies they used were slow and cumbersome. One day, however, after mechanics finished tinkering with it, a brand new 150 h. p. Hisso Jenny, a JN4-H, was wheeled out on the field. It was the pride of the field. Lt. Petersen was to fly it. Pang and Reid decided they could fly it themselves. When no one was looking, Reid spun the prop, and Pang gunned the motor, and before anyone could stop them, were off. They flew until their gas ran out, doing a collection of all the stunts they could think of.

Two days later, Reid got his discharge papers as his term was up. Pangborn was alone and the days dragged on. A week later Pang, honorably discharged, started for home. His ticket called for transportation to St. Marie, Idaho. He never reached there, instead he heard that a man named McClellen had built a knockdown plane powered with a Curtiss OX-2 motor, and had formed the Northwest Aircraft Co., at Spokane, Wash.

McClellen was looking for a pilot to fly the plane for exhibitions he had booked. He made several mistakes. He put the exhausts in such a position they set the longerons on fire. This was soon remedied. He also had built a knockdown plane, one that could be taken down, shipped to its destination, exhibited, then taken down and shipped again. However, Pang signed on as test pilot at fifty dollars per week, and proceeded to give exhibitions, one of which was in Kalmaith, Idaho.

Pang was using the baseball diamond for a field. Among the crowd were a few of his relations. He was giving his best for the crowd when his motor quit. He couldn't reach the field so he set down in a small open space, bounced over a fence and plunged into a dry creek bed. The prop stopped sideways so when the old "McClellen" turned turtle, it did little damage.

The radiator of the plane was an old one used by Cy Christofferson back in 1911. Cy abandoned it after an exhibition flight. Years later, McClellen appropriated the same radiator. One day it boiled over and that was the end of the Northwest Aircraft Co.

Four exhibitions in less than a month had netted McClellen 2100 dollars where the plane had cost only 2000 dollars to build. Pangborn saw the profit of McClellen's short venture and this amazed him. He began to dream of heavy profits to be made out of aviation.

Pang arrived home in July, 1919. He became a promoter. The first thing he tried to promote was his own plane. He poured a tale of heavy profits into the ears of A. B. Annis, banker and druggist of St. Marie. Pang suggested that they organize a company and raise 5000 dollars. The investors were to receive 51 percent, and Pang as General Manager was to receive 49 percent. Although Annis was no aviation bug he agreed to join Pang and 5000 dollars was raised.

The one thing that Pang and other pilots failed to realize was that the public was ready to ride. With the 5000 dollars, he set off for Hagerstown, Maryland, to buy a sport Bellanca Biplane. His idea was to buy a suitable biplane for the purpose of giving exhibitions. He decided to stop off at Lincoln, Nebraska, to see a representative of the Curtiss Company. The agent advised him that he had as many airplanes as Napoleon had at Waterloo. He suggested that Pang go to Los Angeles. Pang wired the Bellanca outfit and was surprised to hear that production on the Bellancas had not yet begun. Pang took a train for Los Angeles. At Los Angeles, he found the agent had moved to Sacramento.

At this time only Curtiss and Bellanca were preparing to manufacture aircraft on a production basis. The reason was that there was no demand for aircraft. The Government had many more planes than pilots and had hundreds of planes stored away that were never uncrated. Curtiss bought back about 2000 Jennies and about 3000 motors and began unloading them for four or five thousand dollars each.

However, to get back to the story, Pang found the agent and found he could buy a ship. Cecil B. DeMille bought the first three for DeMille Field in Los Angeles, O. H. Matley a promoter, bought two more. Pang bought the sixth Jenny sold on the Pacific Coast for 4750 dollars.

Once sure it was his, he wired Ralph Reid, who was trying to make a living in the electrical business in Kalamazoo, Michigan. Reid quit his business and rushed to Sacramento. Both rolled up their sleeves and went to work assembling and rigging the Jenny. They had it up for a test flight a few days later.

It must be realized that in 1919 there were not the airlines, the radio weather reports, beams, beacons, airtrails, or even accurate air-maps that there are today. There were only approximately 150 established airports as against 4000 in the U. S. in 1931. When a pilot took off for a destination he had no idea what he was going to land on when he got there.

Pang's and Reid's first booking was to be Wallace, Idaho. So, without consulting anyone in particular, they loaded their tools and baggage and climbed into their coveralls, and prepared to take off for Idaho. The C.O. at Mather Field was curious. "Where are you fellows bound?" he wanted to know. "Up Siskiyou Pass, across the Sierras for Idaho." "You're crazy," he said, "No ship ever flew over there, yet, you can't do it in a Jenny, she won't climb high enough." "Well, it's time it was done," said the boys, and took off.

Their first stop was Redding, Cal. They were to make several before reaching Wallace, Idaho. They landed OK, fueled, and took off again. As they passed over Mt. Lassen, Mt. Shasta, and other peaks, they ran out of oil. The motor got hot and quit. Pang at the controls was looking for a nice bushy tree to set in when he sighted a clearing, bounced over a ditch and negotiated a safe landing.

They found they were near Gazelle, Shasta County. They refueled and got off again. A sparkplug fouled because of too much oil and they were forced to seek a landing again at Montague, Cal. Just for the hell of it, Pang spun down. People ran five miles to see the accident, or rather the purported accident. They landed safely and stayed overnight. The next morning they were roused by farmers begging to buy rides. The boys were surprised but lost no time in flying 21 passengers at 10 dollars per head. They stopped only when they broke a wing-skid. They also found that lava formations on the ground had nicked their tires. This is, however, the first case of a barn-stormer who carried passengers as part of a daily routine. They didn't expect, when they started out, to find their fortune in passenger carrying. They also made 50 dollars more when Claude Gillis, a local man, paid them to fly him to Medford, Oregon. Reid went by automobile.

They refueled at Medford and took off for Roseburg. They landed in a field that ended against a fence. They broke a prop and the leading edge of one wing. They wired Berkeley, Cal., for a new prop and while waiting took off the wing and repaired it. The prop arrived a week later and they got going again. While at Roseburg, they sighted a forest fire, this paved the way for the establishment of an Aerial Forest patrol.

They flew in and out of Salem and Portland but didn't bother looking for passengers. They then wrote the first air letter in an airplane and dropped it overboard at Cottage Grove, Ore. The finder posted it as directed and Orville Southworth, another flyer, received it in Wyoming.

They then became the first flyers to fly a commercial plane up the Columbia River, landing at the Dalles, Oregon. They landed on a sandlot on the river-bottom. In taking off, they hooked a wing in a bush and spun around landing in the opposite direction. They pancaked, crushing the landing-gear and breaking the paddle. They wired Berkeley for a new paddle. Their original starting capital of $300 and money made along the way, had done a disappearing act. They sent an urgent request to St. Marie for $300, and got it.

The proud Jenny drifted into the old hometown on a boxcar. At St. Marie, they proceeded to repair the Jenny. They imagined they had cracked the fuselage, so they tore off the fabric and sent to Curtiss in Buffalo for a new cover. Curtiss had, in the meantime, moved to Long Island, N. Y. In due time, they had the ship rebuilt but the cover didn't arrive.

Pang's mother, Mrs. Opal Pangborn, desired to see the plane and went on the field. Pang explained the Jenny's intricacies and casually mentioned that if they had the cover they could take off. She set to work and made them a cover of linen. Two days later they were flying again. They painted the Jenny a brilliant red in place of the old olive drab color. A bull charged the plane one day and Reid at the controls, barely got off in time. He remained aloft until the bull was calmed and tied.

A week of passenger carrying in St. Marie netted the company $2500. This was immediately paid in dividends, perhaps the biggest dividends ever actually paid. The stock company sunk soon after.

They barnstormed Idaho and pulled into Coeur d'Alene. The field they chose had a slight downgrade. They got into the field and headed for some big pine trees. Too late to pull out, Pang tried turning, went into ploughed ground and nosed over. They wired Berkeley for a new paddle. They loaded the plane onto a tramp steamer and returned to St. Marie on the St. Joe River. Pang resolved then and there to some day fly into the old hometown.

They rebuilt the plane and, as the winter of 1919-20 was on its way, they shipped the plane to Sacramento. When they reached Sacramento they found that the plane had grown restless and tried flying out through the roof of the boxcar. So they wired Berkeley for a new paddle. They again made repairs and painted the plane a pure white. Pang was becoming widely known as an upsidedown flyer. This type of flying was new to the public at this time.

Woodland, Cal., appeared attractive territory, so they flew there. On the way a wristpin broke and the thrashing connecting rod tore the OX-5 in half. They had to put up $1250 for a new motor.

I Learned About Flying From That!

BY CAPT. ERNEST GANN

Captain Ernest Gann

Over-confidence on a 40-mile flight taught this veteran airline pilot a good lesson.

THE day it happened, I had 152 hours in my log book. With that amount of time I should have known better, but I didn't. I still didn't have my sky-thinking completely in line and I fell right into the same trap that has caught so many other pilots. In my vast confidence I forgot that you can get in just as much trouble on a short flight as you can on a long one.

I forgot the old rule that an airplane isn't landed until it's in the hangar under wraps—it doesn't make any difference if it's been up four minutes or four hours. I never ignored that basic air law again. The smallest mistake made around an airplane seems to have a way of multiplying into a great big snowball that winds up into a real jam.

Ever since I had soloed I had done more cross-country flying than average. I was pretty proud of my aviation and had made a number of two and three day flights out of little Christie airport near New York. On the day in question, I wasn't going very far—just a little 40-mile hop to an emergency field at Wurtsboro, N. Y. I owned an old Waco "A" at the time, equipped with a Jacobs engine. Not many of these were built. They were the usual Waco biplane design, but the fuselage held an open side-by-side cockpit. It was another attempt to get away from the tandem arrangement whereby the pilot had to yell at his passenger or student through a Gosport tube or simply into the prop blast. Although it made the ship very blind on the ground and somewhat so in the air, it was a good idea and is still in use.

The other peculiarity about the Waco "A" was the combination throttle-brake arrangement. Both were operated by the same handle located on either side of the cockpit. To operate the throttle, you simply moved the handle back and forth in the usual manner. To operate the brakes, you pulled *down* on the handle, kicking rudder at the same time. It was a rather tricky setup on first acquaintance—many pilots (including myself) got throttle when they wanted brake and vice versa. You had to be particularly careful not to pull the throttle in its outward arc of motion when landing or the brakes would take hold. The combined duties of this single-throttle handle made for a complicated joint affair attached to the fuselage. Its exact construction I cannot remember and it's unimportant— the lesson I learned from it is something else.

I arrived at the field late. That was my first mistake. Never arrive at the airport later than you originally intend—get there well ahead of time if you can. For some reason or other, the tendency is to make your take-off when you planned, regardless of how little time you've been at the field. A rushed take-off means hurrying things and in an airplane that's asking for trouble.

A friend was to meet me at Wurtsboro airport at 11 that morning. I hadn't seen him for a long time and I was anxious to show him that I was now a good enough pilot to arrive right on the dot. I hadn't yet learned that a dash of humility is a sure sign of a really experienced pilot. I grabbed a few maps, took a quick glance at them (after all, it was only 40 miles), then got the ship out of the hangar. There was a ragged overcast about 3,000 feet but that didn't bother me—remember?—it was only 40 miles. Such was my haste to make schedule that I didn't even bother to call Newark for a weather forecast. I kept telling myself it was only 30 minutes till I'd be gliding into Wurtsboro.

As I ran the engine up and tested the mags, I noticed the throttle was slipping back. The control seemed positive but it wouldn't hold in place. Ordinarily it held through friction and the tight ball joint. Everything else worked fine. It was such a little thing (I thought)—if I were starting on a long flight I'd wait and fix it, but 30 minutes . . . well, I'd just hold the darn thing. Nothing to it. Fix it when I got back. I had 152 hour confidence.

I took off and headed north. At 11 o'clock I was circling the field at Wurtsboro. I could see my friend standing on the ground waiting. I landed and we had a nice talk—too nice. I stayed for three hours. I told him that the throttle was slipping and we tried to fix it with the only tool available, a chipped screw driver from his car. It didn't worry me; it was just a nuisance to hang on to it. We thought we fixed it, but we didn't.

I shook hands with my friend, took off and headed back to Christie airport. I climbed to 2,000 feet and slipped along just beneath the overcast. There was a very light rain and some good dark muck ahead . . . but after all it was only 40 miles. I keep repeating that because I was thinking in terms of 40 miles. I hope you never do. Fat, dumb and happy, I charged right into as fine a thunderstorm as could be had.

Ten minutes out of Wurtsboro it really began to rain. I couldn't fly on instruments even if I'd had them so I let down to 1,500 feet. The hills weren't much below me. Still on the course, I flew into what seemed like a solid sheet of water. The turbulence was terrific. It was all I could do to keep the Waco right side up. I was continuously conked on the head with various sized hail stones—and that throttle! Every bump, every twist would send it scooting back to me. The engine would settle immediately into a nice quiet idle. The Waco and I would respond with a rapid descent. What I urgently needed was a third hand. I had one on the stick, very busy, and the other on the throttle to hold it open. This left nothing to hold the map but my tightly-pressed knees which couldn't stay pressed and work the rudder against that turbulence.

I tried sitting on the map, which held it all right, but if I unfolded it enough for a good look it tried to get out of the cockpit. If you've flown open-cockpit ships in highly turbulent air you know what I'm talking about. I finally put the map between my chattering teeth and did some thinking. Obviously I couldn't keep this up for long. Even with a good throttle and a steady compass it would have been a job to stay on course. I decided I wanted to get out of that thunderstorm and made what I hoped was a 180° turn. That was about mistake No. 20.

The rain got worse, the air became rougher. I was drenched, blue with cold and squirmed uncomfortably in a pool of water. I had long before pushed my goggles up because I couldn't see through them—yet with my bare face hanging out I couldn't see much over the side. The wet map was limp as a rag and kept slipping to the floor. Once it worked back under the seat. I did a hand-stand retrieving it. I took my hand from the throttle for a split second. When I raised my head again the Waco was diving for the treetops. Funny? I didn't think so at the time. I was dressed as an aviator and doing duty as a deep sea diver. I gasped for air and got a mouthful of water. I was literally drowning at 1,000 feet above sea level. With the air speed dropping to 40 m.p.h. every few seconds the water just bucketed in the cockpit.

I called the throttle a number of uncomplimentary things. I should have turned my scorn inward. Only 40 miles! I had already flown that in circles and was no better off. All I knew was that I was slowly but surely drowning in midair somewhere between Wurtsboro and Christie—but where? On a little 30-minute hop I had become a very lost aviator.

By taking a brutal face lashing, I could just peek over the side. A village slipped past beneath. I circled it for what seemed like an hour trying to identify it on the map. Actually it was closer to five minutes, but I was busier than the proverbial paperhanger. I finally managed to tie the throttle in position with my handkerchief—that's quite a performance too—but not in cruising position. It just wouldn't stay there. It held just enough revs to keep me from stalling in level flight. The slightest climb and off we'd go in the start of a low power spin. I hated to look at the airspeed.

The map, despite its government durability, was now a wet mess resembling oatmeal and just about as easy to handle. I finally managed a really good look at it with my stinging eyes and identified the town. I was a good 30 miles off course. I had two alternatives. There was a little field near the town that didn't look too terrible. It was the first I had seen worth considering. I could attempt a landing in it, or set a course for Christie airport. Now that I'm older and wiser, I would have landed and fixed the throttle. As it was, I elected to carry on to Christie.

By the time I had a new course figured out I was involuntarily hedge-hopping again—banging through the rain at a near stall. I was mad and I was scared, and I was kicking myself in the pants—all at the same time.

The only right thing I did was to hold that course once I had set it, through high and low water. I got an occasional bleary glimpse of a lot of strange country. I looked *up* at the top of a few hills as they marched past in clammy procession, but I practically ran into Christie airport. I landed and poured myself out of the Waco. As my feet squeegeed across the grass to the hangar, they mocked every excuse I could think up.

Forty miles, I thought, may be only 40 miles—but they grow long or short, as the case may be. An airplane is the fastest method of transport, but give it time before you start. A few minutes saved scurrying around on the ground, or refusing to be delayed, can turn into embarrassing air hours without half trying.

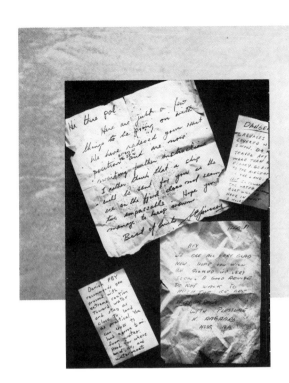

MAYDAY FROM CHARLIE FOXTROT

*This was to be his last transoceanic ferry flight.
He had made 116 others without a mishap.*

BY MAX CONRAD

I CAN'T SAY I learned a lot about flying from this experience, but I sure learned to believe what I already knew. After thirty years of flying, including 116 transocean crossings and a couple of world records, I've picked up most of the well accepted principles of safe flying. And I may have even invented a few. I still learn something new every time I fly, but on this last occasion, as I say, I learned to have confidence in what I learned a long time ago.

I'd accepted a job last December to ferry this lovely looking little Procaer Picchio from Italy to the States. When I got the word that the airplane was ready, it was March first. I'd also accepted a speaking engagement in Iowa on the 20th of March. So it looked like a busy month.

By the time I got everything put together and the paperwork finished, I had put 12 hours on the airplane and found many things I like very much. It has a stick instead of a wheel and it's super sensitive—a beautiful rig with good speed and nice characteristics.

There was just one thing that disturbed me, and that was the engine. Not that there's anything wrong with the powerplant which was in her. I've flown that thousands of hours, but there was something wrong with the installation or the rigging or the components because several times I got a strange vibration, almost a roughness, though not exactly; and once or twice it actually missed. Each time I checked everything completely, but found nothing wrong. I decided to go over to Oxford for a full day's inspection before heading west.

By the time I'd taken things apart, put them together again and test flown without incident, I had five days left before my date in Iowa, and I knew I could make it.

The same afternoon I finished up the engine an intense low moved through England and headed for Greenland. I could catch 30- to 40-knot tailwinds in clearing weather. Conditions were perfect. I left for Prestwick that night. At dawn, with four days left, I headed for Reykjavik, climbed to 8,500 feet and was in Iceland an hour ahead of ETA.

The wind was so great and looked to stay that I checked things once more, gassed up and was pointed on a great circle to Goose Bay in under two hours. Labrador weather was not too good, but Greenland was and so was Boston, where I really intended to go if the wind held.

And it held on the leg to Greenland at almost 40 knots on my tail. I was calling Prince Christian on the East Coast before I knew it. I couldn't raise them at all after several tries, so steered a bit northwest to pick up BW1 and check Goose Bay weather. Although my radios had been working fine, I couldn't raise them either. I kept at it for a while on different channels, all the time continuing on course. The weather was fine; I knew the Boston weather would hold, and I felt alert and comfortable.

The vibration, the roughness and specks of oil on the windshield all happened at the same time. I saw the oil pressure needle go to the stop as I started my one-eighty back to BW1. By the time I got around, the oil was pouring over the cowl, and the windshield was covered. I pulled down rpm to a minimum and figured I had about five minutes' flying time before the engine would seize.

I had just cleared the west coast of Greenland when all this happened, and I knew I could ditch by the shoreline if I had to. But I'd noticed a clear, flat glacier a way back, and I kept S-ing the plane to keep that in sight in case I couldn't make it to BW1.

While still in my turn, I was calling Mayday for the first time in my life. I gave my position and was pleased to hear an answer almost immediately. BW1 had my message, and I told them I'd continue to report my progress every minute. After four minutes the oil pressure was gone, and I knew I'd have to cut the engine.

The glacier stretched out in front of me for several miles. I picked a spot that seemed smooth and free of crevasses, pulled the mixture and started spiraling down. I could see now only through the side vent window. The propeller had stopped in a vertical position, so I touched the starter to bring it horizontal to minimize damage. I cut the switches as I eased onto the ice. I was gear-up, so I slid for at least two or three hundred feet.

I climbed out and brought my gear with me. I spread things around to attract attention. It was about 10 o'clock Greenland time. The sun was shining, and it was not too cold. I had some food but I determined not to eat until I had been spotted, or for two days, whichever was first.

I hoped a helicopter might come for me, but it was a C-54 that flew over, dropped a note, food and warm clothes.

One of the notes warned in broken English: "A good advise, do not walk too much. The ice are dangerous." But another note from a Danish PBY Catalina flying boat recommended that I proceed "with extreme caution" toward the water where the land, ice and water meet.

So I put on some warm socks, and some real fine Arctic boots and a parka that they had dropped, and decided I would try to go down to the shore. Looking over the aircraft, I found it was undamaged. And from it I took four bags of gear, two light ones and two heavy ones. One was a big bag weighing about 45 pounds in which I had loaded food and gear I thought I might need, and the other held a two-man raft, which weighed about 35 pounds.

Then I started out. But as I walked, the snow got deeper and deeper, and I found that after I had gone just 35 or 40 steps, I was rather shocked how my heart started beating. [Max is now 61.] And I thought, "Wait a minute, fellow, you'd better just be careful." So I went back to the aircraft and picked what I thought would be a more direct route to the shore.

The crust was just about hard enough so you could almost get on top, and then it would break.

Well, what looked like a half mile to the edge of the glacier really turned out to be two miles or better. And meanwhile the PBY stayed over me, flying back and forth toward the water. I knew what they were doing. They were showing me the routing, like a pointer, to the water.

And finally I saw a boat coming up through the ice down by the shore. I was surprised how big it was—about 10 times bigger than I expected. I was looking for a little fishing vessel or launch, and this was a large steamship.

But meanwhile the sun had gone down, and I was in the shadows and it was beginning to get cooler. The ship slowly came closer, and I could see the big chunks of ice being pushed out of the way.

Well, I finally had to abandon my two heaviest and most valuable bags—the ones including the candy for the kids—because I still had the problem of getting off the glacier onto the mountain. It was then that I got a panic feeling for the first time, because I was so tired.

But I did make my way to the bottom of the mountain, where there were nothing but round boulders, some big, and some little—all covered with two feet or more of snow. At one time I broke through the snow between two big boulders right up to my shoulders—and my feet didn't hit bottom.

But I worked my way down, holding onto those two bags. And just about the time I got to the water, the ship had worked its way to about 300 feet from shore, and it now inched its way forward again and actually got to within about 75 feet of the shore, when suddenly the prow lifted two or three feet, indicating they had hit bottom.

They lowered the small boat then, and the captain came up to the edge of the ship and called to me in English, and asked how I was.

I told him "Fine, and sure good to see you." Then I said, "Why don't you shoot me a rope, and tie it onto the front of the rowboat. Then get two men in the front and let them kick the ice out in front, and I'll pull the boat in."

And that's what they did, and I pulled the rowboat in. It came in—just like pulling in a big fish. I never even got wet. And they pulled the boat back.

I boarded. The captain took me to his cabin and told me he just happened to be at the end of Greenland at the time of my forced landing. His ship makes the trip several times a year. As it turns out, this was his first command, his first trip out.

Then he told me that the PBY had landed in the water offshore and was waiting for me. We got out into open water, and there it was, just sitting with the engines running. The ship let down a rowboat, I said goodbye to everybody, took my baggage, and we paddled over to the aircraft. I climbed in and we took off.

The USAF C-54 was still with us and followed us up to Narssarssuaq. And when we landed, it was just dark.

Triple Trouble

*Rusty technique is better
than no technique at all—when Fate
plants you in the pilot's seat.*

BY FREDERICK K. PURDUM

AS I looked down at the moving terrain, the field glasses picked out little other than the dark brown of the Korean earth and the equally drab vegetation. There was no sign of the enemy artillery battery that had been harassing our lines with high explosive shells day after day.

"Any luck back there? See anything?"

It was my pilot, Second Lieutenant Russ Matz of Oakland, Calif., speaking over the intercom radio system.

"Nothing yet," I told him through the microphone. "Just keep circling. If it's here, we'll find it."

"Sure. Sure." His voice was cheerful. "But remember what the man said about the third time being the charm." I was remembering all too well. One of the other pilots in our outfit, the observation squadron of the First Marine Air Wing, had asked me just before take-off if I was sure I felt like flying the mission. I had said that I did, and he had cautioned me about running out of luck, since trouble always seemed to come in three parts.

The first two parts had come in the preceding three days. On Saturday it had started, when another pilot and I had crashed on our narrow, bumpy dirt strip that was barely 900 feet long. The light observation plane, an L-19, had run off the end of the runway, cartwheeling into the rice paddy. The crash crew had dragged me out of the twisted wreckage as the plane lay on its back. My safety belt held me head down in the seat, unable to move.

On Sunday, I had been flying an observation mission, searching for what was reported to be a tank. It had been an anti-aircraft emplacement instead. It had found us first, and the 40 mm shells had ripped a pair of holes in the wing and tail of the unarmed observation craft. We had hardly been able to limp back to our home strip and make a landing.

Monday, nothing happened. This was Tuesday. Maybe the jinx was broken, I was thinking—trying to convince myself. Trouble in trios, though. I had seen it before. You may not believe in it. You may say you're above superstitions . . . but you still think about it.

This was one of those situations that happen in a war. Coming out of flight school too late for combat missions in World War II, I had given up my flight status and been

assigned to ground duty, ending up with an artillery regiment. Korea had come. Assigned to the First Marine Division, I had been sent to the observation squadron. In the rear seat of the lightplanes, acting as observer and target spotter for our howitzers and artillery pieces, I had piled up almost 300 missions. It was the first flying of any kind that I had done since the end of the war, even as a passenger.

"If there's an artillery piece up here, they've buried it," I grumbled into the microphone. "Let's give it one more going over, though."

"Roger. How about a little lower?" Matz didn't wait for an answer, but tipped the nose of the plane down, pointing it at one of the Korean hills. I fumbled at the straps beside me and took out the stick, bending down to fit it into the socket which controlled the plane's movements. Then I raised the binoculars to my eyes once more, scanning the terrain in search of the phantom gun emplacement. It had to be there. Intelligence had pretty well pinpointed its position, but that wasn't enough. You didn't throw away artillery shells, hoping for a lucky hit. You had to have a definite target.

I saw the rows of sturdily-built bunkers that held the Communist forces defending their own line. I caught a glimpse of a machine gun crew huddled about their weapon, dark winter uniforms making a poor effort to blend with the muddy, frost-filled ground.

A sudden crack sounded over the steady beat of our engine and I was jerked back against my parachute as Matz pulled the plane over on one wing, banking to head for our lines. Then he banked the other way, going into a steep dive at the same instant. We both knew the sound. Some hopeful hero had taken a shot at us with his rifle. There were no more shots.

"Just some clown with a rifle," Matz reported over the radio. "We'll circle back." He swung the aircraft back to the area we had left, and I raised the glasses once more.

The sudden maneuver left my stomach feeling as though it had become detached and was free floating somewhere in my chest cavity. I had flown hundreds of hours as a fighter pilot and such stunts had never bothered me. Those times, though, I had expected it. Sitting in a rear seat and not knowing what to expect made a difference. Too, I had been used to a heavy plane. Something with body and power. Flying in an "engine-powered kite" didn't impress me as sport.

One pilot had blown a fuse when I had called his pet aircraft by that name. But I had meant it then. Flying over enemy territory in something that carried no guns and was nothing but a repainted sport plane didn't seem to hold much future.

The first tracer seemed to be coming directly into my field glasses. I dropped them, shrinking unconsciously, and heard the flat crack of the bullet as it went past. There were more sounds, then another tracer.

Matz cried out over the roar of the engine and our left wing dropped abruptly. I jerked my head about to glance out the window. The crest of one of the dirty, brown-colored hills was coming up fast.

I don't remember grabbing for it, but the stick was suddenly in my hand, my feet on the pedals and I jerked the plane upright, pulling back for altitude at the same instant. The wings wobbled crazily for a moment, and there was a strange feeling in the pit of my stomach. I started to speak into the microphone, then glanced up front. Matz was slumped forward in his seat.

He was hit.

I reached forward, pulling him back against the seat, jerking the microphone to my mouth in the same instant. I asked how bad it was. His answer was a mumbled blur in my earphones.

I glanced down at the floor. A stream of blood was moving slowly toward me from beneath the front seat. I shouted into the microphone once more, "How bad are you hit?"

His head lifted a little. The words came jerkily, tight with pain.

"Can't move my leg. You'll have to get us down. I'm going to pass out."

"Get your windows open! Get some air in here!" I shouted. In answer I heard nothing but a groan.

The puddle of blood at my feet was broadening. I started to strip off my belt as I spoke.

"Can you get the windows open?" The strain pitched my voice two octaves higher. Russ nodded, starting to fumble with the catches that held the side windows tight against the framework. The belt came out of the loops of my trousers and I unsnapped my safety belt. Standing, I leaned forward to hand him the length of webbing. I shouted in his ear this time.

"Get it around your leg. Cut off that blood flow!"

He was still fumbling with the window catches, starting to nod slowly at first. Then his head jerked forward to rest on his chest. I shook his shoulder.

"Come out of it, Russ!"

I was praying aloud as he fought his way back to consciousness. His head came up slowly and his hand grasped the web belt tightly. I reached over his shoulder, twisting the window latches free with fear-stiffened fingers. It wasn't what was happening, but what was going to happen. My mind was not up here above the enemy, but on that short, dangerous strip back at our headquarters. Nine hundred feet of it, no more. I was remembering the plane that had cartwheeled off the end into the rice paddy. It could have caught fire while I hung there powerless.

The latch open, I pushed out and up, causing the swinging window to catch on the latch attached to the wing. Cool air hit me in the face, and I was suddenly cold. I was sweating. The entire suit of dungarees was dark with water. I reached for the other window, glancing over Russ' shoulder in the same instant to see that he had the belt about his leg above the knee. He was struggling to pass the brass tip through the buckle and pull it tight. The leg of his flight suit was dark red. His groan in my ear sounded like a scream as he said, "It's your plane, Fred. get us down."

"Sure you can't handle it?" The strain was in my voice, and I know he caught it. He knew that I hadn't flown in six years and that this plane was entirely strange to me. He only nodded his head, saying, "Can't move the leg."

I settled back in the seat, hooking the safety belt once more and pulling it tight. Then I looked at the maps I had been using. Up ahead, Matz' head had dropped against his collar. I started to speak into the mike, but changed my mind. If we were going to pile up, being unconscious would be better for him.

It took several seconds for me to find our position on the map. I glanced over Russ' shoulder, checking the compass, then tested the controls gingerly. The plane jerked uncertainly at my touch, then seemed to accept its new pilot resignedly. It settled down to steady, even flight. More slowly this time, I brought the ship into a slow bank, moving toward our lines. I wanted that enemy machine gun as far away as possible.

I noted familiar landmarks below almost unconsciously. My mind was on that air strip we had left only an hour before. I found myself wishing it were a thousand miles away.

We came in between the hills that hid our camp from sight, and I saw the thin ribbon of rock and dirt with the pyramidal tents lined up along its length. I started to cut altitude and speed. As we dropped toward it, the strip seemed to get no larger. It was like a dirty, weathered toothpick that someone had dropped in a gutter.

"Windfall Home Base . . . Windfall Home Base . . . This is Windfall Six. Over." I waited, guiding the plane into the landing pattern. I jerked the controls too hard and the plane almost turned into one of the high hills that flanked the encampment. I righted it quickly as nothing but static buzzed in my ears.

"Windfall Home Base, this is Windfall Six. Over." I could hear that high note of tension in my voice. Panic was clutching at my throat, trying to close it, shortening my breathing. I glanced down at the windsock that hung above the operations tent, checking the wind. It really made no difference. The

rough, rutted rice paddies were at each end of the strip. If I overshot the 900 feet, the plane would cartwheel into that terraced muck.

I tried not to think.

I was coming in, approaching the strip, but there was nothing but static from the tent below that quartered the squadron's radio section. I tried it once more, shouting into the mike. Suddenly the static died and a voice came over.

"Windfall Six, this is Windfall Base. I hear you loud and clear. Over." The radio operator's voice was unconcerned, lazy. I swore. Then I realized he didn't know what had happened.

"Windfall Base, my pilot is hit bad. I'm bringing in his plane." I gave him the code number which I had been assigned. There was a moment of silence, then the voice was no longer unconcerned. It was taut with nervousness.

"Go around the pattern. We'll get a plane up to talk you in."

"There isn't time. Matz is bleeding bad. Get the crash crew on the runway," I ordered.

"Roger."

I eased the plane back into the sky as I reared low over the strip, gaining altitude and starting to pull into the landing pattern. Below, people were gathering along the strip. Even at that altitude, I could make out the red fire extinguishers. As I approached the strip once more, I undid my safety belt hurriedly and stood up to lean over Russ' shoulder once again. I found the right control with my hand near his instrument panel and jerked on it. The flaps dropped and the plane seemed to halt in mid-air for an instant, then nosed gently toward the ground.

It seemed slow as I dropped back into my seat. It wasn't. The ground came rushing up to meet us and I saw that I was coming in too short. The terraced ridges of the rice paddy looked like teeth in a hungry mouth. I gunned the engine, pushing the plane forward a few extra yards, then let it settle. My hands were caressing the stick now. Nervousness was pushed into the background for the moment.

We bounced as we hit the runway and I saw Matz' head snap back against the seat. The safety belt across my lap jerked into my stomach. We bounced again, and I eased on the brakes as we settled on the rough, rocky surface. I caught glimpses of worried faces lining the strip as the plane skidded toward the deep ditch that was meant to drain off rain and snow waters. The gallery. The cheering section. It wasn't like that, but I was thinking it.

I eased up on the brake, then tried it, again. The plane swung toward that flanking ditch once more, but I held the brake on hard. The rice paddy was coming up fast.

Then we stopped.

I sat there looking past Matz' head at the rut-like terraces only a few yards ahead of us. For the first time in what seemed like hours, I took a full breath. The door of the plane was open then and hands were gently lifting the semi-conscious figure out of the front seat. My knees and hands trembled.

"You okay, Fred?" someone asked. I nodded without looking up. Slowly the feeling passed and I could feel the weakness going out of my knees.

But my hand was still shaking half an hour later, when I tried to raise a cup of coffee to my mouth.

I Learned About Flying From That!

Even a hot pilot eventually learns that a flight's got to have some insurance.

BY ROBERT B. PARKE

THE Colonel was a paddlefoot and a nice guy. I was a Lieutenant and an Air Force pilot assigned to fly him from Sebring, Fla., to Atlanta, Ga., for an important meeting. The plane was a shiny AT-6. As we took off, the colonel called me up on the interphone and asked if it was O.K. to smoke.

We wheeled out of the pattern toward the line of midday cumulus clouds to the north and I said sure it was O.K. Soon the fragrant smoke of a good cigar drifted up to my half of the fore and aft cockpit. Two cigars later, and right on top of our ETA, we were on the ground at Jacksonville.

While the plane was being gassed I checked the weather. The Colonel followed me around asking politely about this and that. He'd never been in a plane before. I, on the other hand, had just come back from overseas with a thousand-plus hours and had recently earned an instrument ticket. It was pretty hard not to play the bold pilot role.

A mean-looking front had built up beside the route and was getting ready to smother Atlanta. The weather man suggested we lay over. That didn't suit my plans or the Colonel's, and I took up the study of the weather maps myself. In short order I had it figured.

The front was moving at a predictable rate from west to east in a line several hundred miles long. It seemed reasonable to suppose that in the time it would take us to get up to Atlanta, the front would pass, leaving open both our destination and a flock of alternates to boot. For the record, Atlanta was then open and could logically be forecast to be open on our arrival. I explained all this to the weather man in my pilot-knows-best manner.

Pointing out that the front could just as easily hover around Atlanta, leaving me gasless at night with no alternate, he said he'd have to recommend that operations not clear me. He concluded with a phrase I was to think of many times afterwards: "The flight's got no insurance." The way he said it let me know he meant if any one of a dozen things went wrong I'd be in trouble; the conditions didn't balance the risk.

But I was in no mood for homilies. I re-explained. He said no. I got technical. He was adamant. I insisted. He left. When he came back I relented. Sly as the wolf in Red Riding Hood I announced I'd changed my mind and would clear for Macon, this side of Atlanta. Macon was open and forecast to remain so. The harassed weather man sighed with relief, as did operations. Soon the Colonel and I were off in a cloud of Panatella smoke.

At the first check point I asked for Atlanta weather and found it clear, with Birmingham open as an alternate. I requested a change in flight plan. Flight Service, the clearing agency for airborne planes, saw it my way and soon Atlanta was my destination again. Red Riding Hood was practically mine.

The first clouds swirled around us about dusk. I went on the gauges as we plowed through turbulence and rain with light icing. Finally the Macon beam came in, static-ridden but distinguishable and I checked the weather again. The front seemed to be gliding past Atlanta. Pretty soon a few breaks in the clouds appeared and the Colonel rang up to ask if I wanted a cigar. By that time the cockpit looked like a pool room on Saturday night, so I decided to smoke in self defense.

Twisting in my seat, I reached back over my shoulder for the stogie. I'd taken off my headset for this operation and after lighting up I dialed the Atlanta beam and put back the ear phones. They were ominously quiet. I checked the connections. Just then we hit some more turbulence and the clouds closed in again. Groping around on the floor, I tried to find my radio trouble. When I found it, the cigar went out the window.

The cable connection to the radio receiver was torn loose. I must have ripped it out with my foot as I reached for the Colonel's offering. That added up to an AT-6 somewhere between Macon and Atlanta, 100 per cent on instruments, at night, no radio receiver, and about an hour and a half of gas.

The last weather I'd picked up showed Macon and Jacksonville closing; Atlanta and Birmingham getting better. I figured I must be almost through the front, and the best thing to do was go on and hope something was wide open behind it.

I'd checked the wind and headings at Macon and so had a dependable ETA to Atlanta. But as it ran out the clouds were solid as ever and I didn't relish letting down blind. There is a 2,000-foot mountain just north of Atlanta. I still had enough gas to run up to Birmingham which certainly should be open. I dead reckoned my way up there, found it tighter than a drum and headed back for Atlanta.

My transmitter was still working and I aired my plight to anyone who was interested. Back over where I hoped Atlanta was I searched for breaks in the clouds. I decided to keep this up for 15 of my remaining 30 minutes of gas and then invite the Colonel to join me in bailing out. I wondered idly if the Colonel would smoke on the way down.

At zero hour minus three minutes I spotted a faint glow through the clouds. I knew CAA would have cleared out all flights from below me so I decided to risk a letdown. Starting at 4,000 feet I started easing down. At 3,000 the glow was tantalizingly brighter. I dropped the nose again. At 2,000 the clouds beckoned wraithlike and I continued the letdown, holding my breath now. That mountain was somewhere nearby. At 1,000 feet lights seeped through the clouds and I decided by circling down I'd minimize the chances of hitting the mountain.

I slid lower. Rain was pelting the ship. The needle slipped past 800, 700. I sat tense in the cockpit ready for a fast pullup. Suddenly I broke through. It was raining hard but I could make out lights and buildings. Then dead ahead, like the lamp in the window, I saw a lighted runway. I got a green light as I went by, flashed through a check list, and a minute later skittered to a stop on the parking ramp.

I climbed out, my knees quaking. I'd all but forgotten about my passenger but as I hit the cement the back canopy opened and he popped out, shielding his cigar with his hand. I wondered if he knew how close he'd been to getting his first parachute jump with his first airplane ride.

But he was all smiles, telling me how great it was—and how did I ever find that runway on a night like this? I saluted as the command car whisked him away.

When I got inside I found I'd landed at the Naval Air Station and that sure enough the whole countryside had been alerted to be on the lookout for me. After turning in my clearance and attending to the paper work I thankfully headed for the sign which read "transient officer quarters."

As I passed the weather office, which in the Navy is called the Aerologists, the fellow glanced up and said, "Going back tonight, sir? It's clearing up nicely. Jacksonville is opening and Sebring is clear." It was a wiser I who replied, "No thanks, a flight like that's got no insurance."

Violent electric shocks, ice and finally anoxia dogged this glider pilot as he spiralled up to a new record.

BY DAVID LAMPE, JR.

Sucked Up in a Thunderhead

TOO WARM FOR a flying suit. In slacks and shirt, Derek Piggott squinted skyward. On the point of spilling, cumulus clouds had been piling ominously since morning. Warm and muggy, southern England was perfect for gliding on July 15, 1955—perfect till spikes of lightning began raking cloudbanks a mile from Lasham Airfield.

No use stringing wires up into an electrical storm. Small, loose-jointed, 34-year-old Derek Piggott, Lasham's chief glider instructor, ordered winches and tow cars off the runway. Only the *Tiger Moth* biplane continued kiting the sailplanes aloft.

Imperial College Gliding Club's new *Skylark* sat idly. Piggott got the orange glider onto the grassy runway, and climbed into the pod-like cockpit. He planned to try for a 3,000 meter height gain for his Gold-C, an international gliding achievement badge. And if the weather held, he might reach Diamond-C altitude—a 5,000 meter free climb. Only about five days a year are good for such flying in England; this seemed such a day. Earlier, a jet pilot told Lasham the top of the cumulus reached 30,000 feet. Soggy clouds suck gliders swiftly upward.

Shuddering as the *Tiger Moth* towed it aloft, the *Skylark* felt like an airborne eggshell—a feel Piggott liked. Slouching in his seat he sighted along the tow rope, left hand loose on the control column, body sensitive to the glider's movement. Air tows always recalled combat formation flying. The *Moth* sheered southward from the storm clouds heaped over Lasham's dispersal area. Lightning still snicked nearby as Piggott lifted 1,800 feet into the clear air. To notch the barograph flight-recorder in the cockpit, he pushed the control column hard forward and the *Skylark* dove 100 feet. Piggott pulled the tow-line releasing cable and heeled the *Skylark* on its left wing to show the *Moth* he'd cut free. As his airspeed lowered, the vibration faded and the slipstream's whistle softened to a hiss. He hovered, feeling gently for a thermal updraft.

On the edge of the storm he found it. Tilting the stick, he began an upward spiral. At 3,500 feet, white murk wrapped his glider. He was in a gentle 30-degree rise—a spiral ramp inside an invisible, vertical tunnel. Only the soft slipstream whistle and his own sensitivity told him he was moving—soaring on instruments as he was blown upward on the rim of the storm. Derek planned to cruise northeast through the clouds to compensate for drift. On the instrument panel the glider's variometer had a pair of plastic columns, one marked "up," the other "down," a sensitive little ball afloat in each. The scale between them indicated climb or descent rate in feet-per-second. Sensibly, Piggott wouldn't rely on the variometer in the clouds. Instead he'd check his wristwatch against the altimeter. At 4,000 he did this and pencilled a notation. In five minutes he checked his watch again. The ship was climbing at 14 feet per second—fast but not dangerous.

Ten after four—11,000 feet—air still warm in the snug cockpit. At 12,000, nearing icing level, everything was normal. At 13,000—still going up. At 14,000—trouble! Piggott's body jolted forward. The glider fell and bounced high on an updraft; then slammed downward, then upward again! Piggott steadied the stick lightly, remembering he mustn't over-control. A sudden ominous noise, a squeaking twisting howl, came from above his head where a pair of clamps held the wing in place. The glider sounded ready to rip apart. To keep from banging his head, Piggott braced hard back in his seat and checked his safety belts with his free hand. The beads in the variometer kept a zany up-down dance as the glider pitched in the turbulent air. The altimeter needle swung back and forth. Derek imagined a bail-out from 14,000 in shirtsleeves and slacks! The cockpit was warm; outside he'd freeze.

More noises! Loud splatting on the wing and tail surfaces— hailstones pelting the glider, crescendoed into sound reminiscent of ack-ack barrage. The storm kept lifting the plane and slamming it forward. The pilot had done plenty of storm and instrument flying in gliders, but never like this. Airspeed wavered between 40 and 50 knots, so Piggott put on the air brakes and the bucking slowed. For a moment he was gliding again, hissing upward through the lashing hail, still all in one piece. He retracted the brakes and slid open the small clear-vision panel on the side of the plexiglass canopy. A burst of pea-size hailstones sprayed his lap.

The hail thinned and the *Skylark* soared peacefully again, but not for long. Like flash bulbs set off behind him, lightning played in the white clouds. No thunder roared, and the white lights weren't strong enough to blind, yet Derek began to feel an occasional tingle from the control column and rudder pedals—static electricity, common in gliders, especially during take-offs.

Still no thunder. Then giant shreds of lightning suddenly began flailing the *Skylark's* fuselage! He gauged the jagged bright stalks at maybe half an inch thick, until lashing toward him, Derek saw a blinding bolt like an oversized spark from an atom smasher. It hit the *Skylark's* nose with a pistol noise, entered the cockpit and jumped from Piggott's right to his left rudder pedal! He felt the current run through the soles of his shoes, and guessed they must be damp. A pungent ozone smell filled the cockpit as the lightning continued to thrash soundlessly near the glider. The loudest noise was still the slipstream whistle, interrupted only occasionally by a distant thunder boom. Was being in the storm's core keeping him from hearing it? Sparks kept bouncing off the glider's fiberglass nose and snicking past at what seemed incredibly slow speeds. Flashes now came every few seconds and the control column grip began to feel as if it had been plugged into a live light socket—and English house current is 220 volts! Piggott's arm stiffened and his eyes ached as the current bucked through him.

It happened again and again! The worst electric shocks he'd ever felt in his life, and he could do nothing to avoid them. He must keep his sweaty hand on the control column. No use kidding himself. He was scared, good and scared. The last bolt of lightning hit so hard that he yelped, *"Yow!"*

The *Skylark* spiraled higher and the lightning died but Piggott began a new line of worry—anoxia—the sickness caused by lack of oxygen. A pilot can't tell when he's becoming anoxic, isn't sure when he's tight in its grip. Anoxia feels like a weird binge, but it can kill. First its victim gets light-headed, over-confident, uncertain of his reactions and his plane's response. He does odd, unaccountable things, takes risks he'd never take if he were breathing enough oxygen. Typically Piggott stopped worrying about anoxia. The *Skylark* carried no oxygen gear, anyway.

He wanted to unkink his course, to come out of the spiral, so he shoved over the stick. The glider should have felt to him as if it were actually curving in the opposite direction. But when the compass showed he was flying straight, he *felt* he was flying straight. He shook his head in disbelief and mumbled to himself that this doesn't look right. The compass hadn't been jumpy, even in the worst of the electric storm, and now it kept a steady heading. Soon he was out of the shock area, and headed into new trouble. The control column was stiffening, not yielding when he eased it from side to side, to and fro.

Ice! He could feel the controls champ against the accumulating frost. As he soared higher the controls became more rigid. Soon he needed both hands to shift the column. Finally he took the chance, jolted the controls and the glider leaped as the ice fell away and the column worked again. But at 15,000 feet ice heavily sheathed the *Skylark's* nose and was building up on the leading edges. The trim tab was as far down as it would go, but the glider was nose heavy. Ice wasn't yet blocking the instruments. Or was it? He shook his head hard and squinted at the dials on the panel. Could they all be wrong? He wasn't certain. He was gasping, had been gasping ever since the first lightning struck him. Must have breathed up all the fresh air in the cockpit. Must watch out for anoxia! Poking his right hand through the small clearvision slot he deflected fresh, cold air into the cockit.

The altimeter said 16,000 feet. Or did it say 6,000? Sixteen or six? In a glider the needle that reads feet in tens of thousands isn't often used. Derek swayed his body, dipped his head forward and threw it back, giggled like a child and squinted at the instruments. He was at 17,000—18,000—still going up. He should have turned and dipped down out of the rising damp, but all he could remember was his first flight plan. Occasional hail flurries pelted the *Skylark* but he didn't hear them. He was too intent on the instruments swimming in front of him. "Anoxia," he mumbled aloud. "Mus'n't get anoxic." His eyelids felt sandbagged and his head weighed a ton. He remembered that one of his friends had gone to 17,000 feet to get a Diamond-C. Just to be sure of his Diamond-C, Derek decided, he'd soar to 19,000.

The thermal thrust the glider upward swiftly. He circled into a stronger thermal patch, squinted at the altimeter. The dial read 20,000 feet. Piggott's body became limp, then stiffened in the safety harness. He nearly fainted but shook his head, poked his right hand through the clear-vision slot and deflected in new air. On the ground he'd never dream of climbing so high without oxygen. Derek didn't know that the altimeter wasn't accurate, that its faulty dial showed him to be several thousand feet lower than he actually was! Reaching over, he worked the air brake control lever. The hissing slipstream quieted as his pace checked. Slowly he began to descend. Another lifting eddy caught him. He peered at the instruments, scrutinized them carefully. Might as well go back upstairs. Jus' to make sure. Pulling in the air brakes, he shot skyward. The glider stayed on its even spiral, lifting easily, climbing within the solid white cloud. Piggott's body sank limply back in the seat.

Later he didn't remember having passed out. Only the barograph recorded the story. He had risen so swiftly that the barograph's needle had to lean over backward to record on the revolving cylinder. A six kilometer instrument, the barograph recorded only to 23,200 feet. Piggott's climb carried him upward right off the top of the smoked paper!

How he flew the plane without being conscious, he'd never know. But if the barograph could be trusted—and it was checked and double-checked—he had shot above 25,000 feet, dove at several thousand feet a minute, again tracing a backward curving line on the smoked paper. Consciousness crept back into Derek's mind. He didn't feel himself regaining it because he'd never felt himself passing out. Only the barograph could relate the story.

Finally able to look around, to inventory his position, to squint at the altimeter, he saw he was down to 16,000 feet. The variometer's right-hand bead danced. He was descending. Almost wantonly he slammed open the air brakes and felt himself heel harder downward through the mist. The *Skylark* corkscrewed to 10,000 feet easily, emerged beneath the overcast and slashed through a mixture of rain and hail. Derek breathed deeply, sucking the thickening air into his lungs. His head felt light, giddy, awful.

Alton, the town nearest Lasham Airfield, should be below him. He lived in Alton. The town he saw didn't look familiar. Ah! there was the hospital! Alton! Home! Faraway he saw Friendship Pond. But anoxic, he had to circle round and round Alton before he could remember the way back to Lasham. Several times he tried to veer west toward the field, but each time he had to hover back over the hospial because he felt lost. The rain was slowing as he finally circled at 6,000 feet over Lasham's parked gliders. It was 5:45. He'd flown an hour and a quarter.

If he hadn't been deep in anoxia and shock, he'd never have chanced the victory loops he did before landing. Shaking all over when he skidded the *Skylark* to a stop, he could barely climb from the cockpit. Lightning had pinholed the glider's fiberglass nose and splayed the fabric covering. Slits slashed all over the elevator—hailstones or loosened ice. One side of the rudder was perforated badly. The leading edges of the wings were full of dents.

Lightning burns scarred the fuselage. Where lightning had struck the rudder, it ripped away a large splint and left an ugly burn.

Piggott took two days to recover from anoxia. He got his Diamond-C and set a British glider climb record. And possibly he holds a world record for surviving anoxia.

I, a Pilot

BY NORBERT SLEPYAN

The first dial I use as a flight student is the one on my telephone. I stare at it. It stares at me. I ask myself one last time: Do I really want to do this thing?

As usual, my sensible self says no. The old arguments are put forward: too dangerous, too expensive, a frivolous indulgence, an ego trip, a hell of a lot of work, too dangerous, a wife and kid to support, too dangerous. Me? A pilot? Who often as not spills his coffee in a 747? Whose driving is the laughing stock of the entire family?

Yet I spill my coffee partly because I love being in that airplane—any airplane. I can see a world of things as I can never see them from the ground: Clouds scudding by at eye level as puffs and towers. An endless variety of patterns made by fields, towns, roads, water and highways, all laid out in their own peculiar order, struck by sunlight or shadowed, and the beauty of it all made more so because the whole thing seems to be moving. For me, it is a joy to see what water is like from way up when the sun hits it just so or to see another plane, far off, contrailing clouds of glory.

But my affliction goes deeper still. I clog up the aisle as I stop to stare into the flight deck, gawking at the instruments, the switches and lights and those people in there calmly making sense of them. I love takeoffs and landings and the wonderful sensations of banking, even in the tubular hotels they call airliners.

So, like a kid suffering from infatuation compounded by frustration, I stare balefully and then perhaps maniacally at that dial. My finger reaches, spins the wheel. I have chosen the number via the Yellow Pages: a big ad offering "Learn to Fly in a Country Club Atmosphere," with a large picture of two guys on the eighteenth green flanked incongruously by an airplane. The ad offers everything—"FAA and VA approved, all ratings, aerobatic training . . ." Wow! ". . . FAA examiners on staff, student loans, combine business with traveling, 24-hour charter service, aircraft rental and leasing, credit cards accepted."

The school is at an airport sinfully nearby—a 15-minute drive. A motherly sounding lady answers and I say in my most Steve Canyonish voice, "I'd like to inquire about a first flying lesson."

"Wonderful! What is your name?"

Fast workers, these flying people. I give my name, and it suddenly hits me what I am doing: I am coming out of hiding as a frustrated flying nut. I am committing myself.

After 30-some years, it's about time.

Everything is arranged—time, location, whom to ask for. ("I'm Robbie. If I'm not here, Gail will be.") Visions of beautiful girls (with motherly voices?) taking me into the clouds . . . Raquel Welch guiding me on my first walk-around. I've almost hung up before remembering to ask the price.

"For the first lesson, $25; if you take the whole course, that will be discounted."

Twenty-five dollars. Doesn't seem like too much. *But.* I wonder how much it is at other schools. How do I judge? There are some 15 schools listed in varying degrees of prominence. I've gone for the biggest ad—I, who douse the TV when the commercials come on. I should be able to do better than that.

I call another, less splurgy number, and get a hard, sock-it-to-'em man's voice. I picture the man standing under a windsock, large white coffee mug in hand, enduring my questions with the stolidness that got him through those cold nights on the old mail runs. He breaks it all down, essentially to one fee for the pilot and one for the plane. Pilot, $7.75; plane, $10.50. $18.25. Much better. But this school flies out of a small field that is an hour away.

Another number, another voice. A centipede of prices goes rushing by, for the packages they offer are both endless and unjottable, involving numbers of engine, horsepower, model and innumerable other variables. For my first lesson, I can have my choice of a Piper Cub or a Learjet or anything in between. An F-111 will be a little tougher to arrange, but they can quote me a price. It all seems to average out to $25 to $27.

I call still others, with results that leave me so confused I don't even know whether to be bewildered. In desperation, I opt for big motherliness.

Nine-thirty in the morning. 0930 hours, right? Under an overcast sky, I am driving to the airport. Last night, I made a point of leaving a dinner party early to get to sleep at a decent hour. Nothing but righteous conscientiousness. I *would* rise clear of head and sharp of eye.

Not so. My cat chose last night for a rampage, a symphony of crashing destructiveness that left me in wide-eyed, angry wakefulness at worst, and in light flotation just above dozing at best.

I am resentful, still angry and a little scared.

I pass through the city traffic. I might be going anywhere— to the supermarket, to the drugstore, to grandmother's house. My wife said goodbye as if I were heading off to work. No drama or hint of glory from that quarter. I guess she'll believe it when she sees it.

Suddenly, I am at the turnoff for the airport. There they are. The planes. Rows of brightly colored machines gleaming even under this gray sky. In my head, like a loop of tape, runs Robbie's assurance that this first lesson includes flying. Already I can feel my jaw getting squarer.

Another turn and I am passing among planes and hangars and buildings with large painted names that had been mere entries in the phone book yesterday. A plane comes to life. The roar is insistent. The two men in the cockpit are wearing ties. (Captain Midnight, whatever became of you?) They are going to fly. So am I—maybe. Unbidden, the sour question erupts: Do they let people like me fly when it's overcast? Don't worry, I reassure myself, the school knows what it's doing; when they say you'll fly, you'll fly.

Now my sign comes into view, somehow bigger than the others. I discover that I am driving my car at the deliberate slow speed of a taxiing airplane, and that as I ease into a parking space, I have become ultra-methodical. I run through the checklist. Cut engine. Gear lever to park. Emergency brake pedal depressed. Remove key. Safety belt unfastened. Door open. Climb out of car. Door closed. Proceed toward office.

To the wrong office. From this one, they sell airplanes. I ask for Robbie. The girl at the desk points to a door four buildings down and I toddle off, already with a false start under my belt. Two young men pass me. They are wearing suits. Up ahead, someone wearing a mechanic's uniform is immaculate. Is there any dirt in this place? It seems that in my tattered green windbreaker, pullover shirt, seafarer jeans and walking boots, I am underdressed for the occasion.

I approach the fateful door, a formidable blue one with a chrome replica of a propeller for a handle. The office I enter is resplendent with airplane models hanging from the ceiling, posters trumpeting safety plastered on some walls and informal snapshots on others. Desks, sofas and potted plants— lots of potted plants—complete the furnishings.

Seated on one of the Naugahyde sofas are two men in earnest discussion—instructor and student, from their manner. One of them is dressed sensibly for the demanding, gritty task of guiding a machine through the skies; he is dressed like me. The other is wearing stylish slacks, designer shirt and tie. I

can now guess which one is the veteran.

At the rear of the room is a long counter where planes are rented, and behind it is a pretty girl in a stewardess-type uniform joking with a group of men. Nearer is a smaller desk, also manned by a cute, uniformed girl. She has an I'm-Trying-Harder smile, which is infectious as it turns my way, so I take the three shy steps to the desk.

Shy because I feel ill at ease here. No particular reason to: No hostility radiates, the vibes are cool, all is calm, the laughter is genuine, the girls' eyes smile. But there is a specialness about these people, which perhaps exists more in my mind than in the place. I intensely feel the outsider.

I approach the desk and go through my halting introduction of name, ranklessness and purpose for being there. The smile remains through it all. I ask, "Are you Robbie?"

"I'm Roxanne," the girl says, beckoning me into a small office. She seats herself behind a small desk while I wrap myself around the tiniest chair I've ever seen.

"Let's begin by your telling me something about yourself," she says, as if she is really interested.

I'm about to narrate my childhood when she says, "For instance, are you taking up flying for business, a career or for pleasure?"

Should I tell her? The whole thing? A lump jumps to my throat. Should I really tell her about the years-long war between my desire to fly, on the one hand, and a grisly combination of circumstance, fear, lack of confidence in machines and plain inertia, on the other?

"For pleasure," I say.

"Of course." Still smiling.

Now she transfers to the desk a bulging, bright red briefcase from which she takes book after book, all plastered with the name of the school. My eyes widen. I had fewer books in college: curriculum, procedures, Cessna 150 owner's manual, workbook, a folder containing a Weems plotter and other calibrated goodies, the Constitution of the United States and a few other essentials. These, I am told, will be the course materials I will study as I learn to fly.

"But first," she says gleamingly, "let me tell you about our firm."

I wait out her telling me that in all candor this is the most together outfit of its kind, and that it has finally perfected the challenging art of teaching people to fly. Furthermore, I am assured, this company, in its power and glory, rents, charters, maintains, repairs and spans the nation. And all this only 15 years from its founding. Tomorrow the world.

The essence of the system, I am informed, is that it is "integrated." All the elements—study material, instructors both for flight and classroom, the training aircraft, and various audio-visual techniques—are harmonized so that I will know everything I need to know for each flight lesson. In days past, I am told, flight training and ground school were independent, an aviation mishmash. Now they fit together.

Is that true? How can I know? Still, it all sounds sensible. The training is broken into "blocks" of lessons—28 in a Cessna 150 and 12 in the classroom. The flight lessons consume 35 to 41 hours of flight, approximately 12 of them solo. It's all beginning to sound quite wonderful.

Now she sets before me a sheet with lots of numbers, which has to mean money. My head swims. I've come for a $25 trial lesson and here I am talking with Roxanne about more than $1,000, which is a lot of money for me. After days of dreaming about the wild blue, the day of reckoning has come, and every way I reckon, it looks gray.

The package comes out to exactly $1,069.98, including tax. Roxanne, seeing the look on my face, makes a hasty exit, ostensibly for coffee. Alone, I ponder the future, and it comes up grounded. Bad business times, other needs pressing. My sensible self (back again) rightly screaming prudence. Against this, the self of the glorious takeoffs and steep turns, the idiot self who gawks at DC-9s, is silent. Against reality, his poetry has no answers. Suddenly, he seems almost indecent. I had expected to leave this place feeling heroic; now I simply feel poor.

Roxanne returns with coffee for two, and a possible solution. I've looked at those columns of figures and have seen nothing. She starts to interpret. The course *can* cost $1,069.98, but it needn't cost that much. If I pay before each block by credit card, I get an allowance of $38.98; and if by cash, $48.98. We're down to $1,021. If I pay for the whole course at the beginning by credit card, they allow $51.98; and if by cash or through a student loan, they allow $71.98, which brings the cost down to a low $998. Uh huh.

"You'd be surprised at the people who go through this course," she says triumphantly. "Most of them are nowhere *near* wealthy. You don't have to decide now. After all, you've come for a trial lesson, which would apply to the cost of the course if you go on, and you ought to see how you like it. Pay for it now, and maybe for a second one later; then we can discuss things again."

Terrific! "Okay. Can I go up today?"

"No; too cloudy. Only people rated for instruments can fly." She sees my crestfallen look. "Tomorrow should be better. Come back at 10 tomorrow. You will, won't you?"

"Yes," and she knows I will.

I pull out of the airport without looking back, for I am just beginning to put together the impression this hour has left with me. I have come here . . . looking for what? High adventure? Danger in faraway places? Smilin' Jack? Yes. Part of the beckoning image had to do with leather helmets, a white scarf, crinkle-eyed faces and myself carrying all this and a tough but insouciant air. I went looking for the 1930s and, of course, found 1973: crisp office efficiency, business to the hilt, dress that seems to say, "Flying is not *so* different that we can't look like everyone else while doing it." I might as well have been discussing the purchase of a car.

Yet it *is* different, this flying thing. Navigation, meteorology, radio procedures, flightplanning—the *doing* of it. And the money. So one thing, at least, is up front, dead center and perfectly clear: I am going to be giving my sense of commitment a test I never bargained for.

At 10 the next morning, it is just as gloomy, but Roxanne's uniform is blue and her smile a touch more personal. "Tell you what," she says unctuously, "we may get you up after all. The forecast looks good. Let's start preflight prep."

I am seated at a desk. A tape deck is placed before me.

Installed on top is a ringbound book. Roxanne inserts a tape cartridge. "Play this, and be sure you can answer all the questions."

I flick the switch and a full orchestra greets me as if C. B. de Mille were personally welcoming me into the heavens. Then, a syrupy announcer's voice repeats what Roxanne has already told me about the glories of this wonderful company that's going to put me in the pilot's seat. Responding with Pavlovian simple-mindedness, I flip the pages—which contain the announcer's speech—whenever a little signal beeps its order to do so. Beep—flip. Beep—flip.

Finally, things settle down and I begin learning about airplanes. The beeps go on. The voice talks about thrust and drag, lift and gravity, chords and angles of attack, roll, pitch and yaw, control surfaces and how they are operated. Every so often, there is a pause while the announcer asks questions. These are multiple choice and no great challenge. We go on to airports and landing patterns. Now for instruments, and more questions. More beeps. More information to memorize.

It is 11:30, and I have gone through three of the ring bound books. My head is buzzing with facts, terms, conjectures and the last echoes of the orchestral climax that ends the third tape. Roxanne returns to find me sitting slack-jawed and blinking.

"Through already? Very good. Things are supposed to clear in about an hour, so maybe George, who'll be your instructor, can spend the time familiarizing you with the airplane. Okay?"

George turns out to be the instructor in the designer shirt. About 25, he is a friendly, earnest-looking man with quick, dark eyes and a firm handshake. As he smiles and says hello, he is obviously sizing me up. Roxanne says, "You'd better get this man flying. He's been chomping at the bit for two days now." George smiles. "Let's head out to the plane."

My heart is thumping.

"What we'll do," he says as we walk to the red and white Cessna 150, "is check out the cockpit, and then I'll show you how to do a preflight inspection. It's very important that you personally make sure the plane is in good condition before you fly it. It can mean your life."

He is very soft-spoken, but he makes sure I've gone through the flight prep and checks out what I have studied. He seems satisfied.

The plane is small and sleek-looking, with a rakish tail, fixed landing gear in streamlined wheelpants and an overall sense of zippiness. He opens the door and motions me into the left seat. I bang my knee climbing in, but at last I am in a cockpit, at the controls. A magic moment.

George climbs into the right seat and waits a few moments as I settle in. He shows me how to move the seat forward and back—just like in a car. In fact, what with the carpeted floors, bucket seats and wide-angle rear window, I have a sense of being in a car. But not quite. No car is quite this tiny. George and I are wedged against each other.

George takes me through the instruments again—altimeter, airspeed indicator, artificial horizon, directional gyro—step by step, explaining not only what each instrument does but how it works. I ask questions, many of them dumb, and he patiently answers. As we talk, we loosen up, and there is humor with the instruction. He clearly wants to tell as much as I want to hear, and then some, beyond the required instruction. He apparently likes his work.

We climb out to begin the walk-around.

"You can start anywhere you like," he says, "but once you decide where you're going to start, always start there. It helps you not to forget things."

We begin with the nosewheel and carefully move around the plane, checking for loose bolts, worn brake linings, missing cotter pins, breaks in the skin and other problems. We check the oil and the fuel tanks. We poke around the wiring. Finally, we check out the prop for nicks. And then he makes me do it again, on my own.

"You think you've got it now?" he asks when I've finished.

"I think so."

"So do I. I guess we can take her up."

I have been so preoccupied with the instruction that I haven't noticed how much the overcast has burned away. It is vivid blue up there.

As I climb in, George says, "Well, after all this attention, this plane *has* to be airworthy."

Yeah, I think, I hope I am.

I am aware of a strange feeling. This isn't what I've expected. It's all so low-keyed. George's quiet voice, his precision in answering questions, the matter-of-factness of each step . . . it's as if I were merely being taught to drive a car or run a vacuum cleaner, though more complicated. There have been no wows, no dire warnings, no Steady Son, You're Going Into the Great Beyond. I wonder if George isn't really bored.

Where is the exhilaration I was expecting? The butterflies in the stomach? The tingling in the scalp? I feel a slight glow, and something curious among my viscera, but that is all. Are these people going to take the thrill out of learning to fly? For a thousand bucks, I want some kicks.

George climbs into the right seat. "Why don't you start her up?"

And here are the butterflies, and my scalp is tingling. George hands me the checklist.

Here we go. Seat belts and door get tugged and pulled. My seat is pulled forward a bit until my feet feel solid on the pedals. How solid should they feel? I'll find out soon enough. Radio—off. Reach into the glove compartment and count the spare fuses.

I push the red-handled plunger all the way in. Mixture full rich. I delicately open the throttle an eighth of an inch, conscious of George's eyes and also that we are getting close to *that time*. The carburetor-heat knob is checked at cold. My fingers move over the electrical switches—all off.

Master switch—on.

That time.

No one is anywhere near the plane, but I shout "clear" as if trying to control a rioting mob. I turn the magneto switch to "both" and then engage the starter.

Why am I stunned? Because the engine has actually started? Up to now, I've been doing everything undramatically, in silence. Suddenly the world is full of roaring. Can 100 horsepower sound so loud? The plane is throbbing. All that power!

"Set your throttle," George says patiently, but with a warning note in his voice. Keep your mind on business, that tone says.

I push the throttle in, and the rpm needle swings up to 1,200.

"Too much. Pull back."

The needle is at 1,500 and climbing. I pull back and the engine nearly dies. What the hell's the matter?

"Handle her easy. Move the throttle back in, slowly." And he shows me how to hold my hand against it for precise control.

By delicate increments, the needle eases up to 800, where it should be, and settles there. I stare at it, enthralled. This machine is responding to me!

"Check your instruments," George says. No romantic he. Here I am, falling in love, and he's talking about instruments. But I check. Oil pressure—in the green. Ammeter—charging. Suction gauge—at two. George helps me set the radio to ground-control frequency, says completely unintelligible things into the microphone and gets a similarly unintelligible answer over the speaker.

"What language is that?" I ask.

"You'll learn it," he says enigmatically.

"Let up on the brakes," George says quietly. I do, and the little Cessna slowly eases forward. "Steer with your feet," he says, as, against everything I've learned today, I turn the wheel hard left. "Left *rudder*. Let go the wheel."

I am now beginning to sweat heavily. I press my left foot against the pedal, but the plane turneth not, neither doth it obey. We are heading straight for my own parked car.

"Harder," George says "we don't want to run into that guy's car."

Harder it is—I'm all agreement—and suddenly, we are turning. My car glides from view.

"Now straddle that yellow line and follow it."

Like the waddlingest of geese, we erratically head onto the line. First too far to the left, then too far to the right, too much left, too much left again, much too much right. This plane isn't taxiing, it's bobbing and weaving like a punch-drunk fighter.

"Here, let me show you," and George firmly sets his feet on the pedals. Miraculously, we turn, compensate and are dead center over the line.

"Now keep her there," he says, still the model of patience.

It gets better, and as we near the runway, I'm beginning to get the hang of it, but George takes over and swings us smartly into place behind a Comanche waiting to take off.

As we begin our final checks, my heart is pounding. Control surfaces okay. Instruments the same. We set the altimeter at 30.01 (still a touch of mystery there) as per the magic sounds from our speaker. We note the indicated altitude, and George says, "We'll tack 100 feet onto our readings, right?"

"Right," I say, but not too confidently. We tour the panel and nothing is found wanting. The mixture is confirmed to be at full rich, and the fuel valve, nestled down there in the carpeting, is inquired of and found open.

"Run her up to 1,700," George says, and as I do, the power in the little four-cylinder engine becomes an irrepressible thing. This machine is alive—and I am getting scared. Even when we bring her back to 800, having checked out the magnetos, carburetor heat and engine instruments, that feeling of energy remains.

"Okay, check belts, seats and doors. We're going to fly." I brace myself. I know I should be thinking great thoughts of joys to come, but I'm thinking only of when George will ask me to take the controls.

He talks some more into the mike and gets takeoff clearance. The Comanche has gone, no planes are approaching. "Put your feet on the pedals and your hand lightly on the wheel—just your left hand; the right handles the throttle, and remember, we'll want a little right rudder." I do as he says. My mouth is dry. I pull the wheel back tentatively, it slides out of its socket heavily, without interest. I let it fall back toward the panel.

"Put in the throttle all the way."

I push and push until the black knob will move no farther. The engine is now roaring like a demon—and the plane is moving. We are at 30 miles per hour and accelerating. The ground begins to blur in front of us. Objects zip by. We are at 2,600 rpm and hitting 50. That's less than freeway speed, but it seems much faster.

"Back on the wheel a bit," George says and from the corner of my eye, I notice that his hands are in *HIS LAP!!!* The wheel comes back, with interest now, and I feel a slight tilt. We're past 60. "A bit more." I ease back. Seventy-six mph. According to what they've told me, at this speed the plane should certainly fly.

It *is* flying. Looking ahead, I see only blue, but looking to the side, I see the ground tilting and falling ever more slowly away. Suddenly and briefly, the plane is buffeted. For a moment, I panic. It seems as if the plane is motionless and straining to keep from sliding backward down an invisible incline it can't quite make. I've never had this feeling flying before. Is it because *I'm* doing the flying? The instruments are saying one thing—altitude 2,000 feet, rate of climb 300 feet per minute, airspeed 80 mph, all systems go—and my head is saying another.

This is definitely not what I've expected the first euphoric

moments of my first flight to be like.

But the feeling passes, and, yes, I am revelling, flipping out at being up here and in control—if not in charge—and just beginning the hour.

"You did fine," George says. "Watch out for traffic; always watch out for traffic. Now give me a shallow climbing turn to the right."

He really gets into it, I think, as I turn the wheel a touch right and apply right rudder. The plane begins to turn, or rather, the ground below us does.

"Keep your nose up."

Amazingly, the nose has dropped.

"Remember what we said about turns. The plane naturally tends to nose down because of the increased drag. I want to go to 4,000 feet."

In about 10 seconds, we straighten from the turn but keep climbing. Slowly the altimeter needles move, the rate-of-climb needle stays fairly steady, our airspeed remains at 80, and the engine is at 2,200 rpm. We level off, and now George has me trimming the plane and constantly checking my wing tips against a vague horizon to make sure of level flight.

Then he lets me have a few moments to let it all sink in.

I'm not in the throes of wild delirium, but I feel a deep satisfaction laced with rivulets of concern. This has to be done right, but it is do-able. There are, no doubt, planes all around us, for this is busy airspace, but I don't see any. We are in this vehicle very much alone. What would it be like to be here *all* alone? Below me is the coastline, flanked by a highway I've driven more than a hundred times. From here, all is different, from the light on the beaches and water to the surprisingly comic quality that small sailboats can take on when seen from the air. Off to the right are mountains—not tall ones, more like ambitious hills—and as we head for them, they look gravity-laden, like the bodies of strange creatures too heavy to move. I've seen this from the air before, with other people flying, but never has it looked like this. This time, *I'm* doing the flying.

If I never do this again, say I to myself, I will have done a beautiful thing by doing this at least once—but I cannot in my

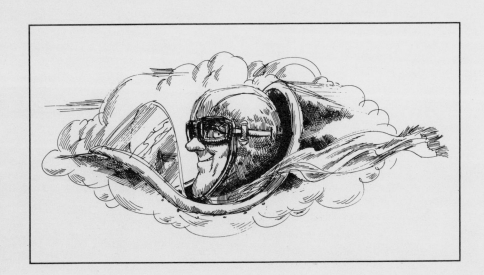

heart support the notion of not doing this again. George brings me back with a jolt. "Do a 30-degree turn to the left."

My eyes skip here and there as I manipulate wheel and pedal; the little ball—keep it centered; altitude indicator—keep your nose up.

"Okay, now I want a 30-degree turn to the right, but use only the rudder."

"What?"

"Do it. No wheel. Only the rudder."

Something seems awfully wrong. I try it, and the plane does a sickening little slide as the tail starts to come around. It doesn't turn. A terrible feeling, like finding yourself in a car skidding sideways.

"Now use the ailerons."

I move the wheel, and the sluggish impostor of a maneuver becomes a respectable turn.

"Just an object lesson," George says.

Now he teaches me to sight on objects as reference points, in order to turn a full 360 degrees and come right back on course; that is, he teaches and I flub. I get back on course, all right, but just as I'm looking over to him smugly, he says, "You lost a hundred feet." And so we do it again, and again.

There are more turns—right and left, 180 degrees, 360, shallow, less shallow.

"We'd better get home," George suddenly says, and as he says it, we are still over the mountains, with the city and our airport part of a bland smudge beneath the right wing tip. I am disappointed to be going back. But not for long, for George is calmly reaching forward, pulling back the throttle and sitting back satisfied as the engine grows silent.

Panic. Sheer, unadulterated horror. Terror and nightmare. What is he *doing*? I'm locked in with a maniac!

The propeller is turning, but not purposefully. I can hear the wind, and that feeling of the plane's being alive has gone.

"Take the controls," my dingbat mentor says calmly. (Maybe he doesn't realize what he's done. Humor him.)

"Keep the nose up. Give me a shallow turn to the left."

Wondrously, the controls respond well, though more leadenly, as we make the turn. Our descent, I see from the panel, is shallow, controlled.

I'm ashamed of myself. I ought to know better. Do the right things and the plane will fly without power, at least for a while. And the engine can be started again. Remember, you're not the first man flying for the first time.

At 2,500 feet, George tells me to push in the throttle. Again power! I know I'll get used to gliding, but it's still a good feeling, this vibration of our obedient engine.

George has the controls as we head toward the airport, which is now plainly visible. There is another transaction with the tower—what *is* that language?—and we enter the downwind leg. We are told to follow a Mooney and report when over—a what? A street intersection? Sure enough, over that

very intersection George reports in again and makes a tight turn onto the base leg. The Mooney is comfortably ahead of us on his final approach. George has me keep my hand on the wheel and my feet on the pedals, but lightly. Another turn and the runway is ahead, but a bit to the right, then a bit to the left, then dead ahead. The wheel goes forward, the engine is very quiet, the flaps are fully extended and again the plane is buffeted as it nears the ground.

"We have a steep approach here," George says, as the ground rushes up.

"Feet firm on the pedals and we keep her steady," he says. "Now come back on the wheel." As the yoke slides back, the plane seems to level in a little flaring motion and then the nose is slightly elevated. There is a bump, and another, and we are suddenly tearing along on the ground, but slowing and swinging off the runway.

"Taxi her in," he says.

As I do, I ponder asking The Question. I've done some dumb things in this hour and have asked silly questions all day. He has me stop the plane in front of the office, and we go through the shutdown. Radio and all electrical switches—off; mixture—full lean; ignition—off, and here's the key. Master switch—off. Now all is quiet but for the gyros whistling down. The silence is sad.

All right, here goes: "George, will I make a pilot?"

"Sure. You've got to loosen up and trust the airplane more, but you have a feel for it. Just keep asking questions. When do you want to fly again?"

Without hesitation, "Tomorrow."

In the office, he logs in my first recorded hour as a student pilot, what I did and how well. "Average." Have to improve on that.

As we say goodbye, I blow what little cool I have and say, "Look, this probably sounds corny as hell, but I'm especially grateful to you. This has been . . . well, a joy and a revelation."

"Not corny at all. It's what happens. I've got an hour like that in my logbook, too. More than one. It keeps happening."

I pull my car out of the lot and into the mundanities of automobile traffic. I know where I still *really* am. Part of me will never come down.

I've got some figuring to do. It won't be easy, but I'll find a way to pay for it. Perhaps at school they want you to think that learning to fly is an ordinary thing. Well, for me, it isn't. For me, it will be another dimension, and worth embracing. And there is the rationale—that it is going to be *worth* the worry and work to know how to fly for the rest of my life. If that seems impractical or even childish, very well; go find a happiness of your own.

Meanwhile, the clouds are returning, but the forecast is good. The blue sky may be hidden for a while, but tomorrow we'll fly. Tomorrow we will *fly*.

Out Spinning with Jules

BY COLBY BLODGET

JULES SITS IN the left seat of his trusty, rented Cessna 150, diligently chasing the rate-of-climb indicator with the elevator, oblivious to the fact that he is 300 feet above his assigned altitude. He has also turned 110 degrees since he began to level off, but he has no time to consider such trivia, since the most important instrument in the airplane is, after all, the rate-of-climb indicator. He finally stabilizes the needle 150 feet per minute too high and sits back satisfied.

I clear my throat. "Jules," I say carefully, "I would like you to perform a departure stall in a left turn, please."

Jules responds to my request by punching the throttle all the way forward. Now everybody knows departure stalls are done at full power, and everybody knows that if you don't look at the rpm indicator, the engine won't go over red-line. Now for that stall: *Boom*! Lots of left rudder; got to coordinate those ailerons. *Clank*! The ball hits the end of the race. Got to get that nose up. Keep that left rudder in there to take care of the torque . . . or is that right rudder? Never mind, keep the nose coming up. There, that's about a 60-degree angle; more right aileron, more back pressure. Why is the prop making those outboard-motor noises? No time to worry about it, airspeed's almost down to . . . Uh-oh. For the sixth time, Jules has just demonstrated his expertise at entering power-on spins.

I look at Jules. He is staring, fascinated, at the spectacle of the Pacific Ocean and the California coastline spinning around in front of his windshield. All thoughts of stall recovery are forgotten with the rediscovery of this ultimate thrill.

"Jules," I ask patiently, "what are you going to do now?"

He doesn't answer. The wheel is held firmly back, and his left leg is ramrod still, planted solidly on the rudder pedal, as if to prevent it from taking any unilateral action to save him. I am losing my patience, but I speak calmly. "Jules, do you think it is a good idea to be spinning down through the practice area like this? After all, you're making it extremely difficult for other aircraft to see you, coming down from this angle."

Slowly he turns to look at me, grins, and asks, "Huh?"

Jules is definitely a challenge. I bite my lip and decide he deserves one more chance. After all, of the approximately 400 stalls he's done, he's only spun 87 times. "Now Jules," I plead, "I'd really rather not go below 1,500 feet. . ."

No reaction. Then, finally, my tone of voice—not the meaning of my words—gets to him and *wham,* aside from that Mach .87 pullout, Jules makes a rather nice recovery.

Although Jules is by no means an average student, he is representative of the kind who force many an instructor into early retirement. He is also the cause of many of the problems that you have with your instructor. A Jules is anyone who has scared the living hell out of him, and you should do everything in your power to keep from becoming one. Remember those times when your instructor grabbed the controls for no apparent reason, or when, tight-jawed and sweating, he told you to do something you were just about to do? These reactions come from a Jules in his past, and you should try to bear this in mind the next time he seems to be an overbearing, overpaid, arrogant prima donna.

My very first Jules was a Mr. Dunley, who tried to get me to fly a glider through a barbed-wire fence. Picture the airport: It has one tree, one hangar, one trailer, assorted gliders and towplanes and three fourths of a runway, the approach end of which coincides with the end of a drag strip. The runway gradually tapers from normal width to nothing, causing many pilots to assume it's not there at all. It is a beautiful day, the sky is cobalt blue, with fluffy white cumulus clouds everywhere, promising a good soaring flight.

Enter Dunley, a 747 copilot with a spectacular female in tow. This jet jock is also a commercial glider pilot and wishes to soar aloft.

"Have you flown here before?"

Negative.

"Are you current?"

Additional negative.

"Well, you'll have to check out with one of our instructors and get current."

Roger.

"Let's see. Colby? Would you like to give Mr. Dunley here a field check?"

"Let's take 81W there, Mr. Dunley."

Dunley says, "I'd like to fly from the back seat, if it's all right with you. I used to be an instructor."

"Sure." My opinion of Dunley rises a notch. Every glider instructor knows that he or she has better control in the rear seat. Since he is about three inches farther back from the stick, he can utilize full aileron and elevator travel. Up front, you keep running into your legs and crotch while vainly trying to duplicate the maneuvers he demonstrates.

By the time we reach 1,000 feet, I know that Dunley is not going to be a problem. Flying tow brings out a pilot's worst, and Dunley's worst is not bad at all. We release at 2,000 feet, and after a few stalls and steep turns, he wants to try some soaring. Fine. He stumbles into a thermal immediately and soon is at 4,500 feet, sniffing and poking among the clouds, exploring their potential. By this time I'm so confident about him that I'm practically asleep. He soon decides he'd better head back to the airport to get current in takeoffs and landings. I agree and go back to sleep.

Suddenly Dunley snaps me back to reality. He has the spoilers out, and a casual glance at the airport reveals that we are now quite low. Since adding power is frowned upon in gliding, I am interested to see how he is going to get himself out of this mess. To my surprise, he flies on obliviously, sinking lower and lower. I can see the duck ponds smiling in anticipation. I sit on my hands, not wanting to preempt him until it's absolutely necessary. He begins to turn final much too early, and he now has my complete attention. Unbelievably, he pulls on full spoilers and begins a slip. A thought suddenly occurs to me: "Where are you going to land?"

"On the runway."

"Point to it."

Dunley points to the drag strip.

"I've got it." I slam the spoilers shut and turn toward the airport.

Dunley emits an embarrassed "Oh..." from the back seat. In all his years of flying the big jobs, he has gotten into the habit of lining his airplane up with a runway just before he lands. This is an admirable trait, to be sure, but it has not prepared him to consider our airport's collage of gravel, ruts, dirt and tar for a runway when there's 3,000 feet of beautifully paved blacktop right next door.

I am now faced with the classic low-and-slow situation. A common, though thoroughly disapproved, method for curing students of landing long is to take them around the pattern and line them up on final one mile out at 150 feet above ground level, then dive down to ground level, trading altitude for airspeed and float up to the fence in ground effect, effortlessly vaulting it at the last moment and making a normal landing. This method occurs to me now, and I only wish I had 150 feet instead of my miserable 50.

The airspeed is relentlessly bleeding off, and the fence is getting closer. It appears that I'm going to stall coincidentally with striking the fence, which, should I survive, will require quite a bit of explaining.

The fence looms closer, in curious slow motion. I dare not look at the airspeed indicator. I wait until the last second, take a deep breath, and yank the stick all the way back. To my amazement, 81W rears up, hesitates, shudders and stalls with a resounding bang into the weeds on the other side of the fence.

As the dust slowly drifts away, Dunley clears his throat. "Sorry..."

I'll admit that I should have known better than to trust Dunley, but his flying and his ATR lulled me into complacency. I learned from him to be suspicious of *any* pilot I haven't flown with before, and if that makes you high-time guys uncomfortable, I'm sorry.

Sometimes a pilot you know quite well can turn into a Jules. That he and his flying habits are familiar merely assures that the change will be quick and unexpected. Sam is an excellent pilot, the best I've ever flown with. He is working on his flight instructor certificate and can do all the standard maneuvers better than I can. I am brushing him up on the things he has not practiced in 10 years, and soon he will be ready to meet his local Fed for the check ride.

This particular evening, we are running through the stall series. Sam has done impeccable power-on and-off stalls, departure stalls, approach stalls and a series of accelerated stalls. I am satisfied and tell him to hurry down to the Long Beach breakwater before it gets dark, so we can do a few pylon eights.

Sam obediently begins to descend, and as we pass through 2,000 feet, he asks, "Is that how you're supposed to do accelerated stalls?"

"Yes, why?"

"Well, I just couldn't seem to get a real stall break banked 45-degrees like that."

I assume my most professorial tone. "Of course not, Sam. You are working with a limited amount of elevator travel whenever you do a stall, and a 45-degree bank in level flight uses up almost all of it, leaving you with very little power to exceed the critical angle of attack. You'll find that if you try to stall power off in a 60-degree bank, you can't stall at all. You just go around in circles until somebody throws up."

Sam nods politely but remains unconvinced. He adds power at 1,200 feet and levels off, saying, "I'm going to try one more." Now, everyone knows it is idiotic to do stalls at 1,200 feet. They are right. Regardless, Sam rolls into a 45-degree bank, stabilizes the airspeed, and begins to haul back. As before, full back elevator merely produces a little extra G and a spasmodic shuddering. Sam frowns and rolls out slightly. Picking up some airspeed, he rolls back in with an extra measure of rudder, and yanks the wheel all the way back.

He gets his clean break. The airplane snaps over into a spin as fast as any I've ever seen, and in seconds, we are spinning down through 900 feet. Never has an airplane been the subject of such concentrated and enthusiastic spin recovery techniques. When the spin stops, we are going through 500 feet. At the bottom of the pullout, the hundreds needle reaches down, gently caresses the two and starts back up.

Sam, who is pretty unflappable, exhales quietly and shakes his head. Then he smiles shyly and says, "Well, that's one way to get down to pivotal altitude."

Next time you're out at the airport with nothing to do, buy your instructor a cup of coffee and ask him about his experiences. He'll probably be more than happy to tell you, and you're sure to learn something. If he's not there, wait around for a while. He's probably just out spinning with Jules.

*Braving the tyranny of time and the grayness of space,
a desperate young woman plunges under the hood.*

IN SEARCH OF LOST HORIZONS

BY DEE MOSTELLER

I HAVE a confession to make. I am a dipso. I am constantly drunk with the desire to fly solely by reference to instruments. Not to break the sound barrier or step upon the moon but simply to fly through cloud . . . no, to possess the scrap of paper that makes it legal. That secret craving has for years been my obsession. My big embarrassment.

Sometimes before I logged the obligatory 200 hours, I told my friends that I was working on my instrument rating. The two hundredth hour crept by, and then the three hundredth and the four hundredth, and no rating. Logged time, even solo, came slowly and dearly, and the price for an instructor to sit by my side could be raised only on very rare occasions.

I began to get edgy whenever someone privy to the number of years I had been flying assumed that I had an instrument rating. "No?" with surprise. "Well, I'm sure you'll have it soon," in hasty consolation. That's like saying to a single girl of 35, "Oh, well, my dear, you're still young . . . no need to rush things."

In 12 years as a private pilot, I accumulated 60-odd hours of instrument time under 12 different hoods, gorged myself on two weekend cram courses, took the written exam four times, flunked it once, passed it thrice and let it expire twice.

I grew cunning; I began to lie. "I'm *getting* it." Not *working* on it, but *getting.* "I've got most of the necessary instrument time logged. And I've passed the written—that's the hard part, you know." Instrument manuals were cached around the house like hidden bottles of Old Crow. I began to nip when nobody was looking, stumbling into a cloud accidentally (Freudian?) and coming out rightside up and pixilated. Once I had to call ATC and ask them to work me up a clearance so I could let down through a thin layer of broken; of course I didn't tell *them* I had no instrument rating. The next day I had a hang-over and the sweaty desire to do it again.

Ah, but you've heard this story before? Maybe even told it yourself? I am not alone in my sin. There are 300,000 private pilots in this country, and only 19,000 have an instrument ticket; yet every private pilot I know is *working* on the rating. Perhaps, you instrument addicts, this story can help you learn to love the ADF and get your license to fly in clouds—or to decide you're never going to do it and be satisfied once and for all with your fair-weather limits.

There are three steps you must take. The first and hardest is confession. I told my story to a friend (instrument-rated, of course), finally admitting that I wasn't getting anywhere. "Help me." There, I'd said it.

My confessor was less then subtle. "Either do it or forget it. Stop wasting time and money. Take some time off and concentrate on it. You'll never get the rating piecemeal like you've been doing."

The second step is to analyze your needs and abilities and determine whether or not to pursue the rating. It had never occurred to me that perhaps I *shouldn't* get an instrument rating. Or that I should at least have a good reason for making the effort. And pride and vanity weren't good enough reasons. When I began to think about it practically, however, I realized that the training would inevitably make me a better pilot, even for VFR flying, and could well save my life.

Another point in favor of an instrument rating: it gives you the ability to pop through harmless clouds . . . to use your license—and airplane, if you own one—more frequently. No more will the thin overcast obstinately squatting 900 feet above the airport cancel your trip. An instrument rating occasionally can lower your aircraft insurance costs slightly, and it often can make life easier for the rental pilot as well: FBOs are more likely to rent aircraft—particularly high-performance equipment—to instrument-rated pilots.

Having found a decent rationale, and having decided that since I can pat my head and rub my tummy simultaneously I could handle it, I was ready for the third and final step: selecting a suitable learning environment, setting up a schedule and allotting time during which to *get* the rating.

For me, the solution was a large flight school with an abundant supply of planes and instructors, somewhere away from the distractions of home, in a place where the sun always shines. The only problem was that a full instrument course, including ground school, takes four to six weeks in a flight academy, and I just didn't have that much time. So I decided to do the written on my own, one more time. I figured I could breeze through it.

The first attempt was a disaster. I never even saw a copy of the test. After dragging myself to the nearest GADO by bus, subway, taxi and foot, I found that I needed a letter from an instructor approving of my "homestudy course" before I could take *any* written exam . . . the first, second or tenth time. New law, went into effect November 1, 1974. The second try ended in a 62 and manic depression. The third time around, I was better prepared, but the FAA threw up one more block: They gave me the exact same test I'd had two weeks before—which is supposedly impossible—and I spent most of the time trying to figure out which questions I'd missed the first try and how I had answered them. Despite these efforts to keep me out of the clouds (I grew paranoid), I passed. With a big 72.

I found a combination of eternal sunshine, professional flight training and nondistracting environment at Flight Safety International, a large academy in the modest Florida community of Vero Beach.

Ideally, one can live in a reasonably pretty dormitory a hundred yards from the runways for the giveaway sum of $2 a day, if you're willing to share a cubicle with one or two other students. Being one of two ladies enrolled at Flight Safety at the time—the other was living as far from the airport as she could get—I settled for Howard Johnson's world of plastic.

Waving my successful test results and logbook with its 45 hours of instrument time, I informed the manager, assistant manager, chief pilot, flight supervisor and three registrars at the school that I was there to get my rating in a week. "I'll fly all day and all night, if necessary. . . ."

They all looked pained, except for the chief pilot, Bush Narigan, who simply smiled vacantly and riffled through my logbook humming tunelessly to himself. He found a few things missing—Part 61 had changed rather significantly in the past few months, I discovered belatedly—and proceeded to lecture me about not fighting a deadline. He allowed that I might fly four hours a day, if I was up to it.

It rained all the first day long, as if the elements were giving me the same message to slow down, but even if it had been CAVU, I doubt that I would have seen the inside of an airplane that day. The morning was spent filling out a thousand forms—this is a real school, not kid stuff—and it was not until midafternoon that I met my instructor, a worried-looking young man who gave me a good hour's briefing on primary instruments and put me in a GAT-1 simulator for 45 minutes

before discovering that I already had 45 hours under the hood.

The second day I got a new instructor, two hours of ground briefing, three hours in an airplane doing basic hood work, 30 minutes of paperwork and a lecture from the instructor about not fighting a deadline. I was getting the message. I was also worn out.

The third day was better. My instructor, the school's flight supervisor and number two pilot, Frank Schwarz, showed me a shortcut to the paperwork, kept his lecture subjects to holding and tracking, and gave me three hours and 20 minutes of flying, including my first approach.

I also learned about the Flight Safety method of grading. You are graded, it seems, on everything you do to or in the airplane, on a scale from one to five. One means you are absolutely superb—better even than the instructor. Two is very, very good; three is good; four, okay; five, unsatisfactory. I got ones on my (not the airplane's) attitude, cheerful little thing that I am, and threes and fours on everything else that week.

There is a great, fast-evolving sense of camaraderie at a school like this. Friendships necessarily are developed rapidly,

and in a short time you get to know a person's first name and ultimate goal in life, though rarely his last name or place of birth. You soon find yourself actually caring whether Art gets his private license, or whether Amy gets along with her instructor, or whether Isaac has soloed yet. You have no trouble telling students from instructors, for the teachers look like authority figures in their neckties, clean shirts and self-assured ways. Besides, they never sweat, and they have a private room where they can sit and drink coffee and laugh about their problem students. The rest of us wander around the hangar looking lost and slightly dirty in our jeans and sweatshirts.

By the fourth day, I and everyone else at the school knew I wasn't going to get my instrument rating that week. Maybe not even the next. I had a lot of work ahead and a few missing requirements to complete. One new necessity is a 250-nm instrument cross-country, during which you must do an ADF, ILS and VOR approach, each at a different airport. I suspect that's going to hang up a lot of pilots, like me, who accumulated their 40 hood hours under the *old* Part 61.

Thursday I had my first filed IFR cross-country, a 30-minute hop up the Florida coast to Melbourne for an ILS and back to Vero Beach for the VOR approach to Runway 11. The VOR into 11 became SOP for all subsequent returns. I also had two hours of night flying with a third instructor, Dick Sharp, who was to take over my second week of training.

By Friday, I was comfortable with the idea that I wasn't going to get the rating, and consequently I relaxed and started to learn more. I found that I was greedy for the learning and the consistency that I had never had before, and I really didn't want to go home. Schwarz, an ex-Marine instructor-pilot made of granite and uncompromise, recognized my new spirit and socked the unusual attitudes to me, along with partial panels, steep turns, holding—until my head ached from trying to determine whether a teardrop, parallel or direct entry was called for—and ADF approaches. I was surprisingly good at those anachronisms in the modern world of ILSes and scored my first (and only) two from Schwarz on the second try. And for the first time since I'd arrived at Vero, I flew the four-hour day that I'd thought would be such a cinch.

At that point, I made two mistakes that nearly cost me getting the rating the second week. I went home for a few days to catch up on some work, which wiped out a few things I had been learning, and I brought some more work back to Vero Beach with me to do "in my spare time." I did not practice what I had begun to preach—stay with it till you get it and let nothing interfere.

Monday I was rained out again—even Florida sunshine needs a day of rest—and Tuesday I got in just a little over three hours, thanks to the distractions of the work I'd brought with me. For Wednesday, Sharp had scheduled The Big Cross-Country, with an ETD of five p.m. The round robin went first to Orlando for an ILS to minimums, a missed approach and an airborne clearance to St. Petersburg. Sharp, being a most kind man, let me take the hood off twice—once to see the runway 200 feet below me at Orlando and shortly thereafter to view the fantasy world created by Disney's dreams.

At St. Petersburg, I botched a backcourse approach and landed frustrated and cranky. Snuffling down a warm Coke and stale peanuts, I decided to give up instrument flying forever. I was beginning to feel rather like an incompetent computer being programmed by a human being I couldn't see. I wanted to revel in freedom in the sky, rocking my wings and flying hither and yon as I pleased. Cold turkey had worked. I got my first two in attitude that night.

The gods provided a spectacular sunset, followed by an enormous full moon to cheer me on. A short leg through a non-radar environment kept me too busy to sulk, so the final stage—an ADF to minimums at Fort Myers and home again to the VRB VOR approach—was a bit better.

Thursday I began to feel pressured again. It was the last day to prepare for the check ride scheduled Friday morning. I was exhausted and vertigo-prone. And absolutely sure that I was not ready. Two short sessions, concentrating on unusual attitudes, partial panel and night approaches, went poorly, but Sharp decided that I *was* ready. He'd let me take his mother or sister flying in real instrument weather, he said, so he was willing to risk his reputation by endorsing me. (Sharp probably doesn't have a sister and knows his mother is smart enough to stay on the ground.) Go home and get a good night's sleep, he said. Does anyone sleep on her wedding eve? Back at HoJo's, I slept two, tossed two, slept two and gave up.

Friday began ignobly with an hour's worth of paperwork to "close me out." Ha! You aren't going to get rid of me that easily, I thought. Narigan breezed past me in the office and casually told me to work out a flight from Vero to Allendale, South Carolina. Aha, a strange little town—probably doesn't have weather reporting—one of Narigan's standard tricks, I'd been warned. I didn't know how to tell for sure on the chart, which Narigan ultimately found out, but the FAA confirmed my suspicions, so I filed a wide-open alternate. The weather briefer must have sensed the tension—I was having a hard time spelling my own name on the flight plan—for he gently reminded me to ask for winds aloft and notams.

Paperwork laboriously done and redone, weight and balance figured, weather briefing recopied in legible block letters, I presented myself to Narigan, who started firing the questions. Flight plan, regulations, procedures, chart symbols, what ifs. Some I knew, some I guessed at, some I stared at. I figured I'd failed before I even got off the ground, but Narigan told me to file and preflight the airplane.

In the airplane my whole flying life flashed through my mind. It's sink or swim time, I thought fatuously. I was trying desperately to keep a grip on myself—reality kept vagueing away—and not to forget the least detail. My clearance came— cleared as filed, godblessem—and we were off. I was so nervous I forgot to turn the transponder on. Miami Center prodded me gently about that, and Narigan seemed happy that I'd caught the mistake early. Hood on at 300 feet—examiners waste no time—and we flew on course long enough to prove that I knew in which direction South Carolina lay. Then we cancelled and went through all those things I'd been practicing for two weeks—slow flight, partial-panel turns, climbs, descents, and then . . .   head down, eyes closed while he went through maneuvers to disorient me (it worked). Okay, you've got it. (I wonder if anyone has ever said to him, "I don't want it, you keep it.")

Then he quietly said, "You're cleared direct to the 270 radial of the Vero Beach VOR, maintain 3,000."

I went absolutely blank. I fumbled with the OBS, turned the airplane right, decided that was wrong and turned it left. I wandered about confused and depressed. I simply couldn't remember the procedure he wanted. I took off my hood and admitted defeat, expecting to be told to go home. He quietly told me to put the hood back on and watch him. Then he

showed me a simple way to find the 270 radial.

I was cleared directly to the VOR, which I could handle, and to enter a holding pattern. I determined the type of entry— direct, luckily—and the holding worked fairly well. So did the familiar, old VOR approach to Runway 11, and I was back on the ground. Narigan emotionlessly said that I had done fine except for the VOR orientation and that he couldn't pass me without it. I would have to go up and practice with an instructor and be signed off again for the check ride.

I was close, so close. But I had to go home that night. I could take the remainder of my flight check with an examiner back where I lived, but I wanted Bush Narigan's signature in my log—for sentimental reasons, I guess. Happily, the school was willing to help. Schwarz was free for a while, so he bounced around with me while I did seven fairly competent VOR orientations. Then I went through another round of paperwork; that's the worst part of flunking a check ride.

Once again, I was in the Cherokee with Narigan, who did the takeoff while I put on my hood. You've got it, at 100 feet. After flying around in circles to get me lost, I was cleared to that foul little 270 radial again. But this time I found it. And the next, and the next. Narigan rewarded me by letting me take off the hood and showing me a shortcut to the procedure.

He also gave me that scrap of paper. My instrument rating.

At the end of my check ride, Narigan delivered me to an airport from which a 727 would take me home. Sitting amid my suitcases and flight bags, clutching my new license in a sweaty hand, I felt an intense desire to tell someone. I had called mother, collect, and she was happy because I was happy and because she didn't need to worry anymore since I was back on the ground. But I wanted someone who would *understand*.

And there he came, strolling through the terminal with that casual self-assurance that all his brothers share, four gold stripes glistening in the harsh lights—the Eastern Airlines captain who would fly me home. I flung myself in his path and begged to be allowed to speak. Probably fearing the worst if he denied me, he granted permission. "I've just gotten my instrument rating," I yelled, and slowly he comprehended. Then Captain Jim Norris and First Officer Wes Hargraves welcomed a little fish to their big pond with an enormous hamburger from the terminal coffee shop.

After inspecting the sweaty scrap of paper, the good Captain wondered, "A private, multiengine, instrument-rated pilot?" He looked at me pointedly and said, "How come no commercial?"

"Well," I said with my mouth full of hamburger, "I'm *getting* it!"

The Salmon River Run

Their airways are valleys twisted by looming mountain walls.

They fly into barely approachable airstrips.

To these pilots, it's all in a day's flying, but it's really the stuff of legend.

BY GEORGE C. LARSON

AVIATION could use a good story right about now, a story of grit and resolve in the accomplishment of a day's labor. We are not getting anywhere with plaintive whining about how difficult it is to hoof it halfway around the world in the front of a Boeing because the jet lag throws your body clock off; nobody is listening except those endless Washington committees, and those proceedings are not fruit for folklore. We need a tale of callused hands and human brawn in contest with nature, doings that will stir our children. Our pilots must be a bit like Frodo in those *Ring* stories: meager lads who are called to higher purpose by forces beyond their ken. They must rise to the challenge, stare death down until it blinks and emerge to shine larger than their beginnings. The machinery must be as near to living as machines can be; it must have within it some force of virtue, a heroism that carries into the far realms of sorcery.

The origins of such a legend are available to us—still fresh and echoing—in the Idaho mountains.

In the early 1930s, the post office at Big Creek, Idaho let a contract for a winter mail route to a man who rode his horse through the snow until the drifts were too deep for the animal to negotiate without exhausting itself; the man would dismount and trek the remainder of the way to the post office on his snowshoes, the entire trip taking a full day. It took him another full day to make it back to his route, which followed Big Creek and terminated at Cabin Creek; this became known as the Cabin Creek Route. By the mid-1930s, it was apparent that more was needed, and an outfit in Missoula, Montana

called Johnson Flying Service had airplanes available. In the winter, Johnson flew mail from Cascade to Big Creek Airport and to the post office at Warren. It also flew supplies to what was then the Mackay Bar Mine, which had the only airport on the Salmon River between the town of Salmon, on the eastern end, and Lewiston, beyond the western extremity. There was a caretaker at Deadwood Reservoir who got his mail and groceries twice a month through the services of Johnson, and the company also flew to Cabin Creek on occasion to deliver mail and freight. In the summers, Johnson picked up the entire operation and moved up the road to McCall Airport, from which it flew for the U.S. Forest Service. In the summers, the mail business fell off when roads opened up to allow surface travel. In 1945, Johnson left Cascade Airport for McCall; it has been there ever since. At about the same time, Dewey Moore's Airstrip, Vine's Ranch, and the Taylor Ranch—all located in the area of Cabin Creek and desiring the same service—plugged into the route.

The mining business began to lose money in those years as the easy pickings gave way to harder times and extraction methods that cost too much; when the Big Creek mines closed, the post office closed down, too. Although a post-office carrier was no longer needed, there were still people in the area who had been reeiving mail service and who were entitled to have it continue; the Post Office Department put the Cabin Creek Route up for bids to the public. Johnson bid on the contract and won, and so began the company's marriage to the carrying of the mail to Cabin Creek.

Next, the Forest Service asked to have mail service extended to its stations at Cold Meadows and Chamberlain, which were manned only during the summer. These lay somewhat north of the Cabin Creek Route and closer to the Salmon River. The post office approved the extensions of the route, and they were added sometime before 1954. About 1952, a Forest Service crew preparing to float a bridge down the Salmon for installation in the area of Campbell's Ferry asked Johnson to survey the area for possible aerial supply during the construction of the bridge. A pilot named Bob Fogg, who had learned to fly with Johnson and moved to the McCall, Idaho operation from the company's Missoula, Montana headquarters in 1943, went into the area on foot and set up a landing site. Whether this was the foot in the door or whether the earlier extensions to Cold Meadows and Chamberlain had done it is hard to say, but soon afterward, the Cabin Creek Route was extended to include delivery to additional sites along the Salmon River's main run through the Payette and Nez Perce National Forests; the new locales were named Crofoot Ranch, Campbell's Ferry, Whitewater, Allison Ranch, Shepp Ranch and James Ranch.

There was now more of the Cabin Creek Route flown along the Salmon River than there was along Cabin Creek, but the name was never changed. The mail run is called the Cabin Creek Route and is considered by the post office to be a Star Route, which means in practical terms that the mail is carried for delivery to people's mail boxes, and not simply shuttled between post offices.

Recently, the post office, as is its practice, put the Cabin Creek Route up for renewed bidding. When the results were in, Johnson Flying Service discovered that, for the first time in the history of the Cabin Creek Route, someone else had won the contract. The amount that separated the winning bid from Johnson's was not very great, but it was enough to end a tradition, and on May 15, 1975, Bob Fogg and two other pilots lifted off from McCall Airport, turned across the valley of the Payette Lake and headed north to the Salmon River Canyon to make the last runs on the Cabin Creek Route. This is the story of that day and of the people and airplanes that are part of the legend of Johnson Flying Service and Salmon River country.

Boise sits like a sunny-side-up egg in one corner of this great frying pan of a valley in Idaho. You cross over the city and head north, following either Victor 253 or a winding, single-track line of the Union Pacific Railroad. The railroad skirts to the east of the Cascade Reservoir, then continues north toward a feather of smoke and a forked lake; the feather of smoke is from the waste burner at the Boise-Cascade mill, and the lake is named Payette Lake. McCall Airport has only one paved runway, and since there is usually little wind, the local pilots like to land to the north and take off to the south, which avoids, for the most part, flying over McCall, a tidy town arranged along the south shore of the lake.

Just as the Boise-Cascade Company *is* McCall, one fixed-base operator dominates the airport: Johnson Flying Service. Want gas, repairs, a charter, advice? The answer to all questions is Johnson. Further, the answer to all questions about Johnson is Bob Fogg, manager and pilot of the McCall operations of the Missoula-based company. Fogg runs the operation from an Alpine-style hangar, one small corner of which is devoted to office space to hold the paperwork, the cash register and the telephone. His wife, Margaret, attends to some of the clerking and admonishes him on occasion to resist doing too many favors for too many people, lest he achieve sainthood—the hard way. There is a full-time mechanic assigned to the station, Don Micknak, with Johnson 15 years, at McCall for six. Fogg himself learned to fly at the Missoula office in 1939, became a Johnson employee in 1940 and moved to McCall in 1944.

In those years, he pushed Travel Air 6000s—six-place, high-wing trucks—and Ford Trimotors for the Forest Service working forest fires, carrying mail as needed by the contracted carrier and running hunters and fishermen to their hunting and fishing. Later, the Forest Service specified the Twin Beech and DC-3 in its contractual agreements; now the Forest Service uses its own Twin Otter.

Fogg says he will make his part of the last Cabin Creek run in a Piper Super Cruiser modified for mountain flying and that takeoff time will be about eight o'clock.

Fogg is lifting cases of groceries into the tail of the high-wing Super Cruiser as I arrive. He figures on a payload of 350 to 400 pounds, mostly because the rough fields are hard on the gear, maybe too hard if the loads ran much heavier. There is produce and fruit and a bag of potatoes in addition to the orange-nylon mail sacks, each with a tag labeling the destination for the contents: Campbell's Ferry, the Mackay Bar Ranch and the James Ranch. The limp sacks, each with no more than a few letters in them, lie together on a large, wheeled cart that looks ridiculously overdesigned for its task, but which is probably called the "mail cart," and, by Heaven, will be used to stack the mail and trundle it to the airplane; such are the routines of the place.

Another pilot, Phil Remaklus, arrives in a medium dither, apologizing for having misunderstood whether he would be flying this morning; he will take a Turbo 206 to the Allison and Shepp ranches. Fogg and Remaklus help each other to load their respective airplanes. There is a fruit tree still to be installed atop the mound in the rear of the Super Cruiser, its roots bound up in a cage of burlap as if it were some drugged animal being shipped to a zoo. Two man-high bottles of propane go prone on the floor of the 206, to be overlayered with groceries—a ham shines up through the pile—and a Formica counter top in two pieces, the larger of which won't fit unless the two men can modify the airplane; they decide to leave it for another way.

Micknak arrives, the pipe smoker, the "Polack," the self-deprecating mechanic who, one instinctively feels, is the foundation upon which the whole operation rests. Remaklus and Micknak discuss the settings on the propeller governor and the manifold pressure readings for the 206; they have been a little off lately, and Micknak has adjusted them. They also discuss fuel and weights. These men fly so far below gross weights that it is never a question of being legally overweight but only operationally overloaded for conditions. They must be their own judges, and no magistrate ever had more incentive to deal with truth, for too heavy means too bad. They check their fuel not just with gauges but with a wooden dipstick scored carefully at each five-gallon mark; since they fly with tanks no more than half filled, it is their best way to know their fuel status accurately.

The passenger, Mrs. Richardson, arrives with a friend, the two of them drinking coffee from steaming paper cups. Remaklus is pulling the prop of the 206 through as Fogg turns to aid his passenger aboard.

The 206 is first out to the runway. The day is clear, crystalline, with the ever-present smoke from the waste-burner on the lake spread out evenly in a thin layer over the town, the lake and the meadows west of the airport; again, it tells as clearly as any sequence report ever did, "winds ◊◊◊◊." Remaklus finishes his run-up, announces his departure over the unicom frequency, then takes off with the muffled snarl that turbochargers have a way of voicing. The sound of his first power reduction is lost as Fogg, right behind him, runs up the Super Cruiser, all 150 horsepower present but a hefty slice not accounted for at McCall's 5,000 feet msl. Fogg lines up and takes off; he holds the tail down—or is it down of its own?—for about 200 feet, and the Piper floats off soon after. Surely there is helium in the wings.

I follow in a floozy, speed-striped 235-hp Charger as Fogg leads; he immediately turns to what later seems to have been the *exact* heading to the entry into the Salmon River Canyon—I never sensed any midcourse correction—and begins a gradual climb that will take us as high as 8,500 feet before he begins his descent into the river's anteroom. I know there are elk below us, because Fogg had told me so earlier, but the elk must have their minds on breakfast in private today, for they are nowhere to be seen. There is the meandering river that feeds the lake—a river now, a creek later, then a trickle when it turns dry—and a topographer's nightmare of knolls, spires, pyramids, spurs, ridges . . . how can Fogg tell one from the other? Moreover, the scale of it is evasive, for there is no telling whether each tree down there is a few years past seedling or a candidate for a mainmast. Without a familiar scale of reference, you are left with only your sense of relative motion to reassure you that you are, indeed, at a safe distance from the upward reaching land. From the safety of the airport and with one's feet firmly planted upon their foothills, these mountains seemed enchanting; now they are threatening, reaching up like vast anemones for us morsels.

At last, after some 20 minutes of cruise at about 80 knots, the Super Cruiser seems to pick up speed; but the velocity is due to descent. Fogg is on his way down to the river.

The Salmon River is called the "River of No Return" on the local billboards and on the travel circulars that ballyhoo trips on rubber boats. The name actually stems from the fact that once you float down the river, the terrain prevents your returning by portage along its course. The river's potential malice should not be gainsaid, though, for although its 80.8 miles drop only 969 feet, sheer walls confine its course, so that when the volume of water increases, the river must rise; when it rises, it is no salmon but a barracuda bent on spoor.

Fogg's first stop will be Campbell's Ferry, which is a well-known river crossing that served miners hunting gold as a jumping-off point to the riches of Thunder Mountain. The mines are gone now, and it serves as a way station and resort ranch serving boaters and sportsmen.

The red Piper has been slowly but steadily accelerating even as it has grown smaller. I remain above the canyon in what is probably the vain hope that if anything goes wrong, I could glide to a safe landing atop the canyon's maw. I check the chart and note that the river is at 2,000 feet, and I am almost even with the rim, according to the altimeter, at between 8,000 feet and 8,400. When Fogg touches down at Campbell's Ferry, he'll be about a mile beneath me.

Down, down, drifts the Super Cruiser, winding now between the walls and shrinking to the size of some winged protozoan from my perspective at the top of this scaled-up microscope. The illusions crowd in: it is at once, for me, like tending a diver who is on a deep descent; there should be bubbles rising from the Piper. My mind wanders to classroom talk about how altimeters measure the proverbial "column of air," and here is Fogg flying at the bottom of one. Surely it must be possible to feel the weight of the gases down there.

Suddenly, the distant speck has turned a tight circle. Something wrong? There . . . houses, a clearing. No, it can't be. This can't be Campbell's Ferry, and Fogg surely couldn't mean to set the little Super Cruiser down into that meager opening in the trees; it can't be done. He is probably just circling in greeting to whomever lives there. Another circle, then he is off upriver again. No, he's turned back. Like a red sparrow, he arcs tightly, and I lose him for just a moment under the wing. By the time I get the Cherokee turned to where I can see him again, Fogg has his airplane turned straight at the canyon wall. He seems to be barely moving as the airplane crosses over some trees—how tall are they?—and I see the shadow of the Piper racing up what must be a steep slope. The airplane and the shadow merge, there is the very briefest of rolls, and he is pivoting around on one wheel, nudged up against a stand of trees, pointed into the center of the clearing. It takes a few seconds to realize that we are, in fact, at Campbell's Ferry. I had not witnessed an emergency but a routine landing at destination!

How—?

There is no time now to try to ponder what has shown every evidence of being an impossible feat. I could try to guess at the size of the airfield—clearing, strip, whatever—but it would be meaningless at this distance. Approach airspeed? I will have to ask Fogg those questions later. For now, I simply circle like an idiot buzzard too scared to chase the mouse into its hole and ill-equipped for the task. All the while, my eyes flick around for flat spots and find none; if anything were to happen now. . . .

What am I doing here? More importantly, what is Fogg doing down there? What would *he* have done if the Piper, or its engine rather, had decided to call it a day? If he had misjudged the approach a bit, how could he have gone around?

About 15 minutes later, the Piper moves down there, scampering uphill, turning in one motion, and running back down. In only two seconds, the shadow and the airplane have split again, and I realize that the Piper has performed another amazing levitation. Fogg turns downriver. Next stop: Mackay Bar Airport, the oldest strip in the territory. I wonder if there had been any ceremony attendant to this last run into Campbell's Ferry? Perhaps a cup of coffee, some kind words?

It is only a few minutes down to Mackay Bar, and the airstrip, unlike that surprising hole in the trees at Campbell's Ferry, is obvious from a distance—a white (probably sandy) runner at a shallow angle to the riverbank. At the downriver end stand some low buildings, next to which is parked an old Cessna, probably a 172. Mackay Bar is situated at the junction of the main Salmon and the South Fork, leaving a fairly wide opening in the canyon for Fogg to hang a U turn and drop into the airstrip. He uses no more than a fourth of the runway rolling. Again, I circle above. There may be more business and conversation here, for there are two mailboxes to be serviced at the Mackay Bar: one for "Buckskin Billy"—his real name is Sylvan Hart—and Rodney Cox, Hart's nephew, and family; and one labeled only "Mackay Bar," for the crew at the ranch—there are six in residence today.

Fogg takes off in the opposite direction from the way he landed and cranks in an immediate right turn to follow the main Salmon to the next stop at James Ranch. Here, the mail will carry the address for Norman Close, who manages the place with his wife and two children. The James place has been described as having one of the more difficult approaches to a landing, but the word "difficult" has a whole new meaning here. The canyon is narrow at its bottom, but it widens a bit more quickly here than it did upriver; at least that's how it looks from my vantage point. Fogg elects to fly over the strip, part of which looks to be under water, then climb a bit before turning back. I can see the shadow from farther out on the approach this time, but I notice that Fogg has it slowed down to what must surely be the near-stall point. Again, that short roll-out and a slow, trundling taxi to the buildings at the far end. While I wait, I am surprised to hear Remaklus' voice over the unicom frequency, announcing his departure from the Shepp Ranch, just a mile downriver from James. Sure enough, here he comes in the 206, appearing to emerge from the very rocks, then banking around to climb along the river toward me. He scoots by underneath me and passes over Fogg's parked Piper without ceremony. Somewhere down there, Fogg is handling out the last sack of mail to the Close family.

"Back in 1954, we were using a Super Cub and a Stinson Station Wagon," Fogg is recalling, "then we had a 180 for eight or nine years, sold the Super Cub. I think it was 1967 when we got the first 206; it was a 1964 model, and we sold it after 6,000 hours. It wasn't until three years ago that we got our first turbo." All high-wingers? "Oh sure. You can't afford the ground float, you know, and the weeds really kill the leading edges on a low-wing. No, it's a high-wing operation all the way."

Fogg reckons he has 22,800 hours. At his peak, he has flown as much as 178 hours in a month, 700 to 800 hours in a year. These days, he figures he is flying at the rate of 350 to 500 hours a year. Though he doesn't say so, it is obviously to his advantage to fly time himself, for he must mind the budget and keep the McCall operation profitable. The office is lined with old photographs of past airplanes working at everything from bug spraying to air drops of equipment by parachute. There are the disasters, too: one photograph shows the Travel Air with a ski driven through the snow crust ("It took a long time to dig 'er out.") and a wingtip buried. Fogg remembers each airplane, how Harry Combs bought the Stinson, how they were one of the last outfits in the country to keep a Ford Trimotor working; it finally left them six years ago, though parts for one still exist up at the Missoula maintenance shop, which arouses fantasies in Micknak—he would like to rebuild one.

Now there is the PA-12 Super Cruiser, good for as much as 400 pounds' payload, the 185 on skis that is capable of toting 1,600 pounds without the seats or three passengers and 200 pounds of gear (on wheels, it does better) and the 206s, which, with 50 gallons of fuel, will haul five passengers or 1,000 pounds for Johnson. Into the really short fields, such loads are reduced by as much as half. Twins are not economical enough to earn their keep at five cents a pound for freight, so there are only singles for the foreseeable future. Micknak says he was skeptical about the turbocharged airplanes for quite a while but that they seem to have proved themselves; he now thinks they are dependable enough to become permanent in the operation.

The pilots and airplanes have a drawnout learning curve in this kind of flying. Fogg says it took him 10 years to learn from his predecessors, one of whom was Dick Johnson himself, who later died in 1944 when a downwash caught him during an elk-counting mission. Most of the pilots who work for Fogg describe themselves as his students, and even the pilot who took away the mail contract with the lower bid was after Fogg to check him out. A check-out with Fogg takes time. You might start by operating in and out of Chamberlain Basin, which, as these airports go, is easy. With time, you will go into fields of increasing difficulty, each time trying it with Fogg observing; when you are ready, he dismounts, and you get to try it alone three or four times, just like first solo. There are those who try it and just can't handle the pressure. Most of Fogg's pilots are, well, "mature," with the preponderance coming out of the retired military ranks. Fogg himself spent only 31 hours at cadet school before he was called back to civilian flying by the Forest Service, which needed his skills badly. Phil Remaklus is an ex-Air Force colonel, Bill Dorris is an ex-Marine and a Fish & Game pilot, John Slingerland is an ex-Marine. In the summers, Bob Franklin, also ex-military, comes aboard, joined by an airline pilot who wasn't identified. The pay? They never talk about it, probably because it is not commensurate with the task, probably because, on their military pensions, it doesn't really matter, and probably because it is the only flying left to find where nobody bothers them—nobody would *dare*—and they find a peace in that canyon that is not measurable in dollars.

Not that they are not conscious of the risks. One said:

"People ask all the time, 'What would you do if the engine quit?' and I just say, 'Say a prayer and kiss my ass good-bye.' " There are no outs, no go-arounds, no "forgiving" by the aircraft, no back doors, nothing. There is only exactness, a Zen-like confidence in the outcome of the maneuver that would ensure its success even if one closed one's eyes. There is judgment and follow-through; nothing in between, for commitment is final. Each man knows what he will and will not do, and there is no bending those limits. One states categorically that he will not go into Campbell's Ferry with a 206, though he knows it has been done; another says he will not go into Campbell's Ferry with *anything*. (It seems like a good moment to ask Fogg how in the world he manages it into the 800-foot strip with its 17-percent slope. "You keep the Piper just above the power-on stall; it's been modified with flaps, you know, and with full flaps and power, you can get down to less than 35 mph." It was no trick; he *was* just hanging there.)

The secret to their flying seems to be that they pick their conditions. Any wind is too much wind, which means they do most of their flying in the morning, and if you want to make the trip, you'd better be there, because nobody will wait around for you. Once into the canyon, they will not proceed farther if they detect even the slightest beginnings of the draft that can get roiling down that wind tunnel and smack them around. There are wind socks at several of the strips and wind streamers at the most sensitive ones. Campbell's Ferry has a streamer that will pick up and move in just a whisper, whereas a wind sock wouldn't. Fogg allows as how he remembers only one go-around. "It was at Mackay Bar. A horse ran out on me and I had to go around to avoid hitting it." He hastens to add, "But I was ready to land, I was lined up okay." One of their biggest worries is animals; if it isn't horses, it's the elk. Or a dog.

They are, to a man, sensitive about the intrusions of Government into aviation. They can recite quite accurately the regulations that govern their operations, including one recent one that requires them to fill out more forms to waive the requirement that they get permission 24 hours ahead before carrying liquid propane or dynamite or any other dangerous cargo. They report their Federal tax collections monthly, their passenger loads daily. They can recite the fuel price rises within the last three months to the day and exact amount. They know all about Mr. Ullman's tax plan.

They disdain towers and controllers and rules, the damned rules. They give themselves their own clearances and checkouts, and live in the knowledge that no man is as qualified to do that as they are.

They despise anyone who purports to tell them tomfoolery in any area where they are more knowledgeable. There is the story of the Sierra Club group that came trucking through to gape at the trees, pointed at a three-year-old burn in one portion of a national forest, and exclaimed, "Oh my, look how the lumber companies have clear cut that hill." The laughter on that is forced, bitter. They do not tell other people how to run their cities, and they do not expect strangers to come in and tell them how to run their woods.

Surprisingly, a respectable number of flatlanders who are new to the area stop in to ask advice before they plunge on. What is even more surprising is that these same yokels will then go out and do what they were going to do anyway, despite the best efforts of Fogg and the others to persuade them to wait until the winds are down or the temperature has dropped. Each year, there are serious accidents in this back country that are brought on by people who are bewitched somehow by the country or by something else. One fellow drove up in a brand-new Jag with a lady in the seat next to him, claimed thousands of hours with an airline, wanted to pay double to rent an airplane and show his little friend what real flying was all about; him they could handle, though it took two hours to say no. There was the one from California who stopped to ask, heard what they had to say, went anyway, dinged it, and then was too ashamed to call them to ask for help. He called another city: "Say, uh, could you ship me a new prop? Oh, and while you're at it, um, might as well throw in an engine and a main wheel . . ."

Oh, the talk!

Fogg: "I remember the time into Campbell's Ferry in our Cessna 180. I lost a wheel. It was not a hard landing, and the gear was spring steel, y'know, so I bounced up a little. Up ahead o' me, I see this blur of something move, and I figure it's some animal and, you know, it's too late to do anything about it now (laughter), so I just went on in. Turns out it was one of my main wheels just fell off from fatigue failure. That cost a landing gear and a wheel, plus a wingtip, prop and bulkhead in the ground loop."

Phil: "I knew a guy flew the Hump said that he'd never seen anything like this, that this was the hardest flying in the world."

Bill Dorris: "We have this one lady flies the whole time with her eyes closed, you 'member that, Phil?"

Remaklus: "Yeah, she won't fly with Bill anymore. I think there was one time, too, when she just crawled in back on top of the mail sacks and just rode the trip lying down back there."

Fogg: "We're well maintained . . ."

Remaklus: "Yeah, the ships look used, but they run good." (Micknak puffs the pipe.)

Remaklus: "Would you (to Dorris) take your 185 into Campbell's Ferry on skis?"

Dorris: "Oh, I've done it, but . . ."

Fogg: "You've got to catch the ledge. Get up there and get turned around onto the ledge. It's a 17-percent slope, and skis is kinda like floats—you ever fly floats?—and then when you go to taxi up top for takeoff, it takes almost full throttle."

All I can see in back of the Super Cruiser is Remaklus' back and his beat-up corduroy hat. We have poked into two entries down to the river and failed twice, are now pressing farther west. The clouds are all cold, wet-looking, not sharply defined but simply separate parts of a ragged stratus layer at altitude. There are groceries in back of me, and I can smell wet lettuce amid the other aromas buried within the fiber mat upholstery and insulation. The Piper had scampered up here quickly, and all that's left to do now is to find a way in and down. We pass the landing strip at Warren, which places us west of the South Fork. At an OAT of 40° F., the carburetor ices up—the humidity.

The residents of the canyon all communicate with each other and with Johnson by a radio net that suffers terrible noise intrusions from atmospheric electricity. Two days ago, a giant thunderstorm had roared into McCall and blown dust up so bad you couldn't see to the end of the main street. The lights had gone out in parts of town. That afternoon, the radio had been nearly unusable for the static. Next year, there will be a single-side-band system installed that will eliminate that problem plus boost effective transmitter power. To the people who depend on them, there is more than an impersonal business relationship between themselves and the Johnson pilots. They call the day before, and someone at Johnson writes down their shopping list. Just this morning, Remaklus has come from the grocery store, where he has picked up an order for the Campbells at Shepp Ranch (no relation to the Campbell's of Campbell's Ferry). He also sought out and picked up a bronze valve for some plumbing need or other.

Suddenly, the Piper whips into a hard right bank. Remaklus has spotted his opening, and just as quickly—right NOW—has made his decision to shoot through a saddle to the north, and then we are going down.

Between two angled walls with some scrub on them, we plummet down to the river. The scrub turns out to be lodgepole pine, maybe 50 feet tall. Remaklus spots an elk all by itself, but I'm blind this morning. Finally, with his help adjusting my sights, I see it, too, a tiny blob of golden fur—the wrong color, the wrong size. I was looking for something many times that big and much darker. (Remember, the scale is way off.) The sun is breaking through the now-thinning layer, and the light catches rivulets of water on the heights that fall so far they turn to haze before they strike anything solid.

"Bear!" Remaklus is looking off to the right, and this time I see him right away, standing alone, looking at us, I think. How the hell is he holding onto the side of that—?

"They drive sheep in here to feed, and every fall we have to help look for the lost ones."

A dirt road appears halfway down the slope of the eastern wall, winds around, descending into a tortuous whip until it disappears again in trees.

There it is! We're at the Salmon River. We are still 2,000 feet above the water and descending, and I can see white water in spots. We turn right to head east to Shepp Ranch, which sits within its own canyon hard by the river. It is upon us before I could have reacted, but Remaklus began the turn in plenty of time, now heads in close to check for horses. Four of them are down drinking at a creek, and he is a little worried about whether they will spook and run onto the landing area. One more turn-around overhead and then he is satisfied; they don't seem restless. We aim back over the river, climb a bit and then turn, sinking fast and getting slowed all at once. Flaps are coming down, power is reduced, and the Piper is settling with a whistle. We are canted off slightly to the right of the runway's axis, and Remaklus waits until he has cleared some trees to correct, shoves in some power to halt the sink a bit, and WHOMP! We are down. I feel like a fool, for it has all happened so fast, and I haven't watched enough of what happened, and my pen is still poised above the kneepad, with nothing to write.

We step out of the airplane, and the effect inside the canyon is as if Remaklus has just flown us into the bottom of a well. The sensation of it all has been more hurried than I'd planned for. I should have carried a motion picture camera and filmed at twice the normal speed.

"You'd fly that a little different in the 206, of course," Phil is saying. "Fogg flies the 206 in here just like the Cub—turns a complete 180 out over the river—but I don't. I turn out and head upriver where there's more room to turn. He's done it more times." Remaklus sounds almost apologetic. My tongue is stuck.

"This is not mountain flying, it is *canyon* flying," Bob Fogg told me.

No, it is much more than that. It is legend, and although it is probably fruitless to try to make legend—for that must happen spontaneously—you will have to forgive the folly of trying anyway. I know that finding the world's toughest, noblest flying enterprise in McCall, Idaho may seem as likely as running into a matador at a Dairy Queen, but you will just have to believe it.

The Compass and the Clock

*A speck upon an endless ocean, a work of science
and of art, Melmoth strikes eastward to test the mettle
and endurance of its maker.*

BY PETER GARRISON

I AM DRAWN by the thought of flying long distances. It is my mountain climbing, my archery, my test of strength. I don't know why it appeals to me so much, but I like to look at a globe, mentally measuring the great-circle routes and translating them into hours and into pounds of fuel. The more remote and the lonelier the route, the greater its magnetism for me. My reason sees such long flights as long sits, merely tedious waits for a machine to tick up its appointed number of revolutions; but in my heart, they appear as the door handles of eternity, glimpses of the inaccessible and the sacred.

When I designed an airplane, it was natural that I would design it for range. It was not *optimized* for range; airplanes optimized for range have long slender wings like an airliner's or a sailplane's. Though graceful, they are ungainly and hamper maneuverability. My airplane would be as compact as a fighter and would achieve its range by means of low drag and a large internal fuel capacity.

The thought of flying around the world was always in my mind; a convenient design parameter was offered in the distance from San Francisco to Hawaii, which I came to think of as the longest unavoidable stretch of open water on an imaginary circumglobal jaunt. (Actually, that stretch *is* avoidable by following the northern Japanese islands into the Aleutians, but the uncertain weather on that route makes it impractical.) From San Francisco to Hawaii is about 2,100 nm; a range of more than 2,800 nm would be necessary to make that flight comfortably.

The airplane I built is named Melmoth. It took five years plus two more for tests and alterations, and it has a paper range of about 3,000 nm on a fuel capacity of 154 gallons. The fuel is carried in two tip tanks that hold 35 gallons each and two integral wing tanks holding 42 gallons each. The tanks can be selected manually, or else a pair of tanks—both tips or both mains—can be selected manually and the fuel system set on "automatic," whereupon it switches from one side to the other by itself every five minutes, keeping the plane balanced.

The engine is a Continental IO-360-A of 195 horsepower, with 210 available for takeoff. It is a six-cylinder, fuel-injected engine; mine has been carefully overhauled so that it uses little oil—a quart every 12 hours or so. It runs smoothly on very lean mixtures.

The airframe is small, less than 22 feet long with a 23-foot wingspan. The cabin seats two, with a large baggage compartment in which a single jump seat can be used by a third passenger of great limberness and endurance. A large all-flying T tail provides a wide CG range.

The airplane is fast and maneuverable when it is lightly loaded, climbing at nearly 2,000 fpm, doing a roll in three and a half seconds, indicating over 180 knots at sea level. It gets fairly good gas mileage: 25nmpg at 115 knots and about 18 at 160 knots.

With this airplane, on paper at least, I could fly 3,000 miles with a passenger and more than 100 pounds of baggage and remain aloft for more than 24 hours. The "door handle of eternity" was within my grasp.

Nancy and I decided at random to cross the Atlantic. A date, remote but not absurdly so, was tentatively set. The scheme seemed thereafter to gain solidity and momentum of its own accord, until it could no longer be stopped. The first preparations were abstract ones. At the beginning of the year, I sent letters to the governments of the countries we expected to visit requesting permission to operate an experimental airplane in their airspace. I ordered a subscription to the *International Flight Information Manual* from the Government Printing Office (it was never to come—though I did mistakenly end up with a subscription to the domestic *Airman's Information Manual*). I wrote to Canada for a detailed account of their requirements for single-engine aircraft departing their shores for Europe. I obtained Ocean Navigation Charts for the European and eastern Canadian areas, and GNC 3N, the Global Navigation and Planning Chart for the North Atlantic.

It appeared that the month of August, during which Nancy

was most likely to be able to take a month's vacation, was as mild a month as might be found for the crossing. The winds would probably be light for the return flight, there would be little chance of icing, and there would be few storms. Disasters were therefore improbable.

You can prepare forever and never be ready, because the first time across you simply don't know the ropes. No one can tell you everything in advance. And so, a well-prepared tyro, I left Los Angeles on the afternoon of August 2 with 20 pounds of charts, a handful of routine permissions to operate in foreign countries, raft, sleeping bags, perhaps 40 pounds of luggage for almost any climatic eventuality, survival gear, camera, a few tools and spare parts, and the beginnings of a sentiment of fear that would remain with me almost until we landed at Shannon.

Gander, Newfoundland. We arrive at nightfall. It is clear, and the lights of the airfield are visible from far out in the gray that grades into darkness. A cold wind is blowing. Nancy dashes off to the terminal to put on something warmer while I wait at the plane for the fuel truck.

The fuelers don't kid around. Gas gets spilled all over the place, but when they're finished, the tanks are brimful. I bring the oil up to 10 quarts, pay the bill and taxi to the light-aircraft parking ramp. The plane is strange with its 924-pound fuel load; the oleo struts are almost fully compressed, the tires, cooled and squashed, bulge out below the wheels, and it takes a lot of throttle to get rolling. I wonder if this lead pig with bumble-bee's wings will get into the air.

The night before you cross, you put in for a weather folder. They make up a forecast especially for your route, altitude and time, including the winds in each five-degree weather sector, photocopies of several graphic synoptic charts on which puffy masses of cloud spread and coil like protozoa on a slide, terminal forecasts in an indecipherable code for everywhere you could possibly end up going, all packaged in a personalized, blue folder. It's a nice system; it means that you have all the weather data handy for consultation should some question—such as the limits of an area of high winds or low freezing levels—arise in the middle of the ocean.

The weather office is on the mezzanine of the terminal building; we go up, passing a little museum of transatlantic flights, which includes a painting of some luckless fellows clinging desperately to the flotsam of their airplane in stormy seas, their haggard and terrified faces reminiscent of the faces of the damned in the Sistine Chapel. In the met office, a beaming forecaster takes our order. On the wall is a huge Mercator map of the world. I like the Mercator version these days, because it exaggerates the length of the Atlantic crossing, which on a globe or a GNC seems quite unimpressive—it is about as far as from Los Angeles to Detroit. Each time I see one, I gaze at that ragged patch of northern ocean over which Greenland hangs like a bunch of grapes and imagine myself vaulting it in a single bound.

The people of Newfoundland seem dour and gray; unsmiling natives direct us to a hotel and a restaurant, and by midnight, we are in bed. I fall asleep easily—I almost always do—with the specter of the crossing hanging like a fog bank off the shores of my consciousness.

In the morning, the airport is zero-zero. Disheveled corpses litter the terminal. The butterflies in my stomach awaken slowly. It is necessary to check out of customs, get the weather briefing, file a flight plan, have a bite to eat and go; this simple series of events somehow devours two hours.

The weather forecast calls for intermittently strong cross-track winds, but the bored and unsmiling met man, friendly beneath the frost, I feel, guesses the net component as zero. No weather at first, then some cumulus buildups, possibly to 11,000, in midocean, combined with a drop in the freezing level to 6,000 or so, then no "significant Wx" till the coast of Ireland, where there is a front with some thunderstorm activity. I am thankful for a mild day but disappointed in the wind.

The international flight plan form is unfamiliar, and it seems to be of the kind on which I always enter everything on the wrong line. It is curious about our survival gear. "We don't like to alarm people," the center chief apologizes, "but it's good to know these things."

I do not feel impatient at all; but finally, at nine o'clock, we get a ride out to the plane. The fog has lifted; there is now a mile or so of visibility under a 100-foot ceiling. The airplane is wet and dripping; a sporadic drizzle falls. Everything seems to contain a strange mixture of the everyday and the unknown. The engine starts up normally, we taxi soddenly, call ground: "November Two Mike Uniform, taxi takeoff, IFR Shannon." IFR Shannon—a little less than 2,000 miles across an ocean of legendary cruelty.

After a lengthy clearance readback, we are cleared for takeoff. We are 400 pounds heavier than the highest weight at which the plane has ever flown before. Full throttle. The prop surges, then settles down to 2,800 rpm. Acceleration is rapid; I realize now that air temperature and the density altitude affect the takeoff more than the extra weight we are now carrying. Here, on a chilly sea-level morning, we will have no problem. I hold the airplane down a little beyond my normal 80-knot rotation speed, then rotate gently. It's heavy all right; I can feel the weight, the lethargy in roll, the reluctance to become well and truly airborne. After perhaps 3,000 feet, the wheels lift off, settle and touch, lift off again, touch again; and then we are 10 . . . 20 feet above the runway, gear coming up, speed already above 100 knots, and into the clouds. At 110 knots, I bring the flaps up, watch the altimeter for a sag; it's there, but not serious. The air is dead-still, and the airplane solid as a rock. With 115 knots indicated, the rate of climb goes to 700 fpm. The rpm back to 2,600, still full throttle, all temperatures and pressures are good, climb is good—everything is okay.

We are in the fog for a few minutes and then break out on top of a seemingly borderless undercast, with blue sky above. We are climbing slowly to 9,000 feet. The sunshine buoys my spirits. I look over at Nancy; she is already settling in with her book for the crossing, adjusting her pillows, her face untouched by any hint of fear. She's going to be fine, it turns out. Perhaps *everything* will be fine.

We level off at 9,000 feet some 45 minutes after takeoff; the coast is now 60 miles behind us. Speed builds up slowly; 120 knots . . . 125—and, to my chagrin, 125 seems to be it. I had anticipated that the plane would be slow with a full fuel load

but not this slow. We are doing less than 150 knots true at 70-percent power, burning almost 10 gallons an hour.

Anxiety wells up in me. All my range figures are predicated on our managing better than 19 nmpg; we are more than a quarter below that figure. I do a series of calculations on the pocket electronic calculator, which I keep aboard as a sort of chew-toy, and breathe easier; even at this mileage, we will make it with ample reserve, and this mileage is bound to improve. Still, this is not a pleasing discovery, because it makes me wonder whether the calculated range of the airplane could possibly be correct, or whether I could even make Hawaii with decent reserve from Los Angeles.

Gander Radar hands me off to Gander Oceanic; Oceanic assigns me a couple of HF frequencies. I turn my attention to the HF. The man at Globe had given me a quick verbal rundown on its operation, which sounded simple, but I worry that I may have missed some fundamental point, as I usually do in such cases, and might be unable to run the thing at all.

It turns out to be so simple that you could work it with no instructions at all. I reel out the antenna until the tuning meter peaks on my primary frequency, set the friction lock, and that's that.

The undercast breaks along a wide, arcing front, revealing blue water on which scattered whitecaps graze like sheep. The air is glassy. The trailing edge of the white wing seems, as I stare at it, to extrude a pebbly-grained, navy-blue sheet. The ocean seems very remote and flat, like a lake, not the towering, tempestuous Atlantic that has propelled so many men and ships into history and legend. Toward the horizon, the water's deep blue pales in the haze, until it merges at times with the murk of a distant overcast or separates itself from the very pale aqua color of the horizon by a strip of thin, white clouds. The velvety, featureless whiteness of the wing and tip tank resembles that of a snowy mountain. From time to time, streets of puffy little clouds pass by, or a grayish, dirty-looking band of overcast lies just above our heads.

At first, everything is uncertainty, and time seems to race by. The first attempt at contact with the HF is unsuccessful, but I easily get a relay from an airliner. They all guard 121.5 over the North Atlantic and use the "party line," 123.45, as a frequency for casual talk. The airline pilot is friendly and efficient. The next leg is 202 nm, yet it seems to be gone before I have finished fiddling with the radios and making my fussy notes and calculations for the first leg. Again, no luck with HF; again, a relay.

I was told that you can pick up the BBC on 200 MHz just out of Gander. I try, but if there is a modulated signal behind all that noise, I cannot make it out. During the first hour out of Gander, I had stayed coupled to the VOR; now, and until an hour short of Shannon, I am dead reckoning, with the wing leveler holding the heading, which changes every 200 nm or so. I am ignoring wind altogether in my navigating. From the whitecaps, I can tell how it is blowing; generally it is a crossing wind. My position reports are based on time and my observed true airspeed. By most definitions, we are lost, because I don't know where we are.

I watch the main-tank quantity indicators and the Fuel Guard as the hours roll by. Fuel returned from the engine's fuel-injection system goes to the main tank on the side from which it came; before feeding from the tips it is necessary to partially empty the mains. My plan is to run 30 or 35 gallons out of the mains, then switch to the tips and empty them completely, and then finish the trip on the mains. After three and a half hours, the time has come to switch to the tips. I cannot wait too long, because if the tips, for some reason, fail to feed, I have to be able to return to Gander; therefore, I have to have at least half the main-tank fuel remaining, which I can always stretch considerably by slowing to a more efficient speed.

Nancy is asleep. I switch the selector to "tip" and watch the fuel-pressure gauge. At first, nothing; then, suddenly, the engine is completely starved and cuts out. Nancy awakens with a start and looks at me wildly; my hand flies to the boost-pump switch while my other hand makes placating gestures. After an eternal instant, the engine surges and backfires, and after gulping and sputtering, it is again on its feet. Nancy is still shaking. "Never do that without telling me first," she shouts into my ear.

Once one tip is feeding smoothly, I cycle the automatic switcher to the other, this time leading with the boost pump. The effects are less startling; in a minute we are getting smooth flow (except for that constantly flickering fuel-pressure needle) from that tank as well. The bubbles have been swallowed; 120 gallons of fuel remain.

The HF has so far failed to raise anybody; looking over my shoulder, I can see the useless plastic funnel trailing behind me, describing wide and random arcs in the air. One Pan Am pilot has reacted with surliness to my request for a relay, saying that if people can't get the proper radios themselves, that's their problem. I explain, in a wounded tone, that I am a tiny single-engine speck in a vast emptiness and how can he be so heartless? He, knowing that other airlines are hearing the conversation, retorts irritably that there is nothing in my N-number to reveal that I am a single-engine plane, but if I had any sense, I wouldn't be out here.

Band of brothers.

Incredibly, the DG has not precessed in four hours. Everything is smooth. Sometimes, when I don't think about the engine for a while, it seems to subside to a soft murmur; then at other times, if I yawn or turn my head or wiggle an earplug, the note of the engine seems to change. I listen to it intently. A prickly sensation moves up my thighs. The smooth hum seems to break apart into a stumbling tattoo, a medley of rumbles, warbles, hisses and snorts with no fixed rhythm; I scan the instruments—everything is normal. I pry my attention away from the engine sound, and it slowly retreats into a featureless, soft hum.

The engine sound is a barometer of my state of mind, which slowly alternates between blissful complacency an anxious, hollow feeling of having gone too far. I wonder at times what I am doing out here, as I look out at the water stretching from horizon to horizon, the rugosity of its surface endlessly repeated, its coloration sometimes mottled as though there were sandbars a few fathoms down; but that is an illusion, for the water here is two miles deep, perhaps three, an awesome weight of liquid darkness through which a pensive drowned man occasionally descends. The depth, the pressure and darkness, the vermin-devoured serpents, the twisted trees, they are repellent and awful to me; I feel as though I were suspended by a delicate thread above a yawning, hungry maw. Again and again I idly retrace the same nightmare path: the leaden depth, glowing worms, the bottomless darkness; the fragility of fuel lines; the inevitability—and therefore the unimportance—of death; the ennuis and chagrins of life that make one sometimes long for death; the sweetness of life, sunshine and breath; to lie in cold obstruction and to rot . . . full fathom five . . . alas, poor Yorick. I think of all the poor devils of past times and present who trusted their fortunes to fragile shells, who traced like tiny laborious insects the web of their travels across this gulf, and of them whom fortune failed, on whom the immensity and indifference of the universe dawned only when they were treading a thousand fathoms of water at night, shouting vainly for help into the gale . . .

And then I dredge myself up from these thoughts into light air and daylight where we are smoothly sailing along. I think of the record—the Canadian inspector said that no lightplane had gone down for mechanical reasons, only fuel and navigational mistakes, since he had been there, which was years and years; and all the hours I have flown this engine without any trouble whatever, over 400 since overhaul, never the slightest problem, not even fouled plugs. I think of how this is only 11 hours, and there is no reason, if I have never had an engine problem in 2,000 hours of flying, I would have to have one now. My spirits brighten and relax then, and I review the

instruments and the navigational progress and the weather and fuel, perhaps read a chapter or two of a book, stretch my legs and shift my position on the seat. After an hour or two, worry starts to come over me again; I think of Nungesser and Coli, Minchin and Hamilton and Princess Lowenstein—Wertheim, Hinchcliffe and Elsie Mackay, Bertaud, Payne and Witt . . . their epitaphs are written on the water. Again, the ocean becomes sinister; like Rimbaud's boat, I long to see some European puddle again; the minutes stretch and the coldness of the ocean again seems to seep into our cozy cell . . .

The OAT has dropped, as forecast, from 18° C. at this altitude out of Gander to 1° here. The fuel flow has increased correspondingly, but the speed is up 10 knots. Our fuel efficiency is gradually increasing. We move, as forecast, into an area of cumulus buildups.

Nancy has been alternately reading and sleeping. When she sleeps, I take her book and read sections of it at random; it is Zola's *Nana*, a sizable tome that she will nevertheless finish in the course of the flight. Sometimes she looks out at the ocean and the clouds. It is too noisy in the cockpit to talk comfortably. Sometimes I scratch her back. Mostly, however, she reads and sleeps. This galls me. Like most men, I would like an adoring female audience for my accomplishments; Nancy is not cooperating. I try to see things from her point of view: she is accompanying me on a possibly suicidal ego trip, from which she gets nothing at all but a cheap ticket to Europe, an economy she could easily forego. I have failed in the two aspects of airplane design that concern her most: noise suppression and comfortable seating. Furthermore, there is

Settled in after hours of overwater flight in a time capsule called Melmoth. Atop the panel are the fuel gauges, gear lights and angle of attack indicator.

nothing to adore. I am not defeating monsters with *kung-fu*; I am just sitting here with my arms folded, twitching occasionally. I tell myself that I should be glad she is willing to keep me company on this gratuitous venture. Since I do not know where I am, and since I have not seen a single ship or even the contrail of an airliner, I have the sensation of being completely alone. Therefore, it does not occur to me, as I begin to climb to clear the buildups, that I am in violation of an IFR clearance; at 11,500 feet, I decide that we will not make it over the tops, so we start down, weaving among the pillars of cloud. This is the first time I have hand-flown the airplane since we left Gander. Nancy wakes up and watches with curiosity as we drop down to 5,000 feet, where we are finally below the freezing level, though still above the bases of the clouds. We pass through a cave of rainbows and showers. Sheets of rain and snow (snow! in August!) finally wash from the windshield the dehydrated remains of a huge bug that we acquired somewhere in Kentucky. The ocean below is blacker and its surface closer and proportionately fiercer. Gradually, however, it becomes calmer; it is interesting to see how we pass quickly across zones of widely differing wind and weather. Almost seven hours out, some time after I give a position report, Shanwick (Shannon-Prestwick, the oceanic control on the other side) advises by relay that we are below the floor of controlled airspace and will

require a clearance to climb back up above 5,500. This seems to me like a laughable absurdity. How pompous of them to imagine that they have any control over what is happening out here! Nevertheless, I request clearance to seven as the temperature begins to rise; I duly get it and climb.

The sun sets. Its slanting rays burst in bundles among stacks of gray clouds, and silver paths and patches, mottled with the shadows of clouds, appear on the water. The glow in the sky lingers for a long time behind us; and suddenly, at one point, I realize that it is behind us and to the *left*. I scan the instruments: wet compass, remote compass, DG—they all agree we are on heading. How can the sun be to the *left*? I finally realize that as far north as we are, the sun crosses the horizon at a shallow angle, and its glow, after it has sunk from sight, continues to move northward along the western horizon.

At last I succeed in making HF contact with Shanwick; but it is unreliable, since the slightest rain seems to blanket the signal, and we are flying in and out of rain and cloud all the time now. In the last light, I have taken the measure of the surface wind; for hours it has seemed to be an intermittent quartering headwind; now it has reversed itself and we have a stiff quartering tailwind from the northwest. The water is rough and frothy. I began to think some time ago about how difficult it would actually be to ditch into this stormy sea and then to climb out of the foundering airplane—which, for all I knew, might do an immediate nose dive for the bottom—into a raft that would be half awash and blowing away, and to help Nancy out (she hates unsteady and slippery places) and to remember all the junk I want beside what is already packed into the raft, loose stuff like the ELT and the flashlight and spare batteries and jackets. And then we would be soaking wet; the wind would be 40 knots and the temperature would be 40 degrees . . .

As the ocean disappears in darkness, I know that if all that would be difficult and chancy in daylight, it would be next to impossible in the dark, and colder still, lonelier and sadder, a miserable way to die without even time to think back over good times or perhaps without even time for a parting kiss . . .

We have been flying for nine hours. Night is fully upon us. Nancy can no longer read; she is awake and looking out, and then suddenly—a light! There is a bright light down on the water. A boat! I know Lindbergh's feeling when he saw those boats, or any mariner's at the first birds he sees as he approaches land. Your heart sings with excitement. Nancy sees more of them to the right, several, a whole *line* of boats! The water around the lights seems peaceful. I do not know how far we are from shore, nor how far north or south of our course; there are thunderstorms and coastlines between us and the BBC, and I can't rely on the ADF indication. We plunge again into cloud; the nav lights' reflections brighten and fade with the changing density of the clouds. It begins to rain, heavily; a white, horizontal rain that resembles fine silver wires in the halo of the white taillight. It becomes turbulent; this is the front forecast to lie off the coast of Ireland. Roaring,

bucking, rattling, we shoot through the glowing darkness. I have not felt so good in days.

Shannon is no go on VHF from 220 nm, as I suspected; I turn on the transponder, but there is no reply light. I relay through an airliner on 127.9; not long afterward, at an amazing distance—about 170 nm—I get Shannon. After a while, he tells me that he has me on radar; yes, the reply light has begun to blink. I hadn't even noticed it. Shannon is clear, unlimited visibility, windy, warm. Where are we? The VOR needle is still waggling around in confusion. "It looks like you're spot on, maybe a mile south of course."

A mile! One part in 1,500. There's dumb luck for you.

Soon the VOR is centered and we are tracking; the transponder is winking brightly. Other aircraft, local flights, European flights, are talking to Shannon, and he to them. The BBC is clear, though schmaltzy. Radar tells us we've crossed the coast, and a few minutes later we break out of the clouds. Nancy points: roads, towns, headlights and, in the distance ahead, an airport.

We land, taxi to light-aircraft parking, shut down. Silence at last. We are 10 minutes late on our ETA—10 hours and 55 minutes altogether. Forty-six gallons remain in the tanks, enough to continue, at best-range speed, to Rome. The gyros are spinning down, the airplane is rocked by a gusty wind. I lift a window; the air is balmy and sweet. I look at Nancy; I do not know what to say. We made it; there is something special about this moment, and then again there is nothing special at all . . .

A little, yellow truck of the Irish Airport Authority drives up, and a uniformed figure gets out. Normal life resumes immediately.

PART III

Personalities

A major element of the romance of flight has been aviation's cast of characters—people of character and just plain characters. And, of course, people of genius. Both within aviation and among the general public, some pilots have been viewed with awe, admiration or simple envy. During the 1920s and 1930s pilots who performed headlined aeronautical feats were as popular as movie stars. Newspapers devoted millions of words to the great flights and to the men and women who made them. Yet even before that, when the "aeroplane" was quite a curiosity, pilots such as Louis Blériot and Lincoln Beachey, among others, attracted large followings. Blériot's fame was launched in July 1909 by his having made the first successful flight across the English Channel, but within aeronautical circles he also was respected as an airplane manufacturer and a force within European aviation. An article by him was therefore an important addition to *Popular Aviation*.

Beachey was a stunter, a showman whose aerobatic feats captured great crowds. It was he who made "loop-the-loop" synonymous with daredevil flying. His death while stunting, in 1915, helped add to the sense of epic heroism with which people then associated aviators.

Lindbergh was no doubt the most famous aviator of all time, though he shunned the role. *Popular Aviation's*

salvos were part of the barrage of publicity with which the modest pilot had to contend after his Paris flight. Certainly, a story about Lindbergh taking *the* Henry Ford up for the auto magnate's first flight was irresistible to the new publication, as was the story of Howard Hughes' revolutionary racing plane published ten years later.

The names of famous—and infamous—people who have filled *Flying's* pages is enormous. Fiorello LaGuardia, Congressman from New York, wrote about Federal aviation policy. So did Harry S. Truman, Senator from Missouri, destined to be President of the United States. Tyrone Power wrote about the most harrowing experience in his life as a pilot. One could go on and on.

The magazine's fascination for exciting personalities in aviation has remained strong. Sometimes these people have been giants in the industry—Glenn Curtiss, Bill Lear, Fred Weick, for instance—and sometimes they have been people who fly who are of interest in other ways.

Of course, there was "Wrong Way" Corrigan. Who could resist him?

And Gill Robb Wilson, whose writings and policies shaped the image of *Flying* in the postwar years. Not including some of his editorials and poems in this book would be like publishing the magazine without its logo.

Early Days with Glenn Curtiss

A story of the pioneer days at Lake Keuka and the various activities that led up to the experiments that made Glenn Curtiss famous in the history of aviation.

BY C. H. QUINN

Glenn Curtiss at Rheims in 1909.

MY INTEREST in Glenn Curtiss was of the most ordinary sort when I first began dropping into Barney's lunch room in Printing House Row. Then, one day, a casual remark I happened to drop about the settlement of the big Curtiss estate opened up a mine of interesting information behind the counter over which Barney dispensed good coffee and fried cakes.

When our first conversation about Glenn Curtiss opened in the most natural manner, I had not the faintest idea that I was talking to a man who had worked for years with the great inventor in his little shop in Hammondsport, N. Y. And the stories of the great aviator-inventor that afterwards drifted piecemeal over the counter with the pleasant odor of cooking food, were ripened by years of thought and reflection in the storehouse of excellent memory possessed by Barney.

Because he had worked with Glenn Curtiss long before the latter began his first experiments on airplane construction, Barney was present at the birth of the great idea. It seems from all the facts in the case, it is most probable that Curtiss himself did not know just when he made his first discovery of principles which afterward enabled him to develop his airplane invention. He was constantly experimenting in his little bicycle shop.

Long before his mind turned to aviation, Curtiss developed the motorcycle, which proved to be a step toward creation of his airplane. One idea led to another as experiment followed experiment. The two men, Curtiss and Barney, often worked till the small hours of the morning perfecting some part of the motorcycle engine which was one of the early efforts of the inventor.

"Glenn had one of the first bicycles that came out with a small gasoline engine on it as a helper," said Barney, telling of this early stage of the inventor's work. "He had it in the window of his store and bicycle shop there in Hammondsport, more as an advertisement than anything else. He had it jacked up and running and of course it attracted a lot of attention because gasoline engines were not common then.

"Watching that little engine started Glenn to thinking. The engine was very small and only intended to help propel a bicycle, merely to make bicycling easier. One day Glenn said to me: 'If those little cylinders give that much power why won't bigger ones make a lot more?' So he went to work to make an engine big enough to carry the whole load. After it was done and attached to a bicycle the first tests proved that the ordinary bicycle frame was not strong enough to withstand the vibration. The engine quickly tore an ordinary bicycle to pieces.

"Then he faced the problem of getting larger and stronger steel tubing. That was not easy. Glenn searched the entire shop for stronger tubing but couldn't find any. Then one day he walked over to the show-window where there was a tandem bicycle on display. It was a new one worth around a hundred dollars and had the kind of tubing that Curtiss wanted, so out it came into the shop and it was not long before he had it all apart and the tubing sawed up in the longest possible pieces."

Then Curtiss began building an engine that he figured would be big enough to run a motorcycle, and carry a heavy man up grades. Back and forth from the bicycle shop to the village machine shop the two workers plodded hundreds of times to get

Glenn Curtiss flying the June Bug when, on July 4, 1908, he became the first pilot to fly more than a mile in distance.

each piece of the engine machined to the desired point.

But the engine did run and it did some other things too. It made noise and how! One of its achievements was to alienate most of the neighborly feeling for Curtiss which is supposed to make life in the small village superior to that in a big city. Sleep at night, when experiments with the new engine were going on, was enjoyed only by residents far from the scene of the Curtiss developments. And they were developments.

The great motorcycle, which later proved to be a step toward the Curtiss Biplane, was set up in the shop on blocks and when tests were made it required a man on each side of the thing to hold it down. It would jump a foot into the air with each explosion. Conversation could be carried on only by shouting and hand signals.

"When the day for the first road trial came," continued Barney, "we had the bike blocked up in a barn near the store. Curtiss filled the gasoline tank—which did not hold much. Of course, he had no carburetor, just a mixing valve as we called it then, a simple device that mixed air and gasoline. With the first explosions, the motorcycle jumped like a wild colt and it took the strength of another man and myself to hold it till Curtiss jumped on and grabbed the handles. The vibration was terrific and his voice could hardly be heard above the din.

" 'Pull the blocks and let her go,' shouted Curtiss and I did. With an ungodly racket the machine tore out of the barn and into the street. The other man and myself jumped on bicycles and followed. Curtiss steered the machine for the edge of the village. Then the importance of one thing that he had left off the machine impressed itself on Curtiss. There was no way to stop the machine!"

So engrossed has the inventor become in getting the machine to run that he had omitted putting on something to shut it off! There was no such thing as a switch on it, and no way to shut off the gasoline flow to the cylinders, so it just had to run till it ran out of gasoline—or something happened.

Curtiss jumped off and the motorcycle ran into a ditch and stopped.

"I could make forty miles an hour on a level road," declared Curtiss after the experiment. This event had far-reaching results in the brain of the young inventor and he was soon launched on experiments with his "wind-wagon," the fore-runner of the biplane.

The wind-wagon was a three-cornered affair, a sort of platform with two motorcycle wheels on the broad rear part and one on the front point. A single shaft at the rear formed an axle for the two rear wheels. On the platform, between the two wheels, Curtiss placed a motorcycle engine of about four horsepower. To the engine was attached a stub shaft that stuck out at the rear and on the stub, Curtiss fastened a wooden propeller.

"Glenn's experiment with his wind wagon with the engine on the back end produced a real sensation in the village," said Barney. "It also threw great clouds of dust when he rode it through the streets. The few residents who had remained his friends up to this time began to desert him as the wind wagon passed from one stage of development into another. After trials with one four-horse engine, Glenn found he did not have as much power as he desired, so he put another four-horse motorcycle engine on tandem fashion.

"That gave him better than eight-horsepower for the three-cornered wind wagon. It also gave him much greater power to stir up road dust and anything else lying dry and loose in the street. Of course, nearly everyone drove horses then and the streets were not paved, swept nor sprinkled. When Curtiss drove his double-engined wind wagon down the street he swept it pretty clean of everything and you can imagine what landed in the yards and houses that he passed.

"Long before this time, Curtiss had exhausted all his money in experiments and all that he could borrow from friends and relatives. So when he got the wind-wagon to working he had no money and darn few friends. He did, however, get some money from making small gasoline engines that he advertised.

"One day I went into the shop in the barn and found Glenn had a big thing built across the pointed front of the wind wagon. It was a square frame on hinges. With a system of levers, he could work the back of the frame up and down. This frame was covered with cloth and was a crude sort of wing.

"What's that thing for?" I asked Glenn.

" 'Oh that's to ease me over the thank-ye-marms'," he said.

"One thing that his wind-wagon taught him was he should put his engines at the front instead of the rear. When he would strike a bump the front end would fly up in the air. It is a wonder that he was not killed in those experiments with that crazy thing, it certainly was dangerous.

Well, he found that his flapper across the front of the wind wagon did help to ease him over the bumps in the road, for by pulling his lever and tilting the frame with the cloth on it he could partially lift the front of the machine off the ground.

"But money did not change him like it does a lot of people," and this is the part of Barney's memory of the great aviator which he cherishes most. "He wanted to help every man who had worked with him in his hard luck days and he never got so swelled up with fame and money that he forgot them.

"After I had left him and went into the restaurant business, Glenn came to me and wanted me to come back and work for him. He made me a tempting offer and maybe I made a mistake in not taking him up. It showed his democratic spirit and disposition to appreciate old friends."

Volumes more Barney can tell of the early days of Curtiss' struggles in the little shop, but workers in the Printing House Row can't take time all in one lunch hour to get them all.

A LETTER AND PREDICTION FROM Louis Blériot

Monsieur Blériot, a renowned pioneer of aviation and now head of the great French organization, Blériot-Aéronautique, predicts and illustrates the trans-oceanic flying machine of the future.

Louis Bleriot seated in one of his airplanes.

DEAR SIRS:

I have today received your letter of August 15, in which you ask me for an article on the past and the future of Aerial Navigation.

It is now a quarter of a century that I have devoted myself to this science, and it would be too lengthy to expatiate on everything that has taken place. Besides, I have too little time to do that.

However, I will give you some important data that I have retained and which have played a great role in the history of aviation.

It's a pleasure to me to refer to the fact that the United States and France have given one another the hand already from the beginning, in order to provide the world with this mode of transportation.

In 1908 I made the acquaintance of Wilbur Wright and attended nearly all of his experiments. I admired his aeroplane and the control of same, without however approving of the basic idea of his apparatus, as I saw the aeroplane in an entirely different form from that which he had adopted.

Already in 1907 I constructed the "Blériot VII," a monoplane apparatus, which has up to the present time remained approximately the prototype of the modern aeroplanes, and of which I send you photographs.

In 1909 the crossing of the English Channel and the competition at Reims gave to aviation its first universal attention, without however adding any really new ideas to the technique.

A very important event which entirely changed the ideas of equilibrium were the experiments of Pégoud, in 1911. It was Pégoud who first looped-the-loop in the air, flew upside down, and made some different experiments such as descents like a dead leaf, which it was wrong to abandon, as these experiments, after all, contained the secret of the use of the sail-spread of the aeroplane for a vertical descent.

Then came the war of 1914 which considerably intensified the building of aeroplanes and enabled me to create the "Spad" which gave us supremacy in the air.

Finally, later, in 1927, I had the joy to receive Lindbergh, a product of your country, who, amplifying the exploit of 1909 came across the Atlantic, to show us the greatness of American aviation.

I could dwell much longer on this question of the past, but I prefer to speak to you of what I consider to be the probable future, as the past should no longer serve for any other purpose than to indicate the way to be followed, to those who analyze it.

In my opinion, it is in speed research that the future of Aerial Locomotion is above all to be seen, and it is in the same adaptation of this speed that we should center all our researches.

The speed should not be dangerous, and what is wanted is not these fire-ball aeroplanes which lift from the ground after a few meters, as has been the case in various races or record flights.

At this time the comparison between the dirigible and the aeroplane is a live question. The feat accomplished by Dr. Eckener in the "Graf Zeppelin" commands our admiration. It would even seem that the dirigible for the moment is ahead of the aeroplane. However, I do not believe I am mistaken in saying that the dirigible will not long endure the battle with the aeroplane, as its inferiority in speed is too great. This will still more be the case in the near future if the present shape of the Zeppelin, too much elongated, is modified, and its elongation is reduced from the ratio of 10 to 8 or 7. It will always be difficult to succeed in giving the dirigible a speed that would enable it to run against violent winds, and the mid-ship frame required for transporting a passenger will always be from 8 to 10 times greater than that required by the aeroplane.

It is through its considerable range of action that the dirigible strikes our imagination. But the greatest distance that separates two points of land is only 20,000 kilometes, and it is not necessary to run over the whole of it without stopping. As for the possibility of remaining stationary, which seems to be reserved for the balloon, it can only with great difficulty be attained, and in a calm atmosphere.

These two advantages are, besides, compensated for by more serious inconveniences, its fragility, and its volume, which requires garages of extraordinary dimensions, its sensitiveness to the displacement of atmospheric currents, and finally, the danger of catching fire in storms, these things constitute, at all events, a reason why the trip in a dirigible will seem less safe than a trip in an aeroplane, and, besides, for land trips the dirigible has now no advantage. Its great present superiority consists in its fitness for overseas flights. But these routes will soon, I hope, be studded with stations like the land routes, which stations I believe will be floating islands.

I have the greatest confidence in the success of these islands, principally islands of the Armstrong system, and when in a few years the infra-structure of the aerial lines will be constituted by these islands, spaced at about 500 to 600 kilometers, the dirigible will have lost its advantage. The aeroplane, much more swift, much less sensitive to the atmospheric elements, will immediately resume its monopoly of aerial transportation. It will, obviously, be necessary that these islands should be numerous, and that they should be inter-connected with

herzian waves, making it possible to go from one to the other without leaving their zone, and it will also be necessary that they be supplied with all the radio-goniometric apparatus and with wireless telephone, so that the aeroplanes will with certainty find them, in case one of them should be surrounded by fog, so that it can indicate to the aeroplane the nearest island that will be completely free.

The equipment to be employed for the use of these islands should have the maximum of strength, and it is in the study of this future aeroplane that I have specialized myself.

I have considered several aeroplanes answering for this purpose, and the form that I visualize as the best one is a plane with several motors accessible during the flight, but provided with wheels or skis for landing on the floating islands. In order to be prepared for the eventuality that, as a result of exceptional circumstances, it would be necessary to alight on the sea, the principal cabin of this plane is constituted of a life-boat completely decked and enclosed and sufficiently solid to withstand any storm. This boat is provided with water-ballast compartments, and a small special motor, and is able to carry all the necessary provisions and wireless telephone apparatus. It leaves the aeroplane, by which it is carried, automatically or through some mechanical means and, as soon as it is released, it has nothing more to fear.

What speed can existing communications of this kind furnish? I estimate that one can count on an average speed of 200 kilometers, which would put France at least 30 hours from America under the present flying conditions, but the flying conditions, especially in aeroboats designed in this manner, can change with the altitude, and nothing prevents one from conceiving that, if this life boat is air-tight, it can be used as the habitation of passengers for navigation at an ever greater height. But the altitude is one of the conditions that can most favorably be adapted to great speeds, as each thousand meters gained in altitude noticeably increases the speed if one retains the same motive power in the motors, which is easy with mechanical compressors.

Thus it will be seen that, as the aeroplane gradually rises higher and higher in the sky, the number of hours that I have indicated will also be gradually diminished, and it is not rash to foresee that some day this number of hours will be reduced by one-half, that is, that it will be possible to go from France to America in twelve hours: it will only be necessary to travel about 13,000 meters high in an air-tight boat.

How long will it take before this becomes a reality? This depends upon the financial effort and upon the technical effort made by the various countries. I am convinced that if America and France come to an understanding in this matter it will only be a few years before it becomes a reality.

I beg you, dear sirs, to accept the assurance of my respectful consideration.

Yours very truly,
L. Blériot.

P. S. I have intentionally abstained from speaking of the prophecies of certain engineers who foretell speeds of 1,000 kilometers per hour and more through the use of rockets, as I want to stay within the domain of possible and certain achievements.

Lincoln Beachey enjoying the fame accruing to a stunter.

Beachey and the "Box Office" Era

BY WILL NASON

NERVELESS was the word for Lincoln Beachey—a stout-hearted airman who probably set more records than any other flyer. Beachey was one who helped pave the way for flying to grow from a daredevil sport to a safe and sane art. Ironically, this came about through his death.

Beachey was a soft-spoken, shy-looking boy, but there flowed in his veins the stuff that heroes are made of. He was flying dirigibles at 17. The following year he enjoyed the dubious honor of being the first aviator to land a power balloon in New York's East River—the result of a crash against a Manhattan building.

At Los Angeles in 1910, 22-year-old Beachey was bitten by the heavier-than-air bug when he saw aerial performances by the famous Glenn H. Curtis. Curtiss taught Beachey to fly a biplane and asked him to join the Curtiss troupe which included Cromwell Dixon, Eugene Ely and Jack McCurdy.

Flying was dog-eat-dog in those days. It was new, it was fascinating and it was box-office. Every pilot was out to pack in the customers and there were no holds barred. Beachey was soon surpassing them all. He thrilled the crowds with his "Vertical Drop" and his "Dutch Roll" and his "Turkey Trot"— all his own aerial inventions. There was national news in his flight over Niagara Falls, down the gorge and under the suspension bridge. In those freshening days all was to be learned, and every trick, every maneuver was a milestone in man's conquest of the air.

Beachey's most spectacular accomplishment and that which gained for him the greatest fame was his execution of the breathtaking Loop-the-Loop at San Diego in 1913. In these outer-spatial days we are prone to look upon a simple aerial loop as child's play. Not so then.

Consider the fact that the consequences of a loop were virtually unknown. Until this time, the feat had been performed only by a Frenchman. Would the pilot remain conscious? Would the plane regain level flight? Would the motor fail to re-start? Would the wings stay put? Faced with these unanswered questions, it required a venturesome spirit to maneuver the airplane into that great sky-circle and see the job through.

Beachey was not satisfied merely to perform the loop. He brought the act to perfection. It was in this way that he became one of the first to analyze and understand the characteristics of the stall.

From 1909 to 1913, brave aviators the world over had died by the dozens. Considering the relatively few qualified pilots, the figures were appalling. Certain factions of representative public opinion had become so aroused over these fatal accidents that the future of flying was, for a time, extremely uncertain.

Worst of all, no one seemed to know just what was causing these mishaps. In the case of Arch Hoxsey, they said, "It was the holes in the air." When John B. Moisant plowed in, "The machine was unstable." Airmen knew that these answers were superficial, but it remained for Lincoln Beachey to supply the facts which cleared up the mystery.

While practicing his loops Beachey found that, during certain attitudes of flight, his machine seemed to lose lift. He also discovered that, should this loss of lift coincide with the attitude of one wing low, he would fall into a spin.

Through experimentation, he learned that a spin could be anticipated. What was more, he found that, with sufficient altitude, a spin recovery could be effected. This knowledge, passed on to other aviators, did much to cut the accident rate which had been plaguing the art of flying.

By late 1914, the novelty of aerial flight was wearing thin. There was hardly an accomplished aviator in all the world who had not put an airplane through practically every maneuver in the book. Attention was shifting to air combat in the European War and flyers on this side of the Atlantic were finding it hard to attract a paying crowd with anything less than the most thrilling acrobatics.

To spur waning interest, Lincoln hit upon the plan of building a small, extremely light racing plane which would be capable of previously unheard-of performance. His creation followed the pattern of the German "Taube" monoplane. It was a sturdy little machine weighing only 600 lbs. with the pilot, gas and oil aboard. Equipped with a Gnome engine and boasting a streamlined construction which placed Beachey in a virtually enclosed cockpit, its power and speed were a thing to behold. It was said that Beachey was the only man in the world with sufficient skill or daring to loop such an airplane.

On March 14, 1915, Beachey took his "Taube" to 6,000 feet over the crowds at San Francisco's Pacific Exposition. To guard against accidental toss-out during his violent maneuvers, he was strapped at his waist, his legs and his feet. Thus prepared, he began to subject the speedy monoplane to a series of acrobatics which were described by one spectator as "simply too terrifying for words."

Coming out of one of his famous loops, Beachey was seen to place the machine in a terrible dive. It seemed that he must have been going nearly 200 mph—far too fast for even those sturdy wings. Locked there in his cockpit, he must have lost his sense of speed. He simply did not know how fast he was traveling.

With a sickening snap, one wing fell off. Then the other. The helpless craft plummeted into the waters of San Francisco Bay.

A coroner's inquest proved that Beachey was alive when he hit the surface. Strapped there in his little seat, unable to free himself, he drowned.

Such a cry went up over the death of this lovable boy who had captured America's heart that it seemed imperative that something be done to prevent the possibility of such a tragedy. The answer was not immediately forthcoming, but Beachey's fatal crash did much to speed the development of the parachute.

Beachey's specially built Curtiss biplane.

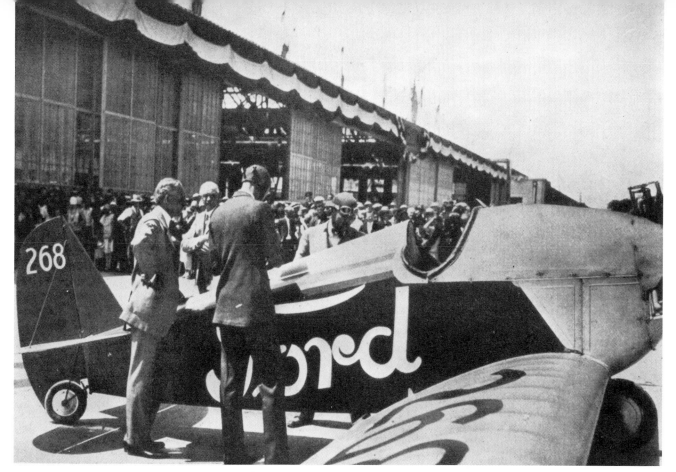

The Ford Flivver Plane, with Harry Brooks, the Ford pilot in the rear. Henry Ford and Lindbergh looking it over.

WHEN FORD WENT UP

BY RUSSELL WILKS

I AM addressing my remarks in behalf of aviation to the adults," said Col. Charles A. Lindbergh, on the platform at Northwestern Field, Detroit, where 50,000 people had gathered to pay tribute to him. "There is no need to speak to the younger generation on this subject, for the younger generation already is a flying one. The adults are the ones I want to sell."

These remarks were made on August 10. Twenty-four hours later Lindbergh made a sale which will become a part of history. He sold Henry Ford and Edsel Ford their first airplane flights. He sold them so thoroughly that they decided to take a second flight thirty minutes after the first flights were over.

While Colonel Lindbergh was inspecting the Ford airport with the motor king on August 11 he invited the latter to ride with him in his famous Spirit of St. Louis, the Ryan monoplane with which he conquered the Atlantic in thirty three and a half hours.

"Mr. Ford," Colonel Lindbergh said, "I have never taken a passenger in my plane. I would like you to be the first to go with me."

Some time elapsed before Mr. Ford answered. During this interim Colonel Lindbergh put the clincher on his famous sale by flying one of Ford's experimental machines. He did not pilot the plane to prove to Mr. Ford how adept he was in the art of flying. He flew it because he wanted the pleasure of flying it and because he wanted to please the thousands who had managed to get into the airport or to find a place along the crowded thoroughfares leading thereto.

That flight, however, made a deep impression on Mr. Ford, just as it did on the other spectators. I am certain of this because I stood by the motor magnate and conversed with him while Colonel Lindbergh was in the air. The remarks he made to me during the young aviator's flight prepared me in part for the startling announcement which he was to make soon.

The plane Colonel Lindbergh flew is known as the "Ford Flivver" plane. It was designed by Otto Kopping, 27-year-old engineer, and was revealed to the public by Mr. Ford on his sixty-third birthday, which was on July 31, 1926. The machine is still an experimental job. It had never been flown by anyone except Harry Brooks, Ford pilot.

Colonel Lindbergh strode out to the miniature mechanical bird and looked down on it as a Gulliver might a Lilliputian. Its wings measured only 20 feet from tip to tip and it weighed only 320 pounds. The gallant aviator towered like a Campanile over it but it was a plane and that was enough for him. He strapped on his headgear and forced his height into the cockpit. At best, his head and shoulders protruded awkwardly but there was nothing awkward in the manner in which he taxied down the field and hopped into the air.

He began climbing and soon the little machine appeared as a silver mosquito in the sunshine which flooded the sky that day. Colonel Lindbergh was in a new plane and, naturally, refrained from spectacular stunts. These were unnecessary, for everyone realized a master was in the air. Spectators thrilled just to see him idle along, a style of flying in which he indulges with much enjoyment. They marvelled when he banked sharply. When he climbed, they climbed in unison to their toes. When he dipped, they dipped back on their heels.

In heart and soul the crowd was Colonel Lindbergh's. No, not Colonel Lindbergh's, but merely Lindy's, for they had know him as a youngster born only 25 years ago in Detroit, and he was still a youngster to them. Maybe they were swayed with the sentiment emanating from the triumphant return of a native son. The day before he had been given the most effusive reception any one man was ever accorded in Detroit. Men, women and children had sacrificed the day for him, just to get a look at him. Perhaps that spirit had carried over to the next day. It was Lindy in the air, their Lindy.

"That boy is a genius," Mr. Ford said to me, as he watched every move of the aviator's plane. "Look at him. Did you ever see such graceful flying? He certainly knows how to handle a plane. Why, he seems more at home in the air than on the ground. He flies as gracefully as any bird. You see, he knows planes through and through.

"When he got into my plane he simply asked if the controls were standard. I told him they were. That was sufficient. You can see that for yourself. I am greatly impressed with his flying. I am also impressed with him as a gentleman. I have never had a guest whom I enjoyed entertaining more than I have Lindbergh. Here he comes down now."

Although I noticed Mr. Ford was following the flight with extreme interest, I did not know at the time what was going on in his mind. No one knew. Mr. Ford was pondering over that invitation. He was wondering if he should break the vow he had repeatedly made—that he would never fly.

As Colonel Lindbergh alighted, we walked out toward the plane. Suddenly Mr. Ford stopped. He turned to me and whispered his startling announcement.

"Lindbergh is right," the manufacturer said. "He says the adults are the ones who must be sold on flying. I am going up with him."

The announcement came with a suddenness characteristic of the world's richest citizen. He had made up his mind. The Lone Eagle had sold him. Orders were given on the field and everything became a hustle and bustle. The crowd did not anticipate what was to take place. The Spirit of St. Louis was brought from the hangar. A special seat was placed in the small cockpit. Mr. Ford placed a headgear over his slightly wavy gray hair and climbed in. Lindbergh followed. They were off.

At the age of 63 the man who had carefully refrained from airplane flights was at last in the air. The crowd cheered as Ford got into the plane. They were seeing the world's first billionaire taking his first flight. Subconsciously they knew in that moment of excitement that his trip would exert a tremendous influence on air travel. They sensed that the confidence planted by Colonel Lindbergh when he made his epic flight from New York to Paris would take deeper root in the public mind as result of Henry Ford's first flight. They knew that others, some felt that they themselves, would soon be emulating the Croesus of the Twentieth Century. Again they cheered when the plane landed.

"How did you enjoy it?" I asked Mr. Ford, as he stepped out of the plane, following a ten-minute flight over the airport and vicinity.

"It was great," he said. He revealed no sign of nervousness after getting out of the plane.

"How would you like to take a flight every day?" I inquired.

"Oh, I don't suppose I'll do that, but I wouldn't mind," he laughed.

The conversation was interrupted at this point by Edsel Ford, who announced he was going to take his first flight, also with the flying colonel. He climbed into the Spirit of St. Louis and took the seat his father had just left. For a similar length of time he was in the air. He was all smiles when the ride was over.

"I enjoyed every minute of it," he said.

"Well, let's get a party and go up in one of our own ships," suggested his father, getting thoroughly into the flying spirit which prevailed at the airport.

At this suggestion a huge tri-motor, Ford, all-metal transport was brought out and made ready for a flight which was to be bathed in fame. It was a monster ship with a wing spread of 70 feet and with motors developing 600 horse-power. It weighed more than two tons. As I looked upon it I could easily picture the air pullman which the Ford organization has announced it will build within the next year. It has considered building air liners to carry 100 passengers and to be driven by six 1,000-horse-power motors.

Colonel Lindbergh and Harry Brooks went to the control room. The passengers started toward the plane. I had not been invited, but the thought of riding in an airplane which was to carry the world's wealthiest citizen and was to be piloted by that noble youth who linked the United States and France with an unforgettable solo flight made me bury all reserve.

"Mr. Ford," I said, "I would like to make this trip with you."

I had only known him a half hour. All I could expect was a courteous refusal and all I could hope for was a most unusual fate. My hope was not in vain. Mr. Ford showed as much interest in obtaining a seat for me as if we had been friends of years' standing. After three conferences with William B. Mayo, his chief engineer, who was organizing the party, he told me I could go.

Thus it was I had the honor of being the only newspaperman given permission to make the trip and the first one to obtain an aerial interview from America's outstanding genius of industry. Thus it was that I had the added pleasure and distinction of interviewing Colonel Lindbergh in the air. Thus it was I made my first airplane flight.

I was glad that it was my maiden flight because I knew I could compare my sensations as they came with those experienced by Mr. Ford. We were in the same ship, so to speak, because his first flight had not lasted long enough for him to feel he was a perfectly initiated traveler of the air.

Our flight was for a period of 40 minutes. During those minutes I had the opportunity of observing and conversing with Mr. Ford as no one ever had. We passed over his various plants in Detroit. We circled over the principal sites of the city. We flew over the Detroit River and the Canadian shore.

The trip as a whole probably was best described by him when he said: "Why, this is just like a picnic."

That was the way it seemed. In the luxurious passenger cabin everone laughed and joked as though they were in the smoker of a Pullman. The party was comprised principally of Ford officials. Major Thomas G. Lanphier, commandant at Selfridge Field, also went along as a passenger.

"There is nothing to it," said Mr. Ford, referring to flying. "It is simple. I haven't the slightest feeling of uneasiness. We are safe up here. I feel just as safe here as I do on the ground."

"It seems to me that every now and then we take a little dip as though we were going down unexpectedly," I remarked.

"Those are air bumps," Mr. Ford explained, giving me the benefit of his first flight experience a few minutes before. "You see, those bumps are encountered more over a city than in other places on account of the heat arising from the city.

"With their exception I can hardly tell we are moving. It seems to me that we are not moving but that it is the ground that moves. It seems to pass back of us. I am not aware of speed. This is the finest ride I ever had."

Mr. Ford undoubtedly was very comfortable, for in no way did he display uneasiness. He smiled and chatted in the same easy manner he had done while we were standing on the ground. Occasionally he would reach across the aisle, touch my arm and point out things below.

"There are some of the ships we have been scrapping," he said, pointing out the window. "I never dreamed that I would look down on my enterprises from away up here. They certainly look differently. So small. There's my River Rouge plant, I have seen air views of it but this is the real stuff, isn't it? Look over here. That is the Highland Park plant. Well, this surely gives one a new viewpoint."

By this time Mr. Ford had come to feel so much at home in the air that he decided to go into the control room, where Brooks and Colonel Lindbergh had been alternately piloting the big transport. The young aviator gave his seat to Mr. Ford, who took hold of one of the two control wheels. He made no attempt to pilot the plane but sat and watched Brooks maneuver it.

Meanwhile, Colonel Lindbergh came back to the passenger cabin.

"How do you like flying this ship?" I asked him.

"Well, it is very easy to operate, but, of course, it is not my plane." His reply revealed his love for the Spirit of St. Louis.

"This is a novel experience for me," he continued. "I am nearly always in the pilot's seat. I don't know how I would feel if I should miss a few flights a day. I have just figured up that during the last five and a half years I have averaged four flights a day. That is a good argument for safety, is it not?

"Flying is not a question of safety, any more. What we need now is an airport in every city. That is the way aviation will develop. Mr. Ford is doing an immeasurable amount of good for aviation. His airport is a splendid one. He is building planes, and today he is showing beyond all doubt his confidence in them.

"But we need municipal airports, too. Every city should provide one and operate it so it will encourage aviation. That is why I am touring the country under the auspices of the Guggenheim Foundation for the Promotion of Aeronautics. I make a plea for airports wherever I go."

While Colonel Lindbergh was in Detroit his plane was kept on display in a hangar at the Ford airport. The bashful boy aviator was at different times the guest of the Fords, Major Lamphier, and his mother, Mrs. Evangeline Lodge Lindbergh, a Detroit school teacher.

Mr. Ford returned from the control room as happy as a boy with a new toy. The renowned manufacturer is always happy in the midst of machinery.

"I think that flying is still 90 per cent man," he told me, "But we have two mighty good men along with us. I mean Lindbergh and Brooks.

"What we must do is to make it more machine and less man. That is what I am devoting a lot of my time and interests to. I expect to continue experimenting and building planes until air travel is comparable with automobile travel today."

"Is that little midget plane that Colonel Lindbergh flew the kind you are going to manufacture for individuals to use?" I asked.

"That will come later on," he replied.

Edsel Ford, seated just behind his father, plainly enjoyed the trip as much as any experience he had had in a long while. Never did the least expression of concern creep over his face.

"How are you enjoying the flight?" I asked him.

"This is the best way of seeing the city I have found. It beats worrying yourself in the congested traffic on the ground."

And, indeed, this mode of travel was vastly superior to that of the automobile in the streets of the city because we were flying at the rate of 100 miles an hour. Our altitude ranged from 2,000 to 3,000 feet. It seemed we had been in the air only a few minutes when Mr. Ford informed me we were getting ready to land at the airport.

All along I had been thinking about the landing. I had heard so much said about how easy it was to go up but how difficult it was to come down properly that I looked on the landing as the climax of the flight. Gradually we began lowering.

"See," said Mr. Ford, "it comes down just as easy as it went up."

And so it did.

The "Mad Major's" Escapades

One of the author's true stories of the World War that contains incidents stranger than fiction. Because of his German name and his unusual practices, Carl Spaatz frequently faced grave dangers on his own side of the line.

BY CARL B. OGILVIE

JUNE had been a remarkable month in the life of Carl Spaatz, ranking veteran pilot in point of flying service, of the U. S. Army Air Corps.

He was born on June 28, 1891, graduated from West Point June 12, 1914, became a pilot of the 1st Aero Squadron June 15, 1916, graduated from several Army service schools during his "lucky" month, and on June 30, 1936, was assigned to duty with the most distinguished unit of our air service—the General Headquarters Air Force.

June, 1937, marks the 46th anniversary of his birth at Boyertown, Pa., and 22 years of service as a military pilot. Carl Spaatz today, slight, wiry, energetic, strong as hickory, is the outstanding member of a small group of distinguished American war-time aces who are still in the military service. Including his entry at West Point in 1910, Carl Spaatz has served our country continuously and with marked distinction for 27 years.

By the public at large, Carl Spaatz is remembered as the commanding officer and guiding genius of the greatest aviation training center in history—the U. S. Army aviation school at Issoudon, France; and as C. O. of the Army's famous "Question Mark," which established a refueling endurance flight of nearly 151 hours, in January, 1929.

In the Officers' Club one will hear the legend retold of the "Mad" Major of the 2nd Pursuit Group going A.W.O.L., making unauthorized flights into Hunland and being suspected as a traitor and spy by the French high command. Treason, the French pointed out, was punishable only by death. It was regrettable that such a brilliant American officer should meet an ignoble end before a firing squad.

No wonder the French, quick to fasten the sting of suspicion upon everyone whose name had a Teutonic ring, were amazed when, instead of a firing squad's bullet through his chest, our government pinned a medal upon the chest of Carl Spaatz. The legend of Spaatz's "treason" is one of the most amazing off-the-record incidents of the World War.

When the United States declared war against Imperial Germany, Spaatz was the C.O. of the 5th Aero Squadron. He

Carl "Toohey" Spaatz as a general in World War II.

hoped to take his squadron to France. To his chagrin he was relieved of his command, ordered to France and given a schoolmaster's job. During the year he was Commanding Officer of Issoudon he trained and sent hundreds of pilots up to the Front. He would gladly have exchanged places with any one of them—even his gold oak leaves for the bars of a pursuit second lieutenant. Spaatz bombarded headquarters with regular weekly requests for transfer to active duty at the Front. They came back endorsed, disapproved.

Within the veins of Carl Spaatz flowed the fighting blood of heroic Pennsylvanian ancestors, who had fought and died at Valley Forge. He champed at the bit of military red tape that kept him bound to a swivel chair job. He yearned to serve his country as he was training fledgling pilots to serve—by fighting in a pursuit cockpit behind yammering twin machine guns.

Fighting pilots were needed at the Front. Didn't statistics show that the Allied planes were outnumbered 4 to 1 by the German Air Force? To hell with the honor and high praise he was receiving for a "schoolmarm petticoat" assignment. He carried his feud to the highest command. But America's early bird, who had seen service in the 1st Aero Squadron on the Mexican Border under Pershing, failed to effect a transfer to line duty.

Instead, in the late summer of 1918, Spaatz was ordered to report at Aviation Headquarters, Paris. The military grapevine intelligence said that the major, with the German name, was to be sent back to the United States.

Also, there were rumors afloat that the American Forces would soon launch a tremendous offensive in the St. Mihiel sector. For weeks Spaatz had been whipping novice airmen into fighting pilots. St. Mihiel was to see the greatest demonstration of American strength since the Stars and Stripes had come to France—an offensive, fighting movement. And Spaatz would be on the high seas out of the war he so ardently yearned to take a man's part in.

One phenomenon of the World War, in reference to transferred American soldiers in France, was that in their traveling to report to new assignments of duty, all routes, by remarkable coincidence, led to the Front, via Paris. Spaatz, ordered to Paris, reversed the tactics. Sensing that the big show would soon be over, he selected a route of travel that went via the Front.

It is related that Spaatz, minus insignia of rank, made his way to the 2nd Pursuit Group at Souilly. The 2nd was composed of the 13th, 22nd, 49th and 139th Pursuit Squadrons. The A.W.O.L. major did not seek a desk job with the Group. The C.O. of the 13th was a graduate of the major's school at Issoudon. Spaatz presented himself to the young captain, demanded a good serviceable pursuit job be put on the line as soon as possible.

Whatever the captain thought of the unusual request he left unspoken. A Spad was put on the line. Quickly Major Spaatz was realizing the fulfillment of his greatest ambition. He was winging over Hunland! No thought was in his mind now that his flight was unauthorized, that he should be in Headquarters at Paris; that he was absent without official leave. Hell, he was where he had begged for a whole long year to be sent—at the Front!

This opportunity—to fight for the country his ancestors had shed their life blood for—was worth any sentence of a general court-martial. No general, no amount of red tape could bind him to a desk now. He was on the wing over Hunland! Then his private war waged against red tape was quicky altered to a war of a lone eagle—Spaatz versus the entire German air force.

Spaatz's mysterious disappearance created a tempest in official circles in Paris. "Mon Dieu!" an excitable French general cried with considerable wind up. Was not the major's name that of a German? Did he not speak German as he did English? Was he not ordered to Paris? Was he not here? He was not! Ah hah! It was all very clear.

France was over-ridden by spies. The major's disappearance was accounted for by this narrow gauge reasoning. The American airman with the German name and tongue had deserted! This meant disgrace not only for Spaatz, traitor or spy (the excited French Army man couldn't readily decide which Spaatz was), but for the entire U. S. Air Service.

The French needed only information from 13 Squadron

that Major Spaatz had taken off into Hunland without explanation or authority to back up suspicions with excited demands and accusations. A major at the Front, flying a a pursuit plane, when he could remain safely at a bomb-proof desk, must be bereft of all reason. The man was a lunatic or a spy, or vice versa.

The mad major—if he dared to return—should be court-martialed. It was a foregone conclusion that he would be found guilty. Oui, Monsieur le General, it was regrettable, but the firing squad would be the only way by which the face, the sacred honor, and the desired discipline of the U. S. Air Service could be maintained.

Spaatz did return. But he only tarried at 13's drome long enough to request confirmation of a Fokker he had shot down in flames, and to have his ship serviced. Then he winged into Hunland again.

The C.O. of 13 had a walking stick of dynamite on his hands. He couldn't continue permitting Spaatz to make unauthorized solo flights; confirmations couldn't be granted to a pilot who displayed every intention of remaining A.W.O.L. To forestall possible military police action the solution to the difficulty was to have the mad major attached to the squadron. An official inquiry could be conducted later.

The fervor and zest with which Spaatz hurled his bristling pursuit job at enemy Fokkers quickly made him the outstanding fighter of Squadron 13. Mounting victories carried him rapidly toward acedom.

Then one of the strangest promotions of the war came to Spaatz A.W.O.L., at the Front when he should have been on a transport homeward bound, he was made Flight leader. A major holding down a 1st lieutenant's job, subjecting himself to orders of a squadron captain. That tells tersely that Spaatz held the job of fighting in the air far above mere army rank. What cared he if a court-martial stripped him of his rank later? He was fighting, in the air, now!

Most of Spaatz's victories occurred too far behind the German lines and in rain and fogged-down skies to receive official confirmation. Let us study the terse, modest and laconic reconnaissance reports of Carl Spaatz as copied in the War Department archives and reproduced, I believe, for the first time:

Sept. 24/18: Leader of flight. Altitude 3,500 meters. Weather fair. Time: 14:45—16:30. 5 enemy airplanes seen. Encountered none. Type seen: Fokker. Later at Lachaussen 4 pilots forced to land near Pont-a-Mousson. Rest of flight approached by 5 Fokkers at altitude of 4,000 meters. Combats rejected by us.

Sept. 25/18: Leader of Flight. Low patrol, 3,500 meters. Thescourt; Chambley. Number of enemy aircraft seen: 1 biplane, 3,000 meters; 2 Fokkers, 4,000 meters. E. A. disappeared north.

The French Air Officer had plenty of wind up by now. Tactically the situation was as follows: The St. Mihiel offensive was over. The Americans had wiped out the sector. The Allies, during that four-day battle, had regained supremacy in the air. Due to poor visibility, pursuiters had wrought havoc on the Boche ground forces by ground strafing. (No mention was made by the French that Spaatz had led Squadron 13 in this dangerous and highly effective work.)

The U. S. Air Service had definitely won its wings. The bulk of the American Army was being moved in to dislodge the Boche from his Argonne stronghold. And the U. S. Air Service had a mad major at the Front, a major with a German name, one who had access to orders for the greatest offensive of the war—and that major was leading a flight, even flying alone into Hunland! His reports showed plainly that he had no intention of fighting in the air. He must be carrying confidential information to the enemy.

It is not a part of official records, but the French military police were instructed to bring Spatz in for questioning.

Then, in the early morning of the first day of the Argonne-Meuse offensive, which was the beginning of the end for Germany, the thing the French were certain would occur, transpired. Spaatz was reported missing—last seen flying into Hunland.

Flying low over Flabse, early in the morning of Sept. 26th, the major's flight ran into a flight of 7 Fokker D-8's, the deadliest fighters the Boche ever put on the Front. Spaatz waggled the wingtip of his Spad, signaling every man for himself. Soon the 5 American pilots were locked in a deadly dog-fight with the 7 D-8's.

Spaatz gunned his motor into a dive that carried him within 10 meters of a Hun single-seater. Spaatz parried machine gun bursts with the Boche; poured a total of 75 rounds into the Fokker, felt his own ship shudder under the impact of countless Spandau slugs.

Snapping his head around the American saw another Fokker on his tail. Yanking the control column against his safety belt Spaatz zoomed up to 4,200 meters. No sooner had he redressed his ship than he was going down in a screaming dive on the pursuing Fokker.

The Hun fought against a superior fighter. Spaatz's deadly fire sent the Hun down out of control. Returning to 4,000 meters the major got on the tail of a third Fokker. Following it down to 3,500 meters he sent withering sprays of lead into the ship. The Boche never knew what killed him.

Spaatz snapped out of a roll, saw a wing sag. Fabric hung in tattered streamers from the wing, fuselage and tail assembly of his fighting ship.

Its pilot had flown through a hail of lead without a scratch, but not the Spad. It was riddled and battle-torn beyond repair. Struts were lead-chewed, wires dangled in the air. It would be a miracle if it could ever regain the American lines.

Filled with the zest of desperate battle, Spaatz was recklessly unmindful that he flew a weakened, faltering plane. Instead of attempting to nurse it back home, he spiralled, looking for more enemy ships. But 4 of the 7 Boche had been shot down. The others had fled with the Americans on their tails.

Then a time-worn climax of fiction stories actually happened to the major. His motor began sputtering. He switched on the reserve petcock. Enough gas for 10 minutes more flying. Spaatz fully realized that he was too far behind the enemy lines to ever get back to 13's drome. He climbed for altitude, headed south for the American lines. The gas supply ran out. A long glide only carried the major to a dead-stick crash in a shell hole of No Man's Land. The ship was so badly shot to pieces that it literally flew apart under the force of the impact.

Spaatz, fortunately unhurt, crawled to the American lines, commandeered a military police car. He identified himself, ordered to be driven to 13's drome.

But you will recall that I related that all points at the Front were preferably reached via Paris. The major found himself in Paris. At Aviation Headquarters he experienced considerable ground-stafing from the Brass Hats.

Spaatz's laconic explanation that he had merely taken time off enroute to Paris to get his bag of Huns wasn't sufficient to explain away a long list of military offenses that had disrupted the hallowed system of military discipline, all of which had given the French high command a terrifying spy scare. Spaatz's future looked as black as the interior of a military dungeon.

Squadron 13 came quickly to his aid. Its staunch members related accounts of the major's heroism, exemplary daring in the face of enemy fire; avowed, to a man, that he had been an inspiration not only to the squadron but the entire 2nd Pursuit Group. Spaatz of Issoudon had taught them to be pilots; Spaatz of the 13th had shown them how to really fight!

The situation was awkward, as always the prospect of facing a firing squad is to the hapless warrior who fights more with his heart for his country than with his head. Then dawned the day when Major Spaatz was marched out on to a military field of honor. Rifles slapped in response to commands. Spaatz stepped forward, head high, shoulders squared. Following is the "sentence for treason" read aloud by a high ranking American officer so all present might hear:

"For extraordinary heroism in action on September 26, 1918. Although he had received orders to go to the United States, he begged for and received permission to serve with a pursuit squadron at the front. Subordinating himself to men of lower rank, he was attached to a squadron as a pilot, and saw conditions and arduous service through the offensive. As a result of his effective work he was promoted to the position of flight commander. Knowing that another attack was to take place in the vicinity of Verdun, he remained on duty in order to take part. On the day of the attack west of the Meuse, while with his patrol over enemy lines, a number of enemy aircraft were encountered. In the combat that followed he succeeded in bringing down two enemy planes. In his ardor and enthusiasm he became separated from his patrol while following another enemy far beyond the lines. His gasoline giving out, he was forced to land and managed to land within friendly territory. Through these acts he became an inspiration and example to all men with whom he was associated."

The officer stepped forward, pinned a hero's reward upon the breast of Major Carl Spaatz—the Distinguished Service Cross—and spoke, under his breath: "What America needs is more mad majors!"

EDITOR'S NOTE. Carl B. Ogilvie, the author, carefully investigates the authenticity of each incident quoted in his articles and therefore, we can assure our readers that Mr. Ogilvie's stories are true stories of the World War days. As we remarked, truth is stranger than fiction.

Howard Hughes and his racer, a breakthrough in design.

A Movie Magnate's Racer

*Flying is just a fascinating hobby to Amateur Hughes.
This dashing young genius has spent a fortune on
airplanes and has added an unique chapter to
American aviation history.*

BY ROBERT C. MORRISON

SO FAR Howard Hughes has become the closest of all humans in beating the sun in its flight across the North American continent.

On a very cold morning in January, a sleek, blue and silver racer was secretly wheeled from a hangar at Union Air Terminal, Burbank, Cal. The engine was warmed up slowly and at exactly 2:14 a. m. Pacific Time, the ship splurted down the long runway and seemingly jumped into the air as it reached the border of the field.

It banked around to the east and was soon lost to sight in the darkness and smoke from California's orange grove smudge pots. Howard Hughes was at the controls and was heading for New York to establish a new transcontinental speed record— which he did in 7 hours, 28 mins. and 25 sec.!

This flight was the greatest of Howard Hughes' series of achievements. One may well remember his movie production, "Hell's Angels." Aviation enthusiasts at that time thought nothing of going to see the picture three or four times. It cost

Howard a couple of million dollars to produce that picture, and just as it was about to be released the "talkies" were just becoming a realization, but "Hell's Angels" was not a "talky."

This fact endangered "Hell's Angels" success, so it was decided to start from scratch and make the film with sound, costing Hughes another $2,000,000 or more. When the film was finally completed and shown to the public Hughes was minus just $4,000,000.

But there was much more where that came from. It seems that Howard's father was a very successful oil man and left him an enormous fortune, and then "Hell's Angels" brought in about $3,000,000 profit.

Howard loved aviation and first caught the "bug" when he was 14. He also had a craving for speed which seems to be the characteristic of every aviator and during the filming of his world-renowned picture there was criticism in the movie magazines on the swift technique he used in getting in and out of driveways with his car. One of his cars was a Duesenberg, but even this could not compare with the speed of an airplane, so up in the air went Hughes.

He was known all over as a rich man's son and probably has held the record for owning more planes than any other single person. Among his assortment have been a Boeing 100, Sikorsky amphibian, Beechcraft, and a Douglas DC-1. But as to his flying ability there was little popular knowledge until he was the unmistakable winner in a sportsman pilot's race in Miami several years ago. His "steed" in this race was his Boeing 100.

The airplane is much the same as our Army's P-12C. However, even an Army pursuit was not fast enough for Howard Hughes. He tinkered with his Boeing for some time to try and get more speed out of it.

Many a prominent airplane factory worked on it. Douglas Aircraft did about $3,000 work on it. Then Lockheed took a hand at it. At Lockheed Dick Palmer had charge of redesigning the Boeing, and finally a few more miles per hour could be pumped from the ship, and soon after Howard won his first race.

Apparently, Hughes was well satisfied with the work Dick Palmer did on his plane as Palmer is now head man in Howard's own concern, Hughes Aircraft Co. After leaving Lockheed to join Vultee, Palmer kept up his contact with Hughes, and the two of them began to lay plans for Hughes' present record breaking speed ship.

All of Hughes' exploits have been shrouded in secrecy, but the development of his racer was more so than anything else he has undertaken. It has been said however that Dick Palmer and Hughes differed as to the design of a racer. Dick thought a racing plane should be long and thin and Hughes though it should be short and fat like a Gee Bee.

To finally decide this question wind tunnel models were made and tested in the wind tunnel at Caltech, Pasadena's noted institute. This was all done at Hughes' expense and under the supervision of W. Curtis Rockefeller, recent winner of the Sperry Award and noted meteorologist for American Airlines.

Finally, the "perfect" design was arrived at, and immediately engineers and draftsmen began the long arduous task of detailing its construction. It was to have the long slim fuselage that Dick Palmer had desired.

At that time Gerard Vultee had as a very competent chief engineer none other than Carlton Strycker who is now head of the engineering school of the Curtiss-Wright Technical Institute for Aeronautics. Carl Strycker took an interest in the plane and was given charge of the stress analysis in his spare time at night.

When a plane is designed to do 350 m.p.h. as this one was, the stress analyst has his troubles. It is said that bending moments of over 1,000,000 lbs. were encountered. Thus the structure had to be built very strong, and the wing was almost a solid mass of wood as a consequence.

The fuselage was of all-metal construction, and since the metal skin was so thick, the rivets were countersunk to make a smooth surface over the entire fuselage.

Carl Strycker promoted the idea of building a keel in the "belly" of the fuselage, and after much discussion the keel became a part of the airplane, later playing an important part in saving Howard Hughes' life, an incident which we will relate later. Though the keel had many good points such as strengthening the structure of the plane, it made the ship heavier and reduced its fuel capacity. However, various parts of the racer were stored away in the keel, such as the Sperry gyro instruments.

The remainder of the fuselage was more or less conventional in design and had in its nose a Pratt & Whitney *Twin Wasp* engine that developed about 800 h.p. normally and packed away 1,000 horses when full out.

The wing was a work of art. It was of all-wood construction with a very smooth plywood covering. This covering was over an inch thick when put on and was then shaved down about a fourth of an inch to give the wing a perfect airfoil section. The plywood was then polished until it was as smooth as glass, was painted blue, and finally waxed. Two box spars were used in each wing panel, but when all the reinforcements and parts were completed to them they were practically solid spruce beams.

Two hundred eighty gallons of fuel may be carried in three or four tanks located in the fuselage. In all the racer weighs approximately 5,000 lbs.

Hughes was naturally born an engineer. Though he did not have much school training along those lines, he has exceptional practical engineering knowledge according to men who have worked with him, and he knows every detail of his ship thoroughly. When his racer was completed he took it out and taxied about the field.

If something did not suit him perfectly he would take the ship back to the shop and with great thoroughness and care would make any alterations necessary. But with all this thoroughness he did not seem to have any set procedure. At times his men did not know when he was going to make a flight but were always to have his ship ready. Likely as not, Howard would get up in the morning, grab any clothes he could find in his room, rush to his awaiting plane and be off for some destination perhaps 3,000 miles away.

This was such the case when he first test-hopped his latest low-wing racing plane, according to an eyewitness. Following several anxious days of testing his ship on the ground, Howard was taxiing along the runway, when without warning he was seen to leap into the air. He was off on his first hop!

When Hughes takes off with the ship he pulls the wheels up when about 25 or 30 feet in the air and then begins to climb fast. On one of his first hops he was in the embarrassing position of not being able to get the wheels down again when coming in for a landing. The hydraulic system would not work, and he yanked an emergency cable to try and pull them down.

When bouncing around in the air at about 300 m.p.h. in a plane as his, trying to pull a stuck landing gear down is quite a hair raising problem. Undoubtedly perspiration came forth from that man Hughes. Fortunately he lowered the landing gear far enough so the plane could be landed safely.

That incident reminds us of the 1936 National Air Races when that daredevil, Harry Crosby, spent the good part of a nice sunny morning careening through the air with his racer trying to get his landing gear down by jerking out of dives until it finally reached the extended position.

The few hours preceding a test hop of a racing plane are usually the most nerve-wracking to the pilot. When he once gets the ship up the climax seems to have passed.

It is said that Hughes, while waiting to fly his plane, would borrow a Great Lakes job and fly it over the countryside so as to keep the "feel" of flying an airplane. He would land, take a look at the progress the mechanics were making on his steed, and would then take off again. On one of these occasions he flew Vance Breese's low-wing racer. It was very humorous the way he hopped in and out of airplanes while waiting for the completion of his plane, but not so to Howard Hughes perhaps.

It cost Hughes in round numbers about $100,000 to build his racer. Soon after completion and test flights, Hughes flew his ship over a measured course at Santa Ana to endeavor to break the world land-plane speed record. He made four dashes and apparently broke the record by a large margin, but the judges declared it unofficial for various and sundry reasons, mainly because of his dive before entering the straight-away which is prohibited.

However, Hughes went up the next day for another attempt. His cockpit enclosure was now missing as it had blown off during one of the speed dashes leaving him in an open cockpit which considerably cut down his speed. It also made the ship tail heavy and at top speed the racer was unmanageable.

Therefore, Hughes had to throttle back somewhat and make the speed dashes at that speed. Nevertheless, on one of four dashes that day he did slightly over 352 m.p.h. to break the existing speed record! His engineers had figured about 350 m.p.h. as the top speed, and that was with cockpit enclosed.

But no sooner had he completed the dashes than the engine of the sleek racer quit. Hughes spotted a newly plowed field down in front of him and attempted a landing with his wheels up, simulating a snappy sleigh ride. Thanks to Mr. Stryker's keel the ship did not fold up like an accordion as I have seen some racers do.

It is always better to land with the wheels up on such a field as otherwise the plane is liable to turn turtle (we unwittingly had the ship change into an accordion and turtle, in the preceding two sentences but we know you grasped the right meaning). Little damage was done to the racer, and it was brought back to Hughes' modern but minute factory at Burbank, California.

From then on, all sorts of rumors flourished. People said that the plane was purchased by the Army to be developed into a pursuit. Others said Hughes would try to break some speed records made by the French flyer, Detroyat, at the 1936 National Air Races.

Finally, the rumors ended when one morning Hughes took his ship out for a test hop, and lo and behold it had a new wing. This wing was long and narrow as compared with the short stubby former wing. It had the N.A.C.A. 23012 airfoil section, the most efficient yet designed. Hughes' ship by these modifications was now a long-distance racer, and Hughes predicted that it would do about 360 m.p.h.! It was to be delivered to the Army.

In the Hughes manner it was delivered. Taking off one morning he flew it to Newark, New Jersey, making his famous record-breaking flight. It was his own record that he had broken which he had established in Miss Jacqueline Cochran's Northrop *Gamma* some months before.

From there our guess is that it went to Dayton. It will undoubtedly be tested for stratosphere flying by the Army engineers, and it is not impossible that speeds of around 500 m.p.h. may be obtained with it with tail winds that are encountered at about 35,000 feet altitude.

With an enormous top speed like that the racer has a very slow landing speed which makes it ideal for military purposes though it will probably have to be throttled back in order to maneuver it in combat. It would take a radius of about five miles to turn the ship when "wide open" without making the wing appear as if it was hinged to the fuselage.

Ordinarily at low altitude it will do 160 m.p.h. Both figures are at cruising speed. So Howard Hughes may take to the stratosphere on his forthcoming flight, and it shall be interesting to watch his progress.

It is fortunate that our industry has such a man as Howard Hughes who is willing to risk his life in the practical development of high speed aircraft.

MAILMAN Overboard!

BY CHARLES A. LINDBERGH

*The author of the report below is star figure in the
world's most exclusive organization—the Caterpillar Club.
Lindbergh is a four-jump member.*

NOWHERE in the annals of aviation can be found a more dramatic first person story than the one written by Colonel Charles A. Lindbergh on November 4, 1926, in which the famous flyer describes the circumstances under which he was forced to abandon his plane over Covell, Ill., the preceding day. From an unimpeachable source, this graphic report has come to POPULAR AVIATION and the editors are glad to pass it along to our readers, since it expresses the innermost thoughts of a tall young mail pilot who was to rocket to world-wide fame a few short months later. It is written by a real pilot in real flying language:

REPORT OF NORTHBOUND MAIL FLIGHT, NOV. 3, 1926
BY CHARLES A. LINDBERGH, PILOT C.A.M. NO. 2

(Editor's Note: Lindbergh's run was then between St. Louis and Chicago with intermediate stops scheduled at Springfield and Peoria, Ill. The route was then being operated by the Robertson Aircraft Corporation.)

I took off from Lambert-St. Louis Field at 4:20 p.m., November 3rd, arrived at Springfield, Ill., at 5:15 and after a five-minute stop for mail took the air again and headed for Peoria.

The ceiling at Springfield was about 500 ft., and the weather report from Peoria, which was telephoned to St. Louis earlier in the afternoon, gave the flying conditions as entirely passable.

I encountered darkness about 25 miles north of Springfield. The ceiling had lowered to around 400 ft. and a light snow was falling. At South Pekin the forward visibility of ground lights from a 150 ft. altitude was less than 1/2 mile and over Pekin the town lights were indistinct from 200 ft. above. After passing Pekin I flew at an altimeter reading of 600 ft. for about five minutes, when the lightness of the haze below indicated that I was over Peoria. Twice I could see lights on the ground and descended to less than 200 ft. before they disappeared from view. I tried to bank around one group of lights, but was unable to turn quickly enough to keep them in sight.

After circling in the vicinity of Peoria for 30 minutes I decided to try to find better weather conditions by flying northeast towards Chicago. I had ferried a ship from Chicago to St. Louis in the early afternoon and at that time the ceiling and visibility were much better near Chicago than elsewhere along the route.

Enough gasoline for about one hour and ten minutes flying remained in the main tank and 20 minutes in the reserve. This was hardly enough to return to St. Louis even had I been able to navigate directly to the field by dead reckoning and flying blind the greater portion of the way. The only lights along our route at present are on the field at Peoria, consequently, unless I could pick up a beacon on the transcontinental route my only alternative would be to drop the parachute flare and land by its light together with what little assitance the wing lights would be in the snow and rain. The territory towards Chicago was much more favorable for a night landing than that around St. Louis.

I flew northeast at about 2,000 ft. for 30 minutes and then dropped down to 600 ft. There were numerous breaks in the clouds at this time and occasionally ground lights could be seen from over 500 feet. I passed over the lights of a small town and a few minutes later came to a fairly clear place in the clouds. I pulled up to about 600 ft., released the parachute flare, whipped the ship around to get into the wind and under the flare which lit at once but instead of a floating down slowly, dropped like a rock. For an instant I saw the ground then total darkness. My ship was in a steep bank and for a few seconds after being blinded by the intense light I had trouble in righting it. I then tried to find the ground with the wing lights, but their glare was worse than useless in the haze.

When about ten minutes' gas remained in the pressure tank and still I could not see the faintest outline of any object on the ground I decided to leave the ship rather than attempt to land blindly. I turned back southeast towards less populated country and started climbing in an attempt to get over the clouds before jumping.

The main tank went dry at 7:51 and the reserve at 8:10. The altimeter then registered approximately 14,000 ft. yet the top of the clouds was apparently several thousand feet higher. I rolled the stabilizer back, cut the switches, pulled the ship up into a stall and was about to go out over the right side of the cockpit when the right wing began to drop. In this position the plane would gather speed and spiral to the right, possibly striking my parachute after its first turn. I returned to the controls and after righting the plane dove over the left side of the cockpit while the airspeed registered about 70 miles per hour and the altimeter 13,000 ft.

I pulled the ripcord immediately after clearing the stabilizer. The chute functioned perfectly. I had left the ship head first and was rolling in this position when the risers whipped me around into an upright position and the chute opened.

The last I saw or heard of the D.H. was as it disappeared into the clouds after my chute opened. I placed the rip cord in my pocket and took out my flashlight. It was snowing and very cold. For the first minute or so the parachute descended smoothly then commenced an excessive oscillation which continued for about five minutes and which I was unable to check.

The first indication that I was near the ground was a gradual darkening of the space below. The snow had turned to rain and although my chute was thoroughly soaked its oscillation had greatly decreased. I directed the beam from the 500 ft. spotlight downward but the ground appeared so suddenly that I landed directly on top of a barbed wire fence without seeing it.

The fence helped to break my fall and the barbs did not penetrate the heavy flying suit. The chute was blown over the fence and was held open for sometime by the gusts of wind before collapsing. I rolled it up into its pack and started towards the nearest light. Soon I came to a road which I followed about a mile to the town of Covell, Ill., where I telephoned a report to St. Louis and endeavored to obtain some news of where the ship had landed. The only information that I could obtain was from one of a group of farmers in the general store, a Mr. Thompson, who stated that his neighbor had heard the plane crash but could only guess at its general direction.

I rode with Mr. Thompson to his farm and after leaving the parachute in his house we canvassed the neighbors for any information concerning the plane. After searching for over an hour without results I left instructions to place a guard over the mail in case it was found before I returned and went to Chicago for another ship.

On arriving over Covell the next morning I found the wreck with a small crowd gathered around it less than 500 ft. back of the house where I had left the parachute. The nose and wheels had struck the ground at about the same time and after sliding along for about 75 ft. it had piled up in a pasture beside a hedge fence. One wheel had come off and was standing inflated against the wall on the inside of a hog house a hundred yards farther on. It had gone through two fences and the wall of the house. The wings were badly splintered but the tubular fuselage, although badly bent in places, held its general form even in the mailpit. The parachute from the flare was hanging on the tailskid.

There were three sacks of mail in the plane. One, a full bag, from St. Louis had been split open and some of the mail oil soaked but legible. The other two were only partially full and were undamaged.

I delivered the mail to Maywood by plane to be dispatched on the next ships out.

He's Laughing at the World

Here is an unvarnished insight into Douglas Corrigan, the man. Perhaps this article will help explain many of his "queer ways".

BY HAROLD KEEN

Aeronautical irony: When Charles A. Lindbergh's The Spirit of St. Louis *was being built at the Ryan factory in San Diego, Corrigan (circled at right) helped work on it. Lindbergh (circled at left) and Corrigan were photographed with the Ryan staff.*

NOW that Doug Corrigan is off the front pages, let's look at him as we would any other human being—let's poke beneath the glamor that clothes a public hero in the first great burst of mob attention.

Doug Corrigan's "wrong way" story has entered American folklore and he himself has become a symbol of many things—of determination to overcome obstacles, of ambition fulfilled, of youth triumphant. To millions of Americans, he is a vivid example of freedom and opportunity for all, and because of him thousands of discouraged young men have gained new hope. If Doug could do it without education, without resources, without encouragement, with only a flaming, unquenchable purpose—then they, too, have a chance.

And he gave the world a good laugh when it needed it. For a time he displaced wars and human bestiality in the average man's consciousness.

That's quite a lot for one "grease-monkey" to accomplish, you say. And you may wonder exactly what sort of a person is this Douglas Corrigan who made the most amazing flight in aviation history.

In some of the cities of his homecoming tour, Doug told the crowds, his vivid smile winning every heart:

"The world's been laughing at me, but I'm laughing right back at it, too, so I guess that makes us even."

Only the few who knew him during the years he worked in Los Angeles and San Diego airplane factories and hung around airports could appreciate that remark.

It was spiced with a deep-rooted irony and told the story, in one sentence, of the lack of sympathetic understanding, the thoughtless jibing he had endured in the long years he had lived to himself, a veritable hermit in the midst of other men.

In that statement is the key to the secret of his "backwards flight" yarn.

Doug was enjoying sweet revenge, but not in the vindictive sense. In the glow of his achievement, the essentially good-natured Irish soul of this hitherto silent sky-man, this loneliest of all "Lone Eagles," found no room for anything more deadly than subtle rebuke to those who once had ignored him, and now rushed eagerly to the Corrigan presence, there to be bathed in reflected glory.

"The world's been laughing at me, but I'm laughing right back at it. . . ."

Life had been none too good for Doug Corrigan up to that fateful day of July 18, 1938. An orphan, he began shifting for himself very early. He was restless, and didn't stay on one job long, even after he had drifted into aviation in 1925 from the building trades, in which he had served as a helper in Los Angeles.

He was 21 when he worked on Lindbergh's *Spirit of St. Louis* in San Diego, and he hasn't denied that then and there he became afire with the ambition he was to realize eleven years later.

But something like that took a wad of money. Lindy himself had backers—his plane was built specifically for a transatlantic flight, and not only that—there was a $25,000 pot o'gold at the end of the New York-to-Paris rainbow.

Doug began preparing. He started putting money aside. You've undoubtedly read of his heroic self-denial—of the meals he didn't eat, the pleasures he didn't seek, the girls he didn't date. He denied himself all these things, yes—but not in the spirit of stoicism. He simply was careless about his meals—he always had eaten a sketchy breakfast and lunch, consisting of a piece of pie or a banana, perhaps both in a moment of indulgence; dinner was slightly more solid. It didn't affect his health, and although it kept his weight down in the 120's, he could hold his own with the burliest mechanic in the plant.

As for the ordinary pleasures—seeing a movie, going to a party, cultivating the friendship of girls—he just wasn't attracted to them. His limited education (he admitted with a note of defiance while in San Diego last September that he had never gone to high school) put him out of reach of good literature. But no one knew airplane engines better than Doug Corrigan. And airplane engines constituted his life.

Under such circumstances, Doug attracted no real friends, because the average person couldn't figure out a fellow like that. Nobody tried to understand him; his sensitive nature rebelled at the thought that others sized him up as somewhat "peculiar"; and he tantalized his questioners with vague, off-the-track answers while he was installing the extra gas tanks in his old Curtiss *Robin.*

The crowning blow came last year, when the Bureau of Air Commerce refused to grant him permission to fly the Atlantic. It is utterly impossible to realize how crushed was Doug Corrigan, and how bitter he felt that walls were always being built around him.

Thus, when he flew the Atlantic, he had crashed through those walls. He had come out into a glorious new freedom. He now laughs back at the world which laughed at him; and those who recall the strange recluse can appreciate the mockery in the "wrong way" story. It was more than the greatest piece of buffoonery of the century; it was Doug Corrigan broadcasting:

"Yep, I guess all you folks were right about me. I never did act like any other normal person and you couldn't quite figure me out. Now I start in one direction and fly another. Just like me, isn't it? What a bum navigator!" Oh, yeah?

This is not just an interesting theory on the background of the story he told in Dublin.

H. A. "Jimmy" Erickson, the flying photographer for whom Doug worked at times in San Diego, points out that with the world at his feet, Corrigan is releasing all the emotions pent up in the years when ridicule was his lot.

Ed Morrow, a fellow worker with whom Doug had "batched" in a small apartment near the airport, recalls how fearful Corrigan was of being scoffed at; and how he consequently assumed a defensive attitude toward all who came in contact with him.

J. J. "Red" Harrigan and Doug Kelley, two of Doug's earliest flight instructors, were amazed at the overnight personality change in their ex-pupil. The once-silent "grease-monkey" had suddenly become self-assured, voluble, a master of crowd psychology and one of America's wittiest men.

Only one thing could explain it. Doug Corrigan had thrown off all the inhibitions that had repressed him; he was the natural little Irishman now, his character enriched by the culmination of a long, tough struggle.

Occasionally reversions to the old Doug Corrigan were noticeable during his swing around the country. For instance, too persistent efforts to pry out information of his past were shaken off. He carried with him several resentments. He resented the inaccuracies about his past life, even about his flight, that marred the press accounts. Repugnant to him was the performance of communities throughout the country which claimed him as a "home town boy," and the attempts of individuals to cash in on the publicity of either having known him or employed him in the past.

He vigorously denied that his uncle, the Rev. S. Fraser Langford, had taught him navigation, or that he had even lived with him for any length of time; he laughed at the stories of his carrying a pole to knock ice off his wings, or that he closed the door to his plane with a piece of bailing wire; he said no airline had offered him a transport pilot's job. There was a long string of untruths in the newspapers that seemed to irritate him.

"Just about every city I've been in has claimed I came from there," Doug commented as he neared the end of his transcontinental trail. "Funny how I got a lot of home towns all of a sudden."

I asked him if he considered settling down anywhere he could call home.

With a poignant trace of the old Corrigan loneliness in his

voice, he answered: "Home? Where is home?"

One of the biggest "kicks" of his life was experienced at the expense of a former boss who had widely publicized the fact that Doug once worked for him.

It happened at the recent National Air Races. The ex-employer was standing alongside Doug when someone remarked:

"It's amazing how you manage to be patient with the demands of autograph seekers."

To which Doug replied: "Years ago, I learned to be patient while waiting for my boss to give me a nickel an hour raise." Just another subtle Corrigan rebuke!

He has become sharp-minded, suspicious of the motives of those who shower praise on him. In San Diego, a Chamber of Commerce official asked him to autograph a photo taken at Lindbergh Field last year, in which Doug and his plane were pictured against a background of two aircraft factories. But the names of those factories were conspicuoulsy splashed across the print, and although the photo was to be harmlessly used in a harbor department booklet, Doug shied away from it. It "smelled" like a publicity gag at his expense.

And shortly afterward, he wanted to be sure that the baseball game at which he was to appear that day was being played for charity; it took the solemn pledge of the Mayor that no individual was going to make money on Doug Corrigan before the flyer consented to go.

There was plenty of truth to Doug's petulant remark to reporters who had asked him who was managing his multitudinous affairs:

"I don't need a manager! I'm the only one who knows what I want to do!"

By the time he reached the west coast and the end of his transcontinental homecoming trip, he was a vastly different Doug Corrigan than the unsophisticated, lovable young man whom the New York crowds exulted over. Adulation in two-score cities had hardened him a bit, had even mechanized some of his gestures to such an extent that he automatically bowed to the right and left and waved his arms to imaginary crowds on empty streets in suburban or thinly settled areas during the usual auto parades.

There were occasional flashes in which he became needlessly aggressive, such as the time he ordered the Los Angeles woman reporter out of his room when she attempted to use his telephone. Or when he warned the photographer who was trying to get a picture of Corrigan in his box seat at a baseball game:

"You better not get in the way! I wanna see this game!"

And could it have been inflation of the cranium when he commanded the police escorting the parades in various cities to lead slowly, so that "everyone could see" him? Or when he telephoned his old friend, Roland Tyce, operator of the Chula Vista airport just outside San Diego, that he was coming over and to spread the word around so that no one would miss the great occasion?

Perhaps all this is being a bit too harsh with the real Doug Corrigan. I don't think so, because it's all explainable and natural when you study Doug's youth.

Don't forget that for the past decade he had been virtually friendless, possibly because of his own eccentricities, also because few persons took time to figure Doug Corrigan out. He had worked as a night watchman in an airplane hangar in return for a cot. He sweated for everything he owned and he had no one but himself to thank for his present position. He drifted from job to job, his restless spirit unsatisfied. He never received more than a modest salary. He sacrificed not only for himself but for his brother Harry, whom he financed through the University of California, and who held a better job at Glenn Martin Co. in Baltimore than any Doug himself ever enjoyed.

As he himself admitted, he never had a real home.

And he had a burning ambition to rise out of his obscurity, an ambition that caused people to observe his strange conduct with raised eyebrows.

Once at the bottom of the heap, Doug now suddenly found himself at the top. It was an exhilarating feeling, and he experienced it to the full. Furthermore, he had a very real reason to laugh and mock and display an occasional arrogance.

What matters most in a public figure, of course, is his impact on the crowd. And Doug's, from the beginning, has been magical. He has an unsurpassed appeal, rising from a fundamentally unspoiled boyish charm. The persons who wouldn't look twice at Doug in the old days were swept along in the gale of laughter that followed in the flyer's wake across the country. Doug's witticisms have become so much a part of aviation legend that they need no repeating here.

It will be unfortunate, though, if his actions on the recent triumphal tour across the country leave a tainted flavor in his particular bit of aviation history. Despite the bitter criticism of certain of Doug's actions, it must be remembered that there is a story behind those actions well worth knowing. Also, much of his resentment and harsh treatment of persons with whom he came in contact during that transcontinental journey was due to his great distrust for those persons and organizations that tried to capitalize on him. For this I cannot blame him at all. There is something about the modern American method of ballyhoo that would make the tamest kitten a raging wildcat. Doug's sudden rise to international fame brought all these hounds down on his head. His only defense was that which you and I probably would have adopted. . . .

Perhaps Doug will return to the obscurity whence he came. Or he may rise to occupy a permanent niche high in the aviation world. But anything from this point on will be anticlimax to the fellow who taunted a world that previously had had no particular use for him.

One Foot from Eternity

BY TYRONE POWER

Tyrone Power says he'll never forget Saipan— and the wartime error that almost cost his life.

The author at the controls of his Marine Corps plane.

WHEN I THINK of Saipan, I think of the Marines and the C-46 I jockeyed around the Pacific during the war. And every time I climb into a cockpit, I force myself to re-live a hellish 30 seconds or so I spent on Saipan while starting a routine flight mission. I re-live those seconds because they taught me a flying fundamental I don't ever want to forget.

During the war I flew a Curtiss *Commando* week after week, month after month. I think I knew its insides better than I knew what was in my pockets.

And—like a lot of other pilots—I got careless.

The hydraulic system on the C-46 is somewhat complicated. Under the conditions we flew the planes, the seals for flaps and gear had a tendency to dry up easily. We had to actuate them frequently to keep the actuating cylinders flexible and to make certain we'd lost no fluid from dried seals.

So, habitually, I'd actuate the flaps while waiting on the line—just to make certain everything was all right. I'd flip them up and down, up and down.

Then, taxiing, I'd flip them up and down again.

In our squadron it was SOP to go through three full cycles on the flaps. First, on the line or in the revetment, again while taxiing for take-off, and once again after final mag check.

That morning on Saipan I was hauled out of bed at 5 o'clock for a rush flight to Okinawa with a heavy load of materiel and four or five men. I picked up my co-pilot—a new one, checked the plane, and we were ready to go.

I wanted to be extra certain those flaps were okay, just in case I should need them on take-off. I made the prescribed check in the revetment and another one while taxiing for take-off. Then, while we sat there on the end of the runway waiting for a bevy of planes to take off, I checked it again. We went through the entire check-off list.

My co-pilot got clearance, nodded to me, and off we went.

The C-46 scuttled down the runway while I kept watching for the prescribed number of inches on the manifold pressure. I knew immediately something was wrong.

Down the runway the plane barreled, trying to pull off, with me pushing hard on the yoke to keep it on the ground. Second by second we were eating up that runway, roaring toward the end, and I had a quick vision of what would happen: We were lurching along at 60 m.p.h. . . . then 65 . . . and faster. The last few feet of the runway would be coming up soon, and I'd either have to chop those throttles right away or there wouldn't be time left to do anything except pray.

I couldn't figure it out, though I was trying to remember in those quick seconds everything that could possibly be wrong. All I knew was that the darned plane shouldn't take off at that point—and yet it was tugging and pulling to get off the ground. We were still accelerating, and the remaining feet of the runway looked like only inches.

Then, off she roared. I couldn't hold her down any longer. As the plane lumbered off the runway, it seemed to stagger. Desperately, I whipped a look around the cockpit. Maybe the trim was off. I thought of a thousand things, discarded them. I caught a glance at my co-pilot, who about that time looked like a lad walking the Last Mile.

And then we both looked down and at the same instant saw what had happened.

I had left full flaps on the plane!

That spanking new co-pilot worked like a high-speed machine. His arm shot down, and he dumped the flaps in a fraction of a second.

There's no need to tell pilots what happens when you dump the flaps—especially with a heavily-loaded plane the size of the *Commando.*

Ordinarily, the flaps would have been set at 15° because any setting beyond that acts as a drag and cuts down forward acceleration. Taking off with full flaps, we became airborne faster (the tail just wouldn't stay down) but forward acceleration was slowed almost to the stalling point.

So there we were. The co-pilot had dumped the flaps. He knew what would happen then, and so did I. I felt like I was piloting a big hunk of lead. We lost altitude—I don't know how much. I was too busy gunning the plane at full throttle and rolling forward with all my might on the trim tab to keep from stalling.

My gear was up by this time, and we were still so low that the prop tips were missing the ground by inches. The red lights at the end of the runway were getting big as cannon balls—and right behind them was a hill. Even at that moment, while I was breaking out in a cold nervous sweat, I thought mirthlessly about the headlines back home: "Tyrone Power's Last Scene—A Smash Hit."

But, almost miraculously, we squeezed over that hill with the old *Commando's* engines screaming. The rest of the ride to Okinawa was anti-climax. But today, whether I'm flying a plane in California or in Rome, I sit for a second or two before take-off and remember that day on Saipan.

Gill Robb Wilson

A memo to the congress

Air power is the single type of power which adapts itself fully to those technological capabilities in which America excels. It compensates for our relative manpower deficiencies. Let us face it: the eclipse of American air supremacy would be total eclipse of our National security.

It therefore is more than disturbing to learn that a year's production of air power has been lost, that priorities for the aircraft industry are still competitive with the civilian economy, and that the trickle of modern aircraft coming from production lines will not greatly increase in 1952.

Air power is the one form of military power which can prevent war by its threat of immediate and awful retaliation. It is the one weapon which immediately compensates for any major shortage of American manpower in field armies. It is the only weapon which lends itself to basing on both land and sea. It is the single form of land-based power which does not require the massing of troops on foreign shores.

Perhaps more important still is the fact that air power provides the only possible carrier for those absolute weapons whose existence can make wars of aggression unprofitable to the point of self-destruction by the aggressor.

The effect of American air power as peace power in the post-war years is one of the clearest pictures of our times.

In a long series of Soviet threats and American counter actions over Trieste, Iran, Berlin, Turkey, Greece, Yugoslavia, the Kremlin backed down on every occasion. On each of these occasions Moscow possessed the physical power easily to overrun its intended victim despite our aid. But she dared not invite action by American air power. In the truest sense, our air supremacy kept the peace.

How then can we fly in the face of these facts by losing that supremacy as we presently seem to be doing? General Vandenburg, Chief of the Air Force, has stated the danger clearly. On their part, naval airmen fully realize that both carriers and planes must be radically improved. The Army pleads for aerial vehicles in its organic structures. The aviation industry is frustrated in its priorities and its ability to produce dangerously compromised by National policy.

In Korea the Kremlin stealthily pushes the war with increasing contributions of arms, transportation and aircraft. Moscow is safe in Korea, free of any risk of retaliation by American air power, since it is our policy not to strike beyond the immediate theatre of action.

Fear of American air supremacy is the single element which has thwarted Moscow in its long list of aggressions. No one can doubt that the threatened eclipse of American air power will be the signal for World War III.

Flying / February 1952

I remember grandmother

THIS piece takes off with Grandmother leading the mission. Her name was Emma Jane Robb. She lived almost a hundred years . . . and I mean lived. She didn't just stay alive. She lived.

Sopping wet, Grandmother never weighed a hundred pounds. She referred to herself as scrawny. Around about 70 she fell downstairs and broke her back. Thereafter she was bent considerable but none of the vinegar ran out.

When I came home from the wars in 1919, Grandmother was my first customer for an airplane ride. She had never seen an airplane. We fixed it up.

"How high would you like to fly, Grandmother?" I shrieked in her best ear.

"About as high as the tree tops," she said. I suggested five thousand feet as an alternate. She agreed.

"But I don't want to scare you," I assured her. She snorted in disdain. Who did I think I was?

Those were the days of open cockpits. First her goggles slipped down around her neck. Then her helmet blew off. Then her hairpins flicked out and her top knot streamed in the wind. Her skirts blossomed around the safety belt. She wouldn't keep her head behind the windshield. When she opened her mouth the prop wash almost pushed her false teeth down her throat.

On the ground again with her petticoats in place and her hair untangled she rendered her report. She said, "Well . . . now I've done one more thing worth remembering."

I didn't understand exactly what she meant at the time. In later years when she could only sit and rock I began to get the idea. Her eyes never ceased to twinkle. I would catch her laughing to herself. Often she would be murmuring, "Count your many blessings . . . name them one by one."

Eventually it dawned on me that Grandmother had made herself a person of great resources. Purposefully and deliberately she had been collecting "things worth remembering" against the day of her need. The walls of her memory were hung with tapestries of wondrous patterns . . . frontier days of girlhood in the west . . . buffalo herds swamping boats in the Missouri . . . hardship . . . toil . . . pain . . . pleasure, met head-on and no ducking . . . flitting around the thunderheads in a flying contraption . . . running to meet life rather than running away from it.

So at the last she had resources and she whipped old age to a frazzle. Her rocking chair was a center stage and she paraded the full years past her with the same eclat as a marshal reviewing the troops.

Today people tell me the romance is gone from aviation. The callow collegian wants to know why he should have any interest in flying. Hundreds of middle-agers assure me defensively, "Oh, I would fly in an emergency."

But isn't the truth of the matter that the romance is gone from us rather than the airplane? The collegian is headed for the quickest buck, and the middle-ager will never know an emergency because nothing will ever be too important to anyone in retreat from challenge.

Ten thousand articles have been written about the ebb of aviation enthusiasm and activity in the land. And they all ask what to do about it. More airports? Cheaper airplanes? Government subsidy? . . . ad infinitum. It's all beside the point. Apathy to aviation is just one of many indications that twenty years' education that "the world owes me a living" has wrought havoc with our individual spirits. We babble about social security and prate in terms of "masses" and promise "extended benefits." And all of it means no more nor less than that we've had a flame-out of the fire that moulded us and gave us power. We've stopped running to meet life because that takes faith and courage and pride. So we have no inner resources to draw on when the chips are down.

Settling the peace after World War II would have been a rough experience. We ducked it. Giving the Kremlin an ultimatum on the first day of the Korean war would have tested our soul. We lacked the guts.

When youth is called to the colors we refer to it as "years taken right out of his life." We shed crocodile tears and are maudlin about our casualties, but how many of us go out to man the ramparts of civil defense?

But why should I get all steamed up? It's no sweat to me. I've got resources. I remember Grandmother.

Flying / July 1953

Armistice

ON THE blustering morning of November 11, thirty-seven years ago, I sat impatiently in a soggy cockpit waiting for other ships of the formation to taxi out through the mud of a French airdrome. We were launching our second strike of the morning.

I was Operations Officer and had plenty of chores on my mind . . . Let's get this fighting business done so a man could get back to his knitting.

But yonder on the line a commotion had developed. What in heaven's name had gone wrong now? A figure detached itself from the group around the operations tent and came dashing toward the starting line. He leaped on the side of the ship, screamed, "Armistice!" and tore back toward the hangars.

I can't remember my first flight in an airplane, or my first solo, or my first combat mission; but I remember that scream of "Armistice!" as though it was uttered yesterday.

If the sergeant had screamed "Peace!" I would have laughed in his face or cursed, as the mood chanced. Armistice I recognized as an apt summing up of the permanent status of human society. I knew there would never be peace and that there never had been peace, for I had seen the naked soul of civilization. In the Vosges, on the Oise, along the Aisne, at Verdun and Soissons and Montdidier, and in the chalk white vineyards of Rheims, I had seen, not the struggle between French and German and English, but between good will and avarice, between good and evil.

I had talked to men of many races and creeds and colors all over the battlefields of Europe. I had heard their ambitions and prejudices and loves and hates. I had played with enough bastard children along the surging map to fill a dozen orphanages. I knew that peace was not a thing of bread and wine, although the wise ones spoke of "economic stability." I knew that there was no such thing as peace except between God and man—all the rest was armistice and armistice alone.

So I taxied back to the hangar line. I didn't feel exaltation or remorse. I didn't feel anything except a great pity for millions of people who thought they had purchased peace and that the world was now safe for what they called democracy. I knew the world would never be safe for any great ideal unless people of good will were strong enough to hold avaricious men in check. And there would always be avaricious men.

Now a simple man and a mere citizen must apply himself to the problems of life in ways which are open to him and in the place where he stands. The airplane in those days was not much but it was the only potential available to handfuls of young fellows who had had eye opening experience with the realities of human nature. If it could be developed to play a great role over the years, it might serve a double purpose in a constructive direction. First, it might be an effective way to hold avarice in check. Second, it might be a method to overcome traditional suspicions by providing better contacts between strange peoples.

We didn't know much about aviation but we knew more than anyone else. I have a great sympathy with the people who laughed at our antics and were skeptical of results. We sure talked fast and flew slow. And probably neither performance made much sense. But finally we got somewhere, not because we were wise or skillful, but because the airplane has the greatest potential since the invention of the wheel.

World War II was the turning back of avarice by air power and the airplane has so far kept the integrity of armistice for a civilization in spiritual conflict. Air power is even declared the instrument of national policy.

Should this be cause for gratification to an old airman? I wish it were all as simple as that. But no, because men of good will again are taking an erroneous reading on the status of human nature—are talking of "peace" and forgetting the periodic demonstrations of how volatile and vulnerable is civilization at its best. The input of airpower is insufferable and the proposed cutback of air power is intolerable to anyone who knows that the word is not peace but armistice.

Flying / November 1955

"You're on your own" was all he said that day long years ago—
So long his name and face are lost in memory's afterglow;
Nor do I recollect of pride or joy or doubt or fright
Or other circumstance which marked that time for solo flight;
The cryptic words alone endure. He said, "You're on
    your own,"
And down through time I've found it so—the test's to
    walk alone.

Not that one choose to draw aside in churlish mein or vein
From common lot of what life holds of pleasure, toil or pain
But that the call's to rise and cruise alone with
    dreams unshared
Or plan alone for some far goal for which none else
    has cared,
Or fight alone for what you hold is worth a warrior's strife
And ask no gain or fame or aught beyond the joy of life.

I owe a quenchless debt to him who bade me seize my fate
And hang it on the faith that I to it was adequate
For when he said, "You're on your own" and sauntered
    on away
I knew that here in four short words was youth's first
    judgment day
—Not wit to learn, not test of skill, not pride to satisfy
But will to walk down life in faith that life is theirs who try.

Flying / January 1961

To each and all some storm is sure!
Think not you'll never meet
By planning every venture with
Allowance for retreat!
In time—some day, some night, some storm
Will intercept your track
Where there is no way 'round or out
Or any turning back.

That you have fortitude to face
Your storm will not suffice
When gust stall snatches at controls
And windshield sheathes in ice;
The stoutest heart is not so stout
That fittings dare not give
Nor will the lightning stroke relent
Because you pray to live.

Is storm some dreadful tyrant then
Who may forbid the skies
To us who love the sun kis't height
Where heart's adventure lies?
Be damned the thought! The darkest cloud
That ever crossed a path
Must yield to him who is prepared
To circumvent its wrath.

There is an art! There is a skill!
And there is knowledge too
Which all must have to meet the storm
That waits them in the blue;
Dream not you can outwit your fate!
Dream not you will not meet
Somewhere, some day, some night, the storm
From which there's no retreat.

Flying / June 1961

Not every tree can rear Sequoia high
in monumental splendor to the sky;

Not every sapling's ponderosa bred
with heritage of girth to lift its head;

Most must endure whatever place they find
bequeathed them by the caprice of the wind;

But when the God of trees—and men—shall note
in judgment what grew up from seed and root,

I ken that he shall grade us every one
not by how tall we stood against the sun,

But by the courage that we had to stand
and grace the place assigned us in the land.

Flying / February 1963

Paul Mantz

A Pilot's Pilot

BY ED MACK MILLER

The right balance of precision pilot and daredevil combined to make the unique and colorful career of a top movie stunt pilot.

STOCKY, GREYING Albert Paul Mantz is a pilot's pilot who once washed out of Army flying school—and went on from there to carve out a career in aviation that will never again be duplicated.

Mantz, who at 57 still has a youthful daredevil imprisoned behind his brown eyes, presides these days like a Graustarkian prince over his little Duchy of Aviation, located just a few citrus groves south of Los Angeles at Orange County airport, Santa Ana, California.

Visiting the headquarters of Paul Mantz Air Services is like taking a high dive into nostalgia for the person who is an "aficionado" of aviation, because it is at Santa Ana that Mantz keeps his priceless winged treasures—such exotic items as a Spad XIII, a Fokker D7, a Nieuport XXVIII, an SE5A, a deHavilland DH4, a Thomas Morse Scout, a JN4D "Jenny," a 1910 Pusher, a Pitcairn mailwing, a Lockheed Vega and a Lockheed Orion, two exact replicas of "The Spirit of St. Louis," and many more wonderful antique aircraft.

One of the few greats of the Old Days still active, Mantz is Mr. Aviation as far as the movie industry is concerned (the "King of Stunt Men," he prefers the term "precision flying"). If you've been going to movies for long, then you've seen Mantz fly, for he has been performing thrilling aerial sequences in movies since the advent of talkies.

But movie flying is only one facet of the fabulous Mantz's life. He's an ace film director (Cinerama), speed pilot (three Bendix speed trophies), personal friend and technical advisor to Amelia Earhart (whom he called "A.E."); retired Air Force Colonel, and Hollywood's favorite charter operator (you name the film personality, Mantz has probably flown him or her to get married—or divorced). A serious guy with a flare for flamboyant clothes and things dramatic, Mantz combines a Teutonic desire for precision with a devil-may-care temperament that has become legendary in a land where legends aren't easily bought.

For years Mantz lived on his magnificent yacht, the Pez Espada, which was equipped with radar, sonar, an automatic pilot, radio direction finders—everything but a swimming pool, and for a time he ran the swank Flying Yachtsman Restaurant on Catalina Island. In fact, for a guy with nine grandchildren, who has flown from Cape Horn to Hong Kong, the legend is anything but static; conversely, it's still growing.

Mantz, who still holds the world's record for outside loops in a stock airplane (set in 1930 when he was running a flying school in Palo Alto at Stanford University), has probably cracked up more types of airplanes than most other pilots have flown—and almost all of them intentionally. And yet his only injury has been a broken collarbone.

Born in Alameda, Calif., only three blocks away from where his then-and-now hero, Jimmy Doolittle, grew up, Mantz took his first flying lesson in 1919—and says he was "scared stiff" every time his instructor did acrobatics.

Since the day he soloed, however, Mantz has lived a life as colorful as any character in a Mark Twain novel (and—why not; Samuel "Mark Twain" Clemens was his grand uncle, and, by dint of paying taxes on the Clemens' estate over a period of years, Mantz is one of the author's major beneficiaries).

In the brassy, booming days of the first talkies, Mantz got his chance as a movie stunt pilot after "Wings" and "Hell's Angels" made America air conscious. Seeing that money, adventure and fun went hand-in-hand in this lusty infant industry, Mantz set out to become "aviation caterer" to Hollywood—and how grandly he succeeded is proved by the fact that he has flown in something better than 250 movies.

He started flying through open hangars in "Air Mail" until the stunt almost became his trademark. For "Blaze of Noon" he spun a plane down to ground level—plus about 10 feet, and for "When Willie Comes Marching Home" and "Suzy" he flew airplanes between two trees, losing the wings in the process. For "Twelve O'Clock High" he flew a B-17 solo and landed it gear up, the only man to ever take one off alone (although, as Mantz is careful to point out, a number of pilots brought their Fortresses home "solo," with their crews badly shot up).

He's been lucky, too, when things went wrong, as they did in "Flying Leatherneck," when 10 cases of dynamite were exploded prematurely, just as Mantz was making a low pass in a B-25. That one "singed his pinfeathers" more than somewhat.

Called by such film spokesmen as director William A. Wellman, "the best pilot in the business," Mantz figures that he has earned better than $10 million in aviation—and put most of it back in.

During all his years as "Mr. Crash" in the movies, Mantz has kept up his charter flying service—and for many years kept on teaching flying (among his students: Veronica Lake and the late Wallace Beery).

Since Hollywood discovered Nevada, his "Honeymoon Express" has been a filmland fixture. In some 30 years of charter operation, he or his pilots have flown everyone from Jean Harlow to Lana Turner to Vegas or Reno to get married.

In fact, the "most scared" he's ever been flying, Mantz says, was once in his Lockheed Vega on the return trip after flying Jean Harlow to Yuma to get married. He was busily cruising through the starry night when all at once he heard a "bang" and something hit him the back of the head. He thought he had been shot—until he found that someone in the wedding party had popped the cork on a bottle of champagne.

He flew Tom Mix's body back to Hollywood after Mix was killed in an auto crash in Arizona, and he transported the body of William Randolph Hearst from San Simeon to San Francisco. He's had more "mercy flights" than he can recall, but one he does remember is rushing a man dying with a brain tumor to Mayo Brothers' Clinic in Rochester, Minn.—and getting him there fast enough to save his life.

You name it, and Paul Mantz has done it. Flying at night once, he saw a house on fire, buzzed the house until the people were awakened and got out. He's gone up into pea-soup fog to lead lost planes to safety—and has had other people help him out of similar scrapes.

His title of "Mr. Upside Down" comes from an inverted stunt he used to do at airshows in his P-12 flying "wheels-to-wheels" formation with Frank Clarke.

In fact, any time he had to fly from here to there, he'd do part of it upside down—just to keep in practice.

Once, while making a picture called "The Bride Came C.O.D." with Bette Davis and James Cagney, he was returning to the movie location in Death Valley and decided to make an inverted pass at the cast assembled below (again in his P-12). It was raining slightly, and a cameraman on the ground told the cowhands who had been hired as "extras" that Mantz always flew on his back whenever it was raining—just to keep the rain out of his face.

This didn't hurt the Mantz legend any—and so "Mr. Upside Down" was born.

His only injury, the broken collarbone, came in the filming of "Central Airport," when he aimed his plane to collide with another on the ground—and leaped out. The only thing wrong was that his plane turned and chased him . . . and caught him!

Mantz gets perhaps his biggest kick out of telling about being washed out at March Field, where his instructor advised him: "Mister, stay out of airplanes or you'll bust your neck" (another of his instructors had been the then Lt. Nathan Twining, later Chief of Staff of the Air Force and head of the Joint Chiefs of Staff).

Military recognition of his prowess came finally, however, during World War II, when General Henry "Hap" Arnold directly commissioned Mantz a full colonel, gave him the Hal Roach Studios ("Fort Roach!") and a supporting cast of many famed actors, including the late Clark Gable, with the project of producing both expert aerial photographers and top-notch Air Corps training films. One of his many unusual wartime assignments was to fly a B-17 up the tricky fjords of Greenland to make a film so ferry pilots could see how to maneuver into Bluie West One and other nightmare fields.

After the war, Mantz became especially famous as owning "the seventh largest air force in the world" (ranking right ahead of China), when he bought nearly 500 surplus World War II airplanes for $55,000 (and got his money back immediately because no one realized the airplanes still held plenty of aviation gasoline—nearly a half-million gallons). For a while he was something for the United Nations to think about, owning, as he did, nearly 100 B-17s, more than 200 B-24 Liberators, more than a score of twin-engine bombers and 137 fighters.

He still has three B-25s and his famed P-51 Mustang. The latter has earned more than $125,000 in prizes with Mantz at the controls, including a $10,000 side bet from oil tycoon Glenn McCarthy on one of the Bendix races.

A special light comes into Mantz' eyes when he talks about "the Bendix," for he claimed this race as his own in the postwar years, chasing his Merlin-powered Mustang across the country to beat the rest of the racing clan three straight times—until it wasn't fun any more.

He likes to fly his specially-equipped B-25 photo plane best now. From his combination pilot-seat/director's chair in it, he has supervised the aerial shots for such flying films as "Strategic Air Command," "Spirit of St. Louis," and "The Crowded Sky," and the Cinerama spectaculars "Seven Wonders of the World," "America the Beautiful," and "Deluxe Tour."

He found two sister-ships of Lindbergh's Ryan monoplane for the film version of "The Spirit of St. Louis" and reconditioned them into perfect replicas—then flew them in the aerial shots.

Mantz was one of the original pilots get interested in artificial rain making. General Electric had been working on the project when he got involved and rigged up his B-25 to drop silver iodide pellets in cloud-seeding experiments.

He's flown nearly every commercial and some military jets and would go in a minute if someone would let him join up with the astronauts.

The ebullient Mantz reminisces with glee about his life-long romance with flying—but he speaks with reverence of another facet of his life, his time as friend, fellow pilot, and adviser to Amelia Earhart.

Before World War II, "A.E." had come to him for advice on equipping and modifying her Lockheed Vega for distance flights. He had shown her how to increase range and speed, the little short-cuts and bits of knowledge he had gained in his own flying experience.

They flew together in their first Bendix attempt, and when Miss Earhart later began planning her 'round-the-world flight in a Lockheed 12, it was Mantz and his technicians who modified her plane, and it was Mantz who checked her out on heavy-gross-weight takeoffs in the Lockheed, flying the first leg to Honolulu with her.

He was watching as she cracked up on takeoff from Honolulu, and talked the ship people into allowing him to put her plane on the tennis deck of the *Lurline* for the trip back to Mantz's shops in California for repairs.

Her next attempt was the last flight of her life, and one that still brings a mist to Mantz's eyes.

The world's "Happiest-go-craziest" stunt pilot has piled up more than 20,000 hours of flying ("In the wild type of aviating he's had to do," says one friend, "it has been 99 per cent terror and only one per cent boredom!") and hopes to fly till he's 100.

Perhaps the secret of his charmed life has been the fact that, behind all the wacky antics, there has been a cool mind that has carefully planned each stunt, each crash down to the intentionally weakened spar, the carefully sawed strut.

"Don't let Mantz' easy-going demeanor fool you," says one Hollywood technician. "This guy has a Univac mind. It's always computing, measuring, calculating—and the answers it puts out are shockingly accurate."

Another good reason for his durability in a profession noted for its early tombstones is the fact that he has always watched his health carefully. Nine at night is "sack time," come what may.

You can see him there thinking as he paces:

"Now, this sequence would open with a wide-screen shot of black sky, panning in to a close-up of the underside of the moon, red and pocked-marked, as we cut in the retro-rockets and ease in for a landing . . ."

Fantastic? Don't sell Albert Paul Mantz short.

"He is," as one Hollywood type says, "always thinking."

The Senator and the Reporter

*FLYING'S man on the spot: keen, witty, perceptive—
and all loused up—tackles Pilot, General, Senator,
Presidential Candidate and all-around cool good guy
Barry Goldwater in the interview of the year.*

BY ROBERT PETERSON

REVOLTING AS IT SEEMS, statistics make it clear that no more than two percent of the adult males in the United States make it big. Quibble all you want about what making it big means. *I know.* It's poising into the local aerodrome, accepting delivery on a new turbocharged Beech Baron and paying for it with a personal check. That's what it is! Sure there are other variations. Big Man on Madison Avenue, Oil, CBS News, Reporters. *Reporters.* Get them. Expense accounts. First-class all the way. Sophisticated. Women. Extra-dry Martinis. *You* know.

Me, I'm not going to make the two percent. I'm Joe Cauliflower, the average guy. Oh, it's not that bad. I work just above the middle level in my salt mine. The ceiling is only five feet, but down there where I used to work, it's only four feet. I can almost stand up straight now. If I keep my eye on the tablet, I'll move up to a higher level and stand up straight for my last five years.

This is no con. A few years back an Uncle named David died in Baffinland. Left me a modest stake. Successfully hiding the dough from my wife, I shoot the wad on lessons and get a private ticket. You read? All day in the salt mines. Suppressed desire to fly with the birds. So I got that going for me. I'm a pilot. Not too good. Low hours. But not bad.

Like every other Junior Birdman, I subscribe to FLYING. After reading it a year or so, I figure I'll tell them about my experiences. Licking the pencil stub on each word, I get off a manuscript. Bingo. Another. Bingo. Pay's not too good but who cares. I'm a writer. Not too good. Low hours. But not bad.

Then one day, I meet Editor Parke. Nice guy. No grass growing on a busy street, head-wise, but personality-wise, he's smooth. I'm going along, talking-wise, at about 180 knots and then into a flat spin. Sure, he says, you can interview celebrities who fly. We'll pay you and *pay your expenses.*

So now I'm a reporter. I tell my wife and everyone else I can corner, conversation-wise. I even start using " - wise" like the syndicated columnists. I'm ready, world. First-class all the way!

All I can tell you is thank God that celebrities are nice people. Senator Goldwater telegraphed:

DELIGHTED TO TALK WITH YOU ABOUT MY 40 YEARS FLYING EXPERIENCE. . . .

In due course, the Reporter, commanded to appear at the Senator's Washington door at 0700, can be found searching frantically in a vast apartment building designed to frustrate the uninitiated wandering its corridors. Three minutes late, he punches the doorbell. The door swings back and there, *My God,* 10 feet tall, orders and decorations in place, in Air Force blue, is the Senator, his appearance highlighted by two stars on each shoulder. The Reporter is struck dumb. What the hell does one say? Where is the journalistic aplomb of an Edward R. Murrow? Plunging mutely along after the General, we climb into a car for the trip to Andrews through the cool dawn of a lovely D. C. day.

Until the car is parked outside the chain link fence at Andrews AFB, The General is The Senator, wise and knowledgeable, putting the Reporter at ease. Inside the fence, approaching the hard stand, with Air Force One and Two to the left, a covey of T-39s ahead, the Senator changes to the General and is dealt with by officers and enlisted men as the latter have dealt with generals since Genghis Khan. Stirring. The General carries his role lightly and effectively.

The Reporter, armed with camera and tape recorder begins the job of being a reporter. What do you do with the camera bag and tape rcorder when you take pictures? Especially when the General said wheels up at 0730 and it's now 0728.

The flight is first-class all the way. The Reporter, microphone at the ready, enthralled by flight characteristics, navcom equipment profligately duplicated, asks innumerable questions. Everything ends. Driving back to Washington, the microphone is poised, the tape spun through the head and the Senator tells one good story after the other.

In the salt mines, we are taught to write thank-you letters:

Dear Senator:

Thank you very much for the great T-39 flight last Friday . . . The tape will be transcribed this week . . .

Am I a reporter? Yes, sir. Sophisticated, suave, expense accounts, casual, debonair. The whole bit is a lead-pipe cinch.

Dear Senator:

It is painful to have to admit that the taped interview (in your car) turned out nothing but virtually unintelligible sounds.

The airborne sequence was totally obscured by skin road . . . the car ambient noise obscured the tape.

Class always shows!

Dear Bob,

. . . I am sure we could get together . . . it will be September before I get East.

I am sorry the tape didn't turn out. I always have great skepticism about these contraptions!

How do you express your appreciation if you are a suave, sophisticated, debonair, casual, dough-footed, stupid amateur of a reporter. Make a novena!

Dear Bob,

I plan to be in Washington . . . We could work an interview one of those days but I cannot take a flight in your Bonanza as my doctor will not allow me to fly private planes until I have made a further recovery.

Still a bit shaky from his back operation, the Senator talked flying with FLYING in his Washington apartment. Seated on a long comfortable sofa, behind a table lined with separate piles each containing stacks of letters, clippings and reports, he relaxed in shoes-off, sport-shirt comfort and talked about his many years as a flyer covering nearly every phase from a Great Lakes trainer to the latest jets. Sen. Goldwater is easy to talk to, fascinating to listen to, direct, and knowledgeable about aircraft. He is also subject to those difficulties suffered by all of us. Visualize him as you are reading, midway through the interview, searching for the newspaper clipping. It was right here; he knows it was right here. Who the hell has been moving his stuff around? Here it is. No. *Here* it is.

Shall we go back to the very beginning, Senator—first flight, how you got interested in flying and that sort of thing?

Sure. Well, I got interested in flying for two reasons. I wanted to fly and I wanted to do some work on aircraft radio. I was an amateur radio operator and my first flight in the instructor-type aircraft was in 1929.

Was that your first flight?

No, I had been up in 1927.

Where was this?

It was in a Fleet over in California.

Were you interested then?

Yes, I was very interested then. But I had just graduated from prep school and I was getting ready to go to college. So, I didn't want to embark on a flying career and have to worry about college. So, I waited until '29 when I quit school and started to fly. And my first flight was in a Great Lakes Trainer and my instructor was Jack Thornberg who is retired now and living in California.

Where was this first flight?

Phoenix. The present site of Phoenix Sky Harbor, but part of my training was done in another air strip when they closed the airport there and made a lettuce field out of it. The fellow that soloed me is a fellow named Irving Dravitz, he's now a senior captain with TWA.

Did you buy this airplane?

No, no, I just—I never owned an airplane until 1940 or '41. My brother and I bought a Waco.

Well now, did you have any particular difficulties in early training?

No. Of course, it was a lot different then.

Oh?

I am glad that I got my license then, because I don't know if I could get one today. All we had to have was 10 hours. And take a written, real tough questions like, "Which wing is the red light on and which wing is the green light on?"

Were there—at this point, were there any tower facilities or anything like that?

No. No, you had nothing.

It was just an open—

Open cockpit. We had no airspeed indicator, no brakes—you just flew it. And you listened to the wires. And if the wires were right, you were at the right speed.

Did you have any particular difficulties with takeoff or landing or turns, or anything that—

No, ah—

It just all went pretty well.

Everything went well. Both of these men were excellent instructors. I think that's more than half the battle. A good instructor can inspire confidence and erase any doubts.

Did you ever teach anybody to fly?

No, I helped; in that I instructed in gunnery, fighter gunnery, during the war, both ground school and air. But that was all. You'd fly in the back seat and tell the boy what he was doing wrong. I've checked pilots out in various types of aircraft, both in the Air Force and in private flying. But as far as starting a man from zero to his solo, no. I've spent an awful lot of time with Air Force Cadets prior to and during World War II. I flew with an awful lot of them on check rides and I taught instruments, I did teach instrument flying for awhile, but not the act of just getting somebody in an airplane and say go. Frankly, I would hesitate now, after nearly 36 years of flying. You pick up a lot of things that you can do and get away with. And I think the instructor always has to be cognizant of the fact that this young guy has no shortcuts. I'd never teach one of my children to fly. My oldest boy is a pilot and he's a hell of a good pilot, but I never gave him a lesson, except instrument after he knew how to fly.

Well, now after you—did you get a license then—after those 10 hours—were you—

I got a license in 1930, I guess.

Do you still have it?

Oh, yes.

Is it a low number?

No, surprisingly, and I never looked this up. It's 18352. Now, did you—what happened after this; did you continue to fly on your own?

Yes. I flew. I'd find time and go. And there was a period there for a few years that I didn't fly because I broke my left knee. I didn't bust it up more than any one guy should, but this left knee was broken in a semiprofessional basketball game and I couldn't get in the cockpit for a long time. I was on crutches and canes for nearly six months. I just couldn't get in. And so, for a few years, I just didn't fly myself, but I did fly with friends, kept my hand at it. And when the war came along, or before the war, I volunteered for duty. I was a reserve first lieutenant in the Infantry.

Now, the last time we spoke, you told me some interesting stories about taking pictures of cadets, and so on?

Oh, yes. When I went on duty at Luke Field in August of 1941, I thought, well, a year would do me some good. I'd get away from the routine I'd been into and possibly the business would be better off if I weren't around it so much. So, I went on duty, I thought for a year. I stayed for five.

Naturally, I wanted to fly and I put in an application for cadet training, but I was then 33 and age and eyes got me. So, being the only photographer on the base and these kids all wanting pictures to send home to their girls, I'd go up in the back seat of an AT-6 and these fellows would just line up; they'd slide right up on the wing and I'd take the pictures, develop and print them and—but I had a deal with them that they had to give me a little time. So, all through the balance of '41, I was piling up AT-6 time and, unbeknownst to any of the brass, I soloed one; in fact, the fellow that started all this is a close friend of mine. Now he lives in Phoenix. A retired colonel—Brignal.

So, in '42 they came out with what they called the service pilot rating and this was a chance for men like myself to get, at least, get flying with the Air Corps. Our flying, however, was restricted to instructing, towing targets and ferrying. So, I guess I had one of the first ones issued. And I went down to Yuma and I ferried planes and I checked gunnery and did a lot of that and then I went into the Ferrying Command and later into the Air Transport Command. I wrote at least, I think, three different letters asking assignment to—first to fighters and then to bombers, and I never could get to first base. And I wound up the war having flown the ocean quite a bit and then after, after I had the service pilot rating about a year, a year and a half, they made it possible to take a test and become what they call a rated Air Corps pilot, which I took, and I've gone through the whole ball of wax. Now I am a command pilot. That's the end of the road.

What was that I recall about the gunnery business in which you developed a new technique and—

Well, it was—let's start at the beginning. We had a fellow assigned to the ground school by the name of Clark who was a genius. We were always talking gunnery and why it was that the best we could do was to qualify five or six percent of the graduating class. Well, he discovered that,

if you aim correctly at the target, it would be impossible to hit it because the rate of the fire of the machine gun was such that the target would go between shots.

Enough spread between—

Yes, that's right. So, we had been reading about what the British call the "Curve of Pursuit." We couldn't get any information on it. We tried all kinds of things. Finally, he worked it out. It took two full pages of yellow foolscap to get the formula out; the angle that you'd have to be to the target at an approach speed of 120 miles an hour to hit it every time. Well, then we got a good break because Squadron Leader Donaldson, who is now retired from the Royal Air Force, came over. He'd been shot down five times, but he knew this Curve of Pursuit perfectly by then. So, we developed teaching techniques for it; developed— my responsibility was all in the ground school field and I made all my mock-ups, made all of my instructing tools and developed lectures and slides on it and then the boys would go out and fly. And our first class, we qualified 94 percent. Well, the Air Corps was really up in the air because at first they thought we were lying. They sent a team down to Yuma to see what we were doing and when we told them that we were using the Curve of Pursuit, they nearly hit the ceiling, because nobody had authorized it. And this colonel said, "Wait, you're crazy; you can't do that." I said, "Colonel, if you come to my lectures for just one day, I'll ride in your back seat and I'll guarantee you I can qualify you at least as a marksman." He qualified expert!

I thought we were all going to be court-martialed. But then this became the standard method of instruction.

Well, now, after the War, did you still maintain an interest in flying?

Oh, yes, yes, when the War was over, I got out in November of '45 and I went back to Phoenix and my brother and I had this cabin Waco and then the Air National Guard got active and I was asked by the Governor to organize it there, but the highest rank they had was captain and I was a lieutenant colonel, so I took a bust back to a captain and organized the squadron. We had P-51s and two A-26s for target towing and a couple of little L-1s and a C-47. Then we went into F-84s and about that time they called the squadron up for Korea and they didn't call me or two others who were senior in grade. And about that time, I was back to—I was a full colonel. And I came back here to Washington and said, "Damn it, I'm going with my outfit," and they said, "All right, you'll get orders." After three months, they wrote and said too much rank and too much age, you know. About that time, I got into national politics and we've always had an understanding out there that, in the Guard, you wouldn't be in politics. So, I went back to the reserve and I was attached to the 4th Air Force for awhile and then Air Defense Command Center at Colorado Springs and then to Deputy Chief of Staff of Personnel.

And flying during that time, of course, in the Air Force Reserve, you have to get 100 hours a year and 20 of it has to be instrument, 10 of it has to be night. You have to have 50 in each six months. Part of it can be copilot and, naturally, you'd like as much first-pilot time as you can get.

Sure.

So, at the present time, I am current in the T-39. That's the one we were up in. That's the only Air Force plane that I'm allowed to fly.

I presume that you've flown the jet fighters?

Oh, yes—

In fact, is this really the epitome of flying?

Well, of course, to me flying a fighter plane is the epitome of flying. It's not just because of the maneuverability, but you're all alone.

Right.

That's you up there.

Right.

You've got to do your own thinking. I remember the first jet that I flew was the F-80. That was back about 1948 or '49 along in there. And I thought that was the end of everything, but I always thought the F-80 was one of the most beautiful ships ever built and I loved flying it. Well, then, we got the F-84s and the T-33s and I flew a lot of— have a lot of T-33 time. And the 84-A was our first airplane—'course that was a real sled; that thing—well, theoretically, it wouldn't fly if the temperature was over 110 on the runway. I've flown—either soloed or flown in—every fighter plane we have in the Air Force, I think. I've flown the F-80, the 84, the 86. I've flown the 94, the F-100, the 101, the 102, the 104, the 106, the 107, and the 105. And, on top of that, I've flown in the bomber-types, the B-47, B-36, the B-52, the B-58 and the B-57. So, I think I've flown everything we've had. I've never taken the 135 off and landed it but I have the 52.

The interview was over. Not that this was the end for Robert Peterson, Cub Reporter. Another letter to the Honorable Barry Goldwater—

. . . If you are to be in New York, I wonder if you could have luncheon. . . . and may I have your permission to include Bob Parke, editor of FLYING?

Gracious to the core, the Senator wrote:

. . . I can change my plans . . . and could have lunch with you . . .

A suave, debonair, sophisticated reporter would find no occasion to bollix this one up. What's tough about it? Call the Senator's secretary, Judy, in Arizona, arrange the date, the place and the time. Ask that Judy confirm it and ask her to ask the Senator to call when he gets in town. Simple. Who could guess that the Senator would arrive in town the night the lights went out all over the Northeast? Callous to the pinpricks of fate, Bob Peterson, Cub Reporter, decided at 1145 hours that the Senator had suffered similarly and could not make this date.

At 1405 hours, telecon:

"Hello."

"Bob, this is Barry."

I don't know any Barrys, so . . .

"Barry who?"

"Barry Goldwater."

"Oh *my God.* Senator. Where are you?"

"I'm at my hotel. I sat on my -ahem- at that restaurant for over an hour. Where were you?"

"Senator—*My God*—I waited all morning to hear from you—"

"Well, it doesn't matter. About one o'clock a man came in from *Newsweek* and I had lunch with him."

"Senator, in my letter, I asked you—"

"No, I called Judy, and she said it was set for 12 o'clock."

"Jesus, Senator—"

"Anyway, what do you want to do?"

Not knowing whether this meant go to dinner or go flying and, being more interested in flying than eating with a highly irascible Senator, the Intrepid Reporter suggested the latter. He agreed for the next day.

(The next day The Senator said he had checked with Judy, that he was at fault and the Intrepid Reporter took a deep breath and let it out slowly.)

So, the scene shifts to Linden Airport, where out on the ramp, polished and swept and dusted to a brilliant shine sat Bonanza N6122V, awaiting its distinguished pilot. Inside the operations room everything was unusually clean and neat and filling with people all of whom had somehow in some mysterious fashion gotten in on the expected arrival of the Senator. A councilman was on hand, the mayor was on his way. Bob Meyer, FBO, was hopeful that the Senator would be impressed by the field and *say so,* for Meyer is in the bind so many operators encounter in the big metropolitan areas. Linden the city, faced with rising costs, casts hungry eyes on Linden the field, whose many acres, sold as plant sites, would earn a bountiful harvest of taxes.

The Senator arrived in a long, black Cadillac that a chartered Reporter had ordered as transportation. FLYING was appalled at the size to which the audience had grown. The Senator may have been equally appalled, but if so, did not reveal it. He met and spoke to the important personages, allowed himself to be the center of photos, but in a manner practiced by practical politicians, kept moving slowly but steadily in the direction of the door to the ramp. Once through it, he headed with long strides toward 22V.

Sen. Goldwater looked the Bonanza over and climbed into the left front seat where he began a familiarization course. Into the back went the FLYING team and we were all set, awaiting the arrival of Jack Olcott who was the check-out pilot for Linden Flight Service and would go along for the ride.

At this moment, on to the field come the reporters and photographers. As flashbulbs went off continuously, the reporters interrogated the Senator.

"Where are you going, Senator?"

"New Haven."

"What is the purpose of this trip?"

"A private flight."

"What do you think of the vote in New Jersey?"

No answer. Nor were there to be any more answers, for Jack arrived, climbed in and slammed the door. The S-Model Bonanza is well insulated. The sounds of tumult and shouting faded.

"It's been five years since I flew one of these babies. But if it's got wings, I guess we can fly it. I had an A-Model. Is there a checklist aboard?"

"Right there, Senator."

"Look, you hold it, please, and we'll run down the list. Is the gas okay?"

"Yes, sir."

The Senator and Jack completed the checklist and like everyone else in this whole wide world, Goldwater had his troubles with the fuel injection engine.

"I had a letter from Gen. LeMay the other day. He said he had trouble starting his new Bonanza. He did solve the fuel injection problem on his Corvette but he hasn't licked the Bonanza yet."

The engine finally hesitated, caught and got down to business. The General called Linden unicom, announcing his intention to taxi: He was *immediately* answered, (not only because he was who he was, but because Linden mans its unicom continuously) with wind direction, runway and the best direction to taxi. Moving out, slowly outpacing the furiously active photographers, he proceeded to the intersection. Again the checklist was collectively followed, interrupted by questions and answers.

"Does the throttle still screw in?"

"Yes, sir."

"The prop?"

Nod.

"Where is the elevator trim?"

"There. That wheel to the left of the throttle."

"What's the best climb-out speed?"

"120."

"Is approach 90?"

Nod.

Another call to the unicom to report that he was about to taxi on the runway. Unicom reported no traffic known, wind unchanged.

Taxiing, a final look around at belts, door, windows, another check on controls. All the way to the grass at the end of the paved runway before turning into position.

"Here we go."

Advancing throttle slowly, the Senator may have been surprised by the S-Model torque but it never showed in 22V's ground path. Right down the middle line, back pressure, and a clean liftoff. Wheels up. The speed climbed to 120 and stayed right there. The Bonanza was bent into the procedure turn; coordinated, the ball stayed where it belonged. The RPMs came down with a pronounced jerk, and were picked up again with the vernier. Throttle back to 25 inches matching the RPMs.

On climb we proceeded south of Newark Airport whose instrument runway has a heading nearly the same as the one we had just left. The Senator sighted a United jet a half mile at three-o'clock level, moving like a huge shark through the air. We veered right, went below him and continued the climb to 2,100 feet, our cruise altitude.

When New Haven Airport appeared right where it should be, the throttle was retarded and we entered the pattern and landed. Touchdown was made to the first faint notes of the stall-warning horn. Very nice.

Sen. Goldwater cleaned up the cockpit and in the process of shutting down left the airplane ready for the next pilot. Even the trims were returned to normal, something many pilots forget and subsequent pilots frequently discover to their discomfort.

As usual after a flight, the group stood around and shot the breeze a bit. Sen. Goldwater has the capacity for getting people to talk, and all of us found ourselves talking about our own experiences. Professing himself satisfied with the flight, the Senator patted 22V and complimented it as a "real fine airplane."

As he walked away across the ramp we climbed back into the airplane. Preparing to taxi out, we saw across the field the TV camera, microphones, lights, photographers and reporters boring in. The General waved back to us, probably wishing he were flying back rather than going through the inquisition that seems to await him wherever he goes. He may have his problems on the ground, but in the air, he's a damn fine pilot.

Far above the Arctic Circle, where the air turns as sharp as glass and the ice is ageless, man is a tender intruder.

The World of Weldy Phipps

BY STEPHAN WILKINSON

AH YES, THE BUSH pilot; we all know him—or think we do: lots of guts and less sense, little respect for the rules and none for the regulations, underpowered and overloaded, here a quick buck and there barely enough to keep his decrepit Norseman in the air. We all know he's on his way out, too, now that the final frontier has fallen and the Learjets are commuting to Alaska and the Canadian sky is smoky with Hercs plodding north to the burgeoning Arctic oil fields.

But wait. Meet Weldy Phipps. Other men have claimed to be "the last of the bush pilots"; if they're right, we'll have to call Weldy the first of the Arctic operators—and if he's not the first, everybody knows he's the best.

Weldy Phipps has a beaten, weathered face—almost the face of an ex-prizefighter, puffy with scar tissue. If you saw him on a city street, a thought as incongruous as an eagle in a chicken-coop, you might say, "There's a man who takes a drop of the sauce now and then." Best you say it to yourself, though, and quietly. Weldy's features have been battered not by fists or whiskey but by a lifetime of flying in the deepest Arctic, by bitter cold and razorous gale, by months-long nights when the sky was ultimate blackness, by endless hours of spark-white sun glaring off ice and limitless, empty sky.

Welland W. Phipps runs an outfit called Atlas Aviation Limited, based at bleak, bitter Resolute Bay, on the northern-

Left: When it's 40 below, hoses become brittle and snap, skin sticks to metal so cold it feels red-hot, and every breath condenses in icicles on your beard. Still, somebody has to get up on the wing and top the tanks. Center: Markoosie, who lives in the Eskimo village at Resolute Bay, is the only Eskimo pilot in the eastern Arctic. He was trained by Weldy Phipps and flies as a copilot on one of the Atlas Twin Otters. Right: Weldy Phipps' tiger-in-the-tank shoulder patch is a memento of the epic voyage of the oil-carrier Manhattan *through the Northwest Passage in 1970. Atlas aircraft helped supply the ship as it forged through the ice.*

most fringes of the Northwest Territory, in Canada. He is certainly the world's northernmost FBO, in fact, and probably Canada's best-known bush pilot. ("Ours *has* been called a bush operation," says his wife, Fran, "but up here, we call it a rock operation." When spring comes—briefly—to Resolute, it's known as "when the rocks come out.")

Resolute is a 6,000-foot-long gravel strip, an assortment of low buildings erected by the RCAF when they built the runway after World War II, a bristle of antennae, a clutter of small tank farms and a tiny Eskimo settlement that should shelve forever the fiction that Eskimos live in igloos; in Resolute, and just about everywhere else, they live in scruffy tin shacks that make the place look very like a South Carolina trailer park.

Resolute, a bare 100 miles from the magnetic north pole and so far north that there's nothing between it and the true Pole but a few more Eskimos, some half-happy scientists and a number of lonely oil-exploration crews, is also home base for three Atlas de Havilland Twin Otters, a single-engine Otter, a Beaver with tires the size of a DC-3's, an Aztec owned by Weldy's chief pilot, a woebegone Atlas Apache with more snow in the cabin than most Stateside airports ever have on their runways, and a Super Cub. It's also a way-station for the Lockheed Hercules hauling fuel-oil for the helicopters, Caterpillars and drill rigs of the oilmen who are almost undoubtedly on the brink of a bonanza that will make the oil sheikdoms of Arabia and the black gold of Texas seem small-time.

When Weldy first came north, in the late 1950s, he did so not to cash in on the oil boom, which hadn't yet started, but because he loved the country—"and because I was curious," he says. Phipps had been a flight engineer in the RCAF during the war, but didn't learn to fly until he was a civilian again. He wasted no time, though, and was soon making a specialty of buying surplus equipment and converting it for high-altitude photo-mapping, a job that showed him much of the Arctic from 35,000 feet, often under a P-38 or a de Havilland Mosquito.

After a decade of work with various aerial-survey and Arctic-specialist firms, plus a stint flying for a small feeder airline, Weldy struck out on his own. He took a four-wheel Cub north—more on a lark than anything else—with a cinematographer who was making a movie about the Arctic. There, he met a geologist who was laboriously roaming Ellesmere Island with dogsled and—when the ice went out—canoe, and Weldy soon had a job flying the surveyor around in his Cub.

The next year, he borrowed $30,000 and bought a used Beaver, and sold his house for operating capital; Atlas Aviation Limited was in business. ("Most wives would have sued for divorce and kept the house," one of Weldy's pilots said. "Not Fran. Within a month of the time he first mentioned the idea, the Phipps house in Ottawa was sold and Weldy had his Beaver.")

Fran Phipps, luckily for Weldy and Atlas and the Arctic, is a Hollywood director's dream of a bush-pilot's wife: Strong and trim, with the sort of make-up-less good looks that become beautiful under the blush of a northern chill, Fran seems to have stepped straight out of a June Allyson movie about the Klondike. Weldy was her flight instructor when they met, eight children ago.

In 1961, Weldy won the McKee Trophy, a cup given annually "for meritorious service in the advancement of aviation in Canada." Sort of a Canadian Collier Trophy, it commemorates the first trans-Canadian seaplane flight (perhaps for lack of anything better to commemorate), and the cup sits off to one side of the living room of the pilots' and mechanics' quarters, a wing of the Phipps home. For a long time, when it was the custom for Weldy to buy a case of beer for the crew every couple of days, each beer-drinker would drop some change into the cup toward the next case. One of Weldy's young daughters told me that the kids were onto the arrangement, too, and used to cop a quarter out of the cup whenever they were thirsty for a Coke.

Phipps got the award for doing the development, testing and certification work to produce the immense tires that are now a standard part of Arctic and Alaskan flying. The Phipps wheels and tires make it possible to land anything from a Super Cub to a Caribou on just about any snow or ground you can walk on without sinking into. Surprisingly enough, they have no appreciable effect on airspeed; the Beaver is faster, in fact, with Phipps tires than it is with the de Havilland-built wheel/skis. "The Beaver's so bloody slow anyway," Weldy says. "Lord, they haven't *made* the airplane slower than the Beaver. Put wheel/skis on it and the thing'll barely fly."

The tires are specially molded by Goodyear (and marked "Phipps Special"); they're not simply adaptations of big-airplane tires, for they'd be far too heavy. The big problem is trying to get Goodyear to punch out an occasional run of 50 or so tires—a big order for Weldy but utterly insignificant in Akron. Most of Weldy's present stock has been to Vietnam and back, for they were ordered by the Army for use on de Havilland Caribous, which utilize a set of dual Twin Otter wheels on each side. Somehow, they were never used, and today, they rest in the rafters of a hangar near the North Pole.

Phipps protects his rights fiercely. "We had the molds for the wheels made in one place, cast them in another and finished them in a third. I'm not about to equip my competitors without getting a price for it."

The deep Arctic is difficult to describe, for it both exceeds and falls short of the popular images. It sometimes gets colder on the farms of Canada's western prairies, and Resolute Bay, far north of northernmost Alaska, has at times been as warm as a Maine summer resort. (They still talk of last summer's beach party, when it went up to 55° F. and everybody went down to the stony shoreline by the Eskimo village and had a cookout.) Polar bears are hard to find and there's not a dogsled in sight; the Eskimos all drive snowmobiles. But when you do hear the story about the bear last seen lumbering across the snow, its arms wrapped around the approach-light pylon that it had torn out of the ground, or when you see from the air a lonely herd of musk oxen—peabrained, primitive heaps of hair and gristle that live out most of their lives in cold that fractures metal—you realize that the Arctic is something else. And when it does get cold in Resolute, it's a searing, wind-driven bitterness that penetrates every layer of clothing and touches your face like a hot iron. Forty below is no rarity, –65° F. is the present record, and a spring day when it gets up to –10° F. is considered absolutely balmy.

Thankfully, though, cold is all relative. If you're a mechanic working barehanded on an engine in –20° and 20 knots of wind, it's like puttering about in a suburban-basement workshop once you get the jug off and into a semi-sheltered shop warmed by body heat and a blower to perhaps 30°. And for the copilot loading and preflighting in the early morning (the pilot is in the office, staying warm), when the wind is up and the temperature down, it's luxury to finally climb into the cockpit, which is probably still so cold you dare not touch metal with bare skin.

The Arctic's real extremes, though, come in its flight conditions, not its temperatures. During the winter, the sun never appears from the beginning of November to the end of January, and for almost three months, the pilots literally fly blind. "Boy, it gets boring, I'll tell you," Weldy admits. "You get so that you're really happy to see that sun come back." The exact hour and minute of the sun's return is, of course, the occasion for a pool, awarded to the closest guess as to the exact moment when the ice begins to move out of the bay; they tie a line from a stake in the ice to a clock on the shore, and the first shift stops the clock.

When the ice does melt and the rocky hills are bared, though, the fog starts—fog so thick you can't see the wing tips from the cockpit. More than one Atlas pilot has had to land his ship on the beach and walk home when he couldn't get into Resolute. Minimums at Resolute are supposedly 600 and one, but if they were honored, little work would get done during the summer. "Ven the ice melt, the temperature and dewpoint usually stay about a degree apart," Jan Backe, Weldy's lean-faced Norwegian Twin Otter pilot explains. "Ven you see them maybe t'ree degrees apart, you run out to the hangar and start flying as many trips as you can."

Even so, the fog can come in with startling suddenness. The Resolute runway is rare for the Arctic in that it has high-intensity strobes on the approach. Weldy considers the fog not too bad if he can let down to 100 feet over them and pick out the strobes one at a time as he comes up the approach. Jan's

approach technique is to make his procedure turn on the southernmost of the two Resolute beacons and let down until he sees the water in the bay.

Smog is yet to afflict the Arctic, but the polar regions serve up every other kind of fog known to temperate man—and a few that aren't: radiation fog, advection fog, up-slope fog, frost smoke, ice-crystal haze, ice-crystal fog and blowing snow, to name the most common. Ice crystals are often present in the Arctic air—invisible motes that glint like cut glass if you get the sun at the right angle; if you do so from an airplane, they can cut horizontal visibility to five miles or less. More spectacular yet, they will adhere to any airborne particle receptive enough to serve as a nucleus for an even larger ice crystal. Make a low pass at a runway, or run a snowplow down it when the conditions are right, and your exhaust pipe will paint a curtain of ice-crystal fog thick enough to shut the airport down.

In the spring and the fall—otherwise the best times to fly—icing's a hazard, for it gets warm enough then even for freezing rain. The ice usually shows up in low-level stratus (it's rare that an Arctic pilot isn't flying in the clear at 5,000 feet), but occasionally it'll be solid right on up to 10,000 feet. The Twin Otters can punch on up through the layers, but the singles aren't so tough. "You take a loaded Otter through a thousand-foot icing layer," says Weldy, "and you may make it through, but you'll be so heavy that you'll sink right back into it."

When there's a crosswind at Resolute, it's a beauty—70 degrees to the runway, like as not, with gusts as high as 85 mph and a steady state of 40 mph considered quite ordinary. Phipps tells of landing his Twin Beech—a truculent airplane even in a dead calm—in a 70-degree, 70-mph crosswind. He says, with a laugh, "That was the only time they had all the fire engines and crash trucks out at the same time at Resolute. Fran was sitting in the right seat, and I had to ask her to lean way back so that I could keep the runway in sight out her window on final." You may not believe it, but Weldy Phipps doesn't exaggerate. He doesn't have to.

A typical Atlas mission, when they aren't flying their scheduled passenger and cargo mini-runs, is like the flight I found myself on with Weldy and copilot John Warner, a short, Scots-accented ex-Shackleton pilot who you might mistake for a cheerful high-schooler if you didn't see him in the cockpit. We headed north in a Twin Otter loaded with three mechanics, a warm extra battery, a big Nelson heater (everyone calls them "Herman Nelsons," and I at first expected an extra passenger) and a stubble-bearded Bradley Air Service pilot who'd had to ski-land his single-engine Otter out in the snow, low on fuel, when he couldn't get into the airport at Isachsen because of fog. He'd put the plane down north of a camp occupied by a hardy band of boffins probing the Polar Shelf, and Bradley had sent their Twin Beech in to pick him up. The Beech had no skis—just ordinary wheels—and Weldy felt that Bradley was lucky it didn't end up with two airplanes down on the ice.

We came in low over the camp—a straggling, struggling collection of tents—and turned due north. The blushing-pink wingtips of the Otter appeared in moments, probably the only sign of human life between here and the Pole: We were north of 80 degrees latitude, a good 250 miles northwest of Resolute, actually over the Polar Ice Cap.

Weldy put the Twin Otter into a steep bank and looked over the tortured snow, slashed everywhere by windswept drifts. (You'd be stunned how rough the virgin snow can be, and the uninitiated won't be aware of it until about halfway through the flare.) He turned upwind, set up a full-flap, power-on descent and the minute the skis touched—perhaps even before—Weldy cranked the throttles back to reverse and we stopped right *now*.

The forlorn single was banked with blowing snow, its cockpit windows frosted white. Within minutes, propelled more by the bitter wind looping around the top of the world than by any more admirable ambition, we had the heavy Herman Nelson out and snorting its dragon-breath up under the Otter's canvas-lashed engine and into the hellhole under the cockpit, to thaw the oil tank.

The temperature was about −25° F., but the slashing wind must have driven the chill down to almost −50. The stagnant twin soon lost every calorie of its warmth—it was going to take us the better part of an hour to get the single startable—but luckily the smaller Otter was slowly coming to life. While the unfortunate mechanics hand-pumped a 55-gallon drum of molasses-like fuel into her flank, we sat in the scarred cabin and brewed a pot of tea. Even that had its moment, though: The gasoline-fired camp-stove went up like a Molotov cocktail the moment it was lit, and I remember a crazy moment thinking that at last we were going to get some heat, as the wooden flooring began to crackle merrily.

Weldy had artfully eased himself over to the after door while the Bradley pilot was vigorously pumping up the little menace; Phipps admitted that he'd rather expected the blaze. (Of such foresight is the good Arctic pilot made; I later noticed Weldy quietly turning the 10-gallon drum of stove fuel sitting on a rack in the rear of the cabin so that the spigot was at the top of the drum, rather than the bottom.)

The Otter finally lit off, backfiring into its augmenter tubes in protest, and we left the pilot with nothing more to worry about than the fact that his oil tank was almost empty and he had only two extra quarts in the airplane; when you add oil to an R-1340, you talk about *gallons*, not quarts. The precautionary landing had already cost Bradley about $600 for Weldy's services, plus a five-hour round trip in their Beech, so the pilot wasn't about to send us back for more oil. . . .

Polar expeditions are another peculiar burden that Phipps must bear, and he considers any year during which nobody tries to ski, walk, flap or jog to the Pole a good one, for he's the gent who has to find them when they go astray or keep them supplied when they don't. One year, some nut even decided he had to parachute onto the Pole, and he hired Weldy to fly him up. ("They weren't anywhere bloody *near* the Pole when he jumped," one of the Atlas mechanics opined to me.) Upon his return to civilization, the chutist embroidered his tale by telling how they were so low on fuel during the return trip that Weldy had to add a quart of cognac to the tanks. Since Weldy's number-one Twin Otter has his initials as registration letters, one story about the expedition was headed, "Whiskey-Whiskey Papa, the Alcoholic Twin Otter"—a legend that was quickly stenciled onto the otherwise-sober de Havilland's cockpit door.

Nonetheless, Weldy has done his work so well that one man he flew north to tackle the Pole with snowmobiles wants Weldy to repeat the stunt at the South Pole this winter. Phipps just may do it, though he mutters vague complaints about not being able to take time away from his business. If he does, he'll fly a Twin Otter down rather than letting it be disassembled and air-shipped, as the expedition leader suggested. "And if I do that," he says with only half a smile, "I might just carry on around the world the rest of the way. Might beat Max Conrad if I did that."

Navigation in the northern latitudes is surprisingly simple, though undeniably crucial. While the magnetic compass is meaningless (variation is 80 degrees west at Resolute, and no indications are to be trusted within 1,000 mi. of the magnetic pole), and though landmarks are hidden during the winter and so numerous and uniform as to be almost useless during the summer—an infinitude of lakes, a limitless dullness of tundra—the primary rules are to trust your dead reckoning, hold the best heading you've ever held, and not believe your ADF until you are sure you're close enough to the station for a firm lock-on.

What about precession of the gyrocompass? All Arctic planes carry a simple device called an astral compass, which looks like a tiny surveyor's tansit, mounted on a homemade tripod bolted to the instrument panel; set the local hour angle and approximate latitude into it with two little knobs, aim the instrument's riflesights at the sun, read off your true heading and reset the DG. During the dark winter months, they take star sights with mariner's sextants.

"With the beacons, and the sun and the moon—and we even have a VOR up here now—I don't see how anybody could get lost," said one of Weldy's young copilots. (Perhaps that's why he's a copilot, of course.) The simple, reliable, largely forgotten technique of ADF navigation is a finely developed art among

Arctic pilots; all of the Atlas ships carry two, sometimes three, ADFs. Weldy's favorite is the ARC Mark 21—an old-fashioned coffee-grinder, "but about as good an ADF as you can get," he maintains.

Fly above the glistening icecap for awhile and you'll begin to notice (it's hard not to) the burnished skeletons of downed aircraft—modern-day mastodons trapped forever by the ice. There's a Lancaster, and you wonder why it never made it to the airport just over the ridge. A B-24, polished by the wind to a luster as bright as the day it left Willow Run. A Hughes 300 helicopter, one pontoon in the air like a beetle on its back. A Fairchild F-27 on the plateau south of Resolute, so clean that it looks like you could get her home with a ramp tug; but her back is broken. (The pilot made his procedure turn off the beacon north of the runway, rather than the south beacon, and that brought him down over rocks and snow rather than open water.)

There are 42 aircraft down just on the island of which Resolute is a part, and whichever way you go, you won't go far without being reminded that the Arctic is rough country.

"You like it up here, do you now?" a Dornier DO-28 pilot from another firm said to me over a cup of coffee at an oil-exploration base near the fringes of the icecap several days later. "You maybe would have loved it if you'd been with us last night. Missed our ETA by almost three hours, and when we finally found out where we were, we were 40 degrees off course—and we'd already allowed for 20 degrees of drift." Maybe there is more to it than dead-reckoning and needle-tracking, but then, that's why they're Arctic pilots and I'm not.

My preeminent concern at the very moment, though, was the cup of coffee, for I was on the ground, briefly, during the second of two days of the kind of back-breaking labor that Weldy's younger pilots have to put up with before they're ready for the comparatively cushy twins. To see what the Arctic can really mean for a bush pilot, you have to fly in a single-engine Otter. ("Bloody pig of an airplane," one mechanic muttered as I hauled myself up to the airy heights of the front seat.) You'll find that the Otters are the trucks of the fleet—literally. You enter the cockpit as though you're climbing up into a Brockway tractor-cab, they fly like trucks, carry loads like trucks and require a teamster's efforts of their pilots.

Jasper Lafrance, a young lad (an oft-heard phrase around Weldy's group, usually pronounced "yung lod") with a cherubic face and a weightlifter's shoulders, took me north in an Otter for two days of fuel-caching on Gardner Island, not far from the permanent polar pack ice. In those two days, we loaded, flew and unloaded 98 55-gallon drums of turbine fuel—six per trip—for the use of an oil company's Hueys and Alouettes out on the seismic-testing lines, where the geologists were probing, blasting and measuring for oil. We fell asleep at night, bone-weary and bruised, in sleeping bags spread on spare bunks in little insulated trailers, amid a womanless, profanity-blue, gripe-ridden routine that must've been vaguely like a base camp in the Vietnam highlands, punctuated by the whap-whap-whap of rotor blades all day and night.

The temperature went down to –39° F. and the barrels got

heavier. Jasper pumped the Otter's skis down to land on the snow near the various caches and back up again to return to the base camp's bulldozed runway out on the Arctic Ocean ice so many times that he finally said the hell with it and left them extended. We heaved the barrels in, strapped them down with a lick and a promise, fired up and groaned off the ground so many times that a tired Lafrance once even found himself set up on approach for a cache of barrels out on the ice only to have them begin to gallop away: It was a small herd of large musk oxen.

The de Havilland Otter is a bull of an airplane, pig though it may also be. Full or empty, the difference in takeoff run seems to be no more than a hundred feet or so. The airplane starts rolling so ponderously that you think it'll never get off, but along about 50 mph, flaps and flaperons drooping, it levitates miraculously and begins to climb out, the big 600-hp Pratt and Whitney clattering like an overworked road-grader. Get it a bit too heavy, though, and it'll lift the same as ever, then thunder on off toward the horizon in ground effect—and no higher. On skis, the takeoff technique is to literally heave the Otter off the ground at 40 mph or so (it seemed to me), to spare it the pounding as it bounces from drift to drift picking up "speed" in its elephantine fashion.

The tiny, compact turbines are the Arctic's best friend, though, and the growl of the round engine is fading into the past, soon to be as dead as a dogsled. Up where lunch can only be a half-hour long, so you can go out and fire up the Wright before the oil solidifies, the PT6 becomes beautiful. It's light, simple and as reliable as a Bunsen burner. The PT6, with its 3,000-hour TBO (1,050-hour hot-section inspection) is ubiquitous in the Arctic. Cover it with a fitted quilt and stick a little electric automobile-crankcase heater inside and she'll fire up in no time the next morning (if you remembered to bring the battery inside for the night, that is).

The only problem is money; Weldy keeps a spare engine on hand, and that means $52,000 tied up in the parts room. If he took the easy way out and had a whole firewall-forward QEC kit, he'd be in for $75,000. The cost of even the tiniest parts is ludicrous: An injector-nozzle tip, of which there are about two dozen per engine, and which is about as complex as a Chevrolet carburetor jet (exactly what it looks like) costs $50, and there's no way that the thumbtack-sized piece of metal with a hole in it can be worth that much. It's small wonder that half the pilots north of Frobisher Bay eventually come shuffling over to Atlas and its well-stocked parts bins when they have a turbine problem. "There are about four guys up here now with nothing but their airplanes," says Weldy in a rare moment of complaint, "and they sit on my doorstep waiting for me to help them out. I don't mind doing it, but I sure as hell don't like it when they take a part that puts *me* down the next day."

Weldy Phipps works with a broad brush, and his enthusiasms, like his parts bins, sometimes get him in trouble. "Dot Veldy, he's something," Jan Backe says, shaking his head in perplexity. "He vas talking on his ship thing again last night." "His ship thing" is Phipps' burgeoning plan to buy his own ocean-going freighter, to haul his own fuel and food and supplies north to Resolute when the ice melts. "And then I can charter it out when I'm not using it." Weldy says with unassailable logic, "and take it to the Caribbean for my vacation in the spring." (He'd do it, too: He'd drop the hook right off the beach at Montego Bay, and put his eight kids and Brutus, his sloppy Saint Bernard dog, ashore for groceries in the lifeboat.)

Jan, who comes from a Norwegian family "dot talked shipping at breakfast, lunch and dinner, and never made a penny at it," is deadset against the plan. "You ask for a lot of trouble ven amateurs get involved," he points out with the proper degree of grimness.

"Well, that certainly includes us," admits Weldy. "But maybe that's what they need, to uncomplicate the whole business."

Even riskier, yet with vastly greater promise, is Weldy's and Jan's plan to acquire a four-engine turboprop big enough to haul large quantities of fuel in bladder-tanks inside the hull, small enough to land on the ice when it's still thin early in the winter, and cheap enough to buy at airline-surplus prices. (They know exactly the airplane they're going to get, too, but if I told you, you wouldn't believe it, so I won't tell you.)

With the oil-exploration madness swinging into a round-the-clock routine, there's enough talent and equipment and money invested in the Canadian Arctic right now that Backe estimates that the oil companies will drill a minimum of 100 wells before they give up—and God knows how many more if they do find oil. Each well requires at least 100,000 gallons of fuel to search it out, set it up, drill it and cart the remains away, and each gallon of fuel has to be *flown* in. The only way that can be done right now is aboard Lockheed Hercules, which go for $4,500,000 a shot, plus a 10-percent-a-year insurance premium. Even Weldy's enthusiasms don't soar that high, though he talks wistfully about wanting to get a de Havilland Buffalo, the T-tailed military STOL turboprop that looks a little like a twin-engine Herc. "It has bastard engines—they were designed for the airplane, I think—that have very low life. I believe they're still only getting 350 hours before overhaul," Phipps says, "and that costs $60,000 per engine. Besides, it isn't civilly certified and probably never will be." Don't count him out, though: He'll probably bang one together in his shop.

Weldy Phipps' two top pilots—Dick de Blicquy and Jan Backe—are said to be the two best-paid pilots in Canada, earning more than even transocean captains for Air Canada. This would put them somewhere above $30,000 a year. They work for their share of the profits, though, and while Weldy straight-facedly insists that none of his pilots flies more than 120 hours a month, the legal maximum, the truth is closer to 190 or 200 hours a month during the busy summer season. Weldy is privately of the opinion that a healthy man might just

as well fly as sit on the ground, as long as he gets a solid eight hours of sleep the night before.

The Atlas crew works as a team, a tight coalition, rather than a company with a boss and employees. De Blicquy, for instance, has bought an Aztec that he flies under the Atlas colors—but to the profit of de Blicquy and whichever Atlas pilot happens to be driving it that day. There are times when it takes trade away from Weldy's Otters, but Phipps encourages this sort of enterprise.

"I'll never go back to airline flying," Backe says. "It's too boring. You're yust a number on a scheduling board. Vith an operation like Veldy's, you're really *part* of something, really contributing to it."

Nonetheless, Weldy has trouble finding good people to come north and work with him. Some sign up, stay awhile and quit. Others take one look and hop the next flight south. The Arctic is a harsh yet pure land that either repels instantly or appeals irresistibly, and for those who can see it only in terms of featureless frigidity or numbing boredom, there is no way to communicate its appeal and challenge.

The world seems full of young men with All the Ratings and 2,000 hours in Cessna 150s, every one of them convinced that fate has dealt them a kidney punch because they can't step straight into the right seat of a golden-route 727 or at least a corporate Gulfstream. If they were truly ambitious, they might be able to load mail and pre-heat engines and bulldog oil barrels for Weldy Phipps for awhile, and if they were truly lucky as well, he might take the time to teach them how to be real pilots. One of Weldy's young copilots, Hans Hollerer, came to him with a lot of ambition but not much time, none of it in anything bigger than an aging Apache; another is an Eskimo—the only Eskimo pilot in the eastern Arctic—whom Weldy taught to fly. Make no mistake, though: Both of them know what they're doing.

Weldy Phipps is a good man, and he has a good life, up there in the barren Arctic. There are mornings when you'll hear him muttering, as the wind tries to tear a rudder batten from his mittens, that anybody who chooses to make a living in the Arctic has to be soft as a grape. But still, but still . . . the Arctic is so clean and quiet that it gives new meaning to those words in our foul land. Weldy has his family around him, working with him, involved in his life and his skills in a way no suburbanite's brood could ever be. His work is his pleasure, for it's all-consuming, infinitely variable—and exciting.

Think upon it. There are those who put up with the frustration of a massive city for the sake of convenience or a kind of transient excitement. There are those who put up with stultifying jobs so they can fill their weekends with golf or skiing. Weldy puts up with the cold.

People Who Fly: Bobby Allison

*If you're Bobby Allison, a racetrack is where you can win
a million dollars; it's also where you can land your Aerostar.*

BY SALLY WIMER

NO DOUBT you already know about the good ole boys, who apprenticed in the stock-car-racing trade by out-driving the law, out-braving revenuers who lay in wait on dark Carolina roads, and who translated 180-degree bootleg spins into racetrack tactics. No doubt you know about the origins of racing Fords and Chevys, cars that *looked* like everybody else's everyday Fords and Chevys but ran like rockets: how the whisky-running good ole boys had at it out in the cornfields, until it became a regular Sunday afternoon event and the rubes started showing up to watch, so that before anybody even knew it was happening, the cornfields became super speedways, with grandstands and scoreboards and infields full of campers, and the whisky runners became super stars, with highly engineered automobiles and pit crews that could change the tires and fill a car full of gas in 20 seconds or less. Nascar came in, stole the sport of the South and presented it to the whole nation, neatly wrapped up in a professional package, so that nowadays, a guy can make a million dollars racing Fords and Chevys—which is exactly what Bobby Allison has done.

Did you also know that, somewhere in the transition from the cornfields to the super speedways, a new kind of driver was born, a driver who'd learned his way around professional tracks by racing on amateur tracks first, and that the new breed came through as businessmen as well as hard-charging, fearless and, yes, sometimes ruthless competitors? Businessmen they had to be, to garner their share of the sponsor money, for stock-car racing was no longer cheap. Bobby Allison is of that new breed—and then some—which may be why he races with Coca-Cola money in Roger Penske-prepared cars.

Being a businessman is also one of the reasons why Allison moves from championship to championship in his Ted Smith pressurized Aerostar.

"I bought my first airplane for business reasons, pure and simple," says Allison. "If you live in the South and die, before you can go to heaven or hell, you have to lay over in Atlanta. But with my own plane, I go direct, and I can race four nights a week, all over the country, if necessary—and that can translate into a lot of wins."

Business may or may not be the reason why Bobby Allison has been known to land his Aerostar on the racetrack itself, park it in the pits and, 500 turns around the track later, take off from the back stretch at a lift-off speed somewhat less than he'd been traveling at all day long. Allison claims that flying in and out of race tracks is a question of saving time, but that he can't do it anyore.

"The FAA doesn't say you have to land on a designated runway," says Allison. "They left it up to me and my insurance man. A track is sometimes 30, 40 miles from the nearest airport, and landing right on the track saves a lot of time. But my new plane, the 601P, is quite a bit more expensive than the others I've had, so my insurance man is objecting to my landing habits.

"There's nothing unsafe about it; why, the back stretch at Road Atlanta (where he landed last march) is a mile long and 50 feet wide." (Actually, it's 4,400 feet long and about 32 feet wide.)

Bobby Allison bought his first airplane in 1967, which was also the year of his first major win.

"When I was a kid," he says, "I had a friend whose father ran an FBO and we'd go out and get an Aeronca and fly all over the place, but I never did any flying myself until 1967, when I bought a Mooney 21 and presented myself and the Mooney to a flight instructor for lessons. That instructor happened to have a girl student he was spending all his time with, so I switched to a second instructor, but he'd push his seat way back in the airplane and just sit there, with his eyes closed. I always thought he was asleep and would ask him what I was doing wrong, but he'd just answer 'nothing,' so I got a third instructor, and after 30 minutes with him, he soloed me.

"I traded in that little Mooney for a Mooney Executive 21, and then I got myself a Baron, but that particular Baron gave me a lot of trouble, so I bought my first Aerostar. Now I'm on my third Aerostar—it was number three of the pressurized 601s to come off the line, and it's equipped with *everything*."

Aerostar 12BA came to Bobby Allison by way of the accumulated million dollars in purse money, the appearance money, the television, sponsor and contingency money and the money from Bobby Allison jackets and other assorted trinkets. Bobby Allison Racing, located in a large, well-equipped shop in Hueytown, Alabama, exists primarily to build Bobby Allison Modified Champion Nascar and STAR cars. (Allison races for Penske only in the big time, in Winston Cup Grand National Championship races.) There aren't many stock car drivers of name and fame who bother with the smaller circuits, but Allison does, and for good reason. It was recently rumored that he'd won some $100,000 at small tracks last year.

At the same time, the Allison business enterprises are growing. He has just bought Birmingham International Raceway, which has been known to draw 10,000 people a weekend, for a long season of weekends. And he wants to build an airport.

"There used to be an airport in Bessemer, Alabama, right next to Hueytown, and I used to keep my airplane there, but they closed it down and moved it 35 miles away." His new proposed airport site is about 5 minutes from Bobby Allison Racing.

The Bobby Allison success story has been a long time in coming. He's been racing 20 years, and it's only been the last five that he's made it in a big way.

"My first big win was really exciting," he says. "It was in Rockingham, North Carolina in 1967. I had been running my own car, without much success, when I got a call at the last minute to show up over there and run a factory Ford, one of Ralph Moody's cars. The whole thing was put together on like a week's notice, and we went over there and won. In 1971, I drove for Moody again and I won eight major races and 11 races total, in like half a season."

Before the 1971 Moody ride, it was said that Allison was broke and through with racing, but before the season was over, he had won over $236,000 in Grand National wins alone.

In 1972, Allison drove for one of stock-car-racing's authentic good ole boys, Junior Johnson, and although he finished second in the Winston Cup series, he won more money than the first-place contender, Richard Petty. That year, he also won two of racing's highest honors: the Martini and Rossi Driver of the Year award, and (for the second year), was voted Most Popular Driver by his colleagues.

This year's second win came at Riverside, California, one day following a five-car crash that netted Allison broken vertebrae and cracked ribs. Not only did he win, but he led the race from the third lap through the finish, some 70 miles later.

"It was the worst wreck I've had since 1969," says Allison. "But you just have to go on through and pick your course. I feel no heart-stopping fear any more; you know it's going to hurt, you just hope it won't hurt too much.

"At first, I asked my parents' permission to race, but they discouraged me from it after a while, so I just went out and raced under an assumed name.

"I had a school car when I was a senior, and I ran that car, a Chevrolet coupe. I drove it to school with the numbers painted on it for a couple of weeks, but it finally got so beat up that I couldn't run it on the street anymore.

Allison grew up in a large Catholic family; there were 10 children. His kid brother, Donnie, has been known to burn up a few tracks of his own, and another brother, Eddie, has been a mechanic for Bobby Allison Racing.

Out of high school, Bobby Allison went to work for Carl Kiekhaefer, in Oshkosh, Wisconsin. Kiekhaefer was then president of Kiekhaefer-Mercury, the large marine manufacturer, but engineering dilettante that he is, he was also racing stock cars—and winning consistently. Old Man Kiekhaefer had a reputation for being a stickler (there are see-through picture windows surrounding all the executive offices at Mercury), and, according to Allison, if Kiekhaefer had three cars running, he expected them to come in first, second and third.

Bobby Allison then returned home to Miami to race. Once, the story goes, he hauled his car up to Virginia to compete, but when he didn't win, he had to sell a carburetor and manifold off the car to get the money to come home again.

"I came to Alabama," says Allison, "because I'd heard a rumor that there was good racing here, and I found that it was true." Before his wife and family moved up, he took an apartment, and the only furniture was the mattress on the floor.

In the annals of racing-car history is the story of a major shunting that occurred between Bobby Allison and Curtis "Pops" Turner, at Winston-Salem, North Carolina.

"It was one of the things that happened in the old days," says Allison. "The general feeling was that Curtis was a little high that night; you know, he had a reputation for being, well, flamboyant. So what he did was to spin me out at the start of the race. He had collided with several cars. I recovered from my spin and worked my way up through the traffic and was about to pass him on the turn when he drove into me, and when he did, I spun him out. So he got all backwards, and I went on to take the lead of the race. He couldn't catch me, so he slowed down and waited for me and wrecked me. He'd cost me the race, so I wrecked him back, and we got into a big destruction derby out in the infield of the racetrack. The race stopped, for all practical purposes.

"Of course, he was in a factory car, so when it was over, he just got out and said 'ha ha'; my gang took my poor little ole wrecked race car back to Alabama and worked all week fixing it. It's ironical, but Curtis and I became very good friends after that."

"When he and Donnie first started coming to the races around here, why they just took over," says retired army officer, Chuck Campbell, who goes to *all* the races. "But when they showed up, they won a lot of friends. It was always 'yes, sir' and 'no, sir,' and they'd give out autographs and help the other drivers. Another thing—they were professional, those boys. I once heard Bobby Allison say he'd rather come in second in a good race than win a sorry race. He's a good engineer. Like I say—professional. And success hasn't changed him."

Bobby Allison arrived at Birmingham Municipal Airport in the longest, blackest Lincoln you ever saw and walked toward the Aerostar on the ramp. "I got that Lincoln from Marty Robbins," he says. "Traded him my old Grand National car and the truck that went with it for the Championship car Marty had, this Lincoln and a cow—black Angus cow; Marty's keeping it on his farm."

You figure stock-car racers to be short, stocky, red-faced (and red-necked) fellows; not Bobby Allison, who looks more like a group vice president than a good ole boy. He's tall and slim with brown hair and just enough gray around the temples. He turned 37 at the end of 1974, but he looks older, not in the gray-green eyes but in the wrinkles around the eyes. From the side, there's a touch of John Wayne and sometimes, a John Wayne tough-guy grimace. Maybe it's because of the broken ribs; they hurt a lot. Bobby Allison has a reputation for being a straight guy; above all, polite, respectful.

The Aerostar was a factory-loaner because his own plane was at the factory for modifications. Ted Smith is designing a wing platform, which will add lift and about 50 knots cruise speed to the 601P and Bobby Allison is letting him experiment with N12BA. (Twelve is the number of his racing car; you know what BA stands for.)

We were cleared out of the Birmingham TCA for Macon, Georgia, where the Georgia State 500 was being run, a STAR race, which means short tracks, not tall names. Bobby Allison was second in the STAR standings, and a win today could mean first place. Bobby would start the 500-mile race, to get the points, but because he was anxious for his ribs to heal before the next Grand National battle, Neil Bonnet, a hot-shot youngster, would finish the race.

"Look at that!" said Bobby. "That's what I like about this airplane. We're climbing at 900 fpm and indicating 192 knots. Look at that! It only took us seven minutes to reach 8,500 feet. You don't waste any time in this airplane.

"My own plane is beautiful," says Bobby. "It's white with red and gold striping." (His race cars are also red and gold). The transponder and ADF are on the left and in the center console is the audio select panel, dual navcom, a King radar unit and the DME. "I used to have RNAV," says Bobby, "but to me, it was just a fancy gadget. Why do you need RNAV when you can get direct routing anyway? Once I was okayed direct from Louisville to Toronto by radar vectors. The controller was a race fan. I do have an RMI, and that may be a gadget, but I do

use it.

"Going VFR, I like to cruise in my plane at around 16,000 feet, or, IFR, at 19,000 or 20,000. I keep the manifold on 29.5 all the time; we're consuming 25 gallons of fuel an hour and truing out at 250 knots, in cruise. The Aerostar gets mushy at about 22,000 feet. But at 20,000—you can move!

"I don't have much use for a jet, because I need to get into the small runways in the small towns where a lot of races are held. But I will say that I'm interested in the Turbomeca Astafan, a little French jet engine. They say that with the Astafan and a change in this windshield, you can get 350 knots cruise at 25,000 feet. And if Ted Smith says it can be done, I'd believe him! Look at this airplane. It's got the same engine in it as other light twins and here we are, at 210 indicated with a full load of passengers and 90 gallons of fuel. I'm a Ted Smith fan.

"I read FLYING'S report ["Pressurized Aerostar," October 1974] and what it said is very true. You can rudder trim it with your feet on the floor. It's not even fair to take a multiengine rating in this plane, because it runs so well on one engine.

"I had just gotten my brand-new airplane, and I went racing that weekend. The following Tuesday, I was taking a doctor-friend flying. We blew an oil seal on the turbo actuator and dropped oil out of the right engine. All the oil wasn't gone, but the turbocharger wouldn't settle down, so I told my friend I had a situation I wasn't sure of, and I was going to shut the right engine down. I told him, 'Please do not be alarmed, because the plane runs very well on one engine.' And that it did. We dropped down to 13,500 feet and indicated 200 knots all the way back to Birmingham. Then I made a back-course ILS approach and broke out right over the numbers. Airplanes really turn me on."

Allison is, if anything, as cool with airplane controls as he is going around the racetrack. You know that every landing is a squeaker. "One night, I was flying with him," says an Allison employee, "and it began to get a little bumpy. Well, I didn't say anything, and Bobby didn't say anything—until we had landed. 'Phew,' he told me, when we were down. 'That was the *worst* thunderstorm I've ever seen.'"

Allison isn't the only racing driver to fly his own plane. Bobby Unser has an Aerostar as well, and Mario Andretti flies a Navajo. David Pearson and Cale Yarborough have Aztecs, and Paul Goldsmith owned a Riley Rocket. Curtis Turner was killed in his twin. Roger Penske travels in his own Learjet, piloted by a full-time crew. "I flew the Lear once," says Allison. "That's quite a plane! But sensitive. And the controls get mushy, up high."

There's a billboard along Route 79, down the road from the Macon Airport, and that's where you turn in for the Middle Georgia Raceway, should you miss the little cardboard sign tacked to the tree, the sign that says "RACE" and points down the dust-bowl road.

Lined up around the pit apron were some 20 local hopefuls, each armed with a late '60s Chevy or Ford, a driver, a driver's buddy, a gas can, a couple of spare tires and a lug wrench. The day's game was, obviously, to whip the two superstars who had shown up: Cale Yarborough and Bobby Allison. In the Bobby Allison Racing pits was a van-full of parts, six or seven sets of tires, quick-fill gas guns and quick-change electric lug wrenches, two drivers (Allison and Neil Bonnett), a full crew and about a hundred hangers-on. Oh yes, and several dozen cases of Coca-Cola.

"We go to every race to win, you know," says Allison. "The big wins are important, they're personally satisfying, but so is a little race like this one, today. Racing has always been a part of my life and if there's not a big race on, that won't stop me from running. Sure, there's some resentment from the local boys, but there's also the possibility that they'll beat us. And they do, you know.

"If Jack Nicklaus goes to the country club in his home town to play golf, it's a plus to have him there. You can see him without having to go a national tournament. It should be the same way with automobile racing."

Allison had hardly started the race when he pitted to let Neil Bonnett take over for the rest of the day. "I would have liked to have gone the distance," he said, "but these ribs. . . . I can't even get up in the morning without rolling out on the floor first." Still, he had spent the first part of this day in and out of, and underneath Neil's racing car, checking everything personally.

Midway through the race, Bonnett shunted Cale Yarborough, or vice versa, resulting in a broken rear axle on Number 12. On the next to last lap, there were about ten cars left in the field. The leader had just gotten the white flag, signaling one more lap, when he, Yarborough and Bonnett got into a tangle from which the leader never recovered. Yarborough came around the track to receive the checkered flag, with Bonnett behind him.

Yarborough moved his car to the front of the grandstands and started signing autographs, scribbling on bubble gum cards, pictures of football heros, hats and crumpled up lunch bags, and talking to the trophy girl a lot.

Unfortunately for Cale, it turned out that he had been a lap behind the leader all along, so he had completed one lap less. The broken-down racing car on turn one took the race and the $5,000.

Meanwhile, in the Bobby Allison Racing pits, Neil Bonnett struggled out of the car to a curious throng.

A little girl, shoved forward by her mother, approached the driver. "Will you sign this?"

He turned around and wrote "Neil Bonnett" across the chewing gum wrapper.

"Hey, that ain't Bobby!" she wailed dramatically.

The stars were bright over Alabama that night, as an Aerostar sped through the skies, taking the real Bobby Allison and the surrogate Bobby Allison home. The day's work was done; Neil had fought a good fight; Bobby Allison's wife and kids were waiting, and he was content.

Somewhere behind him, traveling some 80 knots slower, was Roy Milligan in a Mooney. Milligan is a local boy who races some. In fact, he used to own a car that Allison once raced.

Bobby Allison tuned to 122.8 and keyed the mike.

"Little Mooney, where are you?" he asked.

"Right behind you, Bobby," said a voice in the skies.

Right where Bobby Allison was a few years ago, but not where the businessman/racer is likely to be tomorrow.

At 19, he flew Navy Hellcats over Wake Island. Two years later, he launched a career that led to a group vice presidency in a five-billion-dollar conglomerate. Today, he's a corporate drop-out. Why? To fly again!

BY EDWARD D. MUHLFELD

ROBERT BLOOMFIELD

"THE SLEEK, WHITE and blue-trimmed Cessna 340 banked sharply to avoid the 7,500-foot mountain peak southeast of the field and slid down out of the midafternoon turbulence onto final approach for the Hailey, Idaho Friedman Memorial Airport (altitude 5,300 feet).

Waiting for the Cessna's passengers in the shade of a Phillips 66 sign at the far end of the one-way strip was a powerfully built, ruggedly handsome man in scuffed boots, a worn, plaid shirt and faded jeans; a well-chewed wooden kitchen match was barely visible under a bushy, salt-and-pepper moustache of enviable vintage.

Robert Bloomfield is 50 years old, five feet 10 inches, 182 pounds, but looks much bigger. "I'm short-legged," he says. "If I'd been born with normal-length legs, I'd be six-three and weigh two-ten."

He greets us in a clear, booming voice that rings with authority, yet makes everyone feel welcome, immediately at home. "Ed Muhlfeld, nice to see you again. Richard Collins, I'm happy to meet *you;* I've been reading your stuff in FLYING for years. Boys (to the three teenagers staggering under the weight of backpacks, sleeping bags, fishing rods and other assorted camping equipment that had filled the 340 to near gross), put your gear in the back of that beat-up Travelall over there; it's dirty, but it sure runs good. You two guys come on over here around the hangar and take a look at my Aztec. It's got turbos and it's the pride and joy of Idaho Air Service—that's me—and it's in the air most of the time."

We climb into the Travelall and rattle up Highway 93 from Hailey, through Ketchum, Idaho and turn off on a dirt road that leads to the Bloomfield ranch. The late afternoon sun lights the hilltops and mountain peaks with brilliant greens, reds and purples, while we drive along in the shadow of the

steep mountains.

"Road's in great shape," Boomfield says. "They'll be doing some logging farther up, and the state made them grade and oil the road. Mary Hemingway lives in there," he gestures off to the right with his chin. We look but see nothing but meadow and stands of aspen. The road turns west and picks up a fast-flowing stream. "That's Warm Springs Creek. A few small trout in there, but you've got to go down to Big Lost River to get the good ones."

Bloomfield turns into a narrow drive, and stops the Trav-elall in the middle of a sprawl of seven log buildings. He has threaded his way through an incredible assortment of vehicles: a World War II jeep, a Chevy two-ton truck, a VW bus, a jeep station wagon, a sailboat, three motorcycles, three snow-mobiles, an Allis Chalmers diesel bulldozer and a snow cat. "A snow cat?" "Yeah," says Bloomfield. "They plow the road only so far out from town in the winter. Last year, we had to park about a mile and a half down and come the rest of the way by snowmobile. Pretty tough going, particularly with visitors and groceries, so now we have the 'cat.' Bought it from the Hailey Ski Area people. That thing'll go anywhere." We later learn that Bloomfieldia also includes three horses, two cats, five dogs and 15 pairs of skis.

We are greeted at the main house by Bloomfield's new young wife, Peggy, who is feeding the sixth Bloomfield son (her first), a spunky 10-month-old called Clic. ("That's not a nickname, that's his name.") Red-haired, clear-eyed and soft-spoken, Peggy Bloomfield directs the teenagers to two bunkhouses and me to the guest cabin, a square log structure about three feet from the creek. Inside is an improbably soft canopy bed, a refrigerator stocked with Coors and 7-Up and a potbellied stove.

I could hardly believe this was the same Bob Bloomfield I knew three years ago, back east: the man with the close-cropped hair, white button-down shirts, striped ties, Brooks Brothers suits and Johnson & Murphy shoes. He was a group vice president at Burlington Industries (a $5-billion conglom-erate) and had an enormous house (and attendant sailboats) on the water in Larchmont, New York. He also had an enormous ski lodge (with attendant acreage) in Stratton, Vermont, drove a Porsche and talked about airplanes but never did any flying.

"Flying was the biggest thing in my life when I was younger," he had said. "In fact, on October 5 and 6, 1943, when I was barely 19 years old, I was strafing Japanese beach defenses and airstrip revetments on Wake Island in a Navy Hellcat fighter."

Bob Bloomfield was born in Yonkers, but raised in Larch-mont and New Rochelle, New York. He went directly from Mt. Hermon, Massachusetts High School to the U.S. Navy, enlisting right after his eighteenth birthday, and in May 1952, he was appointed a Naval Aviation Cadet. He preflighted at Chapel Hill, North Carolina, and his platoon commander was a Lt. j.g. named Gerald Ford. "Yeah," says Bloomfield, "It was Jerry Ford, all right. He and I kept pretty well in touch over the years, and when he got to be Vice President, I said to Peggy, 'I ought to call him, because he's gonna be President.' But I never did."

From Chapel Hill, Bloomfield went to the naval air station at Hutchinson, Kansas, for 90 hours in the N2S Stearman, the Navy's well-known "Yellow Peril," and was checked out by "Burt" Morris, better known to movie fans as Wayne Morris. "We had a record winter for snow in Kansas that year," says Bloomfield, "And cold? Jee-zus, it was cold. Try sometime doing a snap roll from the front seat of a Stearman in heavy flight boots, sheepskin-lined, with jacket, pants, gloves and helmet to match; try doing that and still hearing that s.o.b. of an instructor through the gosport."

From Hutchinson, Bloomfield was sent to Corpus Christi, Texas in February 1943 ("a welcome relief") for 20 hours of transition in SNVs (the low-wing Vultee airplane known in the Army as the BT-13). Then he had 40 hours of instrument training under the hood in the SNV (coupled with 20 hours of "hot as hell" Link trainer time) and, finally, 100 hours of advanced training in the North American SNJ.

He got his Navy wings and ensign's commission on June 5, 1943. ("President Roosevelt came to Corpus that Spring. I remember standing in formation on the parade ground as he rode by in that big open car. Some thrill for an 18-year-old.")

From Corpus, Bloomfield was sent to the Jacksonville NAS for fighter operational training.

"In a way, it was the fulfillment of a life-long dream, getting that fighter assignment," says Bloomfield. "When I was eight years old, my Dad took me up for my first flight, and when I was 13, I got another flight as a reward for 'Most Advancement by a Larchmont Boy Scout.' From that time, I knew that's what I wanted—and not just to fly; I wanted to go to war. And I wanted to do it as a fighter pilot."

Bloomfield flew 100 hours at JAX in the Navy's F4F Wildcat fighter, "a good flying airplane, but it sure did a lot of funny things; you had to crank the gear up by hand into a wheel well alongside the fuselage—12 to 13 turns—and that could keep you mighty busy on a hairy takeoff."

From JAX, Bloomfield was sent to the Glenview NAS in Illinois, in August 1943, for a bout with the USS Charger. "She was very famous in the U.S. Navy then, a Great Lakes sidewheel steamer converted to a flattop, with the smallest deck we ever landed on. But everybody had to make eight takeoffs and landings on the Charger, and I mean everybody. There probably wasn't a man in the Navy at the time—in naval aviation, anyway—who hadn't checked out on the Charger. I was lucky; the weather was clear and I did it in one day (in a Wildcat). As soon as you did it, you shipped out."

Bloomfield had married his childhood sweetheart while at Jacksonville; they were both 19. With his bride, he reported to NAS San Diego (Coronado) after 15 days' leave. They'd crossed the country by train: the Twentieth Century Limited from New York City to Chicago, and the Santa Fe Chief from Chicago to Los Angeles. Heady stuff, particularly when you were off to fight a war in a Navy fighter, with bright Navy wings on your crisp, white summer uniform and your beautiful bride on your arm.

"We had a heckuva time finding a place to live in San Diego, but it turned out it didn't matter," Bloomfield said. "Two days after reporting, we shipped out on the USS Kaskiaska, a fleet tanker. Four brand-new ensigns (I was with three other guys

I'd flown with at JAX) headed for 'the Pacific Ocean'; that's all we knew, and I want to tell you, we were scared shitless. That 600-foot tanker was loaded to the gills with hundred-octane aviation gasoline, and we never dreamed that fully loaded Navy tankers ever went to sea without a destroyer escort or air cover. But there we were, alone, zigzagging at 18 knots from San Diego to Pearl Habor. We spent a good part of that six-day trip on deck looking for periscopes."

Strapped down on the cargo deck of the *Kaskiaska,* under canvas covers, were four brand-new F6F Grumman Hellcats: replacements—along with the four ensigns—for the Pacific fleet.

There were no Hellcats in training squadrons at the time; heavy losses in the Pacific dictated that every airplane leaving the Grumman factory be sent directly into combat."So, before leaving port," said Bloomfield, "I went to the cargo officer and told him the greatest bloody thing he could do was to let us get familiar with the airplanes we'd be flying. He said okay, so the four of us spent most of our time under those canvas covers studying the Hellcat cockpit layout and instrument panel. And when we got to Pearl, reported to Cincpac at Ford Island and were sent to Fighter Squadron Six at Maui, we were ready to take our cockpit blindfold test and were in the air within an hour of the time we checked in."

Squadron Six had just returned from an attack on Marcus Island, operating off the carriers *Essex* and *Yorktown;* there were 36 aircraft, of which 24 were fighters and 12 were Douglas Dauntless SBDs. "Butch O'Hare was our squadron commander," said Bloomfield, "and could he fly, let me tell you! We flew plenty at Maui, and then suddenly word came to report to the carrier *Independence.* We sailed for Wake Island early in October 1943 as part of the largest carrier task force ever assembled: six carriers and 400 planes, for the beginning—unknown to us at the time—of the 'carrier war' offensive in the South Pacific.

"My first action was a strafing attack against Wake. We preceded the dive bombers and torpedo planes to knock out parked aircraft in the airstrip revetments. I was excited as hell and scared to death, but it was a great learning experience. At one point, I thought I was really getting shot up; the Hellcat's windscreen shattered, and I was taking what I thought were solid hits all over the aircraft. When we got back to the ship, it turned out I'd moved in too close under my section leader and what I thought were enemy hits were my section leader's .50-cal. ejects. We pried dozens of spent shell casings out of the leading edges of my wings."

Bloomfield went on to fly more than 250 missions in the South Pacific, with six confirmed combat victories. He was at Tarawa and Roi-Namur in the Marshall Islands, and was part of the first carrier attack against the giant Japanese base at Truk. He was torpedoed twice, once while on the *Independence* and later while aboard the *Intrepid.* "Scared? Well, when the torpedo hit the *Intrepid,* I was in the ward room. The only sensation was that the ship had hit something. She was 33,000 tons and pretty hard to shake up."

Later, he flew with Air Group 85 off the carrier *Shangri-la* in actions against Iwo Jima and Okinawa ("that was *kamikaze* time") and against Honshu and other Japanese home islands.

"We were flying Vought Corsairs then, and the Corsair's ordnance load had to be seen to be believed. Full armament was four 20-mm cannon, eight five-inch rockets, a 500-pound general-purpose bomb and even a drop tank. Greatest airplane of the war, and certainly the greatest I ever flew."

Bloomfield's last military flight took place during the Japanese surrender ceremonies on September 2, 1945, "Up Tokyo Bay, over the *USS Missouri,* where General Douglas MacArthur was holding court. We had 1,000 to 1,200 planes, all Navy, off maybe 25 carriers, and, believe me, it was some sight and some thrill. If anybody had told me then that I'd do virtually no flying for the next 25 years, why, I'd have told him he was crazier 'n hell."

But that's the way it worked out. Discharged, home again, a hero, and recovered from a siege of hepatitis, Bloomfield decided against college. "I had no interest in college. I was too impatient; I wanted to be a salesman like my father. Y'know, we are direct descendants of Smith Bloomfield, who landed in Massachusetts in 1632 as a colonist from England. And General Joseph Bloomfield served under George Washington during the Revolutionary War, and later became the first governor of New Jersey. My grandfather was general counsel for one of the large insurance companies. But my Dad? Well, he was first and foremost a salesman—with product, service or girls and, as I recall, reasonably successful in the first two areas; *super* in the third. So when I got out of the Navy, I figured I could get the jump on the competition by not going to college. I'd be a salesman like my father."

Jumping the competition, Bob Bloomfield did, and successful he was. After a brief stint selling fuel-oil metering equipment, he went to work as a junior salesman for an old-line textile company called Pacific Mills. Shortly after Pacific was bought up by Burlington Industries in 1954, Bloomfield was appointed manager of the Rhodiss Mills department of Pacific ("fabrics for men's and boys' slacks and summer suits"). But Burlington already had a similar fabric division called Burlington Menswear. The senior vice president of Burlington Industries also happened to be president of Burlington Menswear. It was Rhodiss Mills' biggest competitor.

"Burlington should have merged the two divisions. It made all sorts of sense, but they didn't see it. With the boss looking over my shoulder, prohibiting me from doing things I should have done because Burlington Menswear was my biggest competitor . . . well, I put up with that for a couple of years and bailed out."

Bloomfield and an associate formed Tri Tex Mills, Incoporated, converters of fabrics for boys' and menswear, and went into competition with Burlington (and Rhodiss). Within two years, Tri Tex had become a "very successful $8- to 9-million business," had forced Burlington to do what Bloomfield had wanted (merge Rhodiss into Burlington Menswear), and Bloomfield himself was lured back to Burlington by the man who was later to become president of Burlington Industries, Ely Calloway.

By 1968, Burlington Menswear had tripled its sales to $90 million and Divisional President Robert Bloomfield became a corporate officer: vice president of Burlington Industries and a member of the 12-man management committee. In 1969, he was appointed group vice president, with responsibility for a group of Burlington Companies.

But having "climbed the corporate ladder, with the top rung at least in sight," Bob Bloomfield was beginning to have misgivings about the kind of life he was leading. "I was doing the kind of things that had to be done to make the company successful: cutting back, closing plants, expanding elsewhere, dropping product lines . . . some of the things that should have been done before I moved in . . . and the company was becoming more centralized all the time. I was an s.o.b. to everybody and a hero to nobody. I began to wonder if there wasn't a better way of life.

"Meanwhile, it continued to bother me that, except for a few highly illegal flights in questionable vintage airplanes in the late '40s, I'd never done anything about flying—the one thing that had been most important to me in my earlier years."

"When I was hired back by Burlington in 1960 as president of a major subsidiary, one of the prerequisites that went along with the job was access to the company planes. We had a Sabreliner 40, then a 60; two GIIs, three King Airs and a Baron. At various times, we'd had a Jet Star and Jet Commander, although we sure didn't need that JetStar for the kind of trips we were flying, like 500 miles between Greensboro and New York City. A Lear or a Citation would've been perfect. I often flew jump seat or right seat. I got a lot of time."

Bloomfield called me in March of 1972 and said, "Hey, what about Flight Safety?"

Well, what about it? I'd had this vague memory of Bob Bloomfield, the World War II pilot, but it was pretty blurred by the image of Burlington mogul Bloomfield, the Porsche driver who won lots of sailboat races and practically ran the ski patrol at Stratton Mountain, Vermont, whose parties and other exploits were legendary and whose wild-eyed son spent too much time chasing my daughter, who was susceptible to things like sports cars, boats and ski lodges.

But Bloomfield the pilot? "Whaddya have in mind?"

What Bob Bloomfield had in mind was a complete change of direction, climate, career, way of life. In short, he'd decided to become a "corporate dropout," and airplanes were to play a significant role in his future plans.

"We'd been spending Christmas in Sun Valley, Idaho for several years," said Bloomfield. "In fact, we bought a condominium there, in Ketchum, actually, in 1970, and I went out there with my family, with this feeling that I might never come back. But my boss, Ely Calloway, who'd lured me back to Burlington in 1960, called and told me about a promotion he had in mind for me, so I went back and gave it my best for another year.

"On vacation that summer, 1971, in Idaho, in the mountains on horseback, I decided it wasn't worth it. So, before leaving for the Christmas holidays in Sun Valley that year . . . well, I resigned."

Starting in March 1972, Bloomfield began flying again, and with a vengeance, as if to make up for the years in the air he lost. From FlightSafety ground school at LaGuardia, where Bob passed his commercial written exam and Peggy her private, they went to FlightSafety's Vero Beach, Florida facility where Bloomfield got his commercial ticket in nine days, taking delivery of a 1959 Piper Comanche from Showalter in Orlando the day he passed his FAA flight check.

"We put 480 hours on the Comanche between May 1972 and October 1973," says Bloomfield, "and we went everywhere. Flew extensively throughout the Rocky Mountains, into many Idaho wilderness strips, and took longer trips to Palm Springs, to Mexico, to Dallas, New York, Chicago, Minneapolis, St. Louis." ("There was never a week we didn't fly," adds Peggy.)

"For a year and a half here in Idaho," says Bloomfield, "I tried unsuccessfully to buy a business. Finally, since my major enthusiasm was for flying, anyway, I decided to go into the air-taxi and charter business. So I went back to FlightSafety, got my instrument rating and bought the Aztec (a 1969 D-model with turbos) from the Charles L. Blair Foundation in Denver. I got my Part 135 Operating Certificates at the end of January 1974, flew my first revenue trip up to Boise the following day, and jeez . . . it seems like Idaho Air Service has been in the air ever since.

"It's funny: this whole air-taxi/charter business started as kind of a fun thing; but all of a sudden it got to be a serious business. A lot of people have lots of reasons to go from Sun Valley to Boise, Salt Lake, Denver, San Francisco, Las Vegas and Los Angeles, and many times it's cheaper for them to fly with us than to go by airline . . . not even considering the time saved, which is usually considerable. We're getting overbooked, and I'm very conscious of the need for pressurization. We'll be trading up soon, maybe to a Navajo, maybe to something bigger."

It's late afternoon in August. We're at 14,500 feet over the mountains of central Idaho. In the rear seats, two teenage Muhlfelds, 15-year-old Spencer Ewald and 17-year-old Tommy Bloomfield are sound asleep and snoring. We had flown into Chamberlain, Idaho's pastoral grass strip and ridden far, far up into the mountains to fish. Bob Bloomfield is flying, hunched forward, left hand on the yoke, and right arm braced against the crash pad above the instrument panel. The ever-present kitchen match is working beneath the bushy moustache as he peers intently at peaks and canyons that look identical to everyone but him.

"We'll go home up the middle fork of the Salmon River. You'll never forget it." He pulls the power back, and the Aztec sinks beneath the ridge line as the altimeter unwinds. "Look at those goats!" We look and see nothing. Twelve-five, eleven, nine-five. The canyon walls reach for us. The river is a silver ribbon not nearly far enough below. Reassuringly, an airstrip, carved into the riverbank, then another, appear ahead and vanish behind us. "There are dozens along the Salmon," says Bloomfield, "and, as you can see, there are plenty of planes down in there right now. The Salmon has some of the best fishing, and some of the best white-water rafting in the world." I relax. Bloomfield handles the Aztec like he was part of it, or it of him.

And I ask him the big question. "Are you happy you did what you did?" He grins: "Boy, am I. Nothing, but *nothing* could get me back East with Burlington or with anybody else. I have no regrets." And I don't think he does either.

Living Legends

Few people have had as strong an influence on the design of general aviation airplanes as this inventor of the Ercoupe, Cherokee, tricycle gear—and the 50-foot obstacle.

BY RICHARD B. WEEGHMAN

Fred E. Weick

A STUDY IN weathered hickory, Fred Weick lends a hulking, diffident presence to an occasional aviation function. That's his public image. A private talk on nearly any aspect of aeronautics, however, triggers a solenoid that unleashes an unexpected current of animation and restless enthusiasm.

Weick, who at the age of 77 has retired from Piper's Vero Beach engineering department, can be credited with some of the major achievements in lightplane design. He is the spirit behind the modern tricycle landing gear, the "spin-proof" Ercoupe and the crashproof agplane.

He also is an abysmal subject for an interview. Touch lightly on any subject and you run the risk of setting off a chain reaction of tangential issues, dates, people, companies and theories, thanks to his encyclopedic recall for events along the eventful Weickian course of aeronautical history.

Expect none of the traditional engineer's cryptic understatement. No laconic Gary Cooper "yep" or "nope" suffices on a favorite subject. Sorting out the flow of ideas with gestures from his big, gnarled hands, Weick attacks almost any category of aeronautics with gusto, in the process becoming so highly charged that the words refuse to tumble out quickly and smoothly enough.

Although his design of the Ercoupe was perhaps his first widely recognized achievement, even before then Weick had added his share of contributions to the history of aircraft development. It was he who developed the NACA low-drag cowling for radial engines in 1928 after making critical measurements of the effects of various cowlings on cooling in NACA's wind tunnel.

As the assistant chief of the aerodynamics division at Langley Field, in Virginia, he also was the guiding force behind development of the W-1, a strange-looking pusher with an unorthodox wing configuration and decidedly unusual performance.

In 1931, to spark interest among his colleagues at NACA in light-aircraft design, he launched a series of seminars to determine what characteristics would be most desirable in an airplane. The desired features were lower landing speeds, good stability and handling and antispin qualities.

Although Weick's research director at Langley wanted to have the aircraft built as a NACA project, Weick preferred to keep it free of smothering bureaucratic supervision. So, with the help of a handful of enthusiastic engineers, he built it in his garage, thereby early betraying an independent spirit.

"If we had made it a Government project," said Weick, "we'd have had committee meetings on every damn thing."

The W-1 indeed turned out to have good stability and control and, according to Weick, was "automatically nonspinning." It also would get in and out of unusually small spaces with some alacrity—"off in 200 feet with no wind to an altitude of 50 feet," Weick recalls.

In the process, he provided one of those small mementos to the vocabulary of aviation: the 50-foot clearance for takeoffs and landings.

Says Weick: "The 50-foot height was my selection, and it stuck all these years. I figured it was the height of an ordinary power line or a tree, plus a little clearance."

Demonstrating a nice insight into how to solve knotty engineering problems with effortless aplomb, Weick came up with what he calls some "fine instrumentation" for measuring the run to the 50-foot obstacle. He simply dropped a paper bag filled with white powder and lead shot when the aircraft passed the 50-foot level; when the bag hit the ground, it provided a marker.

"It's funny how simple a thing can be," smiles Weick.

The W-1 also possessed a mildly revolutionary device called a tricycle landing gear, which is what Weick says he first called it in his SAE report on the aircraft. Actually, the early Curtiss and Wright aircraft originated the concept, but they had fixed, nonsteerable nosewheels, presumably, says Weick, because the Wrights weren't sure how much stability they'd get out of a castering wheel.

The Weick W-1 team adopted the tricycle gear as a way to reduce groundlooping as well as a means of simplifying the entire landing and takeoff process.

The Douglas DC-4 was the first airline transport to incorporate tricycle gear, and the Ercoupe made lightplane history by being the first in its class to do so. The little two-place single was the fruit of a project aimed at developing a commercial version of the W-1. For this purpose, Henry Berliner, in 1936, started the Engineering and Research Corporation and invited Weick to join him.

Along the way, the STOL aspect of the W-1 was dropped, since the new team figured it compromised higher cruise speeds, and they believed pilots were going to land at airports where they could get service. Also, they thought it would be difficult to sell a pusher design.

Although the first production model of the Ercoupe was ready in 1940, the war blocked further development of the project until the postwar boom, when it looked as though the little company were off and running as it turned out nearly 5,000 aircraft during the first year of production. But then, as Weick recalls, "The bottom dropped out, and Berliner lost his shirt and got out." With that, the future prospects for one more imaginative design were nipped in the bud, with the designs for a retractable model and a four-place "three-quarters finished, and a twin nine-tenths finished."

At this point, Weick was called to Texas A&M to set up an aircraft research center, at which he concentrated his efforts on development of the revolutionary Ag-1 crashproof agplane.

The crashworthiness of this airplane and the Piper Pawnee that evolved from it set life-saving standards of lasting benefit to the entire agplane industry.

Next, Weick's relationship with Piper evolved into another lightplane milestone of sorts. Approached by Piper about developing a new low-cost fabric-covered aircraft, Weick told them he believed he could, instead, design one out of sheet metal at no greater cost. With data borrowed from both Piper and Cessna, he showed that all the metal surfaces on the Cessnas were actually cheaper than Piper's "rag" ones of the day.

By this time, Weick had been placed in charge of Piper's engineering and development. He embarked on a quest to replace the fabric-covered Tri-Pacer, which Weick characterizes as "an airplane on a milking stool." The result, with engineering help from John Thorpe and Karl Bergey, was the Piper Cherokee. With fewer than one half the parts and rivets of the predecessor Comanche, the Cherokee was regarded as a major step forward in lightplane design.

Since his retirement from Piper half a dozen years ago, Weick has not been content merely to bask in the sun at his Vero Beach shore home. Instead, he has turned his indefatigable creative efforts in several new directions. Along with Thorpe, Bergey and Jim Bede, he added his design thoughts for Beech's study that led to the development of the company's new, small, two-place trainer.

On top of that, he has embarked on a personal project to develop what he calls a "precision trim control," designed primarily to prevent the stall-spin accidents that statistics show are still plaguing lightplane flying.

Casting a glance at the progress of contemporary aeronautical design, Weick confesses that he is increasingly intrigued by the imaginative efforts appearing in the home-built ranks. But he displays a sense of pique and frustration at what he sees as the mental and legal shackles of today's conventional lightplane research and development.

Says Weick in one last salute to the future: "The young engineers of today spend so much time worrying about making the FAA certification standards that they don't move ahead into new areas."

Living Legends: Wolfgang Langewiesche

BY NORBERT SLEPYAN

In this remarkable man, a hunger for flight has combined with a thirst for knowledge to produce a classic that has survived the weathering of time and assaults by the FAA.

ONE AFTERNOON at three, it snapped. As for so many who fly, for Wolfgang Langewiesche, lightning struck.

Every day, while he sat among the stacks of the university library pecking at data and earning his keep, the mail plane of Transcontinental and Western Air roared low overhead, bringing in its cargo. Langewiesche knew that its pilots would soon lift westward again to set course for country this young German knew and loved. This afternoon, it was too much, and so the young research assistant, accomplished scholar and doctoral candidate closed his books, put on his hat and drove his jalopy through Depression-ridden Chicago toward his dream: to fly, no matter what.

He worked, he pawned, he went without to buy bits of flying time. He made flying his life and ultimately pursued his destiny of revealing the basic mysteries of flying to hundreds of thousands of pilots. He opened a great window to flight with a book called *Stick and Rudder*.

If you don't know *Stick and Rudder*, you are, if not derelict, at least deprived. You may read your Saint-Exupéry, your Gann, your Bach, your Buck. If you have not read your Langewiesche, you are still deprived.

Stick and Rudder has been a phenomenon in two chancy worlds, aviation and book publishing. In the 32 years of its life, it has sold more than 100,000 copies in hard cover. In book publishing, there are two great lodes: the quicksilver best seller that explodes, sells like crazy, spawns movies, paperbacks and book-club selections and then flames out; and there is the true gold, the backlist item, the classic title, the ongoing seller reprinted again and again over the years. Chief among the backlist money-makers is the Bible. *Stick and Rudder* is a backlist nugget.

The first printing of the book was 2,000 copies. Even first-book poets get better first printings today. Such printings usually do not speak well for the future, but in 1976, this work still proudly bears the features of its birth in 1944. In spite of a host of such books since published, it may well still be the bible of aviation instruction.

Langewiesche has a unique combination of talents that have set him apart as a guru to aeronautical novices. When he first came to this country, in 1929, he was a 22-year-old veteran of two German universities and the prestigious London School of Economics. He soon earned a master's in economics at Columbia University and proceeded to the University of Chicago as a doctoral candidate and research assistant in political science. He had a strong head on his shoulders, though his eyes were a bit weak. Economists (the good ones) and pilots (the good ones) have this in common: the saving capacity to face hard facts, be they numbers or forces, and to see them for what they are. Talent number one.

Talent number two: in combination, love—of places and of writing. Born in Dusseldorf in 1907, Wolfgang Langewiesche emigrated to America eager to embrace it, not as a fire-spewing patriot but as a man struck by the magnitude and beauty, the diversity and humanity of this country. America was one place. Another was the sky.

Even as a boy in Germany, he had been fascinated by airplanes. While at the German equivalent of high school, he had been able to work one summer at the Junkers factory. That was in the middle 1920s, when that firm's designers were working on revolutionary combinations, creating trimotored, cantilever low-wing, corrugated-skin models that would become world-known and world-used. As Langewiesche riveted them, he wondered about them and studied them.

That fascination grew in America, but it wasn't borne on the awe generally held for aviators in the Lindbergh era. "I didn't think of it in terms of the heroic," he recalls. "Even the progress of aviation didn't interest me as much as the idea that one could move through the air and look down into people's backyards. One could turn oneself into a little god and see what goes on."

When Langewiesche decided to pursue flying above all else, he had that most cursable, damnable and valuable sort of person for a mentor, a perfectionist instructor. His name was George Adams, and his mission seemed to be to keep Wolfgang Langewiesche, a bespectacled, serious young stu-

dent with a soft, Westphalian accent, confined to the environs of what is now called Midway Airport until he perfected his approaches and landings—while Langewiesche, deserter from academe, pined to be turned loose to see America from the skies. Midway was destined to become the hub of Chicago's busy commercial air traffic, but in 1933 and 1934, as Langewiesche describes it, "flight instruction mixed in easily with the Curtiss Condors that came in and out, the first DC-3s."

The smaller "ships," as they were often called, were Travel Airs, Fleets, Wacos—demanding airplanes not built to coddle ham-handed students, but used to train students just the same. And the students sweated pounds in them.

Each time Langewiesche picked up the moist helmet and goggles draped over the stick of his trainer, by custom left there for him by the previous student, he wanted but one thing: to do the three perfect approaches and landings Adams required to be signed off for cross-country.

Up he would go, with Adams on the ground watching. Again and again, he would fly his pattern, setting up his final glide to follow precisely the path he thought he saw to the runway. He would overshoot; he would undershoot; sometimes he would set the plane on right, but not consistently. The analyst in him would seek the devil out, the demon of imperception, what Langewiesche, the scholar and fanatic, would describe later as Factor X.

"It is a purely psychological thing," he would write about Factor X. "There is something in the air that befuddles a man's mind, his vision, his sense of space." The pilot therefore confronts a grotesque hazard, a potentially lethal chimera: "He cannot gauge decently that which he most needs to gauge: his own motion and height. And so he staggers through the heavens like a drunken man."

Red, as they called him, after his color of hair, finally won the admiration of his instructor and peers for his judgment, skill and tenacity. He went on to do the touring he had longed to do and ranged far and wide, flying some of the newly spawned *light*planes that began to appear in the 1930s: Cubs, "dowager" Aeroncas, Taylorcrafts, Flivvers. He at first shared the scorn of "real" pilots for these seemingly fragile toys, but he couldn't ignore their economy and efficiency. At length, he saw in them the advent of "personal flying" for the public at large.

On one of his long trips, he learned well why Adams had demanded skill in power-off, precision landings. Caught in cloud over a large Georgia pine forest, Langewiesche realized he would have to land. He looked for a landing site and found nothing. With his fuel running low, he finally found a spot with fewer trees than generally prevailed. He "cut the gun," as he described it, and approached just over the trees, expecting the destruction of the Cub and injury to his passenger and himself. He slipped in and gunned the engine briefly. The touchdown was surprisingly okay, but a crash into tree trunks still looked likely. He aimed between two trunks—just as the undercarriage caught in some brush. The Cub came to a halt. Not a nick in the fabric, not a bend in the wood. The clearing was only a bit larger than the airplane, which had to be wheeled out.

He wrote about this adventure and his early flying experiences in *I'll Take the High Road*, which was published quite successfully in 1939. He had actually been taking the hard road—that of a free-lance writer wanting constantly to fly. He had written for many magazines, facing uncertainty all the way.

As he flew, he studied intensely what his airplanes were doing, for his experience was telling him things that his instructors and the current literature had not. He recalls one such time: "Two or three years before Pearl Harbor, monkeying around with a Fairchild 24, I was still trying to figure out how you get into a turn. For a while, I had the theory that the rudder's purpose is to overcome inertia and get the airplane to turn. After it's turning, the airplane doesn't need the rudder anymore. So it was my idea that you give it a shove, which, of course, was wrong thinking. That's not how we do it. The real answer came later. You couldn't find it in any print. These things were not discussed." The techniques of flying were indeed being poorly communicated—and the accident record was proof of that failing.

To most people, flying in small airplanes seemed to be a constant battle against such monsters as stalls, spins and spiral dives. Where the press left off depicting the diabolical nature of these aeronautical struggles, the movies took over in portraying tumbling, screaming, whirling crashes. They often overdid the horror, but too often, pilots themselves approximated their depictions.

The institutional world of aviation tended then to ignore the magnitude of the accident problem. "Selling" aviation was more important. But an insurance man who saw flying more as a way of life than as a sport knew that because industry and Government and the aviation press were neglecting the accident problem, the image and future of aviation were being jeopardized. The man, Leighton Collins, consequently started a pocketsized magazine called *Air Facts,* which was uniquely dedicated to the prevention of mishaps. He didn't shy away from news about crashes. Instead, he analyzed them, searching for lessons that could prevent recurrences. His magazine was not full of romance but of facts about technique.

In 1940, Collins (the father of *Flying's* Richard L.) attended a party at a New Jersey flying club. Contrary to all the predictions, his magazine was surviving. It seems that the smart pilots, especially instructors, were keenly interested in what *Air Facts* had to say. Collins is a shy man, and as so often happens at big parties, this shy person gravitated toward another. Collins found himself talking to a slightly built, curiously intense man who was obsessed with aviation and, in particular, with the basic hows and whys of flight. Collins and this man, Langewiesche, struck up an association that would lead to the latter's attachment to *Air Facts* for many years as a contributor of articles on technique. Some of the articles that appeared in *Air Facts* in 1941 and 1942 were later adapted to *Stick and Rudder.*

Most of the book, however, was drawn directly from notes Langewiesche had been making for years and from further experiments he was able to pursue in the early years of World War II. By 1943, he was ripe for writing it.

"I was full of everything," he recalls, "full of airplanes, full of the United States, full of flying, and I had several books in me. Then came the war, and I had license trouble. I had only a private license—I could get nowhere with a private license and bad eyes. I spent a year as a ground-school instructor at one of the Army flying schools. I had a marvelous time with all those people. We talked a lot, the cadets, the flight instructors and I. It was a very lively period—the year after Pearl Harbor, intense interest on the part of everybody."

He eventually licked the license problem and went to work for Cessna as a production test pilot, bringing his flying hours up to 2,000. As he worked over Kansas, aviation was turning a corner. New designs, new pilots, new demands were appearing, and, as always, instruction was lagging behind. Langewiesche had seen this at first hand: "You should have seen the Army Air Corps manuals. The pilots were excellent, but the manuals—I'm sure they didn't read them. Instructors would tell students things, and what they *said* they were doing was *not* what they were doing." For example, the manuals would demonstrate a turn by keeping the appropriate pedal depressed and the stick leaned to the side until time to end the turn. In real flying, the turn might have ended only at the ground.

In 1943, in a hotel room in Wichita, Kansas, Langewiesche wrote his nugget of a book. Writing a book can be a curious thing. Some writers, especially those with many titles under their belts, see the straightest road to the goal and drive right along it. Others agonize greatly, with many false starts. Most authors drive some and grieve some. Melville or Morrisson, Dickens or Dickey, Jung or Jong, all authors must meet head on their strengths and doubts.

Langewiesche, too: "A writer never really knows what he is doing until he has done it. All sorts of urges make him want to write. In my case, I was then working on things that were lyrical and poetic about flying. There was also this analytical desire to understand and explain the technical side. There was the desire to teach the brand-new person, and then there was the desire to be of use to the person who was knee-deep in flying. So, with all these desires, I wrote *Stick and Rudder.* What I was really doing was to take up the points I felt were not being understood by students and instructors. Cadets had told me that there were things their instructors were having a hard time putting across."

The book was first published in 1944 by Whittlesey House, a general trade subsidiary of McGraw-Hill, which in time absorbed the smaller company and *Stick and Rudder.*

Shortly before the end of the war, Cessna went into subcontracting for B-29s and laid off Langewiesche and many others. Chance Vought snatched him up and set him to flying F4U Corsairs—to his delight. ("You couldn't give one guy more power. Fantastic speed, fantastic strength.") In 1950, Kollsman hired him to help test angle-of-attack indicators. That was particularly fitting, since Langewiesche had long stressed the importance of angle of attack in safe flying. By the mid 1950s, Langewiesche had become a roving editor for *The Reader's Digest* and was assigned to fly about the country to write about America. Not until last July, after 20 years, did he retire from the *Digest.*

Stick and Rudder has known no retirement, which is not surprising, for it is an extraordinary volume, with an extraordinary point of view. In its pages, "Do this, do that" is accompanied by "Here's why you do this, here's how you do that." And the solution to Factor X is outlined thusly—the details follow later: "Flying is done largely with one's imagination! If one's images of the airplane are correct, one's behavior in the airplane will quite naturally and effortlessly also be correct." We now call that concept "positional awareness," but we seldom stress *imagination*.

Much of the literature of flight had involved theory. Langewiesche saw hazards in that: "What is wrong with 'Theory of Flight,' from the pilot's point of view, is not that it is theory. What's wrong is that it is the theory of the wrong thing—it usually becomes a theory of building the airplane rather than of flying it. It goes deeply—much too deeply for a pilot's needs—into problems of aerodynamics. . . . But it neglects those phases of flight that interest the pilot most."

Stick and Rudder does deal with theory, but it also caters to the novice pilot's needs. The foremost need, as Langewiesche sees it, is to know what can bring you to grief. He doesn't conceal the stark realities that might frighten a potential pilot out of buying into aviation, but he offers ways to cope with them and in stark language. "If you ever break your neck in an airplane, your ailerons will probably have much to do with it" is typical of his mode of expression. The book is a combination of blunt warnings, scholarly analyses and encouragement. There is a signed chapter, "The Dangers of the Air," by Leighton Collins, and one feels the same concern that originally motivated Collins working throughout the book.

The book does have some curiously "dated" mannerisms, for it hasn't been changed since 1944, except for the jacket. Elevators are at times called "flippers"; the drawn airplanes resemble 1930s high-wingers, AT-6s, strange one-seat flivvers and even P-39s. Taildraggers predominate. A few of Langewiesche's concepts, including the source of lift, are open to question, but the general soundness of the book is still clear.

Stick and Rudder has had its negative critics, not the least of which has been the FAA. The book maintains that, all things being equal, power controls altitude and the stick (or wheel) controls airspeed. Somehow, the FAA could not grasp, accept or swallow that basic lesson of years of flight experience. Speaking in his quiet, cultivated manner, Langewiesche recalls that "for a while, at the instructor seminars, the FAA was actively fighting me. The confusion was over what the stick does and what power does. Sometimes, instructors would say to the Government, 'There's something wrong with your theory,' and they would quote me. There's a small group that

runs these seminars, and they would say, 'That man is wrong. Don't give people that book.' I was getting a distinct sense of hostility from those people. I have none against them. They have now caved in and teach, in some respects, what I have been teaching, which is that the stick inevitably controls speed."

Among the book's critics, the most negative is its creator. "The book is all confused," Langewiesche says, his face looking a bit pained. "It talks both to the novice and to the experienced student. It's a basic failing, particularly among educators, that the teacher tends to operate on the frontier of his own thinking without realizing that the poor guy is back there trying to work out the first points." Were he to write the book again, he says, he would "clearly focus on what I'm trying to do. At the time, I thought I was doing one thing; I was really doing another. I thought I was writing for the zero-hour student, and I wasn't. Now, I would just call it *About Flying*. I would write the book from scratch, saying, in effect, 'All right, you've never flown. I'll take you through this thing.' "

Langewiesche is too hard on himself. The confusion is not that apparent, and the fact that there is meat in its pages for the person "knee-deep in flying" and for the student makes this a book for new pilots to master and experienced ones to review. As a token of its reputation, more than 8,500 copies were sold in 1975 alone.

Langewiesche's interests in flying and in America have not flagged as he approaches his eighth decade. He still flies around the country with his wife, Priscilla, in their Cessna 182, which he bought new in 1965. He flies an average of 200 hours a year, including at least one trip a year from his home base, Sky Manor, a pleasant airfield in New Jersey, to California, where his son, William, is studying at Stanford. William, 21, is also a pilot.

His interest in America is like his interest in people, keen, penetrating and genial. An interviewer of him finds himself interviewed in return, as Langewiesche's be-lensed blue eyes study him. He wants to penetrate the look of America as he would search an aging person's past so as to understand his face. When he talks about America, Langewiesche seems to look inward, and he quietly but a bit breathlessly describes his current dream:

"I would like, before I die, to write the book that I came over here to write. A book not in celebration of the United States but one that would show why America is *so damned interesting.*"

Bill Lear beside his Waco in the early days.

Living Legends:
WILLIAM P. LEAR

Carrying genius and audacity as naturally as his arms, this unlikely master builder not only marches to the beat of a different drummer but he designed and built the drum.

BY GEORGE C. LARSON

THE OLD SAYING, "Rome wasn't built in a day" takes no consideration of the eventual coming of William P. Lear. Had Lear been there to run the project, it probably would have been completed in 17 hours—and some of it from off-the-shelf components.

That Lear is some kind of genius there can be no doubt. The question is what kind. He says he never got past grammar school, and yet he has an innate instinct for accomplishing projects far beyond his own talents, with results that sometimes amaze even him. He is a loner in some ways, and yet he seems to feel happiest when surrounded by happy throngs of workers who fly through their tasks like so many weavers. He almost never concentrates on one thing at a time, and the fruits of what he calls his "aggressive curiosity" along with a remarkable talent for promotion have led him at a breathless pace across whole fields of disciplines, from aviation through

electronics into consumer marketing and automotive engineering.

He has always been exceedingly generous with the press, and when the time comes to have a photograph taken, he will suggest poses he wants to be seen in. He attaches his name to his products without fail, and this habit has taken his name out of the ranks of industry and into the household. All of this is only partly vanity, for he seems less vain than proud of what he's done. Although he has met with his share of failures, he tosses them off as events of absolutely no consequence, mere ripples in a steady upward-reaching line of progress. Successes meet with much the same treatment from Lear, for the pattern of his accomplishment suggests that he is at his best when he is spawning an idea; once that's past him, the actual production is only of secondary importance. Production is not an idea; it is just production—less fascinating.

This enigma, this curious American giant, is still putting in a full day at his digs that lie just northwest of Reno, Nevada. It seems the unlikeliest place in the world, and yet, in a way, it suits him—off by himself, defiantly and brashly assaulting the impossible. He and Nevada were made for each other.He is now approaching 75 years of age, although "approaching" is not the right word, for when he hits 75 this June 26, it will be at a full sprint.

Lear started out in a time when a young man lived by his wits, and that's been his *modus* ever since. In 1926, he borrowed a vacuum tube rectifier, scrounged a condenser and transformer and built a device that ran home radio sets on alternating current. In a style that he has duplicated innumerable times, he invited all the "bigwigs" (he calls them) in the radio business to his tiny apartment to see the device operate. One by one, they led him out to the hallway to make him an offer. He settled for $1,000 a month. Soon after he went to work, his boss sent him to troubleshoot another company's product that was generating such awful RF interference that you couldn't operate a radio within 12 miles of it. Lear looked at the device, which was made up of several large capacitors. To this day, he claims he doesn't know how he knew what to do, but some instinct told him to place a smaller capacitor across the larger ones. The noise disappeared.

His reputation was made, and he was rewarded with more money and some stock. He came across a dynamic speaker with heretofore unparalleled reproduction quality and instantly assessed its value as a new design around which a home radio could be built. Within months, distributors were lined up with cash in their hands to bid for the available production of the "Majestic" radio set. He dreamed up an automobile radio, but people told him it would be legislated out of existence—too distracting to the driver. He and Paul Galvin, while driving home one night from Atlantic City, New Jersey, coined the name Motorola. Years later, he would design the first automotive stereo-tape player and the eight-track cartridge that fit it. He soon sold that business, but in 1975, the industry he had started sold $4,500,000,000 worth of stereo gear and tapes. He says he has hardly ever sold a business that failed to make the buyer a profit.

Lear sees his greatest talent—and he's not reluctant to discuss it—as the ability to assemble groups of people, fill them with purpose and accomplish some completely new project in an astonishingly short time. His specialty—his "vision," for want of a better word—is conjuring up devices where there were none before. He has an innate faith in his own innter voice, which tells him about certain things he wants, even though they don't exist. Although many people may experience such bursts of imagination, only Lear seems to heed that voice.

"I built the Learjet because I wanted one," he says, and it's true. The car stereo set was another example of desire becoming substance. Lear found a Swiss ground-attack airplane, the P.16, which had experienced a rather spotty career due to systems failures. Lear says, "People thought, 'How could anything that came out of Switzerland be any good?' But it was the best damned ground-attack airplane in the world. The trouble was a lack of systems know-how. They had a hydraulic system on the thing that wasn't pressurized, and as soon as they got to altitude, they had foam instead of hydraulic fluid."

Lear got busy. With no computer to help, he tackled the design of an advanced-performance, complex airplane. There was no wind tunnel, and he'd be the first to say he is no aerodynamicist. He had other attributes, though, principal among them being the unswerving belief that because he wanted a tiny jet that would scooter him around and cost about $365,000, others must want the same thing. "If I'd done a market survey like they do for known products—this had never been done before—the airplane would never have been constructed. The corporate pilots at that time would have said it was way too small, the range was too short, and it flew too high." When he was done, he had a seven-place jet that has become synonymous with the idea of a business turbojet just as the term Piper Cub has come to represent the archetypical lightplane. At airports all over the country, you will overhear "experts" lining the fence who will patronizingly explain to listening youngsters that the SabreStar-125 taxiing over to the ramp is a "Learjet." Part of the magic in its appeal is that Bill Lear's name is permanently attached to that airplane. It's his. It's a Lear. The Learjet.

William P. Lear has never been equaled, will never be. He knows that, and the knowing has set him free from convention. The simple fact that he can bring off the impossible is measure enough, for he has looked around for years now and found no one gaining on him. Part of the reason for that is that nobody else has entered this peculiar race he is running. He is in there, though, setting records for himself and enjoying the cheers.

He is as fascinating now as ever, a mercurial imp who gets work out of people for endless hours and beyond normal endurance by a cruel cajolery that makes all of the sacrifice a kind of tolerable joke. It is hilarious or it is no fun at all. He can count those who have stayed with him—Sam Auld and Hugh Carson—on two fingers of one hand, and he knows it. "People leave me when I sell my companies," he says, by way of seeing past the question, "because I can't sell the buyer an empty shell." That wouldn't be fair at all. He says his secret of getting people moving is that they share in the glory, the success and the profit, and he is a man with plenty of friends. Many of them are Learjet owners. Justin Dart is one who plunked down cold cash in the dark days before there was a Learjet, when it was just talk. Dart—as well as gaggles of other owners—all showed up at the hangar at Reno-Stead last year for a huge birthday party for Lear. "People who own a Learjet will tell you it's their proudest possession," he says, and again, he's right. Never has a jet been built that incited such strong feelings of personal ownership.

Lear is earthy, rebellious, unconventional, colorful, angry, and a bit of an artist. He's always been led by the precept that if something, anything—a machine, a woman, an animal—looked beautiful, it would work and perform commensurately. Because he once lusted after airplanes enough to survive a wreck during his first ride in one and still come back for more, airplanes have been forever affected by his little jet, his Learjet, the needle with a man in it. Everyone's fantasy. But all the rest ignored the voice. All but William Lear.

PART IV

Air Power

During the 1920s, it became evident to many observers that the recently ended "war to end all wars" may have been a prelude to an even more horrible conflict—that the Armistice was just that, a temporary cessation of hostilities. With that terrible vision came a certainty to many, including William B. Ziff, Sr., that victory would come through air power and that catastrophe beyond mere defeat would come through the lack of it. General Billy Mitchell put his brilliant career on the line because of this conviction, and Ziff used the pages of *Popular Aviation* to broadcast Mitchell's warnings. Increasingly, the young magazine covered developments in military aviation and, as the world became enveloped in war, the aerial combat itself. Military affairs remained a major part of *Flying's* contents well after the end of World War II.

How hideous "aerial terrorism," as *Flying* writer Frederic B. Acosta described it, could be was first realized as German and Italian warplanes, flying in support of Francisco Franco, pummeled Spanish cities. The dire predictions that had made by the pro-air power "hotheads" had come true—and much worse was to follow.

The Spanish Civil War revealed to what extent the Germans, Italians and Russians had been developing their air forces. New generations of fighters and bombers battled each other, as these countries used the war—the Germans and Italians fighting for Franco and the Russians for the Republican cause—as a proving ground for their new hardware. Similarly, the Japanese invasions of Manchuria and China served to reveal how strong the Japanese air arm had become. In the Orient, too, bombing was used to terrorize defenseless civilian populations. The Chinese themselves had virtually no air power to speak of. Foreign pilots and planes, particularly Americans flying Curtiss P-40s, offered strong resistance to the Japanese, but even The Flying Tigers were impressed by the air campaign the Japanese were mounting.

This advance in Japanese air strength was deeply troubling as tensions over Japanese expansionism and American objections to it grew. The United States rushed to improve its air power. As this country did so, World War II began in Europe, and within a year the British found themselves facing Hitler virtually alone. Their pilots, including many from the Commonwealth countries, and volunteers from the United States along with escapees from Nazi-occupied Europe, proved the key factor in preventing a German invasion of Britain.

Meanwhile, the United States and Japan were edging toward war—more quickly than most Americans realized. Indicative of that is the irony of an article's appearance in *Flying* proclaiming that the Japanese were such inferior soldiers, sailors and pilots and that Japan itself was so vulnerable that we would easily vanquish such a foe and that the U.S. should initiate preventive warfare. "Stop Japan Now!" appeared in the December 1941 issue, which many *Flying* readers did not see until Pearl Harbor had been bombed and the Japanese had scored several important victories.

During the war, *Flying* published several special issues covering our Army and Navy air arms. The issues were grandiose with many illustrations, including some in full color, and their articles were often written by military officers. They provided *Flying* with comprehensive surveys of most aspects of our air operations.

The magazine's heavy coverage of military aviation lightened considerably in the late 1950s and early 1960s, but yet another special war edition appeared, in November 1966. The subject was the air war in Viet Nam, and the issue was, again, comprehensive but more dramatically written. The Editors felt an obligation to their readers to cover this major event, an event that was to continue far longer than most people had expected—or wanted.

The Startling Truth About War Flyers

Capt. Rogers completely destroys many of our pleasant beliefs about war flyers. There was no chivalry in war fighting,—no ethics. Either you got the other fellow or he got you.

BY CAPT. BOGART ROGERS

EVERY time I hear someone speak of the war in the air— the late war and the French air—as a gallant and romantic business, a modern counterpart of the chivalrous strife of old, I break right out laughing.

It was a cold, calculating, deadly occupation—sans chivalry, sans sportsmanship and sans any ethics except that you got the other fellow or he got you. If you could shoot him in the back when he wasn't looking, or bring odds of ten to one against him, so much the better.

The great aces loved to increase their victories on fat old German planes that couldn't do much to stop them. You find the famous Major Bishop, for example, saying he was "praying for some easy victims to appear."

The only people who fully understand the war in the air are the fellows who fought it. You can't fool them. They know.

A lot has been written about the most famous heroes of the war—Guynemer, Ball, Bishop, Richthofen, Fonck, Rickenbacker, Luke and a host of others. Their chroniclers have penned their doughty deeds in detail. They tell what these fellows did in the air, but, doggone 'em, they say little about what they did on the ground—what they thought, how they felt or why they felt that way.

In a general way, it is intimated they were brave, high-principled young men fighting for king, country, France,

democracy or something of the kind, which had little, if anything, to do with it.

War fliers were a class apart from the rest of the army. They thought, fought, lived and acted differently.

There were more ways for them to die than being killed by the enemy. They traveled at breakneck speed—aloft and below. They were young men, very young, for the most part, and inclined to be wild. Death was always just around the bend. They knew their worst enemy was the law of averages, which would overtake them in time. They knew if the war lasted they wouldn't. So they lived a dizzy pace. Life was to be short—they would make it sweet. "Eat, drink and be merry"—that was their motto.

They were, most of them, brave and daring and skilled in the art of killing, but they thought and did the most peculiar things for apparently the most incomprehensible reasons.

There was, for instance, the odd fact that wounded pilots always tried to get back to their home fields, even though consciousness, or perhaps life itself, was slowly slipping from their grasp.

Bob Graham was just a pilot in a British fighting squadron. He went out groundstraffing one morning during a furious German attack, a nasty job that carried him through a hell-hail of gunfire. When he returned to his squadron field, he taxied up to the hangars, swung his plane into line and switched off his motor.

"Hey, Pete," he called to one of his mechanics. "Come out here and give me a lift—and see if you can find the medical orderly. I seem to have lost part of a leg."

They lifted him from the cockpit, peeled off his flying togs and several pairs of woollen socks and found two very mean shrapnel wounds. The major arrived, took one look and told the medical orderly to rush young Mr. Graham to the nearest hospital for repairs.

"Only," said Graham, "I'll just stop and pack my baggage and take it along with me."

"Never mind your baggage," said the major. "I'll send it along this afternoon."

"Listen, major," Graham replied, "if I was wounded in seventeen places and about to die—which I'm not—I wouldn't move an inch away from here without my baggage. I hate to seem stubborn, but no baggage, no hospital."

I saw him two months later and kidded him about his baggage. I said it seemed funny under the circumstances.

"Funny, my eye!" he said. "Everything I own was in that baggage. I had a razor in there I've had since I started shaving. Somebody would have pinched it–they always do. If I'd been hit in the head by a six-inch shell I'd have come back for my baggage just the same."

The same morning "Binge" Tyrrell, a dashing Irish lad who had received a Military Cross only two days before, was killed. He had been ground-straffing, too. Before his job was finished he fired a green light, signalling his companion he was going home. Half way to his destination he suddenly dived into the ground from a thousand feet. They found him later, a jagged shrapnel wound in his chest. He had died in the air.

If the statistics were ever compiled they would show that scores and hundreds of aviators lost their lives trying to get back home. Squadron commanders used to caution their boys to land immediately if they were badly hit, but somehow or other a lot of them wouldn't do that—and died when they might have lived.

Consider the circumstances and perhaps you can understand the reason. A "dog fight"—a swirling, diving, twisting chaos of planes, all shooting like mad. You're chasing a Hun, another one is chasing you. You feel a smashing blow in the shoulder, a sickening sensation and a peculiar one, perhaps not at all like you thought it would feel like to be hit by a bullet. Your plane tumbles into a dive, but you pull it out and streak for the lines, praying nobody will follow. There you are, thousands of feet in the air perhaps, alone, badly hurt—you don't know exactly how badly. It may be trivial or it may be fatal. You assure yourself it can't be anything serious—you'll go home and find out. You could land as soon as you crossed the lines—but you'd rather go home. Of course you would. You're hurt and you don't want strangers messing around. You want to go back to the fellows you know and like and understand. And your plane—you might crash it landing in a strange field—much better to get back home. . . .

There was no place in the air service for grand-standers. Except with the French, who characteristically made an unnecessary to-do over the most trivial achievements, heroics and theatricals were apt to get what is vulgarly known as the raspberry.

They were hard to put across. There was no audience at all—just the players in the game—and they were exciting and brutal critics. They could appeciate a good show well done, a brilliant bit of strategy or a daring flash of action, but they could see no reason for blowing steam calliopes about it. After all, it was all in the day's work.

It was no place for four-flushers or bouquet collectors, the flying corps. The boys knew a hero when they saw one. A lot of aviators received, for one reason and another, far more praise and glory than their deeds deserved—received it from the outside but not from the fellows they worked with and fought with and who knew to a split hair just how good they were and how much they had accomplished. Inversely, some of the greatest fighters in the service were almost unknown outside of it.

The lads who fought the war in the air resorted to many devices to make a nerve-racking business a little more easy to bear—embraced them unconsciously, for the most part, without recognizing them as the salvation they really were.

Four of these stand out:

A gradually acquired callousness and indifference to anything that might happen.

Strong drink, and lots of it.

A sense of humor, grim and distorted perhaps, that enabled them to push tragedy across the hair-line that separates it from comedy.

An ability to find solace in the doctrine of fatalism.

Callousness and a hard-boiled and unsympathetic attitude were the chief salvation of the air service.

Outwardly nobody was sympathetic. They had feelings, of course, but not very obvious ones. If your best friend was shot down you masked a breaking heart by declaring he was a damn fool who should have had better sense. You tried to make the rest of the boys believe he got just what was coming to a fellow who made a silly mistake—and they pretended to agree. It was his fault and it served him right!

The dear departed were seldom mentioned in public—at least not until enough time had passed to permit their being discussed impersonally.

What else could these boys do? They couldn't go all to pieces every time one of their pals "went west." Others would go the same way and they had to carry on. So they put the blame on the deceased and almost kidded themselves into believing they didn't care.

Only once did I see an entire squadron really go to pieces. It was a most unpleasant affair—like a baby's funeral.

Little Jerry Flynn was the pet of the squadron. "Gee Whiz," they called him—a Canadian kid who was just eighteen, but a captain, a flight commander and a veteran at this new kind of slaughter. The boys didn't like Jerry—they loved him—as you'd love a kid brother. When he would start away on a patrol with his head barely peeking over the edge of the cockpit he looked like a small child—he had little boy eyes and not the sign of a whisker.

One afternoon the whole outfit went on a short patrol over the lines, so short and so simple that everyone got careless.

Three miles over, a dozen Fokkers appeared from nowhere. Jerry never saw them. Before the rest could come to the rescue he was tumbling down in flames.

His particular pal was a chap named Green. They were like brothers—lived together and played together. When Green landed and crawled out of his plane the tears were streaming down his cheeks. He'd been crying all the way home.

"Poor little Jerry," he sobbed. "Oh, my God! They got him in flames! It was all my fault—I let him down—I didn't protect him!" He wailed out more and more worse things than that. If you've ever seen a football player whose mistake has cost his team an important game, you may have a faint conception of how this lad talked and acted. His heart was broken.

Everyone was on the verge of tears. It was too great a tragedy to conceal. It penetrated their calloused exteriors and jabbed at their hearts.

It was at dinner that the whole outfit blew up like a toy balloon. There were more cocktails than usual—it was the easiest way to forget. But for once they didn't work that way—they only made matters worse.

By the time the soup appeared everyone was three sheets to the wind but the teetotalers. There were guests for dinner and champagne was in order on guest nights. They guzzled it like water but it had no effect—the place was a morgue. Nobody talked—nobody dared talk. Jerry's seat was vacant—there wasn't a soul who would occupy it.

It happened suddenly.

A kid named John Trusler grabbed his champagne glass, hurled it the length of the mess, leaped to his feet and started a vivid impersonation of a lunatic passing his entrance exams for Mattewan. He swore and cursed and cried. He cursed God and the Germans. He cursed the war and the army. He cursed his parents because he was born. He told little Jerry Flynn—who he knew was in the room listening to him, who was sitting right there in that chair!—that he didn't have to worry. They couldn't kill him and get away with it.

He cursed far more capably and colorfully than a lad of his years had any right to do. They tried to stop him, and finally pulled him down in his chair. He hid his head in his hands and sobbed horribly. Half of the fellows were bawling. The rest were trying to quiet things down but feeling no better than the weepers.

Finally it subsided. The orderlies brought more wine and the original objective of finding solace in alcoholic oblivion was successfully attained.

The next morning the hard-boiled masks were up again. They had to be, of course. They all told each other it was just another binge. They kidded Trusler for getting on a crying jag.

It was a tough break, they said, and the place wouldn't be the same without little "Gee Whiz"—but, after all, "C'est la guerre," and he had nobody but himself to blame. Why in hell didn't he watch his tail? Any guy who let himself be caught napping couldn't get sore if they shot him down!

Green had apparently regained his bravado with the rest. He went on an early morning patrol and evened things up for Jerry by knocking a Hun down in flames.

But later in the morning he collapsed—suddenly, unexpectedly and completely. His nerves snapped with the twang

of a broken flying wire and they sent him home for a long rest.

Strong drink and lots of it was a boon and salvation to the aviators. Show me a good, stout-hearted, cool, dependable air fighter, and I'll show you, nine times out of ten, a hard drinker. The fellows who drank could stand the gaff. It let them relax, it enabled them to forget and it made them sleep.

You usually found a set-up somethng like this:

A fighting squadron would include some twenty flying officers. About half of them would come under the heading of two-fisted drinkers. A few more would take a drink if sufficiently urged. Two or three might be teetotalers.

The boys whose records showed a lot of hours at the bar were the ones who chased the Hun hardest. Don't get the impression that fliers were habitual drunkards who flew and fought in an alcoholic daze. An overwhelming majority of them didn't. There were rare instances where boys would fortify themselves with a stiff little nip before starting on a patrol, and a few fellows used to carry a flask aloft with them, but they were the exception. It was simply that when the day's work was done the evening's drinking commenced.

Medical men say there is a physiological reason for this—that they can understand why the heavy drinkers might well be the best fighters with the steadiest nerves and the coolest heads. The fliers lived an abnormal life. Death and disaster were the daily diet. The exhaustion of nervous energy was tremendous. They had to relax and sleep and forget. They were in the front line every day, week in and week out. The end of a day's work found them in an abnormal physical and mental condition. They'd have to do the same thing tomorrow. They were tired and jumpy—so they drank—and their liquor made them normal.

The boys who chose to "shun the fatal curse of drink" were under the same nervous strain. They remained strictly sober and were therefore able to think straight. You did yourself no favor by keeping a clear head. The more strangely you could think, the happier you were likely to be. Instead of spending their leisure moments in convivial—you might even say bibulous—pursuits, the abstainers had plenty of time for cogitation. They could mentally figure their chances of coming through with a whole skin—which were awfully slim when you looked the facts in the face. So they worried and fretted and had no way to get the fears and horrors out of their systems.

They began to sleep badly. Some of them used to stand up in their beds and carry on furious battles with the enemy. They'd zoom and dive and rattle visionary machine guns and moan with horror when one of their comrades went tumbling to earth—which showed that these things really did get under the cuticle.

One chap I knew used to lay the whole cast of the show—he was both himself and his enemy and he handled the dual role like a Barrymore. The rest of the boys would rush in and watch these exhibitions with awe and wonder—but they seldom laughed. There was something about somnambulistic aerial combats that rapped your heart and not your funny-bone.

They got the "woofits" and lost their appetites and came to hate the whole messy business and sooner or later they cracked. Sometimes they cracked at the wrong moment—lost their nerve for a fatal instant when a Hun was chasing them up and down the sky—and the commanding officer wrote a consoling letter to their next of kin.

I know of no war flier whose mental or physical efficiency was impaired by drinking. They could stand a lot of it. They were young, they lived comfortably, they could sleep as much as they liked. They were hounds for exercise and spent most of their time out of doors. You could put them to bed at midnight totally out of control and they'd climb out at six in the morning, or earlier, fresh as a morning glory.

I know an indulging lieutenant who would, about once a week, get into an alcoholic tailspin, go to bed and lie on his back shooting holes in the roof of his hut with a six-shooter— "to get a little ventilation." He's a highly respected citizen now. The last time I saw him he politely refused a drink—he doesn't use it any more. But he did then—he certainly did.

And why not? They didn't figure futures, these fliers. There was no future. Life was today—they couldn't tell about tomorrow. They drank and gambled and loved, whenever the opportunity presented itself, and squandered their money and did a lot of things they never would have thought of doing before the war—and that they knew they would never do again if they survived. It was a phase, a life apart from life. It had no connection with the past, nor with the future.

Lack of imagination was another blessing.

The less you had of it the more happy you could be. A lot of fellows who were supposed to have no nerves had nerves but no imagination—or kept it effectively strangled. There were so many things you could imagine might happen. You didn't dare think of them. You learned not to.

There was one almost universal weakness of imagination that helped everyone tremendously—they couldn't imagine themselves dead. Other fellows, chaps who were better fliers and fighters, more experienced, luckier perhaps, could go west by the dozen. The great aces were knocked over one by one. They could be caught unawares and shot to pieces in the air or tumble down in flames or destroy themselves in a crash, but you couldn't quite imagine these things happening to yourself.

It might have been vanity or ego, but it was surely an amazing paradox of thought. You knew your time would come. You knew the law of averages would catch up with you. You knew they'd kill you sooner or later—but deep in your heart you thought they wouldn't. Your brain was positive your days were numbered, but your imagination refused to prove it.

You'll never know the number of fellows whose ability to laugh at tragedy enabled them to see the war through. The fliers laughed at danger—they laughed at death.

Major Douglas Hill commanded a bombing squadron. Seven Fokkers chased him back from a raid one afternoon and shot his machine to ribbons. His observer, sitting directly behind him, was riddled with bullets. Hill headed for the first aerodrome he could find—his companion might still be worth salvaging. A lot of his good friends occupied the field where he landed.

When his plane touched the ground it literally fell apart. The tail surfaces parted company with the fuselage, both tires were flat and one wing drooped alarmingly. He taxied to the hangars dragging the wreckage behind him.

They pulled the observer out, hit in seven places but still very much alive, rushed him off to hospital, and then turned their attention to Hill. They looked at his plane, shattered by a hundred bullets.

They looked at him, white as a ghost and with his eyes sticking out on straws, and burst into uncontrollable laughter. He stuttered when he tried to tell them what happened and they roared till the tears came. The more he stuttered the more they laughed.

Every detail of the harrowing battle, the battle that had nearly been his last, brought new explosions of mirth. The idea of Douglas Hill fleeing for his life from seven deadly Fokkers, his observer unable to protect him, clouds of smoking bullets whizzing by his head, scared pink and positive he would never get home alive, was side-splittingly funny.

They laughed, too, when a comrade at arms hastened home one day shot through that portion of his anatomy commonly known as the seat of the pants. He was listing badly and sitting on his liver, which made it difficult to fly, and he was suffering acutely, but his misfortune was his friends' good fortune—it gave them something to laugh at.

Most fliers, I think, were fatalists. It was the only doctrine that would hold water. If you embraced it, as many did, it was a great source of consolation. You simply decided your destiny was predetermined and inevitable, and ceased worrying about what might happen to you. When your time came it would come—there was nothing you could do to stop it.

Howard Knotts, who flew in an American squadron attached to the British, is only one of thousands of cases in point. He was being run ragged by a Hun who sat on his tail and wouldn't get off. With his head turned, Knotts was watching his adversary every instant, trying to out-maneuver him or escape.

A flurry of bullets passed his face—one of them hit him on the bridge of the nose, so lightly it scarcely broke the skin. Somebody chased the Hun away—or Knotts finished him, I don't remember which—and he came home and repaired the damage with a piece of courtplaster.

Figure it out for yourself. The bullet was traveling some 2,500 feet a second. Knotts was moving at the rate of 100 miles an hour, 147 feet a second. If the bullet had hit three inches farther back—aloha. Figuring the speed of the plane alone, the old man with the scythe missed him by one six-hundredth of a second.

These narrow squeaks happened every day—many times a day. A few of them were convincing proof that you had no more control over your destiny than you had over what Ludendorff ate for breakfast.

With all its misery and violence and destruction, there was a gripping fascination to aerial combat that held these youths in a grasp of steel.

The heart-hurts have healed now. The strain and work and worry are a dim memory—very dim. In the years since it happened the boys have forgotten the bad—they remember only the good.

So you can ask any one of them—they'll all make the same answer—

"It was a great war while it lasted."

Does America Want a Separate Air Force?

THIS is the most critical question in what may yet prove to be the most critical hour of our history: does America want a Separate Air Force?

Which is the road to efficient operation? . . . Which will serve to create that invincible defense force for which America now struggles against time? The destiny of the nation trembles in the balance as the decision is being weighed.

Now we are up against stark reality. This is no cheap political matter, no subject for pork barrel politics. We are in a world struggle, the result of which will determine the course of history for all time to come. The foremost weapon in that struggle is the airplane. Whoever rules the air will rule the continents beneath it.

Along with others who have fought these years through for

William B. Ziff, Sr.

American air supremacy, I believed that the nation would be better served if it had a Separate Air Force, a striking arm additional to the Services employed by our admirable Navy and rapidly expanding Army. I have been influenced by many matters—by emotion perhaps—by what seemed to be inescapable fact, by my love for Billy Mitchell, the great American strategist of the air.

What a great and stirring figure he was, the man of prophecy and vision, coldly honest, but not political-minded. The passionate contentions which centered around his name have been gradually erased by time, but the fiction of contention remains. I am happy to say now that whatever stands in the way of General Mitchell's enshrinement as the great American prophet of aviation is to be removed. His great qualities as an outstanding military genius will be publicly recognized by his reinstatement in the rank of brigadier general, by recommendation to Congress by the War Department. This noble and well-justified gesture toward the memory of a great man will go far to palliate a sore whose roots have struck deep. It will be testimony not only to the greatness of Mitchell but to the stature of those who will make it.

What stand shall the American people take on the question of a Separate Air Force? Frankly, I do not know. We have a great Chief of Staff who asserts that any effort in this direction at this time would be disruptive or even disastrous to the defense effort. Let us recognize the factors for what they are. Since he has gone on record so strongly, a vote for a Separate Air Force will be a vote of non-confidence in his opinion and judgment. It will be a vote of non-confidence in the War Department; it will be a vote of non-confidence in the Secretary of the Navy who as strongly opposes the measure. If it passes Congress, as it now seems likely that it will, it will certainly be vetoed by the President. It may thus create disunity and, with it, resulting inefficiency at a time when the nation can least afford it—even if the proponents of the measure are correct.

On the other hand, the need for a separate air establishment may be so great as to make even so grave a risk as this necessary.

This is the issue which is now to be tried before the Senate. May God grant the gentlemen wisdom!

William B. Ziff
Publisher

Crushing America by Air

Billy Mitchell

BY GENERAL WILLIAM MITCHELL

Former Commander, Air Forces, A. E. F., and Director, Military Aeronautics, U. S. Army

The tremendous military value of air power has burst on the world so suddenly that its realization is difficult for the average citizen. The idea is deeply ingrained into most of us that armies and navies must always conduct our wars, because they have done so from time immemorial. Such, however, is by no means the case today.

The distance from Washington to Richmond, Virginia, is 121 miles. In the Civil War, it took the Union Army about five years to get there. An airplane would have covered the distance in less than one hour. In the European war, the armies faced each other along a 250 mile front. They went backward and forward for about sixty miles until the end of the contest. Men were killed by the thousands by machine gun fire, they were shot up with cannon, stuck with bayonets and devoured by disease. This modern and "humane" warfare could only result in one thing, the utter destruction of all the participants in such a combat. It ceased to be war and was merely a slaughterhouse performance in which no science, art or ingenuity was involved. The European war, except for a few battles at the beginning, was one of the most uninteresting that history gives us.

Armies have now reached a stalemate and the old theory of war, that victory meant the destruction of the hostile main army, is untenable. The idea behind this was that once the army was destroyed, all the vital centers of the opposing country lay open to the invader. Now, however, armies have become mere holders of ground and air power is the only thing that can go quickly and surely to the vital centers of the enemy. If you tie aviation to an army, you are going to stop its effectiveness at once.

Suppose, for instance, that Chicago and New York were the capitals and vital centers of two opposing states. They are about 800 miles apart. Suppose that the military force of Chicago consisted of a great army of one million men with only aviation enough to act as a means of reconnaissance and observation for its infantry and artillery. Suppose that the army of New York consisted of 100,000 men but that they had an air force of 1,000 airplanes capable of flying to Chicago and back, and each capable of carrying two tons of projectiles, air torpedoes, bombs, gas projecting apparatus and guns. Suppose this force were launched at Chicago, with 2,000 tons of projectiles. They would make the trip in seven hours. They could not only force the evacuation of the city and set on fire all the towns and manufacturing plants around Chicago, but could throw gas over the fields where cattle and other livestock lived, destroying all of them. They could drop chemicals on the fields which would destroy all vegetation and even prevent their use for agriculture for a considerable time. The gas would affect all water supplies, even rivers and lakes. The result of such an attack would be absolutely decisive. The armies would be powerless to avert it and nothing could stop it except other airplanes taking the offensive against the enemy at the same time.

The old services put up a howl that this kind of war is inhumane, but can anything be worse than the long protracted torture to which armies and peoples are exposed when a war is carried on in the present accepted fashion? Air power will get a quick decision. The mere threat of its use may be sufficient to force a peace.

Let us suppose that war occurs between a nation on a continent, such as Europe, and one situated on an island or series of islands, such as England or Japan. Both these empires depend almost entirely for their existence on their overseas commerce. As most of their people have been called into the industrial centers to manufacture goods which they sell, there are very few left for the production of food and the necessities of life.

Great Britain is a direct descendant of the Phoenician system. This comprises a series of armed ports throughout the world into which the raw materials of the various countries are gathered. These materials are picked up, put into boats and transported to Great Britain where they are manufactured into articles of trade and then distributed to the world from ships. To break down these trade routes at any place, for instance, the English trade routes through the Mediterranean and the Suez Canal, would be her downfall. Similarly, the breakdown of the trade routes through the China Sea and the Straits of Malacca would have a corresponding effect on Japan.

Air power is entirely capable of breaking down these lines of communication by sinking the ships and causing the evacuation of the ports. An attack by air power on islands such as those comprising Great Britain and Japan would be decisive because the people have no place to go and nowhere to get food except on these comparatively small stretches of land. It is doubtful if a two months' reserve supply of food exists in Great Britain. The targets are well defined, easy to pick up and tremendously easy to hit. Both these nations thoroughly realize this and both have organized great air forces which are entirely independent of the Army and Navy and are charged specifically with the defense of all the air over their respective countries, and with making air attacks against enemy countries.

In all the leading countries, air, land and water are under independent ministries, which are brought into one organization, either a department of national defense with a secretary or minister at its head, or under a committee that handles the whole war-making system of the state. In this way coordination is secured and each branch of the service has its own voice in the scheme.

In England, the Air Commander has charge of all defense arrangements for the British Isles in case of war and the actions of the army and navy within 200 miles of the coast will be coordinated under his direction. In other words, the head airman is the supreme commander of all forces in case England is attacked.

I mention England, particularly, because they received very heavy casualties and damages from aircraft bombs during the war, which brought home to them the realization of what would happen in another war. England started in the World War with an Army Air Service and a Navy Air Service, just as we have in the United States now. When the German airplanes raided England, the Army Air Service would chase them to the shores of the sea, at which point they would have to turn around and go back. The Navy Air Service was then supposed to take up the pursuit but they were never there, so the result was that the Germans got away safely. The British soon put a stop to this state of affairs by creating a Ministry and an Independent Air Force.

In the United States today, there is an indescribable mess about the defense arrangements all along our coast and in the interior. Nobody knows who has charge of anything. There is no air commander and no similarity of instruction between the Air Services of the Army, Navy, Marines or Coast Guard. There is no air force whatever.

From the Navy we have heard a great hue and cry about the Washington Conference for the Limitation of Armaments, brought about in 1921, and how adversely it affected the United States. The fact is that nothing in the history of the world has done more toward the reduction of useless and expensive armaments than the calling of this conference by Secretary of State Hughes during the Harding administration, and the results it obtained. It has saved millions to the tax payers of all the countries involved and has shown people what can be done under certain conditions with an accord between different nations on the subject of disarmament. The

The hulk of the New Jersey, *showing the effect of a 1,100-pound bomb. The ship sank four and a half minutes after this fatal bomb attack.*

influence of air power had much to do with this.

It will be remembered that we sank the battleships with airplanes in peacetime tests in 1921, a few months before the Limitations of Armaments Conference was called by the United States. The board of Army and Navy officers headed by General Pershing that witnessed these tests put in their findings:

"Aircraft carrying high-capacity, high explosive bombs of sufficient size have adequate offensive power to sink or seriously damage any naval vessels at present constructed, provided such projectiles can be placed in the water close alongside the vessel. Furthermore, it will be difficult, if not impossible, to build any type of vessel of sufficient strength to withstand the destructive force that can be obtained with the largest bombs that airplanes may be able to carry from shore bases or sheltered harbors."

This finding had a distinct bearing on the Conference for the Limitation of Armaments. The Navy flooded the country with propaganda that it seriously crippled our national defense. As a matter of fact, it made us much better off because the ships that we were building not only were surface battleships completely at the mercy of airplanes but even at that they were to be equipped with 16" guns, whereas the British had ships projected to carry 18" and 20" guns, and the Japanese would have followed the lead of the British. They would have waited until we had built a lot of old tin ships with obsolete armaments, then they would have come out with new and improved ships and we would have had to repeat the performance. As long as the leading maritime nations can encourage us to build a navy and keep down our air force, they have a great advantage, because they are creating huge air forces that will be effective.

The same conditions continue here. Our country was flooded with navy propaganda about building cruisers, during the last winter; incidentally, the air people were not allowed to flood the country with air propaganda about how easily these things could have been sunk; and how, if the money were applied to the construction of commercial and military aircraft, a much greater asset for the defense of our country would be obtained. This state of affairs exists because we have no separate department of aeronautics and the Navy and Army either prohibit all propaganda about air power, or distort the truth about it.

The Navy, with their lobby in Washington, pushed a provision through Congress for cruisers. These cruisers are steam vessels capable of cruising about 10,000 miles at 13 knots and carrying 8" guns. The paper and ink had hardly dried with the President's signature to this appropriation act, when it became known that the Germans already had in existence a 10,000 ton cruiser capable of going 10,000 miles at 20 knots, carrying six 11" guns in triple turrets with mountings permitting a very high angle of elevation, and eight 5.9" guns behind shields. This ship also carries six 19.7" torpedo tubes on triple carriages, one at each side of the quarter deck. The big guns of this vessel have a loading gear of a new pattern which enables a very rapid rate of fire to be maintained. Their range is 30,000 yards. The ship is driven by internal combustion oil engines which are said to weigh only 17½ pounds per horsepower. One of these German cruisers could stand off and whip the whole force of steam cruisers that the United States authorized last winter. Not only can this single German ship defeat them any place it can find them but it can sail across the Atlantic a couple of times and do it.

What absolute bone-headedness has been displayed! How utterly inefficient the national defense system is which will allow such a condition to exist. The American people have no knowledge of how they are being duped by the Navy Department in Washington, and the people who sell coal, steel and supplies to the Navy. The cruisers authorized have no value whatever in national defense. Have we stopped the building of these worthless vessels and adopted types superior to anything else? No, we go ahead spending billions and all the Navy says is that we must limit the Germans so that they cannot build such ships!

At the present time we spend about one-third of our total national income in arrangements for national defense or payment for past wars. Of the amount actually appropriated for military purposes, the Navy gets about 50%, the Army about 40% and the aviation, which is split up between the Army and Navy, about 10%. Even if the money for aviation were spent wisely, which it is not, it would go a long way, but so much is used up in useless overhead and inefficient organization that the funds are largely wasted. The actual value of air power, the Army and the Navy to a country might be stated as 50% for air power, 30% for the Army and 20% for the Navy. See how the appropriations are handled.

Mitchell during his court martial.

If the United States had a well organized air force, entirely independent of the Army and Navy, it would be completely immune from overseas attack. No ship could approach our coast across the Atlantic or Pacific Oceans in the face of a well organized air force. The air force is the only thing that can go up in the air and fight hostile aircraft. It is the only force that can launch an attack against a foreign country, either north or south of us, or across the Atlantic or Pacific.

In 1923, I conducted some maneuvers with what aircraft we had left and what airmen remained of the efficient war trained personnel, along the Atlantic coast, to show what could be done with even the primitive aircraft we had then. Having received authority from Congress to have two battleships placed off Cape Hatteras, to be used as targets, I landed our small air force of fifty or sixty airplanes on the island of Hatteras. This strip of land is a sandy fringe along the Atlantic Ocean. It lies 30 miles from the mainland, a distance which no cannon could reach effectively. It is protected to the north and south by deep inlets, Hatteras Inlet on the south and Oregon Inlet to the north, across which an army could not move in the face of the most rudimentary of aircraft. I landed 2,000 pound bombs, 1,100 pound bombs and 600 pound bombs, and all supplies necessary for the operation of these aircraft, from vessels in the water. This was to show that aircraft could be supplied by submarines, if they used an island and held control of the air. We also landed supplies from aircraft to show that we could be supplied through the air. In the old days, before the advent of aircraft, we used to think that a small island was a dangerous place to be caught, unless complete command of the sea were assured. But now aircraft can seize an island and operate from it without fear of interference by surface vessels, because they can sink them hundreds of miles away.

From our air base on Hatteras, we flew out over the sea and sank the battleships. One 1,100 pound bomb sank the New Jersey in 4½ minutes, tearing the whole top of it off. The other vessel was sunk in seven or eight minutes, after a decisive hit was registered. Some of our airplanes flew 179 miles to make the attack, which they launched from an altitude of 11,000 feet, securing around 60% of effective hits. The fire of great projectiles from aircraft is the most effective and the best directed fire we have ever known, as far as missile throwing weapons are concerned. It is more accurate than cannon or rifle fire.

Having completed these tests and shown how an island could be used, I assembled our little air force the following day at Langley Field, Virginia, on the shores of Chesapeake Bay. The day following, I took these sixty ships (including some twenty-five of the logy old Martin Bombers) and flew from Chesapeake Bay to Bangor, Maine, landing in a field which we selected and which had never been used before by aircraft. This maneuver was executed not only to show how we could defend the whole coast of the United States from Maine to Florida in one day, but also to show how this same air force with which I landed at Bangor, Maine, in the summer of 1923, having flown in one day from Chesapeake Bay, a distance of 700 miles, could

have kept right on and gone to Europe. By putting auxiliary gas tanks into these planes, which would have weighed no more than the bombs we carried, the radius of action would have been doubled.

Carrying out the same journey of 700 miles in the direction of Europe, we could have gone the first day from Bangor to Cape North in Labrador; the following day to Cape Farewell in Greenland; the day after to Iceland and the day after that to either the Scandinavian peninsula or the British Isles. With these old wartime aircraft, we could have gone right over to Europe! The following year, our Round the World flyers flew from Europe to America by the same short route. It is the one the ancient Norse mariners used when they went from Europe to America. With modern aircraft, this route can be used, islands seized and occupied as air bases easily. Airplanes capable of going 300 miles an hour can span this distance in a day.

An even closer bridge exists to Asia. There are only 52 miles of water separating Alaska from Siberia. It is along these lines that future invasions will come because they are the shortest lines and aircraft will fly direct across them, using comparatively small islands for their bases of operations and striking directly at the great vital centers, cities and transportation lines.

The United States is the most continuously self-contained country that the world has ever seen. In case of war, it can easily exist without ocean borne commerce. We raise everything that is absolutely essential to our maintenance and supply. Therefore, even if a Navy were effective as a means of controlling a sea, it would have comparatively little to do for us because we can very easily get along without it. Countries such as England or Japan are unalterably tied to their sea lanes of communication for existence. We in this country are the least tied to them of any nation on earth.

Do the Army and Navy give any consideration to the avenues of approach and attack by aircraft as I have outlined above, from Europe or Asia? They do not. These things were not done hundreds of years ago, they cannot find any precedence for them in their textbooks, and therefore what is the use of studying something that cannot be found in the pages of some old musty volume and then copied and modified a little bit to meet existing needs? In this day, constructive minds can employ the armed forces of a nation and bring about strategical results which would lead to world empire.

If Germany had known the power of aircraft and the submarine at the beginning of the European War as they are known today, she certainly would have had world empire. All of this knowledge has come about in the last ten years. It is just as absurd to match an army and navy against air power as it was for the slow heavy-armored infantry of Europe to oppose the mounted horsemen of Genghis Khan the Tartar.

All our old conceptions of navies and military strategy have to be changed, so great has the influence of air power become. It is a distinct move for the betterment of civilization.

Sunny Spain
A Bloody Proving Ground

Savoia-Marchetti SM. 75s pound nationalist targets over Spain.

BY FREDERIC B. ACOSTA

Here is the story of what ruthlessness in the air has done to Spain. The author, an international journalist and cousin of famed Bert Acosta, tells of the holocaust.

IT is the airplane—the intensely developed war weapon since the World War—that is spreading death and desolation over all Spain. The plane has revolutionized war. It is not a mere military weapon like a cannon, a rifle, a battleship. Now it is considered, as the air demonstrations in Spain prove, as the most effective arm of the new technique. It is the most persuasive instrument of the dictator to carry terrorization and frightfulness not only to the enemy camp but to distant regions as well.

The ruthless and methodical bombing and extermination of women, children and the aged of Guernica, the ancient capital of the Basque provinces in northern Spain, by German warplanes two months ago, shocked the civilized world. This murderous bombing and machine-gunning of civilian population in the Basque towns by the rebel planes provoked indignation and disapproval among the military opinions of the world.

I shall cite here a graphic example of a thorough job of annihilation of both life and property made by a modern plane bomber. I refer to the total destruction of Guernica in northern Spain by German warplanes fighting on the Spanish insurgent side.

The havoc wrought in this event exceeded the imagination and shocked both military and non-military world—it was typical of the ruthless and inhumane methods employed by the German and Italian aviators serving under insurgent General Franco.

The facts, as nearly as I have been able to find, are as follows: When the squadrons of bombers set out for Guernica from the German airdrome at Deva in Northern Spain, their crews had orders to carry out the most methodical destruction of life and property since Louvain was razed in the World War.

First, they dropped the explosive bombs to wreck the houses and buildings, then dropped the aluminum incendiary bombs to fire the debris. The inhabitants, who survived explosion and fire, were then machine-gunned as they tried to escape. A captured German pilot told the following: *"My orders were to machine-gun from my plane anything moving."* And so the German crews have had fun swooping down and machine-gunning the defenseless civil population.

The destruction of Guernica and the bombing and burning of Durango, Eibar, Arbacegui, Arteaga and three or four other places, were accomplished by three types of German bombers, of which the Junkers Ju-52 and the Heinkel 51 are known to have been in action since last year, but the German light bomber He-111 first appeared in Spanish skies early last April. In Spain the great Junkers bombers are called *pajaros negros* after the blackbirds which the Spanish know so well.

The real truth is that the civil war in Spain is, in a sense, a European war on a small scale. The Russians and the French are actually fighting Italians and Germans, though in a minor way.

Even here in America we nourish fears that these small scale operations in Spain, today, will inevitably drift into large-scale operations in the future. Some observers believe that in two years there would be another European conflagration of the deadliest sort. Unhappy anticipations of the next war are felt keenly in large countries of Europe. Great Britain has made a sudden decision to build a formidable air force. France is reforming and expanding her air armada and is spending billions of francs for this purpose. Russia, unquestionably, has a superior air force at the present time.

Spain, at the outset of her trouble over a year ago last July, was the least of Europe's air powers. She had only 162 planes, some of which could not match the best of the German's or Russian's. When the rebellion began, the Spanish government had difficulty in keeping her air force intact. More than half of that number had either been flown to the rebel camp, had been wrecked or sabotaged during the first few weeks after the civil war started.

Then, in a twinkling of an eye, there appeared in the Spanish skies every class and make of the latest and best European military planes fighting either for the government or the rebel side.

Russian and French pilots, flying their respective national planes, swiftly and efficiently, scoured the enemy's trenches and camps while, from the rebel lines, one could see three and four-motored Italian Breda and Caproni bombers, German Junkers and Heinkel bombers and the light bomber He-111 roaring into battle.

French war planes in Spain are the best known. They are the Bréguet, Bleriot-Spad, Nieuport, Dewoitine, Loire, Hanriot and Bloch. In the language of a French technician, they are carrying the best work of France's warplanes into the great test.

The Russian "Chato," according to a foreign military

Natural enemies: The Junkers Ju. 52 (top), used as a bomber by the German Condor Legion on the side of the Franco-led rebels, and the Polikarpov I-16 Russian fighter, one of the several models representing various nations on the side of the Republicans.

observer in Spain, is the fastest and most maneuverable pursuit ship.

Modern fighting planes have been greatly improved in speed and swiftness. The speed range of modern bombers is between 200 and 250 m.p.h., while a pursuit ship flies between 250 and 350 m.p.h.

German warplanes are reaching the Spanish rebel's territory in larger numbers. Two or three months ago, Germany established a direct airline to Burgos, Spain (city held by rebel forces). This Nazi airline, though primarily organized as a business venture, is merely a subterfuge for passing warplanes from Germany into Spain.

Messages from various sources point to the departure of a continual stream of Italian airplanes for the Spanish rebel forces.

These foreign flyers on both sides are deadly earnest and seriously determined to fight the enemy. Beside the Russian and French pilots on the loyalists' side there is a great number of British and American flyers serving the Spanish government. The Italian and German aviators compose rebel General Franco's air force.

The rule is for these foreign aviators to fly and fight in the planes from their respective countries. The Italians and Germans have their own airdromes; they have their respective mechanics, their own commanders, wear their national uniforms and eat in the exclusive messes.

As a foreign writer puts it: "The foreign pilots serve both sides, under secrecy, and they obey only the voice of their dictator."

Generalissimo Francisco Franco and his German and Italian advisers believe that aerial terrorism is of great importance to insure victory. They believe that, by bombing industrial centers and the civilian population, they undermine the morale of the inhabitants and thus will cause the loyalists to surrender earlier.

The rebel plane squadrons freely bomb the parks, market places and other centers of civilian population where noncombatants are likely to be gathered. The rebel aviators avoid attacking munition plants and trenches where anti-aircraft defenses are effective.

The inhumane bombing of defenseless civil populations gives you the impression that there is no sure defense against a campaign of terror from the skies, against the attack of superbombers in vast murderous flights. The civilized world recognized it in the destruction of Guernica and even Gen. Franco deplored it.

European military authorities were greatly disturbed by this appalling event. London, Paris, Berlin and other great continental cities are laying foundations against air assault and military experts are scheming devices that might prove to be of much defensive value.

In this regard, the military defensive technique has developed quite as rapidly as the airplane. At the present time no military device can be used as a sure defense against attack by modern bombers and super air squadrons.

In the event of war in Europe, each country will become one vast network, listening in by means of a highly technical device known as "mechanical ears" while searching the sky with powerful lights.

The "mechanical ears" are ready to pick up and amplify the faintest sound from a hostile plane coming to work its job. The searchlights, projected miles up in the sky, will reveal any moving and flying object and make the anti-aircraft guns ready to shoot.

The principal defense against attack from the air is a well organized home airforce, swift and speedy, prepared to take off instantly when the ground-intelligence picks up signs of approaching enemy planes. This powerful defensive home aircraft, always ready to fly, will meet the enemy and engage him in combat before he can release his bombs.

Foreign military observers in Spain have noted with surprise the effectiveness of the German anti-aircraft batteries in General Franco's rebel forces. These ground guns have time and time again saved the rebel troops from defeat.

The German anti-aircraft guns have equipment similar to that of the present-day American and British anti-aircraft batteries. It is an apparatus that electrically calculates the speed and direction of oncoming hostile airplanes, and which sights and fires the guns.

Up to the time of the bombing and destruction of Guernica, old capital of the Basque province in northern Spain, the city had only one anti-aircraft gun for its defense against air raids—not one battery but just one old gun, quite out of date and always breaking down.

Military observers believe that if Guernica had five or six modern anti-aircraft guns, this ground defense might have kept the German bombers at a considerable altitude so that they would have worked less damage. They would have, at least, prevented the Nazi Junkers and Heinkels from swooping down and machine-gunning the civilian population, and it has been the machine-gunning from these German planes that has caused most casualties among women and children.

Yet despite the provocation of ruthlessness and the inhuman cruelties of foreign aviators serving under rebel leader General Franco, the Spanish government does not attack civilians and endeavors to avoid unnecessary destruction of life and property. The secret of the present civil war in Spain is hidden in this distinction of policies and attitude between insurgent General Franco and the Loyal Spanish government.

No Basque, no Madrilenos, no decent-thinking Spaniards will forget or forgive General Franco's wholesale butchering of the civilian population as in the Guernica massacre. Doubtless he knows this well but does not care. The loyal Spaniards will continue to fight for their government against the domination of alien dictatorship and tyranny.

The Loyalists seem to be content with checkmating the enemy planes and defending the territory. But, if the Russian and French aviators see a good opportunity to fight Germans and Italians up in the air, they do so with a relish.

And as this is written, it looks as though ruthless and inhuman destruction of life and property will continue in that death-ridden country, and *the airplane as an instrument of a new technique, produced and developed by modern science of war, will continue to spread terror, frightfulness, devastation and death in sunny, bloody Spain.*

Everyone has been talking about the Flying Tigers lately.
Here, for a change, those phenomenal Yanks discuss themselves.

JAPS ARE THEIR SPECIALTY

BY RUSS JOHNSTON

"SLASHING, twisting, turning, Tiger flyers of the American Volunteer Group hurled themselves recklessly against the enemy. Applying every unorthodox method of aerial combat, these wild men of the skies completely bewildered and put to flight Nippon's best airmen."

Bunk! That's what the American public has been led to believe about the Flying Tigers of Burma. It makes good reading, sounds nice in headlines, stimulates the imagination but upon investigation it falls flatter than a pancake on a hot griddle.

Now that Uncle Sam has stepped into the war and diplomacy is no longer necessary, the true story of these "madmen of the skies" can be told and it proves conclusively that these defenders of Burma were neither madmen nor reckless fighters.

I have talked to these flyers over a cool glass of beer in a quiet restaurant far from the battlefield. I have visited them in their hotel rooms, watched them at work on their new jobs back in this country and can report that there was nothing fantastic in the story of the Flying Tigers who, like the men on Bataan, simply out-flew, out-fought and out-maneuvered the Japanese. It would make a swell story if I were to follow in the footsteps of the press and dish out a wild yarn about the devil-may-care flying of these American heroes, but I'm going to tell it to you just the way it was told to me.

"We whipped the Japanese in nearly every encounter because we *out-thought* them, not because we *out-fought* them."

In that simple statement we find a tribute to their leader— quiet, efficient, Col. Claire L. Chennault, now Brigadier General Chennault. It is also a tribute to the men who designed and engineered the Curtiss P-40C's that were used in the Burma fighting. In it is a tribute to the extensive and thorough combat training of the United States Army and Navy and Marine Corps. Above all it is a tribute to America because here, for the first time in action, were American pilots, in American

planes, fighting under American command. If there's any doubt in your mind as to the outcome of our present war with Japan let these Flying Tigers, now home from Burma, tell the story.

"In that first savage Japanese raid on Kunming, hoping to catch us on the ground, only one out of 27 attacking bombers escaped to tell the story." That's what Larry Moore, blond, 22-year-old staff secretary to General Chennault, told me the other day when I visited him on the set at Republic studios where he is working as technical advisor on "The Flying Tigers."

But better still, let's get the story from Walter Pentecost, who had complete charge of the engineering section of the AVG. He was the man who inspected every airplane, saw to it that they were fit and ready to fly, supervised the repair, overhaul and maintenance of the warbirds.

Walter is 30 years old, but he looks older. He was the first man of the AVG to set foot on Burmese soil when he landed at Rangoon early in 1940 to supervise the unloading and assembling of the first P-40's.

"Don't let me hear you folks say anything against the P-40's," was his quiet warning. "There are a lot of things that can be improved about them, sure, but stacked up alongside the Jap "0's," (the famed Japanese Zero fighters)"they're head and shoulders above them."

"Take leak-proof tanks, for instance. How the Japanese ever thought they could fight a major war without them, I'll never know. A couple of bursts in the Jap tanks and they go up in flames. Thank God our boys had bullet-proof, self-sealing tanks that swallowed the Jap bullets.

"Then the matter of armor plating. I've looked at dozens of wrecked Jap planes, both fighters and bombers, and I have yet to see a piece of bullet-proof steel. The armor plating on our P-40's is about as perfect as could be expected. Time after time the boys bring them back full of holes but thanks to this protection, they have nary a scratch on themselves."

Of course leak-proof tanks and armor plate are heavy. So is the sturdy dural with which the ships are built. It's this added weight, plus the low horsepower of the early Allisons (about 1,050 h.p. for take-off) which gave the Japs superiority in maneuverability and climb. When the Tigers were able to open them up and leave them open for awhile the Curtiss ships were faster than the Zero's and the P-40 in a *dive* is a thing to behold.

The AVG boys would hide in wait behind the clouds until the Japs showed up on schedule (as they always did), then, poking the nose at the ground and opening wide the throttle, they'd come tearing down. With pointed sharkteeth gleaming in the sunlight they'd be on the Jap bombers before they had time to man the guns.

Most of the Tigers told me that in this dive they'd aim for the Jap motors. A hit in the powerhouse with even the lightest ammunition seemed to explode the Jap engines in mid-air. Then the AVG boys would haul back on the stick to climb again for precious altitude.

The superman stories about the AVG, fortunately, are all a myth. These men were using standard U. S. combat training planes with special adaptations worked out for them by General Chennault. There isn't an AVG man to whom I've talked who wasn't utterly devoted to this slim, greying general of the Burma front. Chennault knows flying, knows Burma and he knows the Japanese. He had been U. S. military advisor to China's Chiang Kai-shek, had seen the Chinese coolies move the factories of China on their backs and had later witnessed the bombing of these reconstructed defenseless factories by Japanese planes.

It was Chennault who planned the bases, commanded the operations and instructed the pilots in the art of shooting down the Japanese bombers and fighters. It was a job that had to be done slowly, carefully and with fine regard for international law.

While a defenseless China was being bombed, the Curtiss planes were shipped to Rangoon, assembled and flown to training fields in northern Burma. England, at that time, was not at war with Japan though war-clouds were smouldering on the horizon. Strict neutrality had to be observed. No flights from Burma could be made over Japanese held territory. After weeks of planning and training, the AVG squadrons were moved to Kunming, in Yunnan province, to begin operations.

How many P-40's were there? Not enough. But let Walter Pentecost answer that more fully.

"The original order was for 100 planes. A lot of these were diverted from original British orders and arrived in Rangoon with the British insignia on them. One slipped off a loading crane and fell in the harbor. The salt water didn't do it any good so it was dismantled and used for replacements on other planes. Ninety-nine planes finally went to the front."

Walter believes that when the complete history of the war is written, it will be shown that these 99 planes saved China during the dark days of 1940 before the United States and

Back from a battle with the Japs, these Tigers discuss the fight. Pilot at right wears a U.S. Army insignia.

Britain entered the war. Thanks to these few American planes—less than three complete squadrons—the Burma road was kept clear of Japanese bombers and the Chinese industrial cities of Kunming and Chungking were protected. In the all-out Jap assault on Burma, the heroic rear guard fighting of the few planes and pilots left slowed up the Jap advance long enough to permit the forming of the defense of India.

The word "tiger" does not have reference to the carnivorous beasts of the jungle. The name comes from the flashing tiger shark whose faces have been painted around the air intake on the Curtiss planes. Did the AVG boys design the wicked, flashing teeth and evil eyes of these jungle sharks? "No," says 23-year-old Ken Sanger of the AVG group. The British Tommies on the Libyan desert must be given credit for the design. These "shark faces" had pounded General Rommel's tanks with cannon and machine gun fire long before the first P-40 was unloaded on the dock at Rangoon. The AVG boys saw a picture of one of the Libyan P-40's in an English aviation magazine while in Rangoon and adapted it to their own ships. "We took a few liberties with the design," Ken told me as we were having lunch in Hollywood. "We made the teeth a bit larger because we had heard that the Japs had poor eyesight."

Don't let a quip like this make you think that the AVG Boys look down on the Japanese flyers. They treat them with a healthy respect both in the air and on the ground. Nearly every fatality in the Flying Tiger group can be traced to a momentary disregard of instructions or to a hot-headed attack by a single flyer.

According to Pentecost the AVG attacks, again contrary to newspaper reports, were always made in pairs. One ship would make the attack, with the other pilot covering the attacking plane. In this way the leading flyer could concentrate on his target freely, knowing that his tail was being covered by a fellow pilot. In the event of a miss by the leading plane the other pilot could get in a burst before the target swept by.

Ask Paul Green about this. Paul dived in at 10,000 feet without waiting for his team mate to provide cover. He got his bomber but a Navy Model 97 fighter shot his tail off. Paul threw back the greenhouse and climbed into the blue at 9,000 feet. He yanked the rip cord on his chute and floated lazily with the Jap machine-gunning him all the way. His shroud lines were cut, his parachute looked like a sieve, but Paul didn't have a hole in him. For 8,000 feet he climbed the shroud lines and swung like a monkey to evade the murderous Jap passes.

AVG Flyer Cressman wasn't so lucky. The Japs murdered him in his chute. After that, General Chennault issued new orders and henceforth when a Flying Tiger bailed out he was followed down by his team mate who, by circling the descending chute, saw to it that his buddy was not harmed. Bill McGarry was seen to land this way in enemy territory, gather his chute in his arms and run for the brush. He didn't get back to Allied territory and it can only be assumed that he became a prisoner of war.

One of the questions most frequently asked these returning heroes is their reaction to this murderous Japanese practice of machine-gunning defenseless pilots who had been forced to bail out of their planes.

"Did you men retaliate by shooting parachuting Jap pilots?"

The answer is "No." The AVG flyers seemed to feel that their job was to prevent the planes from reaching their objectives. The destruction of the plane was what counted most. Besides most of the Japs who bailed out near the Chinese and Burmese bases were taken prisoners.

There is a strange psychology about these Tigers. They went to Burma to fight for the Chinese whom they believed to be in the right. But most of the boys will admit it was the $600 per

The Curtiss P-40 with the familiar shark's-mouth nose.

by remote control by the top turret gunner who was in a position to see the attack.

Several of the AVG boys have said that the Japs apparently are not instrument men. Even when closely pursued they have been known to fly *around* clouds instead of through them. The shot-down Jap bombers prove to be full of blind flying instruments and whether the Japanese pilots lack instrument training or hesitate to trust their instruments, the Americans don't know.

Larry Moore, who went through some of the heaviest raids on the air field at Kunming, says that one squadron of Japs would bomb with unerring accuracy while the next day another squadron, under the same conditions, would miss their objective entirely. These were the ones Larry said you had to be careful of if you were caught on the ground. As soon as the Chinese cry of "Jin Bao!" would be heard, the ground forces would rush away from military objectives and throw themselves into specially prepared slit trenches. The bombs that missed the air field often fell too close to these shelter trenches for comfort.

Larry tells of being caught on the ground in that first raid on Kunming by unescorted Jap Army 97 bombers. The Kunming communications had not been thoroughly tested, with the result that Larry only had time to pull on his shoes and dash across the Burma Road into some holes on the side of a nearby hill which had been constructed, he thought, as raid shelters.

After the first run across the target which resulted in no little damage, Larry was hauled out of his place of hiding by a fellow employee.

"You're lucky those boys were good shots," he was told, "they usually miss the target completely and plaster this hillside. These holes aren't raid shelters, they're bomb craters."

Where do the Flying Tigers come from? About 75 per cent came direct from fighter squadrons of the U. S. Navy. These men had all had at least five years of standard Navy training and are, for the most part, graduates of Pensacola. About 10 per cent came from the Marine Corps and the other 15 per cent from the U. S. Army Air Forces. It is untrue that any of them were released convicts or renegade commercial flyers, as has often been reported.

They are nearly all, with few exceptions, between 25 and 29 years old and are cool-headed, crafty and determined. They had been well trained and found it easy to adapt themselves to actual fighting. The stories of single-handed missions and slashing, twisting, upside-down attacks are not true. An AVG flyer wouldn't have lasted three minutes in that kind of combat.

Many of the veterans of America's first war with Japan have returned to this country under orders. They are at present working where it will do our country the most good: in aircraft plants telling designers what they need most, at material centers conducting tests and on training fields teaching teachers in the fine art of "blowing the Japs to hell." To that end let us hope they are successful and that, when the war is over, America will hold a special place in her heart for the "wildmen of Burma," who gave the Japs their first taste of Yankee fighting: the Flying Tigers.

month and $500 per plane bonus they were paid that caused their intense resistance. After Pearl Harbor was bombed it was different. America, their own country, had been attacked. Their own friends and relatives were in danger. On December 8, the AVG fought Japanese over Toungoo with renewed fury. They got 21 planes that day. On December 10, the salary and bonus were withdrawn and the Flying Tigers became merely American pursuit pilots assigned to a deadly serious business. The next day they got every plane of a 12-plane flight.

As has been the history of the war in the air in other zones, by far the greater number of AVG planes were lost through operational accidents and through lack of replacements than through enemy action. When tires wore out or were damaged in combat, for instance, the whole plane became a loss as far as active service was concerned. It had to be pushed off into the trees and used as a salvage ship.

How are the Japanese ships? Well, we've already written that the Army Model "O" (the AVG boys still haven't learned to call them "Zeros") is more maneuverable and climbs better. We know also that the ships have more horsepower than our P-40's. They aren't as well constructed and they have no defensive armament. The Jap bombers (Mitsubishi 96's and 97's of both Army and Navy design) are faster than the Zeros, but still lack armor plate and bullet-proof tanks. The Jap engines are all air-cooled and are *not*, to the best knowledge, supercharged.

The Japs have one wicked little weapon that the Tigers learned to stay clear of: a tail stinger which the boys called a "dust gun." The Mitsubishi 96's and 97's were originally designed without thought of a tail gun position (you don't need one to bomb defenseless Chinese) with the result that the AVG boys flew right up on the Japs and blew their tails off.

One morning four of the boys sneaked upstairs to intercept a squadron of bombers headed for Toungoo. They came in at 9,000 feet and made the attack in the usual manner. But one of the Tigers noticed a row of holes appear on the front windshield. Luckily, no one was hit and one of the Jap bombers was shot down. Later in the day a searching party brought in the wreckage and the men found an ingenious rear gun mounting which sprayed bullets in a 30° arc. The gun, mounted in a space too small to accommodate a man, was fired

Superiority of the RAF over the enemy is due partly to superior aircraft, but largely to its manpower— among the finest in the world.

THE MAN IN ACTION

BY FLIGHT LIEUTENANT H. E. BATES

ON a bright December afternoon a *Stirling* returning from a raid on the docks at Brest came into its aerodrome to land. Its pilot, an Australian, touched down with an exactitude that seemed to indicate that nothing extraordinary was wrong. He was within a day or two of his 23rd birthday. His second pilot, contrary to the popular supposition that the pilots of bombers are for some reason middle-aged men, was a boy of 19. Among the crew was a young engineer, who was badly wounded. The floor of the plane was greasy with oil and blood and from the port outer engine a stream of petrol was pouring out like a bright fan in the wind. As the plane touched down it began to disintegrate. First the port outer propeller fell away; then the whole of the port inner engine. Then the complete port wing fell off and, falling, caught fire. The huge tyres of the *Stirling* were punctured and flat, and finally the whole aircraft tilted violently to port, flinging the starboard wing high into the air.

A typical man of action is Squadron Leader A. C. Deere, who was shot down three times during fighting at Dunkirk, had a bomb explode beside his plane on an aerodrome, ran into a Messerschmitt in a 700-m.p.h. air collision, then had his own plane cut in two in still another unexpected meeting in the air. Shown below: The Handley Page Halifax, one of Britain's mainstay heavy bombers.

The task of carrying a dying man from a plane which seemed at any moment about to blow up would, in civilian life, have been front page news; complete with pictures, personal interviews and possibly framed certificates. Fortunately the plane did not blow up. Half an hour later its crew with the exception of the wounded engineer, now safely in hospital, were having tea, and of that extraordinary episode, and of the earlier episode in which the *Stirling* had beaten off the violent attacks of 10 Messerschmitts, there was nothing in the next morning's newspapers.

In this story and behind it, in ten thousand others of which the public will never see a record because there are so many of them, there is to be found a fraction of the picture that is being made, every day, every night, over sea and land and in all latitudes, of the men of the RAF in action. The title of these pictures is not heroism, since the RAF, with its vital genius for the rejection of normal speech and the coinage of new, does not know the word. Nor are they ever likely to occupy the place in popular literature and art long held by "The Charge of the Light Brigade." On the contrary they are, and will probably continue to be, taken very much for granted. For since the day when the British made the incredible and ghastly heroic blunder in the Crimea and somehow got it immortalized as a national epic, the standards of heroism, like the standards of speed and science and plain inhumanity, have undergone great changes. They have changed so far that they have, in some cases, ceased to mean anything. This has become very true of flying in general and, within the last three years, of the RAF in particular. One simple result of this is that the work of the RAF, which an earlier age would certainly have immortalized in verse, even if it was bad verse, is today recorded in language as plain, matter-of-fact and perishable as a cardboard box. The work of the RAF—words like deeds, exploits, feats of valour and so on have no place in its vocabulary—lies hidden too often behind colourless communiques and factual broadcasts that we now take as much for granted as the kitchen stove.

For this flatness, the lack of the heroic touch in the history of its action achievement, the RAF is itself partly responsible. The men of the RAF are largely inarticulate; they would prefer anonymity; and perhaps the worst of all sins in the RAF is the process known as "shooting a line." The now immortal father of all line-shooters was a fighter pilot who was heard to declaim, with fine pomposity, in one of those unfortunate silences that sometimes fall on general conversation "There was I, flying upside down." This remark has since become the standard by which any personal flying achievement in the RAF, whether heroic or not, has come to be judged. To shoot a line is now embarrassing, boring, comic, or in plain bad taste. In this way the RAF has elected to speak even of its most exceptional achievements in terms of understatement. To find the pilot who will tell you the story of a combat with simplicity, directness and lack of embarrassment is in consequence a rare thing.

But what the RAF *does* and what the RAF *says* about what it does are necessarily different things. The fact that the achievements of the RAF have not so far been recorded with outstanding objectivity, force and imagination is due to very simple things. Pilots, for the most part, cannot write; writers,

for the most part, cannot fly. The day on which we get a fusion of these qualities in one person we may be given, even though in a limited way, a real picture of the RAF in action. It will be limited not only because one pilot cannot speak for ten thousand but because the fighter pilot differs from the bomber pilot and the bomber pilot in turn from the coastal pilot and all three in turn from the ferry pilot as much as the *Spitfire* from the *Stirling*, the *Hampden* from the *Sunderland*, the *Liberator* from the *Kittyhawk*. These types are not only temperamentally different from each other; their daily life is different; their actions and reactions are different; only their final aim, the destruction of an enemy, is the same. To understand them, to assess the differences in their lives and their performance in action, one must live with them for a long time.

Of all flying types the fighter pilot is, to the public, the most attractive. The reasons are again simple. His actions are seen as a triumph of individuality; he flies alone in a high-powered piece of mechanism which is capable, in his hands, of evolutions at great speed, of great beauty and spectacular effect. He is engaged in a dangerous, apparently wonderful and often fatal occupation. Like the bull-fighter, he works near to and often in line with death. To the public this near-fatal occupation, whether of bull-fighters, tight-rope walkers or fighter pilots, is fascinating to contemplate. For of all pilots the fighter pilot is most likely to be seen—and during the Battle of Britain was constantly seen—in action. The coastal pilot is invisible, far out at sea; the bomber pilot is invisible, far out in darkness. The *Spitfire* pilot flies in the sun, turning his plane like a silver fish many thousand feet up, and fascinates the world below.

This is one part of the picture; the finished, pleasant, spectacular part, seen by the public. There are two others. One is the daily and almost certainly boring routine of the aerodrome and the dispersal point, which the public never sees; the other is that of the action itself—which the public also rarely sees and which the combatant will rarely discuss,

In the Middle East, the crew of a Douglas Boston attack bomber approaching their plane. The view is through the nose of the aircraft.

the action involving emotions, reactions and tension, for which the public has no remotely comparable experience of its own. The picture of life on the aerodrome is simple. On duty throughout all the daylight hours—as few as eight in winter, as many as 18 in summer, the fighter pilot's problem is often not how to kill Germans but how to kill time. His life, while waiting for action, may be intensely boring, dull, directionless. To counteract these things he reads, plays cards, revises navigation, talks shop. His action, even when it comes, is packed into a hundred minutes, of which perhaps 10 or 15 form the vortex in which he destroys or is destroyed.

From the thousands of stories by the fighters in action it is not possible to select one which will typify the fighter pilot. It is simpler to select a man who has crowded into one life the experience of 20 pilots and to record of him simply—"This is not typical. It belongs to the highest and rarest achievement. But it happened."

Of all fighter pilots Squadron Leader A. C. Deere, D.F.C. and bar, is probably least typical. He was born in New Zealand; joined the RAF in 1937; won service boxing championships; played football. In four months of the summer of 1940, from May to August, he destroyed 17 enemy planes. He was shot down seven times; bailed out three times. He collided head-on with an enemy machine; a pupil pilot cut his *Spitfire* in two. One plane of his was once blown 150 yards along the ground by a bomb; another blew up three seconds after he left it.

His action began in May, 1940. He had volunteered, with another pilot, to escort a small training-type aircraft to Marck, the aerodrome at Calais. He was to land there and pick up a squadron leader who had forced-landed. The Battle of France was in progress; Calais was surrounded by the enemy; the aerodrome was a kind of "No-man's land." When the trainer landed, the passenger was nowhere to be seen. After waiting a few moments the trainer took off again and was just airborne when a dozen Me. 109's attacked the escort planes. The trainer, forced back to land, hit a hedge. Then followed what Deere himself has described as "a grand shooting match with the Me.'s"—the words are typical—leaving Deere and his fellow pilot still flying and the wrecks of Messerschmitts lying all about the beach, the aerodrome and the town. The action seems to have been a fine example of the gay ferocity for which fighter-pilots have become famous: Deere claimed two certainly destroyed and one probable, his companion one certainty and two probables. The pilot of the trainer got away.

From that moment Deere seems to have had no rest. His patrols over the coast of France began at 3:45 a.m.; often he made two before breakfast. In a few days his squadron lost six pilots; and soon Deere, although only a pilot officer, had become its leader. His adventures began to be fantastic. When he pursued a Dornier from Dunkirk to Ostend and both the Dornier and his own machine crash landed, Deere was knocked out and his machine began to burn. As he scrambled clear, half conscious, his machine exploded. Two days later he collided with an Me-109 in mid-air. Each pilot seems to have thought that the other would give way. They passed each other once, turned and came together again. Deere lost a little height in the second turn and the belly of the Me tore along the fuselage of his plane, running the hood down on his head. The

propeller was snapped off by this impact and the engine partially torn out. Deere, blinded by smoke and flames, could see nothing and could only hold the plane in a 100 m.p.h. glide which took him gradually over the English coast. There his craft hit a concrete anti-invasion post, ripped off a wing, skidded through two cornfields and finally burnt up.

A week or two later he chased a Heinkel over the Channel towards Calais. Shooting down the Heinkel, he was immediately "jumped on by a swarm of other fighters." A bullet ripped the watch from his wrist; another scorched an eyebrow. His aircraft was full of holes, and over Ashford, 20 miles inland and at 800 feet, it began to fall to pieces. Those were the days when shoppers in that market town in the heart of the Kentish sheep and hop and orchard country would look up 20 times a day—I myself among them—and watch the battles in the blue summer air. They must have watched with horror that day as Deere bailed out of his disintegrating plane, his parachute not opening until he was perilously close to the earth. Yet Deere was flying again next day.

Two days later he had his oil tank shot away and had to force-land without his air speed indicator; soon afterwards his rudder was shot away and his engine set on fire. He bailed out again and landed in a plum-tree. Next morning his aerodrome was dive-bombed. A bomb fell immediately in front of him as he sat in his plane, blew the engine of the *Spitfire* out and sent the plane itself hurtling upside down with Deere in it, for 150 yards. He was helped out by a fellow pilot and ran for shelter. Afterwards he was put to bed and was still in bed, next day, when another raid began. He got up from his bed immediately, went up in his *Spitfire* and shot down a Dornier.

Shortly afterwards he was teaching tactics to a pupil. Something went wrong; the pupil collided with the teacher, cutting the *Spitfire* in half. Deere was caught in the wreckage of his plane and could not jump. The plane fell for several thousand feet before he could get clear. Half of his parachute harness had been torn off; the rip cord dangled out of his reach. Deere closed his eyes and waited, knowing there was nothing he could do. Suddenly his shoulders were violently jerked by the parachute opening of its own accord. That day Deere was in hospital for the third time.

A large part of these adventures cannot be called heroic; they are examples of fantastic misfortunes which may and do happen to any pilot, any day, anywhere. Deere's career is remarkable because he brings to such a succession of events, any one of which might have been his death, an indomitable stoicism; the spirit of apparently careless fatalism which is an important part of the fighter pilot's make-up. Above all he brings to them, as almost every pilot does, a passion for flying. This passion, which in the lives of young pilots who have often known no other life than flying is a dominant quality, is well illustrated by the experience of a *Hurricane* pilot, Sq. Ldr. J. A. F. MacLachlan, D.F.C. and bar. MacLachlan is a young, exuberant, magnificent personality. In France he shot down six dive bombers in four days; over Malta he destroyed several Italians. He himself was finally shot down over Malta by an Me-109: a type for which the *Hurricane* of that day was far too slow. In that action MacLachlan lost his left arm and spent most of his time in hospital not only in an agony of physical pain but in an agony of mental pain—the agony simply of fearing he would never fly again. Yet in 15 days from the time his arm was amputated MacLachlan was flying solo again; an enterprise most people would have been extremely chary of undertaking in 15 weeks. His joy at being able to fly again was precisely the joy of the ordinary man at being alive at all: not to have flown would have been synonymous, for him, with not being alive. Today MacLachlan flies with great skill, daring and success as a night-fighter, using a specially constructed intricate artificial arm in place of the arm he lost. He is an exceptional example of the fighter pilot in action: of the supreme individualism which even such physical disability as the loss of an arm (in the case of the famous Bader the loss of both legs) will not crush. It is this individualism, enabling them to undertake extreme risks because they know that failure can affect no one but themselves, that sets fighter pilots apart from all the rest.

This individualism is not dominant, on the other hand, in the pilots of bombers and coastal aircraft. In planes carrying crews the attitude of the pilot is very different from that of the pilot flying alone; his position becomes much more like that of the captain of a ship. His word is, in fact, law; he is in absolute command. He must also not only think of the safety of others; he is aware that he depends on them—his navigator, his radio operator, his engineer, his gunner, his observer—not only for his own safety but for the success of his operation. They too are aware of the same thing.

The successes of these men are not individual; they are the successes of units in which individuality is sunk. For this reason, and also because the flights of bombers and coastal planes are often long, dull and uneventful, they are much less publicised than those of fighters. It is only on such dramatic occasions as the sinking of the *Bismarck* and the pursuit of the *Scharnhorst* and *Gneisenau* that the work of Coastal Command suddenly lights up the headlines. But even in such a magnificent feat as the *Beaufort* torpedo attack by Flying Officer Kenneth Campbell V. C. on a battlecruiser at Brest it is

significant that the crew, all of them sergeants, are officially given the very greatest commendation. All of them died in that attack, and there are no medals to their names solely because the V. C. is the only British award which may be given posthumously. From this it is not wrong to assume, perhaps, that all of them share in that award.

Flying Officer Campbell, a Scot from Ayrshire, came to the RAF from the University Air Squadron at Cambridge. Flying a torpedo-carrying *Beaufort*, he began to distinguish himself very early: so that he was once pointed out to the Air Officer Commanding by his station commander as "a young man who before long is going to win the V. C." He was very successful and tenacious in attacking shipping; he sank a 3,000 ton supply ship in moonlight, made a direct torpedo hit on the port beam of a 6,000 ton vessel off the Dutch coast. He once attacked strong gun positions on the Dutch coast and was very much shot about in consequence, by two Me's. His wireless operator was wounded, the hydraulics shot away, the gun turret put out of action. He himself was wounded and lost a good deal of blood. He crashed on return to base, his aircraft very badly damaged. But he was very soon out on operations again.

The whole attitude of the bomber pilot in action is very similar. Long journeys in darkness, intense cold, frost-bite, flak that has been described by one of them as filling the sky like the paper streamers of a ballroom, the difficulties of taking-off and landing in darkness: those are a few of the routine problems of the job. There is a popular notion that the pilots of fighters are all very young, wild and volatile, and that the pilots of bombers are older men, dour, cautious, steady. This is not true. There are plenty of instances where the pilots of *Stirlings,* for example, are boys of 19 and 20. Moreover it is not the bomber pilot but the fighter pilot who, in the periods of preparation for action at least, strikes one as being cautious. It is the bomber pilot who, off duty, strikes one as being volatile. The reasons are again quite simple. The fighter pilot must always be ready; his orders to fly may give him less than five minutes to reach his plane. His recreations must be sober,

In the troubled skies over England, a Spitfire circles below its enemy in an engagement with a Dornier Do. 17 bomber.

rather limited, never excessive. The bomber pilot, on the other hand, is briefed some hours before he flies. Once his trip is over it is unlikely that he will fly again, except in an emergency, for a day or even two days. And since his trip has occupied six, eight or even more hours, he has accomplished, in one night, the equivalent of three, four or more fighter operations. He is entitled to an equivalent rest. The cold, the vigilance, the danger, the nervous tension of those hours all produce a considerable and sometimes violent reaction. In consequence the bomber pilot, off duty, tends to behave something like a hilarious schoolboy released from unpleasant detention. His life, on operations, is one of violent extremes.

The strain of this violent way of living can be seen in the bomber pilot's eyes. They give the impression of not being correctly focussed. Strain and cold make them rather old and glassy. The eyes of the fighter pilot never give the same impression. Theirs is much more an expression of exhilaration. Their triumph when it comes, is a personal one. But in a bomber there is far less likelihood of a personal triumph. A pilot relies very much on the skill of his navigator; their co-operation is extremely and necessarily close. He relies on his rear-gunner for protection, for information about hostile aircraft and for warning of ascending flak; it is through this information that he can take evasive action. He relies equally on his radio operator, his observer, his engineer. There have been many cases where these men have accomplished outstanding performances, even cases where a navigator, when his pilot has been wounded or killed, has piloted and successfully brought home a plane. But the greatest performances on bombers are probably all collective; the crew are indivisible; their comradship, morale and unselfish tenacity are not equalled in the world.

It is in fact hardship rather than heroism which, in a survey of bomber exploits, seems more remarkable. The story of a *Wellington* crew composed of a pilot officer and five sergeants who spent 57 hours in a dinghy before being rescued is astonishing but not, as an example of fortitude and discipline, unique. The objetive of this *Wellington* was Berlin; the cloud over England thickened into inpenetrable masses over Germany. It was a night of great difficulties: so that in all we lost 37 bombers. Thirty miles from Berlin the *Wellington* was hit—"we heard a crack and everything in the aircraft shook"—but the flying capacity of the bomber was unaffected. The captain, Pilot Officer L. B. Ergeland, took the plane on over the capital, dropped all his bombs except his incendiaries and then turned back. On the way home the bomber was hit again and almost simultaneously the second pilot, Sergeant D. A. S. Hamilton, and the rear gunner, Sergeant A. V. N. Fry, reported that the plane was on fire. The captain, opening the door behind him, saw an appalling sight. The fuselage was full of smoke: flames were rising through the floor and creeping up the sides of the plane. The racks of incendiaries, not dropped over Berlin because of bad visibility, were burning furiously along the whole length of the plane.

The fight against that fire went on for a long time. From the ground every ack-ack battery in its course fired at the blazing bomber. The pilot tried to jettison the burning incendiaries but the electrical switch would not work. The wireless operator sent out repeated messages. The front gunner carried the orders of the captain to the rest of the crew. The second pilot beat at the fire with the burning window curtains which he had ripped down and extinguished with his hands.

While the crew were fighting the fire in the interior of the bomber, flames had spread along one wing, burning part of it away. The plane had all this time been losing height. It crossed the coast finally at 1,000 feet. And because he liked the chances of the sea rather than the chances of enemy territory, the pilot flew on. Twenty-five minutes after that decision the petrol ran out. The flaming aircraft had been flying for nearly three hours. The cases of the incendiaries were still burning.

When the bomber came down into the sea its back was broken. It sank almost immediately; the flotation gear which keeps an aircraft on the surface for a few precious minutes had been burnt away. The crew scrambled into the dinghy. The pilot had an injured leg, a face badly cut by being thrown forward against the instrument panel. The rest were unhurt.

They waited for dawn and it was a lovely day. The sea was very calm and the sun very warm and about 10 in the morning *Hurricanes* flew over the sea in the distance, not seeing them at all. A plane came over and dropped Very lights, but it was too high for them to identify. They paddled all that day and then, taking turns, two at a time, all the following night. The next day the weather was very bad. The dinghy, though always seaworthy, shipped plenty of water, and they baled with tins and their hands. It was colder too and they rationed their food, which was biscuits and chocolate, and their drink, which was rum and water—enough for six days.

That night the wind came up even more strongly. But when the sun came up they saw a buoy go floating past them at "a devil of a speed with the tide." That buoy encouraged them enormously. They saw the coast too in the distance and began to paddle frantically, "like ding bats," for fear of being carried away from it again. They began to be taken ashore very fast and finally they could see people. They waved and the people ashore waved back. They paddled to within 20 feet of the shore and then the seas finally drove them in to the outstretched hands of the rescuers.

The files of bomber operations are full of such stories. The individual details change from story to story. The effect remains the same. These stories are rarely publicized; yet they are the solid foundation which make the great spectacular raids on Augsburg and Brest, Lubeck and Rostock, Skoda and Trondheim, Cologne and the Ruhr, possible. Behind the courageous who are decorated stand the courageous who are undecorated. Who is to choose between them? The whole world has heard of the great exploit at Augsburg of the crew of a *Lancaster;* only a little group of people has ever heard of the Australian pilot with whose story this article began. But between the crew of a triumphant *Lancaster* and the crew of a disintegrating *Stirling* there is really, in the final assessment, no difference at all. Their status is one with that of all fighter pilots sweeping the coast of France and of all *Beaufort* crews torpedoing ships; one with all night fighters intruding over France and all *Swordfish* piling themselves up in the decks of battleships. They have a common and in the end an anonymous name. They are the men in action.

SPITFIRE

*"Just the sort of bloody silly name they would choose,"
said its designer. Little did he know the Air Ministry's
brainstorm would become one of the most evocative
words in the history of flight.*

BY JAMES GILBERT

WE ENGLISH ARE, except in the rheumy eyes of our dreamier politicians, no longer a great power. Our empire, the foundation of our wealth, has crumbled and disintegrated; our factories grew tired, till they now offer merchandise often shoddier than that of nations that once sought only to imitate our goods. Our military no longer deters any but the meanest savages. Our economy is enfeebled by ruinous taxation and the inequitable distribution of such meager wealth as remains to our citizens. Ours is, however, still a lovely land, as gentle and happy a place to dwell as any on earth, and this we have come to trade on, taking in summer migrations of tourists in millions—our late and traditional enemies among them—because we sorely need their money. We are Britain; no more Great.

Our decline was gradual, taking almost 100 years. Yet there was, only 32 years ago, one short summer as glorious as any in our history, when we stood alone against a cruel and vicious enemy, and held him, as no other nation had been able. It was his first defeat, and the beginning of his end, and our finest hour. We were not aware of it then, however; the Battle of Britain seemed to us to be just another jolly battle, such as our tumultuous history was often punctuated with. It was our then-new prime minister, an ancient warrior who had himself ridden out, as a youth, in the last cavalry charge ever made by the British Army, who spelled out for us in thunderous rhetoric the titanic quality of the struggle whereat we were engaged. Winston Churchill to the House of Commons, on June 18, 1940:

"Hitler knows he will have to break us in this island or lose the war. If we can stand up to him, all Europe may be free and the life of the world may move forward into broad, sunlit uplands. But if we fail, then the whole world, including the United States, including all that we have known and cared for, will sink into the abyss of a new dark age, made more sinister, and perhaps more protracted, by the lights of perverted science.

"Let us therefore brace ourselves to our duties, and so bear ourselves that, if the British Empire and its Commonwealth shall last for a thousand years, men will still say, 'This was their finest hour.' "

The British Empire and its Commonwealth didn't last the decade, and "broad, sunlit uplands" is a metaphor only *that* extraordinary old aristocrat would have dared use, but in one thing he was unquestionably perceptive: It was our finest hour. And for it, we had, in just barely sufficient quantity, the most magnificent airplane we ever made—the Supermarine Spitfire.

In all humble honesty, we were jolly lucky to get the Spitfire at all. Its begetters, Supermarine's, were far from scions of our military-industrial establishment, but essentially a boat-building yard; *flying* boats, lovely things, cat's cradles of taut piano wire, drumming fabric wings and varnished mahogany hulls. But Supermarine's were blessed with a genius of a chief designer, one R. J. Mitchell, and by good luck, he and his directors and the British Government of the day (back in the 1920s) had grown interested in competing for a ridiculous silver trophy offered by a French industrialist for international seaplane races. Supermarine's success in these Schneider Trophy contests was total, for the Supermarine racing seaplanes won the contest three times running (against the best endeavours of the French, the Italians and the Americans), thereby retiring the trophy forever under the rules with which it has been set up. These planes, despite their huge floats, also took the world airspeed record for *any* type of plane. Thus Mitchell and his team, though they had never designed a fighter, knew more about high-speed flight than anyone else, for they had spent a decade wrestling with its problems: the problems of working with highly stressed Napier and Rolls-Royce liquid-cooled racing engines whose pistons and valves might last the course or might melt; of cooling the coolant at no increase in frontal area for radiators; of snuggling the pilot's cockpit in the shadow of the engine so that it too, would cause no extra drag, while still allowing its occupant to see out; of thin wings and monocoque structures; of carving and filleting and melding the airplane's members sensuously together for minimum interference drag; of control-surface flutter, in those days a quite new and altogether terrifying problem.

Mitchell was the speed king, but pure speed was hardly sought by those concerned with fighters in those days, other qualities being more highly named: high rate of climb and a low landing speed; robustness, and simple handling to match the mediocre skills of your average squadron plot; and above all, maneuverability and the talent to turn very tight. It was an era when an ordinary private airplane (the Beech Staggerwing, for example) might well be faster than most of its military contemporaries.

One point had been generally overlooked, even at first by Mitchell: While maneuverability can decide the outcome of a duel between two fighters that are evenly matched, superior speed gives you *command,* since you can enter or break off combat at your will. The more maneuverable but slower of two planes may *evade* the faster, but it will be lucky indeed to *attack* it.

When the board of Supermarine's decided to go in for fighter manufacture, Mitchell's first design was nothing like his sleek racers. It was a tubby, crank-winged, fixed-gear, open-cockpit design not too different from the Germans' wretched Stuka. It had a Rolls-Royce steam-cooled Goshawk engine, and attained a modest 200 knots—140-odd less than the last of Mitchell's racers. This first Spitfire met the Air Ministry specification to which it was built, but had little else to recommend it, least of all its engine—one of Rolls-Royce's failures. The production contract went to a rival, and a biplane at that: the Gloster Gladiator.

Mitchell thought some more. Rolls had a new V-12 engine coming along; it, too, was beset with problems—the cylinder jackets kept cracking on test runs, and the double-helical reduction gear washed itself out completely and had to be replaced with straight tour gears—but this new Merlin engine might do. It was a complex unit, made up of 11,000 pieces, 4,500 of them unduplicated, but it was expected to give almost 1,000 horsepower; it was slender in height and width, and it was supercharged, to retain its high power at high altitudes.

Around this engine, Mitchell designed a new fighter, this time devoted not to some Air Ministry specification but to his specialty—speed. Its fuselage was even more graceful than those of his racers, tapering away fore and aft from the point of maximum cross-sectional area at the cockpit. He used a monocoque structure for the rear fuselage, and made the tail surfaces integral with it—something still unusual in those times. A little canopy sliding on rails gave the pilot complete shelter—which his seaplane pilots had not enjoyed, for the belief died hard that a pilot needed the wind on his cheeks, even at 350 knots. He chose a double-ellipse planform for the wing, in the interests of low drag; I don't know where he got the idea for this, though Ernst Heinkel in Germany came out with an exactly similar wing for his He-70 at around the same time; by a weird coincidence, Rolls-Royce bought an He-70 to serve as a flying test-bed for the Merlin. Mitchell had used a wing shape rather like this for his early S.4 racer, but had later abandoned it for straight taper and rounded tips on the S.5- and S.6-series airplanes; but now he decided to go back to it. Whatever his reasons, it is the one thing that distinguishes a Spitfire at a glance from all other planes, and it gave the Spitfire a structurally thick wing root combined with an aerodynamically thin airfoil. Look at a Spitfire head-on; it almost disappears, so slight is its frontal area. Yet Mitchell's skill managed to combine low drag with light structural weight; a Mark I Spitfire weighed only 6,200 pounds loaded for battle, and its wing loading of 26 pounds per square foot was appreciably less than that of the Me-109. It was in the powerful interest of light weight that both Mitchell and Messerschmitt chose narrow, outward-retracting landing gear,

with all the problems of ground handling that narrow gear brings: Narrow gear meant that only the inboard foot or so of the main spar need be made heavy enough to carry the stresses of landing. (Neither of these two designers of the outstanding fighters of early World War II, by the way, had ever designed a fighter before—nor had their Soviet equivalent, Alexander Yakovlev. Yakovlev and Messerschmitt had made their reputations with sport-planes.)

The prototype Spitfire, raw metal silvery overall, was trundled out for its first flight one chilly morning in March 1936.

There was little enough that test pilot Mutt Summers could find to fault in the new bird. They bulged the cockpit canopy to give pilots a better chance of spotting the 109s that would soon be on their tails; they replaced the fixed tail skid with a swiveling wheel; and they altered the shape of the rudder, which Mitchell had at first gotten a mite *too* light. The Air Ministry swiftly issued a new "specification," which did little more than define what Mitchell had given them, and then they ordered 310 of the new fighter—which threw Supermarine into a total tizzy; no one had ever asked them for 310 of

anything, let alone something as complex and hard to make as the Spitfire. They subcontracted madly, and then found their subcontractors weren't up to it either. Spitfire production remained painfully slow for years; by the war's outbreak, the RAF had only nine squadrons of them, and by the time they were really needed, when the Luftwaffe set upon England, only 11 more. Happily, they had Hurricanes in modest abundance.

As for Mitchell, he never saw a production Spitfire, for he died in 1937 of cancer. He was only 42.

The first planes shot down by Spits were two Hurricanes from another RAF squadron, by hideous mistake. Spitfires first tangled with Messerschmitts when sent to the defense of the British Expeditionary Force being desperately evacuated from Europe at Dunquerque; results were "inconclusive." Then there was a pause, till the Luftwaffe attacked Britain in earnest in mid-August 1940.

The Germans expected it would be all over in a week; by late September, it was, but the Germans had lost. They had been unable to destroy more than one radar station; unable to knock out any fighter base for more than a few days. They had lost nearly 2,000 planes—twice the RAF's losses. The real ratio of destruction may well have been even more one-sided, for when a Heinkel or Dornier crew went down alive, it was for capture and imprisonment, while a Spitfire pilot who bailed out might well be back in action in another airplane almost as soon as he had gathered up the silken folds of his parachute.

When the Luftwaffe switched to night raids on British cities in October 1940, Fighter Command could relax a little, for there was virtually nothing they could do about night raiders: Airborne interception radar had yet to be perfected, and the chances of your spotting an enemy bomber above the exhaust flashes from your Spitfire or Hurricane were negligible. The Vickers Supermarine plant at Southampton was well bombed, destroying the two prototypes of Mitchell's last design, a four-engine bomber, so the company could devote its best efforts to new marks of Spitfire. The Battle of Britain had been fought by Spitfire Is and IIs, armed with eight 1,000-round-per-minute .303 Browning machine guns (an adaptation of an American Colt design built in England under license). With the Mark V came a change to four .303s and two 20-mm Hispano cannons—just as soon as they could persuade the cannon to feed without jamming. More Vs were built than any other mark. Succeeding models of Messerschmitt and Spitfire were evenly matched, though the introduction of the first Focke-Wulf 190s gave the Luftwaffe a clear edge.

To match the 190s, the Mark IX Spitfire, with a new Merlin giving 400 more horsepower, was developed. A Mark IX would do almost 350 knots at 20,000 feet, and take just over five minutes to climb up there. With the Mark XII came a completely new Rolls-Royce engine—the Griffon, with rotation opposite to the Merlin, so that it swung the other way on takeoff; your first takeoff in a Griffon Spitfire kept you on your toes. There were unarmed high-altitude reconnaissance Spitfires and armed reconnaissance variants; lightweight extended-wing variants with Rube Goldberg pressure cockpits, which (if their guns didn't freeze up) could have a stab at intercepting the German photo-reconnaissance Ju-86Ps as they staggered along at 45,000 feet. There were variants with clipped wings and modified engines for chasing German pulse-jet buzz-bombs at 2,000 feet; dive-bomber Spitfires with shackles for a 500-pounder; Spitfires with so much power it took a five-bladed propeller, or even two counterrotating propellers, to use it. The final Spitfire was the Spiteful, or, in its naval version, the Seafang, with 2,500 hp and a new laminar wing like the Mustang's; but it was too heavy and uncouth, and in any case, the war was over and the jet had come along.

Jeffery Quill, now a senior sales executive with the British Aircraft Corporation, worked on the Spitfire program from the earliest days to the shutdown of production. Quill joined Supermarine as a 23-year-old boy-wonder test pilot from the RAF in 1936, and watched his boss, Mutt Summers, make the first flight—heard him after he had shut down the engine say delightedly, "I don't want anything touched." Only three weeks later, Quill got a chance to fly the prototype; would a lad of 23 be allowed to fly anything so valuable today? Quill forgot to raise the flaps after one landing, and made the next takeoff with them full down. "Never mind," said Mitchell to the shamefaced youth. "I wanted that done sometime anyway." At that time, the airplane had no name, but was simply

"Mitchell's baby Schneider racer that folds up its feet." Mitchell and his pilots were enormously unimpressed with the name their masters selected. "Sort of bloody silly name they *would* choose," said Mitchell.

Quill test-flew Spitfires throughout the war, and must have flown just about every one of the 33 variants. He even shot down three 109s while seconded to RAF squadrons to see how the Spit held up in battle.

The final Spitfire was a monster with twice the gross, two and a half times the power and nothing like the delicacy of handling of the early marks. "There was," says Quill, "the inevitable conflict between 'design' and 'production,' but on the whole, it was resolved as successfully as I think was humanly possible." Not every service pilot agreed. "It was galling," Quill remembers, "to be told something like, 'what you want to do with the Mark So-and-So is double the tail-plane area.' As if we didn't know!"

Mark VIIIs and IXs were generally reckoned to be the nicest to fly, so let's, in imagination at least, go fly one of those. Strap your chute on before you climb aboard, to see that it is adjusted right. It should be tight, with the seat pack dangling uncomfortably behind your thighs. Don't worry, it's comfortable enough once you're settled in the airplane's bucket seat. The cockpit is strangely roomy; you sit straight up, as though in a long-legged chair, and your feet on the rudder pedals seem way below you. They are: Messerschmitt pilots flew from much more of a reclining position, and blacked out from G a little *after* you. The cockpit is filled with pipes and tubes and taps and dials and switches, every one of them most beautifully constructed, as though you were in a vintage Bentley. Every instrument, every control, seems twice the size of those we fly with today. The stick is a gigantic column, hinged twice—once at the bottom for fore-and-aft movement, but only 10 inches or so down from your hand for aileron movement. At its top is a heavy circular spade grip, from which the gun button, burnished brass in a scarlet surround, glares at you like a malevolent eye. The throttle is another huge handle angled out from the cockpit wall toward your hand like the head of an umbrella. At your right elbow is the gear lever, quaintly labeled "chassis," and likewise huge. Gear and flaps are hydraulic, with a CO^2 bottle back-up for emergencies, while the brakes are pneumatic, worked from a cycle-type squeeze-grip lever on the stick. They are notably ineffective; the airplane can leave the ground after only a hundred yards run in any breeze but needs 900 to stop easily. If the brakes *were* more powerful, though, you'd go right over on your nose. The pneumatics also fire the guns.

Two vast knurled wheels on the left-hand cockpit wall control elevator and rudder trim. The rudder-trim wheel's plane of rotation is also fore-and-aft, and you must rotate it 90 degrees in your mind's eye before it makes sense. The cockpit has no floor, so anything dropped will not be seen again until the next time you slow-roll.

The instruments are big and clear; flight instruments—the usual ones—in the center with lesser dials scattered haphazardly around.

Messerschmitt pilots sat on their fuel tanks; you are hardly better off, for your tanks are ahead of the panel, right in front of you. There is armor plate behind your seat and behind the firewall, and there is a sheet of heavy glass affixed to your windshield, but those tanks are still quite vulnerable. The Spitfire's other great weakness, should you get shot up, is the engine cooling. It uses a mixture of glycol antifreeze and water, piped and pumped to a radiator under the starboard wing; one bullet anywhere through that maze of piping and your engine is going to seize in a very few minutes. (The oil cooler is the bulge under the *left* wing.) The radio is another disaster area; it is a T.R. 9D, or a T.R. 1133, full of vacuum tubes, among the first VHF aircraft radios ever made, and dreadfully unreliable. It was not unusual for a Spitfire squadron to go into battle with only half the radios working. But the 1133 model did have a useful device whereby your voice could be used to trigger the transmitter, so you needed to press nothing to speak. (Another bit of circuitry cut out *that* gizmo whenever you fired the guns, so their racket didn't switch on the transmitter. Very clever.)

There are some fine automatics in the engine controls, too: an automatic mixture control, for one, which gives you a simple choice of "normal," or "weak" for long-range cruise. There's also a throttle-mixture interconnect, so that opening the throttle also enriches the mixture. You have an automatic

boost control that prevents you from exceeding nine psi and blowing up your engine. And with later Merlins, you had a two-speed supercharger that could be set to change gear by itself as you climbed or descended through a certain altitude.

For starting, the "trolley-ac" is dragged up—a low cart laden with lead-acid accumulators to boost your Spitfire's own battery. Crank the starter, let four prop blades slide by, then *contact*. And then, oh lordy the noise! It's a bark, a rattle, a clatterous roar that you seem to feel in your gut rather than in your ears. A big V-12 is the loveliest-sounding type of engine in the world—you don't have to tell Ferrari or Lamborghini owners that—but a 1,700-hp, completely unsilenced, open-exhaust V-12 is a feeling like you could work miracles.

Be careful taxiing; keep it brief, for you can boil the coolant, particularly in a long drag downwind; no air going through the radiator. She rocks on rough ground, on that narrow gear. Line up. Ready? Feed the power in slowly, and you'll find you've always got ample right rudder. You will be off the ground in 150, 100, maybe as little as 50 yards in any breeze. Yes, 50 yards. If you were lucky enough to share a base with a USAAF Thunderbolt unit, you used to take off, retract your gear, do a slow roll that battered the daisies and pull up into a steep

Stages in Spitfire evolution: (Top, left) the Supermarine S.6, Schneider Trophy winner and world speed record setter. Another version (top right), the Mark IX carrying a four-bladed prop and differently tapered wings. The Mark XII (Bottom, left) lost the graceful curving wingtips of the classic Spit and took on the 1,500-hp Rolls Royce Griffon engine.

chandelle. *They* needed 600 yards just to get airborne.

Close the door, the hood and the radiator flap, and climb away at 4,000 fpm.

The Spit is curiously easy to fly, while at the same time very light on stability. Keeping the ball in the center takes effort, and it's easy to rack up horrendous amounts of G, so light are the elevators. The engine noise is at all times quite shattering. If you've only flown government-approved production airplanes, you will never have met anything like the Spitfire, with its marvelous engine bellow and nearly neutral stability. At times during the War, Quill and his merry men felt they'd got the Spitfire's stability too light, and put a bobweight in the elevator circuit. Then they'd relent, and tell the squadrons that as long as they watched the aft loading, they could take the weights out. Don't worry, said the squadrons, we took them out long ago. But you could pull the wings off a Spit; people did. The brave men in the pressure-cabin Spits chasing Ju-86Ps at 45,000 feet found the lack of pitch stability very wearing; and in a low-level beat-up, you have quite positively to *hold* the Spitfire the right height above the ground.

The stall is another thing you'd never get past any certification authority: sharp, with a real wing drop. But for warning, there's a sharp nibble of buffet, particularly just before a high-speed stall, when there's a sudden clattering judder just in time to warn you to back off. If you knew what you were doing, you could usually out-turn an Me-109 in a Spit, even though he had automatic leading-edge slats. You can really pull a Spitfire round amazingly tight. Getting away from a 109 on your tail was a problem; SOP was "everything in one corner," forward and left with the stick, and full left rudder to throw him off his aim. But your engine would cut with the negative G, while the Me, having fuel injection, came right on after you. But you had rudder trim, which the 109 didn't, and you had much better elevator effectiveness at high Mach. A Spitfire could be safely dived faster than early British jets: One intrepid British test pilot got one up to almost .9 Mach, at which point the experiment ended as the entire propeller reduction gear left the aircraft with a colossal bang. He got the rest of the airframe down in one piece—on the main runway at Farnborough, yet.

Deliberate spins were *verboten* by the RAF. This caught out a number of people who accidentally spun in combat at low altitude; it was easy to think you'd stopped the spin before you really had, and start to pull out, and then spin again, often the other way. When you stop a spinning Spit, the rotation first of all speeds up as the nose goes down. Then you must wait till at least 130 knots shows on the ASI before starting to ease out of the dive, and I mean ease out. The whole performance takes about 2,000 feet. Here is an airplane you'd need some time to get used to. I'd hate to have gone into battle with a lonely 10 hours On Type, as used to happen in the Battle of Britain.

Aerobatics in the Spitfire are glorious fun. All that power, and those light controls! Rolls are best barrelled a bit, unless you enjoy the engine spluttering and coughing and choking. You start a loop from 260 knots, and use a good deal of sky, and pull pretty hard at first, then lightly over the top unless you want to snap out. Upside down, those strangely shaped wings seem to fill the sky, like huge, outstretched hands.

Afterward, coming in to land, you'd best fly a curved approach, else the wing hides the threshold. Flaps-and-gear limit speed is 150 knots, and then you will be utterly astonished at how slowly you cross the fence; as slow as 70 knots with a mite of power. After crashing around at 300 or 350 knots, 70 feels so slow you can hardly believe it. You can wheel it on, or else three-point it, stick right back in your gut, but something half-hearted in between can lead to some mighty porpoising. Three-point, you can run the tailwheel along the turf for fully a hundred yards before the main gear touches.

It was, remember, an order for 310 Mark I Spitfires that threw Supermarine into such consternation; how dumbfounded they would have been to know the eventual total production— 20,000! How many Spits survive? I saw a detailed record recently of 98, most of them rusty, empty hulks doing gate duty on concrete plinths outside RAF stations. How many fly? Hamish Mahaddie was able to get a dozen airborne for the Battle of Britain movie, and there are maybe half a dozen more; the Tallmantz collection has one; so does the Confederate Air Force; and a lucky man in Chicago named William Ross has one, immaculately dolled up in its original wartime colors. One hears rumors of more: There's this dump outside Damascus, and then again there's this mysterious rancher in North Australia who's supposed to have a completely unmarked Spitfire that occasionally buzzes airliners. And what did the Burmese do with all the Spitfires they bought from the Israeli Air Force? Serviceable Spitfires very rarely change hands, and then only upward of $25,000. Merlin engines are less scarce, happily, for I've seen $36,000 quoted to overhaul one. Most Spitfire owners keep several in reserve, and cannibalize.

Pierre Closterman, who now runs Cessna's factory at Rheims, in France, flew Spitfires during the War, and later wrote about it. I'll let him have the last word:

"An essentially reasonable piece of machinery," he called the Spitfire, "conceived by cool, precise brains and built by conscientious hands. The perfect compromise for all the qualities required of a fighter, and ideally suited to its task of defense."

On Sunday, December 7, 1941, even as some Flying readers were reading this article, the issue of war or peace was being decided by the Japanese themselves at Pearl Harbor.

STOP JAPAN NOW!

American sea and air power can halt Japanese aggression immediately if used now, the author contends. He advocates the U. S. call Nippon's bluff.

BY JAMES R. YOUNG

JAPAN's threat in the Far East is perpetuated by an Axis plan to keep our Pacific Fleet from being released for immediate convoy duty in the Atlantic. America *can* and *must* call Nippon's bluff and free our Naval squadrons in the Pacific, if we are to get the necessary aid to Britain in time. This is the hour to act. Japan is vulnerable to an attack by our Army and Navy air units operating from bases in Alaska, the Philippines and even China. Almost single-handed our air power, as it now stands, can cut the lifeline of Japan's only real menace—her navy—destroy her crowded cities, demoralize her army and render the nation worthless to the Axis.

Just as the German war machine smashed through the Low Countries and pinched the Balkans the Nazis plan, by using the Japanese as a stumbling block to keep America silent and busy with "peace negotiations" and afraid to use force against the Land of the Rising Sun. In this way, Japan can become the key pivotal point whereon the Axis can whirl to threaten the United States. The first part of the plan is to eliminate Russia.

Already Japan is organizing White Russians in Harbin, Tientsin and Shanghai in preparation for a puppet regime in Siberia. As we ship planes and vital war supplies to Russia through Siberia, they are being immobilized in a sense by increased Japanese forces numbering nearly 1,000,000 in Manchuria. When and if the Soviets should be overcome by the so far victorious Nazi machine, America will be in a tough spot. Japan will be free to go south and, jointly with Germany, bludgeon China through a pincer move which will be followed with a push around to India, joining in Africa the German thrust through the Balkans. The only safe action for the United States to take to prevent such a move is to strike *now* at Japan with the full force of our air, sea and land forces.

If we permit the totalitarian countries to control the Far East, we are inviting Japan to threaten our Pacific Coast, just as Hitler's futuristic plan is to attack our Atlantic shores and disrupt Western Hemisphere solidarity. But, if we weaken Japan now and, at the same time, the German force is defeated in Europe, then the democracies can re-establish in Europe and the Orient.

Right now the Orient is an international boiling pot. If Stalin should quit or sue for peace with Germany, we would immediately face Japan. Knowing the vacillations of the Kremlin in the past, I would not be surprised if, because of our indecision to act in the Far East, the Soviet would turn, under forced collaboration with the Axis, against the United States and Great Britain. This would put China out of the picture. The Philippines would be joined with Thailand to Axis Japan. Singapore would become isolated and our channels for rubber, mica, tungsten, manganese and tin would be cut off by Japan's southward knifing. If we permit Japan to continue we will find ourselves in a perpetual state of agitation. If we do not take the initiative, we will be forced to "bail out" with our Navy in the Pacific, leaving that ocean under Nazi-Japanese domination.

Our job is to thwart Japan's encirclement. We must give her "Hobson's choice" and, if the Japanese do not accept, we must drop our compromise policy and face the dilemma with firmness and force.

There are three trends of thought in America today: (1) Japan cannot attack America. (2) The United States cannot defeat Japan. (3) By using our combined air power and fleet, we can eliminate any further "mug wumping."

The possibility of a Japanese attack on our shores was minimized by Russia's turn against the Axis. Until that time, Japan was in a position to encircle the Philippines. Now Japan is encircled on the north by the Soviet and on the south by strengthened American, British and Dutch forces—and by China in the middle.

Virtually, Japan is caught in a web of modern citadels of air and sea power. She is 1,500 miles from Dutch Harbor, 500 miles from Russia's Komandorski Islands. Her southernmost navy post is Camranh Bay, 2,500 miles away. Her South Sea mandated islands in another direction are 2,000 miles distant. Less than 1,300 miles away from her industrial centers near Tokio, at Alaskan air bases, are U. S. Army Air Force bombardment groups, standing ready for action. And facing her Gibraltar of the Pacific, Formosa (largest island in the Japanese group), are strong U. S. land, sea and air forces in the Philippines. Japan could not withstand the highly developed and speedy striking ability of America's new Pacific might. She would quickly crumble.

Actually, Tokio is a military aviation officer's dream of pandemonium. An attack on water reservoirs and any of the baker's dozen major hydraulic and thermal generating plants will cause complete disruption. Fifty suburban electric inter-urban railways depend on several centrally-located electric power houses.

Rivers—narrow, shallow and rapid—are useless as supplemental navigation, when railways and bridges are blown up. A poorly equipped 200-mile-long street car system, using burdensome double trolleys, and trying to carry 3,000,000 Tokio passengers daily, will be paralyzed by the military expedient of wrecking suburban electric power plants. Also two million bicycles jam the narrow streets.

Tokio's suburbs, which have mushroomed thousands of armament factory smokestacks, can be decommissioned by enemy planes. Cutting water pipes connecting the reservoirs which feed the city's fire plugs, homes and industrial water tanks will create uncontrollable havoc.

Trains run on narrow gauge rails which—with curves, tunnels and mountains and allowing for landslides and earthquakes—reduce express trains to a 35 m.p.h. average. The British-type small freight car, the size of an American caboose, transports freight at an average of 20 m.p.h. More freight cars move from sundown to dawn on the Norfolk & Western than shuttle in one week over any parallel Japanese trackage.

Much mileage is electrified—particularly the main line from Tokio to Nagoya (the center of aircraft production)—and industrial Osaka. Bombers will aim at the one main line and at transformers and transmission lines which web Japan like neon signs on Broadway. Four railroad lines lead from Tokio. The main line—the Tokido—winds through 20 tunnels (the longest is five miles) to the end of the islands, 700 miles away. Twenty passenger trains run every 24 hours, taking 20 hours each. Imagine the composite facilities of New York City limited to a double track to Chicago, over one major railroad bridge a third the size and quality construction of the George Washington over the Hudson! North from Tokio, two lines carry traffic 400 miles over plains and through mountains. Through the backdoor of Tokio is another railway penetrating a mountain chain. A flight of bombers will find, in one glance, three vehicular bridges and approaches of the mile-wide Shinagawa River south of Tokio—the Empire's sole traffic arteries connecting highly industrialized sections of Yokohama where automobile, rubber, chemical and electric manufacturing plants stand in open, reclaimed land.

Bombers over Osaka will intercept the flow from the Yodogawa reservoir, set fire to giant cotton and rayon-spinning mills, wartime chemical plants and industrial alcohol units. Japan's war productivity centers at Osaka, 50 miles closer to Vladivostok than Tokio.

Westward 250 miles is the arsenal terminal of Japan, the coal mine center—unprotected by mountain ranges— concentrated near the seacoast 600 miles from Vladivostok and bordering the Pacific. A combined air attack from the north and carriers from the south could practically disable the region in one trip.

One effective attack from an American aircraft carrier, bombers from Siberia, or a squadron from China could cripple

Japan's transportation system. Night attacks on any part of Japan will leave her practically helpless. If bad weather prevails they will not be able to patrol the island and return to any base, due to the limitation of their flying fields. One open spot in Japan could be a focus point to navigate and bomb with considerable success, due to the small area and the cities being so closely adjacent to each other. In other words, it will almost permit her enemies to do blind bombing and still be very effective.

Important naval bases also will be vulnerable to bombardment, in view of the fact that antiaircraft defense cannot function until the attacking forces are over their objectives. Such airports as Japan has could be bombed completely out of commission, as they have few such bases and cannot move freely to newer fields, due to the condition of the terrain.

Then the Japanese are not up to par with other countries in night and blind flying operations. They have been lax in this phase of aviation; only in the last few years have they installed night-flying equipment on certain air routes. American military and Naval aircraft could protect themselves easily against any pursuit action of present-day Japanese airplanes.

The island of Japan—being limited in size and its major cities and industries located so close together on the shoreline—does not permit them to locate antiaircraft defenses sufficiently distant from those cities to endanger an approach by enemy aircraft. Even if their antiaircraft defense is effective, the enemy will have already reached his objective before the Japanese can attack.

The concentrated effect of an American and Soviet bombardment of Japanese rail heads will seriously handicap Japan; they have not yet developed main highways between their major cities. Where they have done so, they have made only single roads. All shipbuilding industries are located in the main ports; Japanese naval bases and shipyards are extremely vulnerable because of the lack of antiaircraft defense.

Weather conditions in Japan—particularly in winter—are a major obstacle. Because of the few airports, heavy snows would handicap them in sending up aircraft to fight off bombardment. Any formation of airplanes attacking Japan, carrying only a small number of incendiary bombs, will be so effective over the helpless country that they will be able to carry all the gasoline necessary to permit long-range operation. Weather conditions are such that the center of Japan, when completely overcast, will permit enemy aircraft to take off from distant places, attack in comparative safety and return to their bases while Japanese defensive aviation remains grounded.

Production of planes in Japan is variously estimated. The generally accepted figure is 250 a month. Their plan was to increase the output in 1942 to 360 planes. But the lack of essential materials cuts their production figure from a normal 250 to about 150 per month. Lack of materials and worn and overworked American precision tools have slowed the aviation industry.

But, most important, she is totally deficient in every essential war material—iron and steel, aluminum, modern machine tools, copper, cotton, tin, wool, scrap, oil and gasoline, lead and manganese. Her war machine at home—

which might be called her fourth front—would be paralyzed. She has no reserve material, is economically exhausted, financially ruined and paralyzed from America's freezing order.

In aviation, as in the auto industry, the Japanese have always had something of everything from everywhere. A big handicap is the development of carburetors to contend with notoriously bad Japanese weather conditions. They have French carburetors, American piston rings, German and Italian cylinder heads, Swedish ball bearings and all kinds of machine tools—plus American cotton, Dutch rubber and imported aluminum.

One cannot estimate Japan's accumulated stocks and reserves, but in a searching, conservative analysis—and having observed the world's greatest armament producing power—I believe it is suicide for Japan to start a shooting match with any organized force.

Japanese manufacturing industries are operated for the national defense and all companies are closely supervised by the Army and Navy. There are eight large aircraft companies in Japan. Largest is the Mitsubishi Aircraft Works, which produces pursuit and bombardment airplanes, engines, propellers and aircraft accessories. Nakajima Aircraft Company is another major producer. The Kawanishi Aircraft Company manufactures seaplanes. The Aichi Aircraft Company, Showa Aircraft Company and Tokyo Gas & Electric Manufacturing Company also are among the Big Six.

For a number of years the Japanese have developed aircraft through adapting a type purchased through a manufacturing license from some other country. These licenses and designs have become obsolete by the time they have gone into production. However, all of this has been an educational factor with them and they recently endeavored to design their own airplanes, particularly fighters. They have had some success, but these airplanes are limited in range and usually have excessive landing speeds, necessitating almost perfect ground facilities. They are far behind with regard to the development of armor plate and bullet proof gas tanks.

Their bombers are a combination of various adaptations

from German Junkers and Heinkel types. The majority of Japanese engines are of foreign design and low horsepower. It has taken them considerable time to tool up to producing these engines, as well as airplanes, and today are more handicapped than ever in their attempt to change their tooling for the manufacture of newer types of airplanes and engines recently obtained from Germany.

Japan's air force is believed to total 5,000 planes. Several hundred, or as many as a thousand, are perpetually grounded for repairs, because they cannot now import spare parts. They do not have materials, patterns or the metallurgical ability to make parts. Of the 5,000 planes, 3,000 are navy and 2,000 army. The total first-line fighting planes in their army and navy is about 3,000, including about 1,000 trainers. The remainder are transports and other types. Distribution of Japanese army and navy planes places approximately 400 in Manchuria, 700 in North and Central China and 300 in South and Indo-China. With all their aircraft planning, however, they are frightened at news of the United States' program. They look eastward and worry about our two-ocean, $10,000,000 Navy of 338 large and small warships and 40,000 skilled men. With our expanded fleet including 12 aircraft carriers, 15 battleships, 54 cruisers, 197 destroyers and 74 submarines—thereby increasing our fighting ships from 338 to 691—our Navy will have tonnage which Japan cannot hope to equal. Also we have a Navy training program of 6,000 pilots to be attained by the end of December. Again, the Japanese have noted with worried brows that our expanded air force will include 17,000 men. As the Japanese look south they find that, almost overnight, we have joined with the British and the Dutch to treble our reinforcements in the Pacific.

Expansion by conscription of flyers has been part of the Japanese program at numerous new schools. The first had an enrollment of 400 five years ago and now is training men by the thousands. But Japan never will equal American airmen in quality or quantity, mainly because of inherent physical handicaps.

Japanese pilots have no fear. In temperament and psychology, the Japanese are like the Germans; they have a fanatic and religious devotion to their leader and they believe their bombing missions are holy acts of a manifest destiny. They will deliberately crash themselves into an objective. In such suicide attacks they plunge with the thought that their spirits will be enshrined in the temple of Imperial worship.

Japan's best pilots are in the navy. This may be due to the fact that many of them have been trained abroad and the physical and the educational requirements bring a higher type than to the army, but in either branch of the service they are not natural flyers any more than are the Chinese. They cannot adapt themselves from a life of work in the rice fields to flying in the air. There is an instability in their functioning and their reflexes simply do not correlate under pressure at great altitudes.

Nor can they get away from book learning. The Japanese think and operate in groups. They memorize mechanical equipment, but do not take to it naturally. In military tactics they are all right as long as everything is going according to the book, but when they encounter a discrepancy or problem, and they have arrived at the last page of the manual, they are lost.

Most Japanese youths come from the country where, despite a government campaign, mothers still carry children on their backs. This handicaps Japanese pilots with eye trouble since the head, from infancy, is held with the eyes skyward.

Lack of blind flying training, as I have pointed out, is a major handicap. As the terrain in Japan does not permit many emergency landings, they would lose many planes in wartime through lack of blind flying ability. American and European aviation experts who have been in Japan, and who have been stationed as exchange officers with Japanese air regiments, have noted in particular that, while they can fly almost any type of airplane—providing everything functions normally—they have extremely limited knowledge of the mechanical operation of their equipment with respect to limitations and safety factors. The percentage of pilots who could be considered first class is extremely small and consists mostly of men who have been fortunate enough to get by for a number of years and who have acquired a fairly good knowledge of the points necessary to make a good flyer. The younger conscripts are extremely inefficient. Considerable question is raised by competent observers as to Japanese ability to handle armament. In addition to bad marksmanship, they do not seem able to coordinate speeds.

Their ability to land airplanes runs from one extreme to another. They either land far in excess of the speed required, or they make stall landings. This is due to their lack of knowledge of the speed limitations required for landing. They never seem to hit a happy medium and in no way do they appear to have a finesse to coordinate mentally exactly what they themselves and the airplane are going to do.

I had never given much thought to the physical condition of Japanese flyers until the day I received a cabled request from a Sydney, Australia, newspaper. The paper asked whether the increased vertigo of Japanese aviators at altitudes could be attributed to the lack of eating beef. This was a new type of request for a foreign correspondent and I was interested in the follow-up:

Basically, the Japanese diet is raw fish, sea weed, bean curd and uncooked vegetables. Naturally such food is not substantial for anyone in active work. I found that the Japanese navy in particular had had to establish a special diet for its aviators, based on a combination of rich vitamin-producing Japanese and American-style food. (The Australian inquiry was a bit mercenary; if it could be established that vertigo in Japanese flyers was increasing, an enterprising Australian beef salesman expected to open a new market.)

So Japan's downfall will be the poison of her own reasoning to the effect that she is invincible. The answer in dealing with such a nation is to take Vladivostok, Dakar and Martinique and do the arguing afterward. If we find we do not need the bases, we can hand them back. A hermitage empire, helpless, isolated, Japan cannot maintain a major military assault. Her present leaders are blind if not cognizant of the high industrial mortality from airplane attacks.

But as the late ambassador to Washington—Hirsohi Saito—remarked: "Japan has crazy men."

Artist's conception of a Russian I-16 in deliberate collision with a Nazi bomber.

*A "suicide" tactic of World War I has been developed
into a scientific maneuver by Soviets.*

The Ramming Russians

BY LIEUT. ROBERT B. HOTZ

ONE of the specialties of Russian airmen in their battle against the German air force is the tactic of ramming enemy planes. The sacrifice of a dying pilot in a damaged plane by a deliberate collision with his foe is a relic of World War I; but the Russians have developed ramming as a definite tactic from which both pilot and plane may escape undamaged.

Ramming was developed by the Russian airmen after they observed that frequently German multi-engined bombers escaped after being hard hit and seriously damaged by Russian pursuits. Often the pursuit pilot scored heavily, killing part of the bomber's crew and disabling one or more motors. However, these attacks usually exhausted the pursuit's limited ammunition supply, permitting the bomber to limp back behind its own lines. Ramming is designed to destroy these crippled planes. It takes a combination of skillful piloting and utilization of the crippled victim's lack of maneuverability to execute a successful ramming operation with a minimum of damage to the attacking pilot and plane. More often the attacking plane is damaged and the pilot bails out.

Soviet flyers employ three types of ramming, according to Maj. N. Denisov in a recent U.S.S.R. embassy bulletin. The most dangerous is the direct blow. Hitting the enemy plane with a part of a Russian plane and clipping control surfaces by slight propeller contact are also used. The latter method calls for the greatest skill but offers the best chance of survival.

Major Denisov points out that the propeller clipping method calls for an approach from the rear with the attacking plane's speed adjusted to that of the enemy. As soon as slight contact is felt the attacker must drop away to avoid crashing with the enemy plane as it falls. If the ramming flyer is too slow he may easily become entangled with the stricken plane and dragged down with it.

American Air Force observers abroad report numerous examples of the Russians' ramming tactics and there are accounts available from Soviet flyers who have rammed German bombers and made successful landings. Here is the account given by Jr. Lieut. V. Talalikhin, who was awarded the Order of Hero of the Soviet Union for his exploits:

"On the night of August 6, 1941, when Fascist bombers made one of their attempts to break through to Moscow, I was ordered to take off in my fighter and patrol the approaches to the city. I soon spotted a Heinkel He-111K at an altitude of about 15,000 feet. Swooping down I managed to get on its tail and attacked.

"With one of my first bursts I put the bomber's right engine out of commission. The plane banked sharply and set its course for home, steadily losing altitude. I continued to attack the enemy and gave him about six bursts, following him down to about 7,500 feet, when my ammunition gave out. What was I to do? I could have followed the bomber farther but that would have been useless. With only one engine it could still fly quite a distance and perhaps escape. There was only one thing to do—ram the enemy.

"I decided to chop off his tail with my propeller and opened my throttle. Only about 30 feet now separated the two planes. I could clearly see the armor plating on the bomber's belly as I approached from behind and below.

"At that moment the enemy opened fire with a heavy machine gun. A searing pain tore through my right hand. Immediately I gave my plane the gun and the whole machine, not just the propeller, struck the bomber. There was a terrifying crash. My fighter turned upside down. I unfastened my belt and drew up my feet, crawled to the opening and threw myself overboard. For 2,400 feet I fell like a stone, not opening my parachute. Only after I heard the roar of my plane to one side did I pull the ripcord. I landed in a small lake and made my way to shore."

Pilot Mikhalev of the Soviet Fleet Air Service was credited with ramming a Henschel Hs-126 in one of the first appearances of this new German aircraft on the Russian front. Mikhalev dived on the plane after exhausting his ammunition. His propeller ripped the Henschel's stabilizer and rudder. A flying piece of wreckage struck Mikhalev on the shoulder but he managed to bring his plane down safely. The plane crashed and burned.

Pilot Vinogradov did his ramming the old-fashioned way. Fighting a single Nazi bomber over a vulnerable Russian target he exhausted his ammunition without getting a decisive hit. Meanwhile a bullet punctured his gas tank and his ship burst into flames. Vinogradov hurtled into the Nazi bomber and both planes were destroyed.

Another Soviet pilot who rammed and lived to tell about it is Alexandrovich Kiselev. He escaped with only a scratch on his cheek after bailing out. His plane was lost.

"It didn't come off very well," said Kiselev in describing his ramming. "I am sure it is possible to ram an enemy ship without losing one's own machine. I was a bit excited and I suppose that is why I muffed the job.

"My ammunition ran out. The enemy had hit my oil tank and radiator and my engine was just about giving its last gasp. I didn't want to let him get away so I went at him from below to get at his tail with my propeller. It was possible to calculate the movement so as to clip him with the tips of my propeller. But a stream of oil messed up my windshield and I couldn't see very well.

"Just as I was approaching him the suction of the air whirls caused by the Nazi plane swept my machine upward. I got mad then and rammed him from above, digging into his left side. I knocked my face against my stick. If I had figured it out properly that wouldn't have happened.

"The enemy plane disappeared. My own plane went into a spin. I tried to pull out but it was no use. I took my feet off the controls, stuck my head outside and was knocked back into my seat by the air blast. I pushed off with one foot, counted to eight, ripped and floated down."

Lieutenant Katrich of the Soviet Air Force relates another ramming incident:

"At about 10 a.m. I was told that an enemy plane had been sighted heading for Mocow. I took off at once and soon spotted a vapor trail at about 18,000 feet. The enemy was above and ahead of me. I put my oxygen mask on and picked up altitude. I drew up to within 300 feet of the Nazi plane. I sprayed him from stem to stern. It was only then that the Nazi crew noticed me. The cabin gunner returned fire. I gave them another long burst until I saw flames streaking from their port engine. After the third attack my ammunition gave out and their tail gunner was silent. The left engine was burning but the plane continued to fly. The pilot was apparently counting on my fuel supply being exhausted. It was then I decided to ram him.

"I had thought a lot about ramming. The first reports of ramming by our flyers interested me but in most of them the planes had been lost. I thought it would be possible to ram without sacrificing one's own plane and here was a chance to test my theory.

"I approached the bomber from the left of its stern and aimed my nose at its tail, calculating my attack so as to clip its stabilizer and rudders with the tips of my propellers. My calculations proved correct. There was a slight jolt. I throttled back and banked. When I came out I saw the enemy gliding sharply downwards. I glided after it. The Nazi pilot made several attempts to level off. By gunning his motor he managed to fly level for a few seconds before dropping off again. He finally lost control and dove into the ground. The ship burned. I landed at my home airdrome. My plane was undamaged except for a dent in my propeller which caused heavy vibration."

One of the most spectacular instances of ramming which throws an interesting sidelight on the combat psychology of Russian airmen was told by eyewitnesses at the airdrome over which the battle occurred. Sgt.-Maj. Nikolai Totmin took to the air as his home field was attacked by eight Junkers Ju-88 dive bombers escorted by a pair of Messerschmitt Me-109 fighters. Totmin set one bomber's port engine afire with his first burst but was attacked by the Me-109 fighters before he could finish the bomber. Totmin banked sharply to battle the fighters. One Me109 followed the bombers but the other stayed to take on the Russian.

Totmin and his Nazi opponent went into a tight circle trying to turn inside each other. The Nazi went into a quick climb and Totmin followed him. The Nazi then turned to attack and Totmin banked sharply to bring his plane hurtling head-on at the Nazi. Both planes sped toward each other but at the last moment before collision the Nazi heeled his plane over. At that instant Totmin banked in the opposite direction and drove his plane into the Nazi's wing.

Totmin's plane staggered under the shock and both planes spun earthward. Totmin twice tried unsuccessfully to bail out but the air pressure forced him back into the cockpit. The third time he got out but he was only 120 feet from the ground and his chute didn't have time to open. He fell not far from the wreckage of the plane he had rammed.

I FLEW FOR ISRAEL

A 20-year-old Chicagoan fought for the Israelis.
This is his dramatic story.

BY ROBERT J. LUTTRELL

WE CLIMBED steeply from the Aquir airfield near Tel Aviv, our old C-46 shadowing the dun and green coastal strip of Palestine. Off the starboard wing lay the blue haze of the Mediterranean. Ahead of us were the Egyptian lines, nearing rapidly as we approached 10,000 feet.

I lay flat on the belly of the cabin, peering through a window cut in the floor, sighting with my "bomb sight," an astrocompass. (Other bombardiers used German bomb sights.) When I figured we were one minute away from the port of Gaza, terminal of the Egyptian lines, I poked my head through the cargo compartment door.

Our mechanics, standing beside the open door of the plane, took the appearance of my head as a signal for action. Pushing, heaving, rolling, and kicking, they started dumping fragmentation bombs through the doorway, to fall on what we hoped were the enemy positions.

The mechanics were our "bomb-kicker-outers." I was the "bombardier." And that's the way Israel bombed the Egyptian invaders in the early days of the Palestine war.

A year ago March, I never even dreamed of flying for the Israelis. I'd flown only once, as a passenger in a commercial American plane. By profession I was a ship's officer in the merchant marine. The idea of becoming an aerial navigator, of navigating a plane across the Atlantic, of smuggling arms, ammunition, and planes from behind the Iron Curtain to Palestine, of going on front line raids with jeep parties of Israeli commandos, of dodging "ack-ack" over the Negev, of sniping in Jerusalem, never even crossed my thoughts.

They wouldn't have become fact for me if I hadn't stuck out my neck during a bull session at the Officers Upgrading School, U. S. Maritime base at Sheepshead Bay, Brooklyn, N. Y. At 20, after four years at sea including plenty of convoy action, and my third mate's papers, I was sitting for a second mate's license, in March, 1948.

The argument turned to Israel and I sounded off. After all, in three years in the forecastle you learn a lot about humanity,

A Curtiss C-46 Commando, such as the author describes.

and I believed the Jews had as much right to their homeland as did men of any nation. Twenty-four hours later an agent of the Zionist underground asked me if I had the courage to back up my convictions with action. It didn't take me long to decide, and in even less time I was deep in the cloak-and-dagger atmosphere of the Jewish underground.

They signed me up to ship as second mate on the old presidential yacht, the S. S. *Mayflower,* which was being converted into a blockade runner for displaced persons from Europe. But let's get one thing straight right now. I was no mercenary. Many of the gentiles among the 6,000 Gahal, the overseas volunteers in Israel, signed up for money. That's all right, too, but I didn't. I signed because I believed in a cause. I asked only $100 a month, to the amazement of my new bosses, to be sent to my mother in Chicago. By ancestry I'm Scotch-Irish, but since I was going to Israel as an idealist, I told the underground I was Jewish and I was accepted everywhere as a Jew. It helped me to know some mighty fine men a lot better that way, too.

The *Mayflower* deal fell through after a couple of months. Federal agents impounded the ship and the crew was dispersed. (Later the ship sneaked away and is now in the Palestine trade as the S. S. *Mala.*) I spent several weeks with these new friends of mine in the *kibbutzim,* the youth camps run by various leftist and conservative young Hebrew organizations in New York and New Jersey, even helping shift heavy crates which I was told held TNT, pistols, or ammunition. Later I was broke again and trying my old shipping line for another berth, when again an underground agent found me near a subway entrance, slipped a note into my hand, and vanished. Once more I was with the underground.

This time I was on the books as an aerial navigator. As a hobby I had studied celestial navigation, which proved handy background for the intensive 10-day course I received. This was given me at night in the Yonkers home of a New York lawyer who had been a navigator and instructor in the Army Air Force during the war. Checked out by him, I flew commercial to Panama to meet for the first time the members of *Lineos Aeros de Panama Society Anonyme* or, as we called it, *LAPSA.*

LAPSA posed as a new airline surveying a route across the South Atlantic but, curiously enough, no *LAPSA* plane ever returned to Panama. In Palestine *LAPSA* became the Air Transport Command of Israel. When I reached Israel in May, *LAPSA* was operating eight C-46's, and two four-engined jobs. Later we had three four-engined transports, 12 C-46's and numerous C-47's. Most of these were purchased, with American Zionist funds, as war surplus in Burbank, Calif., and elsewhere. Other groups bought planes in England, Europe, and Africa. In time, in addition to *LAPSA,* Israel had a real air force which included Messerschmitt and *Spitfire* fighters, three B-17 *Flying Fortress* bombers, and such planes as Piper *Cubs,* Lockheeds, Beechcraft, *Lancasters,* and *Norsemen.*

From Panama, we hopped in stages down to Natal, Brazil, and you can be sure that on the first leg, Panama-Dutch Guiana, I didn't tell the two pilots, radio operator, or mechanic that this was the first time in my life I had ever tried to navigate a plane—first time that I'd ever flown as a non-passenger, for that matter.

Our C-46 was a crate, in every sense of the word, held together by faith and baling wire. They wanted to put in new engines at Natal but there weren't any. Patched up by experts, with extra gas tanks aboard, we made the long 1,885-mile hop to Dakar, our nerves shattered by repeated misfiring. From there our route was Casablanca, Algiers, (where we came down with one engine afire), and Catania, Sicily. "Jerry," the chief Mediterranean underground agent from Rome, was at Catania to meet us. He instructed us about another field at Ajaccio, Corsica, which we later used as a refueling point, and sent us on to Brno, Czechoslovakia. Other agents briefed us on radio procedure and we made the last jump to Zatec, a small town 35 miles northwest of Prague, In Israeli code Zatec was "Zebra," the secret air base behind the Iron Curtain, the aerial arms smuggling terminal.

Ten minutes from "Zebra" we tuned in on a mobile field radio and came on in, after a rocky crossing over the Alps had put the finishing touches to the raw nerves of the pilot who had also picked up dysentery in Africa. By this time he'd had it. Cursing every plane in the world, he went on by train to Rome and sailed for home on a slow boat. We loaded up another plane, leaving our ship, the RX-131, behind for repairs, and took off for the south, refueling at Ajaccio, and then flying on to Aquir, the big air base outside of Tel Aviv which was to be my home base for 7½ months.

We ATC men lived in luxury in the Park hotel at Tel Aviv, the same hotel used by the foreign correspondents. We weren't supposed to talk with any of them, and we didn't, but they got so nosey the big brass moved us to another inn, not quite as fancy. Still, we lived well, except that the food wasn't too good. Before every mission we would unobtrusively bring down our bags and take off for Aquir, wearing our cotton khaki trousers and shirts, the usual mid East "uniform." Actually, we were outside of the Israeli Air Force, deliberately so, jealously hanging on to our status as civilians. Most of us in the ATC were Americans and we didn't want to run the danger of losing our citizenship. Some of the other crews were English or South African. There was one crew of handsome Swedes, who'd been flying D.P.'s out of Europe for years, and one Russian crew, our only Communists.

If the front line fighters, the Hagganah, or Jewish Defense army, or the Palmach, which corresponds to commandos and were all tough well-trained volunteers, needed some air support, frequently we would be asked to bomb from our transport planes as we started the long haul to Czechoslovakia. That's when we'd load up with fragmentation bombs to be heaved through the doorway when we passed over the Egyptian lines. It was a big laugh to be called a bombardier for such a crude job but the title made a big hit with the girls in Tel Aviv—and that's what counted.

The real bombing was done by the B-17's, although sometimes the crews of the old British crates also did a bit. But B-17's gave Israel quite a thrill. They were the planes which took off secretly from Halifax, Nova Scotia, landed at the Azores, then "disappeared" at sea. Actually, they flew into Zatec and then on to Ramat David, the former British fighter base which had been enlargd by the Jews so it would take bombers. On their first mission each B-17 hit an Arab city: Cairo, Egypt; Damascus, Syria; and Amman, Trans-Jordan. On their return mechanics sprayed on a new paint job, altered the gun positions, changed the wing markings, and off they flew again to fool the Arabs into thinking that Israel had a full squadron of four-engined bombers.

In those early days the Jews only had a few fighters—ME-109's and some *Spits*—all of which had to do double duty. I remember one morning when an Egyptian C-47 came over Tel Aviv and dropped bombs on us, exactly as we used to do. It stooged around for 45 minutes until an ME-109 could get off at Aquir. The fighter shot down the Gyppo in a matter of seconds. We transport "bombers" felt sympathy for those Egyptians crashing in their poor old transport.

When the ATC acquired its *Skymaster* and other four-engined transports, they were used almost entirely to bring in boxed-up ME-109's from Zatec. Not far from the Czech airfield is a former Messerschmitt factory, now in production again, and these planes were bought from that plant. The Zatec field had also been a German jet fighter base in the closing weeks of the war and there were rumors in Israel that the air force was going to get jet jobs. Most of the pilots were boning up on jet manuals of the U. S. Air Force which were smuggled out of America, translated into foreign languages and republished. But up to the time I left Israel no jets had arrived.

More *Spitfires,* however, were coming into Israel. One flight of 15 took off from Zatec but had a tough job navigating, and some were forced down in Greece where the pilots were interned. When they brought in the next flight, three C-46's acted as guide ships, mothering the flight down the Mediterranean, doing all of the navigation, aided by an Israeli corvette which stood off Haifa to serve as a rescue vessel in case anyone had to ditch.

Most of our flights between Zatec and Tel Aviv were uneventful, except for an emergency landing we made, fully loaded with arms, at Treviso, Italy, where we were jailed for five days and interned a month, until "Jerry's" door-opening dollars got to work. At Zatec our field was well protected by Czech soldiers and often Czech army brass would drop in for a visit, or an inspection. We were all under the direction of a very smart underground agent who also supplied the crews liberally with Czech kroner.

We had to close down Zatec in a hurry last fall when an Israeli deserter spilled the works to the United States, which complained to the United Nations, then busily engaged in trying to stop the war. The crews got all the planes away except a four-engined job which had been damaged, but we lost all of our warm clothing. After that, continental operations shifted to another Iron Curtain country, Jugoslavia, which gave Israel the use of a small base at Nicsic, in Montenegro. Down there we had no radio guide and we navigators had to be careful to avoid the Jugoslav fighter bases. Sometimes the "Jugs" buzzed us in their Russian *Yak* fighters, but gave us no trouble.

Our hardest work came when the fighting in the Negev desert flared up, and the Egyptians cut the main supply road to isolated garrisons and communities. We dropped everything else and pitched in running an improvised air lift, all night long, every plane making at least three or four flights. It was only a short air haul of about 25 miles from Tel Aviv, but it took about 40 minutes since we had to reach 5,000 to 7,000 feet before running the anti-aircraft gauntlet.

The Egyptians had quite a lot of 40 mm, ack-ack and many fine German 88's which the British had captured in the western desert and given to Egypt. But they were miserable gunners. If they hadn't been we would have had serious trouble. As it was, most of our planes were hit at some time or other by light stuff. At the desert termini of the air lift, what we called Dust Bowl No. 1 and 2, unloading was often done by front line troops who'd break open the boxes, load up with grenades and "ammo" and race the mile or so back to the firing lines. We had to be alert for Arab snipers who would infiltrate the lines every night to shoot at the crews and planes. One night I saw two Arabs killed while they were laying mines at the end of our runway.

We tried to be off and away by dawn, but sometimes I'd stay over a day to go out with the Palmach hunting Egyptians on Jeep raids, much as one might organize a rabbit hunting expedition. I never bagged any. Nor did I make any scores in Jerusalem where I went twice to visit the front, then held by the Irgun Zvai Leumi, a former terrorist outfit, absorbed, temporarily at least, into the Hagganah. My favorite sniping spot in Jerusalem was the Church of Notre Dame, then a legitimate Arab target.

It was down in the dusty Negev, however, that we found the body of a pilot from Chicago, a gentile mercenary, who had been forced down. The Arabs had hacked his body to pieces and strewn it over the desert. It was in the Negev, too, that I saw my most horrible sight, in the village of Negva, just after its recapture by the Palmach.

A Palmachnick, his eyes burning in fiery anger, led me down a basement stairs. In one corner were piled the headless bodies of 35 Jews, including two women, all members of the farm. Their heads were stacked in another corner. As I stumbled outside away from the grisly sight, I met another Palmach, a captain, who was cleaning reddish stains from a curious curved knife. He too had seen the evidence of wanton massacre.

His reaction had been swift and instinctive, in keeping with the ruthless eye-for-an-eye warfare of Palestine. The captain herded the first 18 prisoners taken by his men into another basement, took a knife from one prisoner, and then slit every Egyptian throat. There had been two women in the lot. He gave me the knife for a "souvenir."

And that's the way it sometimes was in Palestine, before Dr. Ralph Bunche succeeded in arranging an armistice between Israel and Egypt, on a pattern which gives promise of being followed by the other Arab states which were at war with the new state of Israel. But the peace talks flowered after I left. Like many of the other ATC personnel, I decided to come home when it appeared that some of the big wheels in Tel Aviv were starting to organize a rival air line which would be in a better position to operate commercially after hostilities were over. By this time, too, our job was just about done. Israel was a state, a victorious state. It was time for an "old" sailor to look for a new berth.

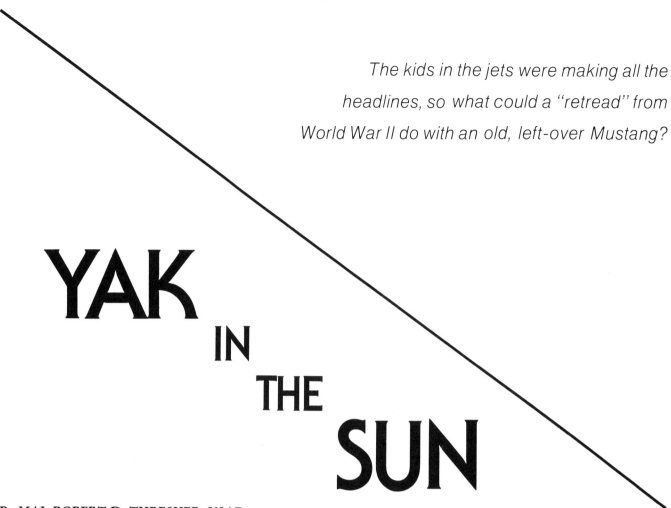

The kids in the jets were making all the headlines, so what could a "retread" from World War II do with an old, left-over Mustang?

YAK IN THE SUN

By **MAJ. ROBERT D. THRESHER, USAF**
as told to **ELIOT TOZER**

I WAS HUDDLED OVER the oil stove when "Dad" Flake, Lt., USAF, shuffled in. He snugged down the tent flap carefully and came over beside me.

"Ground support again," he said. The lines on his face were tight. "All the way to Sinuiju."

So we had it again. Ground support. It was all we'd had since we'd hit Korea six months back. We never climbed up to where an airplane belonged to do our fighting. Because we were flying left-overs—P-51's. But then there was a kind of ironical justice to it. Because we pilots were strictly left-overs, too. Retreads from World War II. Most of us had been in the 9th Air Force but in different squadrons.

My tour was typical. I hit England in May, 1944 in a *Thunderbolt* squadron. By July 19, I was stationed near Caen. I was lucky enough to fly 107 missions. But the 8th Air Force had done such a magnificent job on the *Luftwaffe* before we got there that I saw only four Nazis. And I never got near enough to tag one of them.

So when we got to Korea, we figured we were in. But the nod went to the new young pilots. Jet-trained, they copped the F-80's. We flew Ground Support.

And now even that was turning sour. Dead winter in Korea

and the North Koreans were holed in. Buttoned up for probably six months . . . as they had been since we'd arrived. And now doubt was beginning to gnaw at us. Maybe if we did run into a bogie we wouldn't stack up any more. Time was running out on us and we wondered. And the newspapers from home didn't want to remind you, but there you were, a hot fighter pilot with a second chance, and you never made any kills. While the guys in the jets—the kids—were making headlines every day.

Dad Flake turned down the oil burner. "Take-off's at 0700," he said.

It was quiet as I dragged myself out of the Jeep in front of my *Mustang*. I gave it a quick "walk-around," my flying boots crunching the powdery snow. I jammed myself into the cockpit and struggled with the safety belt and shoulder harness. My arms were bulky in my fleece-lined, sateen jacket.

I looked over at Dad. His Packard was trailing billows of blue smoke. And he was cussing. I grinned. I signalled for the chocks to be pulled and blasted the *Mustang* out onto the taxi strip and braked her toward the runway. She felt good. She felt really ready.

Dad finally came beating down the taxi strip, blowing a fan-shaped cloud of snow behind him. He hit his right brake hard and swung out onto the runway in front of me.

Dad put his hand up and I moved into position on his right wing. I heard Ross and Ausman of the lead element pour the coal to their planes and scoot down the runway. With six .50 calibres, six five-in. rockets, and two napalm bombs aboard, the *Mustang* picks up her skirts slowly.

Then Dad had dropped his hand and I was adding throttle to stay with him. We broke ground and I had my gear up. Prop pitch back. Throttle back. She was in fine fettle. Too fine for the mission we were saddled with.

We were at 4,000 feet when I saw the panels that marked the Bomb Line, the line inside which we dropped no bombs unless we were under Forward Air Control. Then we were over the front.

I picked up the *Mosquito*, a souped-up AT-6 that was our spotter, and directly underneath him saw the tank column we were supposed to support. The *Mosquito* went into a long, sharp dive. I heard Bill Schuman say, "He says he's got a road-block. Give it a squirt."

The *Mosquito* pulled up sharply to point out the road block he'd found. I saw Ross rock his wings, then pull up into a fast climb and roll over. He bore in fast, firing all the way. There was no flak. Ausman followed Ross and Dad was right on Ausman's tail. Still no flak.

But I decided to go in fast anyway. I added throttle and pulled up to start high, flicking on my sight at the top. I was at 280 when I smoothed out my approach. The road block grew big in my sight, but I waited. Maybe this was the day.

The road block was big now, clear in my sight. I squeezed, and the tracers drifted out in front of me. They centered squarely on the road block. Still no flak. By the time we'd joined up, the tank column had come up. I saw the lead tank sniff at the road block, then shift gears and go through. Nothing there at all. Tension crept down my back.

The *Mosquito* hovered over the tanks for a moment, then headed northwest. Ross broke to the northeast to stand off and take a looksee. Then Forward Air came in again. "He's got some ox carts heading north, Give 'em a whirl."

I cursed. The day was going sour all of a sudden. A dead road block and now ox carts. Ross and Ausman peeled off. I saw Dad lift his wing . . . then big orange balls of tracers were snapping across my left wing.

I yelled "Flak" and broke right.

In my haste somehow it came to me that an F-80 had un-covered a nest of antiaircraft in this area just the day before. And now the AA was centering on me. But as I unwound from my turn, a shape hurtled by me and I knew I hadn't been in flak at all. I had been in the gunsight of an enemy fighter.

"Bandits," I screamed.

Like any good ex-*Thunderbolt* pilot, I shoved my throttle to the firewall and snapped into evasive action. When Dad heard my yell, he chopped throttle. I pulled away like he was standing still.

For the moment I was alone. I put my hand up to shade my eyes. The sky seemed empty. I rolled into a sharp left turn to close up on Dad, but by that time, he and the other element had found a bogie. They were far to my left. And the three of them

were high-tailing it down a valley.

As they bored in, I saw that they'd bounced a *Yak*. At 7,000 lbs. maximum gross, the *Yak* is strictly a kite. It has the climbing and turning edge on every *Mustang* that ever pulled off an American strip. And the *Yak* that the boys were chasing was showing them how handy such turning radius can be. I nudged my throttle forward to get in on the kill. This could be what the whole day had been pointing to. But I never caught up with the other guys.

The voice of the *Mosquito* crackled into my headset. "Look out, No. 4, he's right on your tail."

I snapped my head around and saw the yellow winking lights of a *Yak's* two 12.7 mm. machine guns and 20 mm. cannon. He'd been sitting up there all this time in the sun and now I was directly in his line of fire. The turn sucked the blood from my head and for a moment I grayed out. I knew that the *Yak* could turn inside me, but in my panic, I held her there, hoping that somehow he wouldn't be able to follow through after his run.

And he wasn't. I rolled out to see that he'd gone across the top. And, taking advantage of his speed, he'd pulled up into a tight loop. That was all he needed. He had altitude advantage. If he could come out of the loop right, he'd be perched on my tail again, only this time with speed to burn.

I swung sharp left to pass directly beneath him.

He must have been watching me because he kept his stick back too far and stalled out. He spun all the way down from 6,000 feet. I rolled back into a turn, hoping desperately to come out behind him. Actually, he was spinning down the axis of the circle I was making.

He recovered and I thought for a moment that I should try to tighten my turn and jump him while he was floundering out of his spin, but he snapped her over, nursed her back to flying speed and then—kite that the *Yak* is—he went into a climb.

This time he slacked off at the top of his loop and followed through gently.

So that was to be the pattern. I was flying a tight circle on the horizontal. He elected to fly his sparring circle on the vertical, snapping at me as we passed each other. We held our turns. I looked for a break and as he picked up speed, I saw that we were getting closer each time we closed on the south side of my circle. More and more rapidly the two planes got out of phase. I was able to kick my rudder and snap a shot at him each time now.

He saw it. He knew that after another pass I'd be coming in behind him as he began his climb. It unnerved him and he wobbled at the top of his next loop. He fell unevenly this time. I cut wide briefly.

At the bottom, he recovered and began his pull-up but luckily my timing was good. I wrapped up my *Mustang* and while I was still in my turn began firing. I saw the tracers converge and pour into his wings. I eased rudder to the right to keep him in sight. Then a puff of blue grew on his wing. He wobbled and fell out of his turn. Slowly he rolled over and went into a long glide.

I watched him go down. A *Yak*. A left-over from the last war just as I was. But the kill meant that now my wings stood for something. The old guy still knew how.

The Air Force Covers a Retreat

'It was a nightmare,' said the

tired young lieutenant.

'Our fighter planes killed Chinese in carload lots—but still

they kept coming . . .'

By ROBERT FRANKLIN

I SAT on a hill overlooking the river, directing the fighter planes." The young, tired Air Force lieutenant paused to puff on a cigarette and gulp some hot coffee.

He continued: "The F-80's were pouring lead into hundreds of Reds as they tried to ford the Chongchon. None of them made it, but they were stupid enough to try. We call it the 'Red River' now because the water was crimson with Chinese blood."

The lieutenant, on duty with the army as a forward tactical controller, had just escaped capture by advancing Communists.

"When night came they tried again, and this time they made it. They overran the command post. They were everywhere, behind us, in front of us, on our flanks. It was a nightmare. We fought with everything—guns, knives, even bare fists. We killed them by the hundreds but no matter how many we killed, more seemed to take their place. We fell back for four days. Every day fighter planes killed them in carload lots. Still they kept coming."

The scene was a regimental command post just north of Pyongyang, Korea. The date was November 30. For five days a Chinese Communist army of 200,000 had been savagely attacking the United States Eighth Army, bent on destroying it. The five-day-old withdrawal was to last until December 13, and during that time, the United States Far East Air Forces was to be largely responsible for an Army's safety—to see that it remained intact and reached its destination with a minimum of losses.

Master of double-envelopment and infiltration tactics, the Reds threw the bulk of their strength against the right flank. If they could succeed in breaking through, the Eighth Army would face annihilation. The sheer weight of Communist troops caused that flank to collapse on November 28.

The late Lieut. Gen. Walton H. Walker selected a new line to the south. The new position was to stretch from the east to west coasts of Korea, just north of the 38th parallel, 110 miles south of the Chongchon River.

It would be a hard trek—narrow, twisting roads, nothing wider than cow paths through ice and snow-covered hills. Half-frozen rivers, spanned by battered bridges, and in some cases no bridges at all, crossed and recrossed the path of the retreat. There was a staggering amount of heavy artillery and equipment to be removed to safety. Faced with these obstacles, General Walker ordered the withdrawal.

Since June 26, planes of the United States Far East Air

Forces had been pounding the North Koreans without let-up.

In the early days of the war, air-ground teamwork enabled a numerically inferior UN army to hold the small perimeter around Pusan until sufficient strength was available to go on the offensive.

In mid-September, with the Inchon landings and the breakout from the perimeter successfully accomplished, the sweep northward was rapid. Fighters and bombers blasted a way for the ground forces, until by late October UN elements had reached the Yalu River, the natural border between Korea and Manchuria.

Peace seemed but a matter of days. The fleeing North Korean army had been destroyed. More than 140,000 prisoners of war languished behind barbed wire. Another 200,000 Communist casualties accounted for the remainder of Kim Il Sung's ill-fated army.

Then it happened.

The Chinese Communist government had threatened to intervene, but with the landings at Inchon and the drive across the 38th parallel accomplished, the military moment had passed. Good strategy or not, the Chinese crossed the Yalu River into Korea and, as General MacArthur said, "A new war was underway."

After heavy attacks on United Nations forces, the Chinese broke off contact just as rapidly as they had started. For days an unnatural quiet covered the battle fronts—the Eighth Army line in northwest Korea and the Tenth Corps zone of operations in the northeast sector of Korea.

Something was afoot. Having made their point, were the Chinese withdrawing to the Yalu River, or were they preparing another large scale attack? The UN high command didn't know—but they ordered an all-out attack on November 24. Advances up to 20 miles were made the first day without contact with the enemy. Some resistance developed on the morning of the 25th, and by the 26th the Chinese had hit back all along the line.

A new kind of war faced the Air Force. Previously it had provided an aerial umbrella for a comparatively small defensive force, and later had paved the way for an offensive army. Now the Air Force had the job of giving protection to an army that was retreating under constant attack.

During the first week of the Communist offensive, fighters and light bombers of the Fifth Air Force flew 2,350 separate sorties, 1,360 of these in direct close support of the ground forces. The remainder included armed reconnaissance strikes against targets of opportunity, and B-26 attacks deep into North Korea against communication lines.

Over the battle lines, the Air Force made it miserable for the Reds. So concentrated were the fighter attacks around Kunu-ri on December 1 that one group of Chinese surrendered to U. S. infantrymen after night and day pummeling by airplanes.

That same day an encircled U. S. infantry battalion called for air support. Responding F-80's killed more than 200 Communists and eliminated seven road blocks, permitting the battalion to rejoin its division.

"It reminded me of Braddock's march through Ohio," commented one of the pilots. "The Chinese were marching toward our troops three abreast. I kept pumping lead and firing my rockets right into them and they dropped all over the place—except those who weren't blasted. The ones we didn't get, the infantry took care of."

That same day, December 1, the Far East Air Forces flew a record-breaking 807 sorties. At the end of the day's operations, 2,050 enemy troops had been killed or wounded.

In addition, more than 480 buildings used by the enemy for supply storage or shelter were rocketed and strafed, 125 trucks blasted from the roads, several tanks shattered, and five ammunition dumps exploded under *Shooting Star* and *Mustang* attacks.

Superfortresses of Major General Emmett (Rosie) O'Donnell's Bomber Command ranged throughout Korea, hammering interdiction targets along the Yalu River and tearing up rail lines along an already battered transportation system. Every avenue of resupply had to be neutralized. Incendiary bombs saturated large stores of war materiel at Huichon, on the main rail line, 80 miles north of the former Communist capital of Pyongyang. Another dozen smaller areas were left ablaze.

With some units of the Eighth Army now in Pyongyang, the northern defense arc spread from Sukchon to Sunchon, a little more than 20 miles north of that city, but still 85 miles from the new defense line. The withdrawal continued to be orderly. Combined air-ground attacks were taking a 10 to 1 toll of the enemy.

At the time of the all-out attack on the Eighth Army, the Tenth Corps, operating in the northeast section of Korea, had been advancing with little or no opposition. The 17th Regimental Combat Team of the U. S. Army had reached the Yalu River and had stopped there, a few hundred yards from Manchuria. South Korean divisions were racing up the east coast in the direction of Siberia.

To lessen the possibility of being outflanked, Maj. Gen. Edward M. Almond, commanding the Tenth Corps, brought his offensive to a halt and ordered the most northern elements to begin a withdrawal southward toward the coastal port of Hungnam. The First Marine Division had reached the Chosin Reservoir, 50 miles northwest of Hungnam, by the time the order was received for them to prepare for possible action to the west.

On November 28, the Chinese threw seven divisions around the Reservoir, an attack that was to produce one of the most amazing air operations of any war. The coming battle produced the now famous remark by Maj. Gen. Oliver P. Smith, First Marine Commander: "Retreat, hell, we're just attacking in a different direction."

Enemy attacks were repulsed, one after another. Close support aircraft operated so close that one tactical controller, after requesting a napalm attack on a group of Communists complained, "Hells bells, man, I didn't mean *that* close!" The flaming jellied gasoline had dropped 30 feet in front of him, wiping out the enemy pocket.

The overwhelming enemy strength was too much for even the hard fighting Marines. Chinese units succeeded in enveloping them and units of the Army Seventh Division that had returned from the north and joined the First Marines.

Enemy troops honeycombed the mountains on all sides of the UN forces and the Reds established countless road blocks along the highway leading to Hamhung. These units had to travel this road if they were not to be destroyed. The withdrawal was started.

Every mile of that battered, frozen road was under the guns of the enemy, but every yard of the surrounding mountains felt the brunt of the most intensified aerial close support of the Korean War.

The retreat to Hamhung was a three-phase operation. Phase one started November 30. The objective was the small town of Hagaru-ri at the southern tip of the reservoir. A distance of but 15 miles, it seemed like a trip to the moon for the embattled troops. At that point a link-up would be accomplished with other elements of the First Marine Division, and phase two— a 10-mile trek to Koto-ri—started.

It took four days for the last man to filter into Hagaru-ri. During that entire period, Air Force, Marine and Navy fighters and bombers poured machine gun fire, bombs, rockets and napalm into Communist troops on the surrounding mountains. Resistance, although remaining strong, had lessened considerably by the time phase two was under way.

With the main highway of Humhung cut off by numerous road blocks, the problem of resupply was solved by transports of the Combat Cargo Command. Day after day, Fairchild C-119's *(Boxcars)* and Douglas C-47's *(Skytrains)* rained ammunition, food, clothing, gasoline, even drinking water to the column. More than 1,000 ton of supplies were parachuted to the Marines and soldiers in four days.

As the main column continued toward Hagaru-ri, one of the most dramatic airlifts in aviation history got underway. Operating from a 3,000-foot emergency strip at Hagaru-ri, transport after transport landed, unloaded ammunition and supplies, quickly reloaded with wounded and frost-bitten GIs and headed for Japan.

On November 30 a doughty engineer battalion had started contruction of the strip. It was no easy task. Both the elements and the enemy made things difficult. So hard was the frozen earth that bulldozers' blades would not penentrate it. By

welding steel "teeth" to the blades and using dynamite to blast the more stubborn blocks of ice covered ground the battalion hacked out the strip. The entire operation took only 24 hours, the men working under constant enemy mortar and small arms fire.

For six days, from December 1, the C-47's landed from dawn to dusk, removing hundreds of casualties, the majority frost-bite cases, to safety. With the last casualty evacuated, the second phase of the operation was started.

By December 7, all elements closed in at Koto-ri, where the aerial evacuation was repeated once again. This time the strip was but 2,600 feet long. The same rugged engineer battalion had lengthened a lightplane strip so it could accommodate transports. With casualties much lighter (less than 400) the operation was completed in only two days. Phase three was started December 9.

Thirty miles from their destination now, the Allied troops knew they were going to make it. Just a few more miles up the road that wormed through deep gorges and over the wintry mountains and then it was down-hill and across a small plain to Hamhung. A combined Army-Marine relief column was headed north to relieve pressure on the retreating forces.

One obstacle still remained in front of the withdrawing troops. The Chinese had blasted a highway bridge that spanned a deep gorge. It seemed an impossible task to get the thousands of men across the wide gap. Actually the problem was licked before the main body reached the area. Patrols had discovered that the bridge had been blasted. A quick call and within a matter of hours, C-119's of the Cargo Command parachuted an eight-span bridge to the column. That same engineer battalion wasted little time getting the bridge across the gorge. The link-up accomplished, the tempo of the march stepped up. By December 11, the entire command was safely in Hamhung.

"I've never been in an airplane in my life," said one weary Marine, "but I love them. They sure raised hell with the Reds all the way down that damned road. Sometimes I could have reached out and touched them, they were flying so low. A lot of us can thank the fly boys that we are still alive."

Tan Son Nhut/Saigon

By FRANK L. HARVEY

Tan Son Nhut Airfield in Saigon is a steaming cauldron in which the ingredients of war are mixed and stirred and blended in preparation for a multitude of missions in support of the Vietnam war. The field is dominated by a single 10,000-foot strip, and during the morning and evening rush hours it is close to a madhouse. Tan Son Nhut is the busiest airfield in the world—busier than Kennedy, busier than O'Hare.

Everything is designed to clear the planes off the runway in a few seconds. Spaced along the main strip are three feeder strips angled into the runway. Sixty planes are sometimes waiting in these three lines, engines running. Often two planes take off together—either in formation or in tandem, one after the other. As one lifts off, it turns sharply to make way for the one behind it. Takeoffs and landings are often taking place simultaneously. The FAA wouldn't like it, but it sure gets planes up and down in a hurry.

Pipers, Cessnas, Beeches, DC-3s, Caribous, Grumman HU-16s and other smaller aircraft line up at the 5,000-foot turnoff while C-54s, C-118s, Connies and the like wait their turn at the 7,500-foot mark. Heavy fighters F-100s, F-102s, F-104s, F-4s, F-5s—and lumbering jet transports such as the Lockheed C-141 lock their brakes and spool up to takeoff power down at the far end.

A lightplane may land long and wheel off quickly around the banked racetrack curve at the end of the runway while a jet comes in hot behind—landing short and popping its chute at the beginning of the strip. A C-130 full of ammo turns on short final and squeezes in front of a Pan Am 707 on a long, straight-in approach.

Emergencies occur almost hourly. A squadron of choppers, down to their last drop of fuel, flail toward the strip, waiting for their engines to cough out. An F-105 touches down gingerly on battle-damaged landing gear. A B-52 limps in all shot up, and they lift out the wounded pilot after he shuts it down and rush him to surgery.

A Kaman HH-43 helicopter lugging a large red CO_2 dispenser on a cable under its belly stands by constantly for just such emergencies. As the wounded ship comes on final and

The Huskie H-43 Kaman helicopter working as a fire suppressor.

touches down, the chopper flies alongside ready to douse any blaze instantly. Crash trucks and a mobile derrick are always at the ready too. If a crash occurs, the trucks race out, spray foam on the damaged plane and the mobile derrick muscles it off the runway. With the wild stackup that's usually overhead, the field can't afford to close down even for a few minutes.

Tan Son Nhut is the biggest American air base in Southeast Asia. Around the field are rows of ramps and concrete parking areas where planes are outfitted for every conceivable kind of mission. In addition to being a marshaling yard for the goods our cargo planes bring in from overseas, Tan Son Nhut is an operational base for search and rescue, defoliation, in-country supply redistribution, air defense fighters, helicopter salvage work, bombing missions, day and night reconnaissance, VIP air taxis, spotters, medevac, troop transport—and a host of other activities.

The largest planes in the Air Force, giant Douglas C-133 Cargomasters, are working hard out of Tan Son Nhut. There are clusters of various sorts of helicopters, which are used for every conceivable purpose. There is a row of handsome little North American T-39s, including Henry Cabot Lodge's personal VIP jet. Cessna 310 "Blue Canoes" dart in and out on their never-ending ferry flights back and forth across Vietnam.

One sees light transports, such as the de Havilland Dove and the C-123, and the Army Beavers and Caribous. A strange-looking French STOL propjet with two boxy kite-like tails floats out of the field buzzing like an angry bumblebee, yet hardly seems to be moving (its mission: supplying the boondocks). The pilot of an F-101 Voodoo, just back from dodging flak over North Vietnam, waggishly waves his fuel probe up and down at you in greeting.

On another corner of the field, C-123 Providers, which have been modified to spray defoliants on jungles and rice paddies, sit in line amid the smell of defoliant chemicals. These "Ranch Hand" planes (code name for defoliation operation) carry spray racks under the tail and wings. While spraying, they have to fly low, slow and vulnerable. When you get close to them you see they are grimy with spray and peppered with

neat little riveted patches to cover the bullet holes.

Sitting incongruously in this crowd of military hardware are the familiar Pan Am and Air France 707s, DC-6s of Vietnam Airlines and swing-tail Canadair 44 turboprops of Slick and Flying Tiger Airlines.

On the main strip, a four-engine C-141 cargo plane touches down and rolls quickly to the major unloading area. As it brakes to a stop beside a mountain of supplies, a huge flatbed vehicle trundles up to the back door. The door opens and the palletized load rolls out of the airplane onto the truck to be carted away and stacked, or shifted to another part of the field for transhipment.

As soon as the C-141 is empty, a crew of mechanics climbs into its cavernous belly and sets up stanchions and litters five high, transforming the huge jet into a medevac plane. Right afterward, a fleet of ambulances arrives from the local hospital specializing in preparing the wounded to travel; the C-141 is quickly loaded with wounded men, their nurses and attendants. The C-141 taxis out and leaves for the Philippines.

Minutes later an air alert sounds. All traffic gives way to a squadron of F-102s that race out of their Armco-clad antimortar revetments and scream off, in formation takeoff. The all-clear is called and yet another senator flies in to have cocktails with

The McDonnell F-101 Voodoo landing with drag chute.

(continued on page 209)

PORTFOLIO

Over the years, Flying *has followed the course of avia-tion, civilian and military, in words and pictures, some of the pictures being in color.*

Many civilians in the late 1930s made a transition to military flying in the Boeing-Stearman Kaydet primary trainer (below), and many of them first became enamored of aviation because of The Spirit of St. Louis *(shown in replica on page 194). Others admired the rugged bush pilots' delight, the Noorduyn Norseman (page 196).*

For many years, the magazine's orientation was largely military, focusing on World War II planes such as the Spitfire (page 198) and the Boeing B-17 (page 199) shown here in the modified peacetime version flown recently by Robert B. Parke. In more recent years, the magazine covered the Viet Nam war, portraying in color bombing runs by the Republic F-105 Thunderchief (top, page 200) and the Boeing B-52 Stratofortress (bottom) and perhaps pointing to the future in depicting the VTOL Hawker-Siddeley Harrier fighter (page 201).

The basic thrust of Flying *has been civilian, with much attention paid to business aviation, as embodied in the Ted Smith Aerostar (below) and the lighter but highly practical Cessna 206 (page 203). The ordinary has been accompanied by what to many pilots are exotic sides of business flying, such as flying the mail at night in venerable Beech 18s (page 204).*

However, it has not been all work and no play. Though Popular Aviation *inveighed against air racing, which in the 1920s and 1930s took a high toll of lives,* Flying *has since taken new looks at the sport and has caught the color and thrills afforded by such planes as Darryl Greenamyer's Conquest I (page 205, top), with glances at quieter sports such as soaring and ballooning (bottom).*

Always, of course, there has been the simple direct appeal of a photograph beautiful in its own right, whatever the subject, as in the case of this Aerostar caught on top, above the California coastline by photographer George C. Larson (page 206). This is part of the Flying *tradition, which is that of being colorful, as a glance at* Flying *covers over the year will show.*

One of the logistics mainstays in Vietnam, the Lockheed C-130 Hercules cargo and troop carrier.

General Ky, get the complete picture of what's going on in Vietnam, after which he'll hurry on to Bangkok or Hong Kong for R and R.

Premier Ky's famous Coup Squadron of black Skyraiders, with their swish insignia and Vietnamese markings, stand jauntily in a row. They are parked behind the General's headquarters, a low building with lovely light purple shutters, ringed with sandbags, manned machine guns and multiple layers of accordion barbed wire.

Armed choppers orbit steadily over the field, engines clattering, doorgunners sitting, black visors down, with their hands on .50 cal. machine guns, watching for the first sign of a mortar attack or a VC infiltration, ready with the most hellish firepower ever assembled in so small a machine. These are the "Huey Hog" gunships, known among the VC as "The Muttering Death," and in this writer's opinion the most savage machines of the Vietnamese war. They patrol Tan Son Nhut and the other American air bases around the clock.

The newcomer to Tan Son Nhut is soon struck by the type of Americans who inhabit the place. A hot shooting war seems to float the takecharge types to the surface. There are very few flabby nearsighted officers or petty by-the-book noncoms. In fact, you might think you had somehow stumbled into the Green Bay Packers' preseason training camp. Everywhere you

look there are big guys with clippered heads and deep tans. They wear dark jungle fatigues with huge pockets in the baggy pants, Special Forces canvas boots with deep mud cleats and either turned-up Aussie jungle hats or dark loose-crowned baseball caps.

The Officers' Club bar at the base is jammed, starting at 10 in the morning. The noise level is only slightly less than the chipping bay of a steel mill. At times, in order to hear what your companion is saying, you have to put your ear a few inches from his mouth. He then shouts. The barroom conversation is not wholly dedicated to women, stateside ball scores and how may days, hours and minutes until a hitch is up. They also talk shop. Fighter-bomber pilots argue the merits of low-level lay-down versus high-angle dive bombing. Air Commando guys (they fly the paradrop and shortfield supply runs) discuss such matters as going to full power with water augmentation if you get in a bind coming up a canyon and have to pole-vault over the ridge, or how to bounce yourself 30 feet at the last moment, to clear the trees, by dumping full flaps on the C-123.

The Air Force's twin-engined C-123 Providers, some modified to add GE jets and tail braking chutes, are designed to be STOL. You could get an argument on it. Some say, urgently, that the only way to get a grossed-out C-123 in the air on a hot day is with a hydraulic jack.

(continued from page 192)

Nonetheless, the C-123 is the best we've got for small-field supply runs. Just about every C-123 ever built is here. Our jocks have shoehorned them out of plenty of back country jungle strips. Some Air Commandos say they've gotten them in and out of 1,200 feet, which is taking your life in your hands. In '67 the AF will probably start outfitting with Army Caribous. The payload is only half as big, but the Caribou is more STOL, and a lot thriftier on the old adrenalin supply.

Across the street from the Tan Son Nhut Officers' Club is the 7th Air Force Information Office, headed (when I was there) by Col. William McGinty, USAF.

McGinty is the kind of information officer you find sitting at his desk, looking disgustingly healthy and bright-eyed, when you drag in at 9:30 with red eyes and blue fingernails after a hard night "interviewing" at the club. He informs you at once how glad he is to report he was able to clear all those little requests (Oh God, *what* requests?) you made at the bar last night. Hastily you move your cotton lips to say you didn't really mean—but Mac is ahead of you.

"At 10 sharp," he announces briskly, "you will be making a high-speed low-level attack in the back seat of an F-100. At two in the afternoon I've got you aboard Maj. Ralph Dresser's defoliation plane. Ranch Hand is going to spray a real hornets' nest down in Delta. And tomorrow, if I get final clearance— I'm sure I will!—you can go on a night patrol with the Marines, and they assure me—" At this point, if you have any sense at all you slump over and simulate a heart attack.

Some newshawks, of course, prefer to get the "curt word" at Lt. Col. Jack Keeler's "Five O'Clock Follies," as Keeler's everyday briefing in downtown Saigon is referred to by the press. (As far as anyone knows, nobody has yet been wounded while sitting in Col. Keeler's air-conditioned room, watching him point at a map.)

One of the places the Information Office will not encourage you to visit is the mortuary at Tan Son Nhut, where all American dead are prepared for shipment to the States. Such visits are verboten. Nobody is supposed to really get killed in this war, and if they do, let's whisk them out the side door quickly and be about other business. But I had smelled the C-123 death ships on the line at Tan Son Nhut, and I had seen men quicken their pace during the hot Saigon day, as they walked by the big mysterious building where the American dead are processed.

With the help of an independent-minded noncom, I was able to get inside the warehouse-like building. I saw the man-sized aluminum containers stacked high against the wall. I saw the ambulances back up tight to the door and unload the bodies of American soldiers. I saw the white tiled floors, the tilting tables, the grim but necessary tools. I spent an hour there. The dreadful task was being handled as well as is humanly possible.

One of the last things I did at Saigon before going south to Can Tho, and the war in the Mekong Delta, was to visit the Tactical Air Control Center at TACC, which is the major nerve center of the whole air war. TACC in Saigon assigns all missions over South Vietnam. The direction comes out of a grand-style war room, with the usual impressive array of radars, radios, elaborate phone hook-ups, plexiglass grids, flashing lights and battle maps, all manned by skilled operations directors.

The TACC briefing itself, to be honest, was a drag. Important as hell. But alphabet soup. My mind was already on the live action in the Delta, and anyway, I tend to go to sleep in flip-chart briefings where they have these big, meaty titles like OVERALL INPUT TO THE MAGNAFLUXOMETER—MOD 3. Old Hermann Magnaflux himself would have to struggle to keep his lids open. I generally nod politely and murmur, at appropriate intervals, "Really? Why, that's *fantastic.*"

After furious attempts at concentration I can now tell you that the TACC is backed up by the DASC (Direct Air Support Center) and the DASC is backed up by the FAC (Forward Air Controller) and the FAC is backed up by the ALO (Air Liaison Officer) and they are all assisted by the TASS (Tactical Air Support Squadron) and the whole ball of wax is a JAGOS (Joint Air Group Operation System). The TACC is composed of a staff of VNAF (Vietnamese Air Force officers) and their opposite numbers in USAF (U.S. Air Force). So much for alphabet soup.

PART V

Airways

For all its romance, glamour and excitement, aviation was boosted in the period between the world wars mainly for its transportation value. In their hopes, the boosters "saw the heavens fill with commerce," and no publication had that vision more clearly than *Popular Aviation*. Lindbergh's flight to Paris proved more than any other before it the potential reliability of the airplane, and subsequent "route-proving" transoceanic and transcontinental flights deepened that impression. The advertisements that filled *Popular Aviation* seemed to carry to a single theme: For any young man with spunk and ambition, aviation was the way of the future.

The many uses of airplanes were covered, and while the inherent hazards of committing oneself to the sky were not concealed, the editorial emphasis was on the safety and even the luxury of aerial transportation. By the 1930s, for instance, carrying the mail by air, a very dangerous way to make a living in the preceding decade, was reduced to the mundane. By the time World War II broke out, many people were predicting that small planes—then epitomized by the Piper Cub—would become so easy to fly that they would become automobiles of the sky, family necessities.

Even more promising to many people than light-planes was the helicopter, which needs no runways for takeoff and landing. Fanciful illustrations in ads and various publications, including *Flying*, depicted cities with airspace dotted with commuting helicopters. Stanley Hiller's helicopter, for instance, seemed to be an aerial Model T in the making.

By war's end, it was thought, a big economic boom would be in aviation, considering the many thousands of pilots trained in the Services plus the expected inexpensiveness of small airplanes and helicopters. It was not to be. Most of the new pilots simply gave up flying; the helicopter proved to be beyond the capabilities and interests of most pilots; for most people, the economics of taking up flying were wrong. Even airline piloting had become precarious as jobs came and went. Aviation—and *Flying*—lost the healthy glow they had expected to retain. Another ingredient was needed.

That ingredient proved to be a mounting general prosperity accompanied by intense business competition in the 1960s. Business flying no longer was simply a luxurious image builder for a company or corporation. Decentralization of industry and other businesses had outstripped the airlines' abilities to provide truly convenient service except for major cities and towns. The company airplane(s) became a necessity, whatever the cost. The advent of business jets provided airline-swift flight. The Learjet became as well known and ubiquitously identified as the Piper Cub, and often as erroneously, for a wide number of "bizjet" models had become available, just as the many types of lightplanes, singles and twins, being produced began to turn the classic Cub almost into a rarity.

General aviation aircraft were being used for many functions—flying cancelled checks and the mail, crop dusting, commuting and recreation. While the helicopter has not yet become Everyman's car of the skies, it has been taken on for many uses, from transportation and dusting to hauling and specialized industrial tasks. General aviation has continued to grow, spurred on by reduced airline service and the threat of fuel shortages. Many corporations now run fleets of planes—mini-airlines.

All this growth was bound to compound the problems of aircraft safety. The air traffic control system became overloaded both with airplanes and procedures. In its early days, *Flying* favored Federal involvement in aviation safety, but in recent years the apparent inability of the Government to reform the current cumbersome traffic control system has caused the magazine to levy harsh criticism at the Department of Transportation and particularly at the Federal Aviation Administration. In 1968, the magazine published a long and thorough analysis of the traffic problem, "Air Traffic: the Clearing Skies," which was hailed as an important and statesmanlike contribution to possible reform. The very existence of the problem, which still is a headache, is an indication of how far aviation has come from the magazine's founding, when the airways were uncrowded and the "argosies of magic sales," of which Tennyson wrote, were still few and far between.

AIRWAYS

Air-Mail Schedules

Air-Express Schedules

Air Travel Information

†New York to — NATIONAL AIR TRANSPORT — Chicago

Westbound	Close P. O. Day	Close P. O. Nite	Leave Day	Leave Nite	Phone	Eastbound	Close P. O. Day	Close P. O. Nite	Leave Day	Leave Nite	Phone
New York, N. Y.	10:00	6:30	11:00	8:00	Lexington 8044	Chicago, Ill.	6:30	6:30	8:00	8:00	Harrison 8234
New Brunswick, N. J.			12:15	9:35		Cleveland, Ohio			11:00	12:15	Lakewood 9006
Cleveland, Ohio	2:30	12:00	3:35	1:30	Lakewood 9006	New Brunswick, N. J.			4:45	4:45	
Chicago, Ill.			7:00	5:35	Harrison 8234	New York, N. Y.			7:00	6:15	Lexington 8044

*†Chicago to — BOEING AIR TRANSPORT — San Francisco, 1943 Miles

Westbound	Close P. O.	Leave	Phone	Eastbound	Close P. O.	Leave	Phone
Chicago	6:30	7:50	Wabash 8084	San Francisco, Calif.	6:05	7:00	Kearney 2041
Iowa City, Iowa	8:40	9:40	425	Sacramento, Calif.	8:15	9:15	
Des Moines, Iowa			Maple 707W	Reno, Nev.	10:00	10:30	195
Omaha, Nebr.	10:30	12:35	Atlantic 9301	Elko, Nev.	12:15	12:40	1553
North Platte, Nebr.		2:00	29	Salt Lake City, Utah	3:00	4:05	Wasatch 5569-3321
Cheyenne, Wyo.	3:30	4:45	656	Rock Springs, Wyo.			
Rock Springs, Wyo.	6:30	7:05	415J	Cheyenne, Wyo.	7:30	8:00	656
Salt Lake City, Utah	9:00	10:20	Wasatch 5569-3321	North Platte, Nebr.			
Elko, Nev.	10:45	11:15	1553	Omaha, Nebr.	10:30	12:45	Atlantic 9301
Reno, Nev.	1:00	1:45	195	Des Moines, Iowa		1:45 Deliver	Maple 707W
Sacramento, Calif.		2.45		Iowa City, Iowa			
San Francisco, Calif.		4:30	Kerney 2041	Chicago, Ill.		5:45	Hemlock 8180

*†Rt. Cam-3, Chicago to — NATIONAL AIR TRANSPORT — Dallas, 1003 Miles

Southbound	Close P. O.	Leave	Phone	Northbound	Close P. O.	Leave	Phone
Chicago, Ill.	3:45	5:45	Harrison 8234	Dallas, Texas	7:00	7:45	5-7113
Moline, Ill.	6:50	7:20	Moline 738	Fort Worth, Texas	7:00	8:15	Prospect 2169
St. Joseph, Mo.	10:10	10:20	St. Joe-6-0349	Oklahoma City, Okla.	9:00	10:10	Walnut 4214
Kansas City, Mo.	10:50	11:18	Leeds 1090	Ponsa City, Okla.	10:40	11:05	Market 6185
Wichita, Kans.	12:45	1:18	Market 6185	Wichita, Kans.	11:15	11:57	
Ponsa City, Okla.	1:40	2:05		Kansas City, Mo.	12:15	2:05	Leeds 1090
Oklahoma City, Okla.	2:00	3:05	Walnut 4214	St. Joseph, Mo.	2:17	2:40	St. Jos. 6-0349
Forth Worth, Texas	4:00	5:15	Prospect 2169	Moline, Ill.	4:00	5:15	Moline 738
Dallas, Texas		5:35	5-7113	Chicago, Ill.	6:30	7:20	Harrison 8234

*†Rt. Cam-2, Chicago to — ROBERTSON AIRCRAFT CO. — St. Louis, 278 Miles

Southbound	Close P. O.	Leave	Phone	Northbound	Close P. O.	Leave	Phone
Chicago, Ill.		5:50	Maywood 4010	St. Louis, Mo.	3:30	4:15	Avery 2725
Peoria, Ill.		7:25		Springfield, Ill.		5:20	
Springfield, Ill.		8:15		Peoria, Ill.		6:10	
St. Louis, Mo.		9:15	Avery 2725	Chicago, Ill.		7:30	

Rt. Cam-4, Salt Lake City to — WESTERN AIR EXPRESS — Los Angeles, 600 Miles

Westbound	Close P. O.	Leave	Phone	Eastbound	Close P. O.	Leave	Phone
Salt Lake City, Utah		9:10		Los Angeles, Calif.	6:30	7:35	Trinity 6754
Las Vegas, Nev.		2:25		Las Vegas, Nev.		10:40	
Los Angeles, Calif.		5:25	Trinity 6754	Salt Lake City, Utah		3:20	

Rt. Cam-5, Elko to — WALTER T. VARNEY — Pasco, 435 Miles

Westbound	Close P. O.	Leave	Phone	Eastbound	Close P. O.	Leave	Phone
Elko, Nev.		11:30		Pasco, Wash.		6:00	
Boise, Ida.		1:15		Boise, Ida.		9:20	
Pasco, Wash.		4:35		Elko, Nev.		12:30	

Rt.-Cam-8, Seattle to — PACIFIC AIR TRANSPORT — Los Angeles, 1,099 Miles

Southbound	Close P. O.	Leave	Phone	Northbound	Close P. O.	Leave	Phone
Seattle, Wash.		11:45		Los Angeles, Calif.	11:00	12:01	Vermont 4279
Vancouver, Wash. .... }				Bakersfield, Calif.		1:45	
Portland, Ore. .... }		7:00		Fresno, Calif.		3:30	
Medford, Ore.		9:30		San Francisco, Calif.		5:30	
San Francisco, Calif.		1:15		Medford, Ore.		9:30	
Fresno, Calif.		3:15		Portland, Ore. .... }			
Bakersfield, Calif.		4:45		Vancouver, Wash. .... }		12:00	
Los Angeles, Calif.		6:15	Vermont 4279	Seattle, Wash.		2:00	

AIRWAYS — Continued

Rt. Cam-12, Cheyenne to				COLORADO AIRWAYS			Pueblo, 199 Miles
Southbound	Close P. O.	Leave	Phone	Northbound	Close P. O.	Leave	Phone
Cheyenne, Wyo.		5:30		Pueblo, Colo.		**4:00**	
Denver, Colo.		6:55		Colorado Springs, Colo. ...		**4:50**	
Colorado Springs, Colo. ...		7:50		Denver, Colo.		**6:00**	
Pueblo, Colo.		8:30		Cheyenne, Wyo.		**7:30**	

*†Rt. Cam-9, Chicago to				NORTHWEST AIRWAYS			Minneapolis, 377 Miles
Westbound	Close P. O.	Leave	Phone	Eastbound	Close P. O.	Leave	Phone
Chicago, Ill.		5:50		Minneapolis, Minn.	1:30	**2:30**	
Milwaukee, Wis.		6:50		St. Paul, Minn.	2:10	**2:40**	Cedar 5693
La Crosse, Wis.		9:30		La Crosse, Wis.	3:30	**4:00**	
St. Paul, Minn.		11:30	Cedar 5693	Milwaukee, Wis.	5:15	**6:00**	
Minneapolis, Minn.		11:40		Chicago, Ill.		**7:00**	

Rt. Cam-1, Boston to				COLONIAL AIR TRANSPORT			New York, 192 Miles
Southbound	Close P. O.	Leave	Phone	Northbound	Close P. O.	Leave	Phone
Boston, Mass.	5:15	6:15		New York, N. Y.	3:00	**6:00**	Ashland 7750
Hartford, Conn.	6:15	7:35		Hartford, Conn.		**6:35**	
New York, N. Y.		8:15	Ashland 7750	Boston, Mass.		**7:50**	
(Hadley Field)							

Rt. Cam-6, Detroit to				FORD AIRLINES			Cleveland, 91 Miles
Southbound	Close P. O.	Leave	Phone	Northbound	Close P. O.	Leave	Phone
Detroit, Mich.	9:50	10:40		Cleveland, Ohio	3:30	**4:30**	
Cleveland, Ohio		**12:10**		Detroit, Mich.		**6:00**	

Rt. Cam-7, Detroit to				FORD AIRLINES			Chicago, 237 Miles
Westbound	Close P. O.	Leave		Eastbound	Close P. O.	Leave	Phone
Detroit, Mich.	2:25	3:15		Chicago, Ill.	7:55	9:30	
(Central Time)							
Chicago, Ill.		**6:00**		Detroit, Mich.		**12:00**	

Rt. Cam-11, Cleveland to				CLIFFORD BALL AIRLINE			Pittsburgh, 123 Miles
Southbound	Close P. O.	Leave	Phone	Northbound	Close P. O.	Leave	Phone
Cleveland, Ohio		**12:15**		Pittsburgh, Pa.		**2:30**	
Youngstown, Ohio		**1:00**		Youngstown, Ohio		**3:15**	
Pittsburgh, Pa.		**1:45**		Cleveland, Ohio		**4:00**	

†Express
*†Express and Passenger
Light face type—A.M.
Bold face type—P.M.

Time used is local time, whether Eastern, Central, Mountain or Pacific.

Incomplete data will be filled in as rapidly as authentic information can be compiled.

The above schedules of airways information, while carefully compiled, are not guaranteed to be correct, but every effort and precaution is used and will be, from now on, in keeping this schedule up to date. It is proposed, from month to month, to correct and incorporate such further information as may be necessary in keeping pace with the air transport progress of the United States Air Lines.

An attempt has been made to print a schedule which is readable and may be understood by all who have use for air transport, whether air mail, passenger, or express service and is published for the purpose of promoting air transportation and educating our readers so that they may intelligently answer questions regarding aircraft operations when such information is requested.

The present air transport companies now operating and listed above, are primarily trunk lines, but there is a new phase of air transport, heretofore seemingly neglected, springing up throughout the country, namely, Aerial Taxi Service. We hope, in the near future, to be able to print authoritative information

on this class of service wherever it prevails with the proper kind of equipment, so that at any time anyone seeking air travel information may refer to POPULAR AVIATION and there find the answer.

Passenger service heretofore neglected by the air transport companies is now available from New York to Boston over the Colonial airways, Chicago to Dallas and intermediate points via National Air Transport, Chicago to San Francisco and intermediate points over the Boeing Airways and from Salt Lake City to Los Angeles over the Western Air Express and from Los Angeles to Seattle via the Pacific Air Express. Phone numbers have been given wherever possible so that people desiring information regarding air travel and transport between any two particular places can be secured through these offices.

We would appreciate comments from our readers as to how we can further make this schedule better suited for their particular needs, as it is the ambition of POPULAR AVIATION to assist in bringing the industry closer to the market.

An Eaglerock providing transportation
and advertising at the same time.

Go Up, Young Man—Go Up

BY ROBERT S. CLARY

TODAY, J. Harold Smith is a third year student in the Charlotte, N. C., high school. Tomorrow, he may be the Henry Ford of Aviation. By 1948 he may be America's largest manufacturer of popular priced airplanes . . . turning out five thousand $500 airplanes a day.

Within twenty years, 226,948 young men like him—keen eyed, forward looking, sky minded—will undoubtedly be filling highly important, highly paid positions in the aeronautical world. Their combined salaries and wages will doubtless exceed $379,284,935.00 a year. And in all probability they will be making and selling every twelve months some $2,934,488,639.00 worth of airplanes!

Two hundred and eleven of these young men, perhaps, will be presidents of giant corporations engaged exclusively in the manufacture of pleasure planes and air transports. Another 211 will be vice-presidents. And 11,347 will be important executives. Their job will be to make, sell, operate, and service each year the 4,157,830 flying motor cars that may be necessary to take care of new demand and replacement of the 24,452,267 airplanes of all types that experts predict will then be cruising the heavens.

This eventually inevitable consummation spells Opportunity . . . opportunity for tens of thousands of men like Harold Smith, men with vision and soaring ambition, men who realize that their future in flying is almost unlimited.

The future of flying is assured by two factors: First, everybody wants more money; and second, Time is money. Because of these two universal truths, any industry whose operations are predicated on the saving of time invariably grows to colossal proportions.

Moreover, this is the distribution age. The world's chief problem today is no longer mass production. It is mass distribution. And all distribution depends on two elements . . . transportation and communication—both of which are provided to an unprecedented degree by aviation. It is obvious therfore, that the young man who is thinking of his future can profit by thinking seriously of the future of aviation.

Geography itself insures the industry's development. For thousands of miles of desert and mountain and sea are insuperable obstacles to commerce until man's genius provides steamboats, railroads, motor cars, and airplanes.

Just a little over 100 years ago, Robert Fulton piloted the first successful steamboat along the Hudson River while crowds on both banks hooted and jeered. Yet today in the United States there are 567 shipyards, repair yards and boat building establishments employing more than 62,000 men, turning out $213,232,382.00 worth of equipment per year . . . and receiving an additional $112,548,461.00 for repair work. Liners, freighters, tankers, ferries . . . water craft of every class and purpose . . . daily cruise to every corner of the world that can be reached by water. Why? Because passengers, mail, express and freight all demand transportation.

Leaving out passenger traffic altogether, for the moment, the American public pays every year a freight bill of more than five thousand million dollars! And freight revenue for the railroads alone in 1925 was $4,546,670,891.

Let us glance a minute longer at the railroads, to obtain a further insight of the potential demand for faster, and therefore more economical, transportation. If you study a railroad map of the United States, you will see that this vast nation of ours is a perfect network of surface transportation lines. But more important still, you will see a map that looks almost exactly like our airline map will look in all likelihood by 1948.

Another conception of the potential demand for aircraft

may be formed from the fact that the United States alone now requires 64,370 steam and electric locomotives, 2,349,752 freight cars, more than 60,000 passenger cars, and approximately 1,769,000 railway employes.

Aviation does not require for way maintenance the annual purchase of 170,564 net tons of steel track, nor 77,474,983 cross ties, nor literally billions of board feet of switch ties, crossing plank, fencing, poles, bridge ties and piling. The airplane's tracks are already laid down by Nature, and maintained by the same beneficent agent. The tiniest village and the greatest city, regardless of location, are focal centers of a hundred rustless sky-roads . . . radiating in every direction like spokes from the hub of a motorcycle's wheel. *Truly Nature has loaded the dice for Aviation.*

Small wonder that our new infant industry of aviation is now lustily attracting the whole world's attention. Small wonder that enthusiasts now confidently assert that "Aviation is unquestionably destined to eclipse the steamship, railroad, and automobile not only in importance but in magnitude and in profit as well."

Until very recently, any airplane flight created enough excitement to jam all traffic on a busy street; but now the whirring of an airplane's motor receives only an interested glance from the hurrying throng. Why? Because aviation is now an accepted and established mode of modern travel. Its comparatively new experimental work has already covered twenty years, and exacted a total of many lives. But practical perfection has just arrived, and the front page of nearly every newspaper depicts daily the thrill and romance of new feats of progress.

Transcontinental passenger service is now in successful operation daily from coast to coast and from border to gulf. Mail, express and package freight are being transported every hour in the day—safely, speedily and economically by airline. Flying records for speed, reliability, distance, and payload are being shattered with increasing regularity. The aviation wonder of today is immediately surpassed by the achievement of tommorow. And all this is transpiring before our very eyes. A gigantic new industry which seems destined to become the greatest the world has ever known, is just beginning to kick off its swaddling clothes. Right now it offers every reader of POPULAR AVIATION a wonderful opportunity to get in on the ground floor and grow up with it. It offers to make you an integral part of a marvelous movement. It offers to share its huge profits with you. It offers you excitement, glory, honor and fame. It offers you opportunity!

That this husky infant of the automotive industry will eventually surpass in growth its powerful parent (which in three brief decades has become the largest single industry in the United States) is indicated by the fact that there are over a hundred airplane manufacturers in the United States and Europe, many of which have recently doubled and trebled their production without being able to meet the demand for new planes.

In addition to the army, navy, and marine corps, aircraft manufacturers, air mail contractors, and a host of other airline operators offer ample opportunity to the man who will qualify himself for any one of the hundreds of jobs connected with aviation that are multiplying so rapidly. And the U.S. Air Commerce Act at present specifies no particular educational requirements. This means that anyone who can read and write intelligently, and is gifted with a reasonable amount of common sense, can now get into aviation and with the proper training become successful.

Moreover, one need not necessarily fly in order to obtain the rich rewards that aviation offers. For there are some eighteen men in "ground" positions to every pilot in the air . . . positions incident to the manufacture and repair of planes and equipment, and to the maintenance of hangars and airports.

Compensation for such work of course varies, but is considered eminently satisfactory; and many excellent positions await trained men. Higher paid positions include airport superintendents, repair depot managers, general foremen, builders, inspectors, and similar jobs that pay from $50 to $125 per week.

As for the pilots, there are four types: Private, Limited Commercial, Air Mail, and Transport . . . in addition to military and naval. The air mail pilots receive, as a rule $65 to $150 a week; and are usually paid on a guaranteed weekly compensation basis, plus a bonus of about 5¢ a mile. Some have been known to make as high as $1,000 a month, and average about $18 for every hour of actual piloting.

Aerial photography and surveying is another big-pay profession that is proving decidedly lucrative. Greater New York was recently surveyed from the air, and required 3,000 miles of flight—covering an area of 325 square miles and necessitating 2,000 separate film exposures. Stunt flying, too, offers a profitable field. Numerous big money prizes are being constantly offered and awarded for noteworthy flights. And even barn-storming, or everyday "gypsy" passenger flying, has been known to pay up to $500 a day.

Still another profitable activity for independent pilots is crop dusting for the purpose of destroying insect pests. The charge made for this service is generally $1 per acre; and as an ordinary plane can cover 75 acres a minute, you can readily see that one's earnings are limited only by the amount of contracts he can sell.

But perhaps the greatest need is for office and technical executives. In addition to the wide range of high salaried inside position, there are many sales openings. Sales managers, distibutors, dealers, and salesmen are urgently needed. Just as soon as the right men can be found and trained, the aviation industry will provide as many "selling" jobs as does today's motor car industry.

One must not lose sight of the fact that all new industries require untold pioneering hardship . . . a singleness of purpose, a stoic disregard of long hours, dirty work, and insufficient remuneration. This prospect, however, will not deter the man of real vision. For he will fix his eyes steadfastly on the ultimate goal. He will literally hitch his wagon to the star of aviation's magnificent destiny. And he will stay with it till he reaps his reward.

Of course there will be many isolated instances of meteoric success. There will be other Lindberghs who fly overnight from obscurity to fame, who take off today in penury and land tomorrrow in the very lap of fortune. But for every one of these lucky ones, there will be hundreds and mayhap thousands who will wing their way to success only through stern self discipline and grim determination to get into aviation at once and *stay with it through thick and thin*.

What the distribution of men, mail, money and merchandise *must* have, to obtain maximum efficiency, is Speed . . . speed in time utility, and speed in place utility. And aviation meets this need better than any other transportation medium ever devised by man. This one reason destines the aircraft business to become as great if not greater than the automobile industry, in order to serve adequately an ever-increasing population that must be fed and clothed every day in the year.

Often you have heard it said that the airplane today is just where the automobile was a few years ago, but that is not true. The airplane of today is far ahead of the automobile of yesterday. If you doubt this for one minute, compare the pictures of yesterday's airplanes which illustrate this article, with the modern planes shown throughout this magazine. Few realize it, but there is in the United States today a greater number of factories turning out airplanes than there are factories in the United States making automobiles! The business of making and operating airplanes has already attained the proportions of a $100,000,000.00 industry annually!

Twenty years ago who would have predicted that by 1928 forty million Americans would be taught to drive automobiles? Is it, then, unreasonable to predict that within the next twenty years at least 10,000,000 must be taught to fly airplanes?

Aviation schools in existence today will require hundreds of pilots and literally thousands of ground men to teach their proportionate share of millions. Aircraft manufacturers in the country will require hundreds of test pilots and flying salesmen and literally thousands of ground men to supply the inevitable demand for planes. And within the next two decades the mere making and selling and operating of aircraft may make more millionaires than now exist in any other two industries combined.

For flying fulfills a universal want and need, and its eventual universal adoption is inevitable. Occasional accidents cannot deter it. A hundred million Americans will not "storage" their motor cars merely because a few crash up. They will not forsake business merely because a few fail.

Of course the best way to learn to fly is by actually flying. But for those who cannot start their practical flying or engineering training at once, there are a number of excellent home-study courses available. Theory as well as practice is essential in all industries—particularly in aviation. And no one can know too much about the basic principles of flying and aircraft.

J. Harold Smith has never been up but once in his life, but he *has* built models. He *has* read everything that he could get his hands on, on the subject of aviation. And already he knows far more about aircraft than did Lindbergh ten years ago. When he takes up actual flying training immediately upon graduation from high school next summer, he will have a tremendous advantage over those who will have to learn everything from the beginning. For he already knows aircraft nomenclature as well as he knows the alphabet. Futhermore he has learned the theory of flight and hundreds of other technical points so thoroughly that his rapid progress is assured.

Typical of the training in engineering which he may eventually take is the course in Aeronautical Engineering provided by the Daniel Guggenheim School of Aeronautics at New York University (located in University Heights, New York City). Here he can devote four or more years to learning the fundamentals of airplane theory and design, aerodynamics, airplane engines and installation; principles of radio communication, trade and transportation, internal combustion engines, advanced applied mechanics, propeller design, applied thermodynamics; materials and methods of aircraft construction, elements of navigation, instruments, meteorology, and other allied subjects.

The entire cost of this training approximates $330 per year. Aeronautical graduates of the classes of 1924, 1925, 1926 and 1927 were immediately placed in suitable positions. Of course, no position can be guaranteed to graduates, but every help will be given in placing qualified students . . . and classes will be restricted to such numbers as can be readily absorbed by the aeronautical industries.

The average flying course begins with dual instruction in straight and level flying; and includes turns and glides; spirals; approaches and take-offs; landings; eights and forced landings; tail-spins; spot landings; acrobatics; vertical banks; side slips; forward slips; fish-tail landings; hurdle landings; solo; and cross country flights. With this is combined detailed instruction in service and maintenance on modern type aircraft, an opportunity to study assembly and factory methods, and a broad ground-course foundation. The complete course requires about six weeks and includes twelve hours of actual flying instruction and a complete ground training.

Some flying schools offer three distinct classes of training: (1) *A primary flying course* for those desiring to qualify as private pilots. . . including thorough dual training and complete instruction in straight flying, vertical banks, spins, stalls, acrobatics, cross country, forced landings, and solo. (2) *An aircraft industrial course* which includes aerodynamics, airplane construction, repair, maintenance, rigging, engine overhaul and maintenance, aerial navigation, meteorology, parachute work, airdrome rules and regulations . . . qualifying for examinations as airplane mechanic and engine mechanic. (3) *A special advanced course* to graduates of the regular flying course who desire fifty hours' training, to qualify as an industrial or limited commercial pilot.

A few aviation schools are in a position to offer students a 250-hour course which qualifies them for a transport license. There is little question that such courses will become more and more popular. Experienced pilots are being hired by all the commercial operators and it will not be long before the supply of men leaving the army and navy will not be sufficient to take care of the needs of the operators. Naturally they will have to turn to the graduates of the schools. Men with hours in the air will be in demand, but they will have to be qualified for the work to which they will be called.

It is not unreasonable to see a remarkable future for the well trained man, and the day of his reward is close upon us.

500 MPH. Can't Be Done!

Present type planes would burn up at 500 m.p.h. The terrific heat caused by the air resistance would melt steel, claims prominent engineer and author.

BY JOHN B. RATHBUN

A SPEED of 500 to 1,000 miles per hour has long been the theme of the air-dreamer. To this end, many weird inventions have been proposed from time to time for attaining such speeds. All of these trick inventions, however, show that the actual facts in the matter have received but little attention at the hands of the romancers.

Even the most amateurish of the speed fans has recognized the necessity for tremendous power plants, but even with the wildest of imaginations he has not come anywhere within shooting distance of the truth. Above 500 miles per hour, the laws of air flow and resistance change radically, and it is this new resistance that we must overcome if we are to rise much above the present maximum of slightly over 300 miles per hour.

Just as a comparison, the speed of sound through the air is approximately 1,080 feet per second, or very nearly 750 miles per hour. At flying altitudes, due to lowered air density and lower temperatures the speed of sound is reduced to 500 m.p.h. approximately. A modern high power service bullet travels more than twice as fast as the sound wave, approximately 2,250 feet per second or 1,500 miles per hour. These two existing examples will give us an idea as to what will be required of this super-speed airplane in the way of power and performance. Very fortunately a great deal of data has been gathered by scientists on the trans mission of sound and upon "exterior ballistics" or the study of bullet flight, so that we will have some actual figures to work with in estimating our high speed futuristic airplane.

It has been repeatedly demonstrated by bullet flight measurements that there is a tremendous increase in the air resistance when we reach the velocity of sound, or 500 miles per hour. The slow motion pictures taken of bullets in flight confirm these measurements and indicate a new arrangement of the air about the bullet. Instead of flowing more or less smoothly over the surfaces, as with the present airplane speeds, the air is torn violently wide open by the bullet and so highly compressed that the air shows as a dark liquid streak in the photographs. It is believed in some quarters that the air is compressed to liquification at the nose of a bullet having a speed of 3,000 feet per second. Now this upsets all of our nice plans that we were making for our around-the-equator tour, and makes it necessary to revise all of our equipment. If the pressure at the nose of our ship will be sufficiently high to liquify air, then we must give up all thought of windows or any thought of viewing the scenery.

And then there is another sad fact that arises to spoil our dream. This is the knowledge that the friction of the air on the sides of our super-soarer will quickly raise the temperature to dull red heat, making it necessary for us to don our asbestos pajamas and carry a refrigerating plant along so we won't burn up. At ordinary flying speeds of 100 miles per hour, the heat due to air friction is hardly noticeable, but at 500 miles per hour where the resistance is increased about 900 times, we can reasonably expect to become uncomfortably warm at the end of the first hour, and still warmer at the end of the trip if we succeed in lasting that long.

A meteor from the outer spaces enters the earth's atmosphere at a speed of about 1,600 miles per hour, or about three times our estimated speed. If any of our readers still remain uninformed as to what happens to this meteor after a brief visit of a few seconds, we can tell them positively that it is heated white hot and converted into vapor by the air friction, not withstanding the fact that its temperature was about 460 degrees below zero a few moments before. And still they talk of 2,000 miles an hour in the Sunday supplements as though it were merely a matter of adding more engines and putting on more propellers!

A projectile flying at the rate of 2,250 feet per second lands at its destination in a very heated condition, sometimes in a semi-plastic state, and it was not the heat of the powder that did all this. To conceive of flying at this speed for more than a few seconds at a time is inconceivable for the reason that the shell of the ship would soon be white hot—always providing that we could assemble enough power to accomplish this feat. An air duct becomes very warm when the velocity of the air passing through it is only 100 feet per second, and yet we are considering the advisability of sliding through the air twenty times as fast!

And now to the matter of power, which as the poet says, is something else again. To travel through the air at 500 miles per hour requires a power increase of approximately 960 times the power at 100 miles per hour. If we are using 100 horsepower in our present plane at this speed, then we would be compelled to install 96,000 horsepower, and I hardly think that we would have room for this engine, let alone space for the storage of the gasoline and oil. To attain such a speed means a radical change in the motive power department, and it is likely that such attempts can never attain their objective until we have learned how to avail ourselves of the tremendous energy that now lies latent in the atom. By breaking up the atom into its electrons we may be able to obtain sufficient power but in no other way.

FLIGHT 14

What actually goes on "up front" in a big transport? This airline pilot, a former Army pilot who wrote "Cadet Days of Randolph Field," takes you with him.

BY MacDONALD HAYS

THE place: the Kansas City airport terminal. The time: any day at 2 p.m. "Attention please: First call TWA, Flight 14. Eastbound for St. Louis, Indianapolis, Columbus, Dayton, Pittsburgh and New York. All aboard, please. Give your name to the hostess as you board the plane."

The Kansas City passenger agent lays down his microphone. Passengers file out of the terminal building and enter the plane. The captain gets the "all clear" signal from the ground, they taxi out and take off. As simple as that.

But let's look behind the scenes at the preparation necessary for Flight 14, then board the plane and ride to St. Louis (the first leg of their trip) in the cockpit behind the captain and first officer.

First of all let's follow the captain and first officer after they arrive at 1 p.m. on the field—which is one hour previous to take-off time. The captain goes immediately to the meteorology department to check ceiling visibility and air temperature along the route. The chief meteorologist enters with the latest weather observations, which are made every hour along the route and a forecast sheet covering the next four-hour period.

"Captain Smith," he says, "You will have turbulence below 5,000 feet to Indianapolis with high scattered clouds above 10,000 and lower broken clouds at 8,000. I recommend 7,000 feet as the best altitude. From Indianapolis to Pittsburgh you will encounter a cold front at Columbus (he indicates a blue line on the air mass analysis chart), with lowered visibility. Your best altitude from there on in to Pittsburgh is 5,000 feet."

The cockpit of a DC-3, where a pilot's work seems never done.

Captain Smith checks the previous weather map to discover which way the bad weather is moving. He hands the weather sheet to his First Officer. "Here, Jones," he says, "make out the flight plan for 7-7-5-5-5."

Jones, a tall, alert-looking chap translates this to mean climb and cruise at 7,000 feet to St. Louis, 7,000 to Indianapolis, then 5,000 Indianapolis to Dayton and Columbus to Pittsburgh. Jones walks to a shelf outside the operations office and selects a sheet on top of which is written "Flight Plan." He writes the courses and distances between points on the route, direction and speed of the winds aloft at 7,000 and 5,000 feet, with temperatures at that altitude, then with his calculator figures what loss or gain of ground speed the wind will give at their chosen altitudes.

TWA captains and first officers are equipped with engine and performance charts made by the Wright engineering laboratory and TWA engineers. For example the True Air Speed vs. Horsepower vs. Temperature Chart gives 1,200 hp. necessary at 7,000 feet to make a true air speed of 184 mph., with outside air temperature at 40°. Winter schedule time between Kansas City and St. Louis is one hour and twenty-four minutes.

Jones figures from the meteorology sheet containing the wind velocity and direction at 7,000 feet that they will have a tailwind of 10 mph., which will make their ground speed 194 mph. and should put Flight 14 in St. Louis three minutes ahead of schedule.

He finishes his flight plan—on which is given the horsepower, estimated ground speed and gasoline consumption per hour for all legs on the flight through Pittsburgh.

Next in the sequence-before-flight is the making of an Airways Traffic Control form on which Jones writes the approximate time Flight 14 will be over Higgensville, Columbia, New Florence, Weldon Springs and St. Louis—the check points over which radio reports are given. If the winds aloft are correct this forecast check point time is accurate to within two or three minutes.

Jones gives the ATC form to a radio operator who puts it on the teletype as an aid in coordinating this time with that of other airplanes in the same vicinity (called traffic). Captain Smith and the flight superintendent check and sign Jones's work and now we're ready to depart.

At this point I can hear some of you saying, "So what?" What good is a flight plan or an ATC if the winds aloft change so that more or less power is required to maintain a schedule?"

Well, the winds *do* change. But here's the point. The flight plan gives the captain and first officer a good basis upon which they can figure and calculate while in the air. It's an added safety measure. For example: suppose Jones has figured on using 102 gallons of gasoline per hour. That means they will use 143 gallons to St. Louis. Now it's a Civil Aeronautics Administration rule that airliners must land with a least 70 gallons of gasoline—or a 45 minute supply in the tanks, so 143 plus 70 equals 213 gallons, but to be on the safe side we carry 350 gallons and should land with 200 gallons of gasoline remaining—enough to take us anywhere within a 400 mile radius.

It's 10 minutes before schedule take-off time at 2 p.m., so Smith and Jones walk up to the cockpit. You can stand up now and watch Jones go through his regular before-take-off routine. He picks up a check sheet titled, "Before starting engines" and calls out: "Battery cart on—gasoline checked, radio checked . . . ." He and Captain Smith check each item.

He picks up his mike, "Flight 14 to Kansas City. Radio check." TWA's radio operator returns: "Kansas City to 14. Transmission normal, time 57, five seven (minutes past the hour), altimeter 29.80, two nine eight zero."

Passengers and baggage are already on the ship, and Captain Smith gets the "all clear to start engines" signal from the ground. He starts the right engine, then the left. Jones shifts the electrical load from battery cart to airplane battery. Okay—let's go—but no; first Captain Smith gets a salute from the passenger agent which means "all clear to taxi."

Jones changes radio frequency to the control tower frequency and presses his mike button: "TWA Flight 14 to Kansas City tower. Ready to taxi out."

The tower operator returns, "Okay to taxi out TWA 14. Wind north-west three, NW 3, traffic taking off north. Your clearance: Flight 14 is cleared from Kansas City to St. Louis tower to cruise 7,000 cross Weldon Springs 2,500 no essential traffic reported."

Jones repeats the clearance as the plane stops at the south end of the north south runway for the routine engine run-up. Out comes the check list with items to be checked before take-off: "Flaps up, trim tabs set for take-off, seat-belt sign on, carburetor air cold . . . ."

Captain Smith sets the parking brake, runs each engine up to 30 inches of mercury, tests the mags. Jones presses his mike: "TWA Flight 14 ready to take-off," receives the tower's okay and we begin to roll.

You sit back in your "jump seat" and fasten the safety belt. The throttles go to the full forward position and as we leave the ground Captain Smith says: "Up gear." Jones unlocks the safety latch on the floor of the cockpit, pulls the landing gear lever to the up position and in the same motion picks up his mike: "TWA Flight 14 off at zero three and changing over."

The control tower repeats and Jones shifts to TWA company frequency and repeats our time off the ground. The company radio operator at Kansas City contacts St. Louis and gives our time off and also estimated total time to St. Louis.

Now we're 1,000 feet in the air heading on a course due east and Captain Smith pulls back the throttles to 30 inches of mercury and synchronizes the props at 1,900 rpm. We're climbing at 120 mph.

When we reach our cruising altitude at 7,000 you see Jones pull out his Wright power calculator. He figures a moment, glances at the outside air temperature indicator and says: "28 point 3 for 1,200 horsepower. Captain Smith moves the props to 1,850 and the throttles until the two manifold pressure gauges register 29 point 3. Now he moves the gas tank selector from the positive tank (left main) to the left auxiliary tank and glancing at the gasoline gauge we see there are 85 gallons in this tank. The mains hold 210 gallons and each auxiliary holds a maximum of 201.

Well, now we can sit back and study our map for a moment. Straight ahead is a town with a small lake bordering on the east. Directly over the town the captain's thumb makes a sidewise movement—okay—we're over our first check point—Higgensville. Jones picks up his mike: "Flight 14 to Kansas City."

Kansas City replies: "Kansas City Flight 14. Go ahead." Jones makes his report using the standard procedure.

"Flight 14 over Higgensville at 24, two four, 7,000 contact (able to see the ground), estimate New Florence at 45, four five, temperature 35 . . . Jones." Kansas City repeats the report.

Captain Smith and Jones seem to be busy every minute checking oil pressure, moving a lever to increase oil temperature to 150° or checking air temperature in the passenger cabin.

"Jones," says Captain Smith, "You'd better let down a quart of water. I see the steam pressure in our boiler is dropping."

Jones leaves his seat, lets down the required amount of water and notes a steady rise in steam pressure up to 25 pounds. When you ask about the heating system Captain Smith explains that the passenger cabin is heated by a water jacket surrounding the engine exhausts and that steam generated by a quart to three pints of water is generally sufficient for an hour and a half flight. A gauge over the co-pilot's seat shows the exact temperature in the passenger compartment at all times. Air entering the nose valve of the ship is heated to 70° and passes into the passenger cabin.

Captain Smith glances at the ground, then wags his thumb. We're over New Florence and Jones contacts St. Louis.

"Flight 14 over New Florence at 44, four four, 7,000 contact, estimate Weldon Springs at 10, one zero, St. Louis at 21, two one, temperature 33, gain 22 . . . Jones."

Our speed is slightly more than Jones had estimated since we've picked up a tail wind of 22 mph. You settle back thinking that's all the work necessary before our arrival at St. Louis. But

not Jones. He reaches behind his seat and pulls out the airplane log—Form 76.

This log gives a complete record as to functioning of engines and airplane for each leg of a journey. Over his shoulder as he writes we see:

Engines—ignition—radio—oil pressure and temperature . . . normal. Altitude—7,000, air speed—175 mph., horsepower—1,200, fuel per hour, 90 gallons.

Jones replaces the log. When we reach St. Louis he will enter our total time between Kansas City and St. Louis, and the ground crew will enter the number of gallons of gasoline and oil with which they service the ship.

Jones makes his report over Weldon Springs, then when we're five minutes from St. Louis, picks up his mike and says: "Flight 14 to St. Louis. Over The River and changing over."

"Okay to change over. Altimeter 29 point 94, two nine nine four."

Now we leave the TWA radio frequency and contact the St. Louis tower on 278 kc.

"TWA Flight 14 to St. Louis tower, over The River, go ahead."

"St. Louis tower to TWA Flight 14. Okay, over The River, the wind west nine, runway number one, local traffic. You are cleared to the field."

Smith and Jones keep a sharp watch on two small yellow *Cubs* circling the field. Jones gets out his "Before-landing" check list. . . . "Gas on left main, full rich, props 1,850, tail-wheel locked." Captain Smith checks each item and then slows the ship to 150 mph. "Down gear."

Jones pushes down the lever which lowers the landing gear and, when the landing gear strut pressure reaches 750 pounds, locks the safety latch and repeats: "Gear down and locked, pressure 750 pounds, green light on."

We swing to the east of the airport, descend to 1,200 feet, line up with the east-west runway and slow to 110 mph.

"Down flaps three quarters," says the captain.

Jones lowers the flaps, returns the flap lever to neutral. "Flaps are down three quarters."

We touch the ground. Captain Smith pulls back the control column, says: "Up flaps"—then pushes the propeller controls ahead to full low pitch, returns the elevator trim tab to neutral, unlocks the tail wheel and we taxi slowly to gate No. 3 where a mechanic in white coveralls stands with hands over head.

Jones presses his mike button, changes to TWA radio frequency: "Flight 14 on the ground St. Louis at 24, two four."

The figure in white below us extends his arms horizontally—we come to a stop. Jones pushes the carburetor controls to the "full lean" position, Captain Smith reaches overhead, cuts the ignition switches and sets the parking brake.

As you leave the ship you see Smith and Jones head for a telephone in the operations building where they will give their estimated time and altitude on the next leg of the flight—to Indianapolis.

Smith and Jones wave to you as they pass on the way back to their ship. The first leg of a routine flight has been completed and the last thing you hear as you leave the terminal building to search for a taxi is: "Last call for TWA Flight 14, eastbound for Indianapolis, Dayton, Columbus. . . ."

Limitations of the Helicopter

BY GROVER LOENING

*Difficult to fly now, helicopters won't have controls
simplified for general use for five to 10 years, author
believes.*

THE Buck Rogers type of predictions on helicopters must be discouraged. There isn't anything in the cards to indicate that grandma will be doing her neighborhood shopping in her latest helicopter or that the tired business man will be flying in his own little model to his chosen suburban community.

Advertising artists have gone overboard on the future of the helicopter. Picture magazines have simply let their imaginations run riot on the subject. We must call a halt to this sort of thing or Americans will have to face a period of disillusionment at the very time when faith in the future of the helicopter is needed to the greatest degree.

The true picture today, faced honestly, is that the helicopter will be limited to usage by professional pilots and professional aviation companies for the next few years. It is not at all a vehicle to be placed in the hands of the public. The way things stand today, the helicopter is even more of a professional apparatus than is the airplane.

Igor Sikorsky's eminently successful pioneering has given us our real start in this field. Developments are going on apace. There are some 70 or 80 different designs being worked on, out of which 10 or 12 will be successful. Three or four of these designs now are flying well. Some of the successful designs include those of Piasecki, LePage and Young.

The War Production Board now is granting materials and designating laboratories to several young pioneering engineers who have good new ideas in this line. The Army, Navy

Left: An early Sikorsky helicopter hovering before authority at a time when some visionaries saw the helicopter as everyman's car in the sky. Right: Inventor Igor Sikorsky flying one of his creations.

and Coast Guard are placing strictly military orders and developing strictly military uses for the helicopter. These cannot be disclosed except in a general way. The National Advisory Committee for Aeronautics is proceeding with an ambitious research program to make sure that no stone is left unturned in the interests of safety for the helicopter pilot and in the success of the technical development of this field in the United States.

The net result of a long, intensive study of the helicopter is that—for the present and for many years to come—it will be hard to fly. It requires the co-ordination of five controls, and for that reason is far less suitable for the use of the ordinary layman than is the airplane.

Any notion that everyone is going to fly helicopters as soon as the war is over is nonsense. The amount of skill necessary to fly these machines correctly and with safety is greater than that required to fly an airplane and is very much greater than that needed to drive a car.

This serious limitation is not insurmountable by any means. In due course of time the three airplane controls used for balancing can be placed on an automatic pilot mechanism. Engineers and designers eventually will develop a governor that will assure the proper rotor speed without any manual action on the part of the pilot. In somewhere between five and 10 years, in the author's estimation, the control of the helicopters will finally become sufficiently simplified to make these craft available for more general public use.

Another serious limitation on their use in the near future is the fact that helicopters are fundamentally limited in the speed they can attain as at present designed. The advancing blade, since its speed is always the sum of its speed of rotation, plus the speed of the whole machine advancing through the air, more quickly reaches the limiting airflow breakdown

speed of sound than would an airplane. This speed of sound is around 1,100 feet a second and trouble in shockwaves has been run into by airplanes at around 800 feet a second. Since in large sizes it is impractical to develop the rotors to give the necessary lift without having a top speed of hovering rotation alone of much under 400 feet a second, this leaves a forward velocity of about 400 feet a second which would be equivalent to about 270 m.p.h. This is definitely in the high bracket and it is not likely for many years that helicopters will be developed which will fly much over 200 miles an hour. By that time there will be many airplanes flying around 400 miles an hour.

A third limitation of the helicopter is a definitely poor performance at increasingly high altitudes. As the helicopter is envisioned at present, if it is efficient at sea level, it is not an efficient machine in rarefied air.

There also are quite serious limitations on size but this might be met by the development of multi-rotor machines.

Many of these limitations are due to preserving the requirement that a helicopter must at any time with engine dead be able to become an autogiro and autorotate to a safe landing. Present helicopters can do this. However, as the art develops, we may find that this requirement is a deterrent to development that is not warranted. Possibly the development of twin-engined power plants or of parachutes or some other emergency device might allow a heavier disc loading that would be far more efficient than the light disc loading required to meet autorotation requirements.

We have been taking a rather dark view of the whole situation, however. Because these limitations exist, it does not mean that helicopter development has too restricted a future. On the contrary, when handled by professionals, its professional future immediately after the war will be wide and extensive.

The success of helicopter evolution is indicated by the turbined Bell 214B.

Stanley Hiller, Jr., 19 years of age, taking his Hillercopter off from a San Francisco street intersection.

The Hiller-copter

BY WILLIAM FLYNN

Teen-age inventor constructs helicopter with twin rotors and simplified controls.

A CO-AXIAL rotor helicopter, whose twin blades revolving in opposite directions eliminate the necessity of a tail rotor to aid control, has been added to the list of aircraft that may provide wings for personal flying.

It is the "Hiller-copter, NX-30033," the invention of Stanley Hiller, Jr., 19-year-old Berkeley and San Francisco, Calif., youth. He has been employed by Henry J. Kaiser, famed Pacific Coast builder of dams, seagoing freighters, and Fleetwing airplanes, to design and build the revolutionary type of craft. The Navy Department requested deferment of his scheduled Selective Service Act induction to permit him to continue his scientific experimental work.

The first public demonstration of the *Hiller-copter* proved that it could fly—up and down, sideways, forward, backward and chase its own tail in a tight circle, right or left. After the flight, Grover Loening, veteran airman and designer and aircraft consultant for the War Production Board, described

the craft as a "remarkable achievement," and Hiller as "a very talented young man."

Hiller became interested in helicopters three years ago when he saw pictures of the Sikorsky helicopter in flight. Study and investigation convinced him he wanted something more than the conventional helicopters could provide. He decided to design one himself.

"The earlier helicopters were unwieldy and complicated and took too long to learn to fly," he explains.

Son of Stanley Hiller, Sr., San Francisco shipping executive who flew planes of his own design before World War I, young Hiller began his helicopter experiments with a fund of experience in the field of construction and design.

He began tinkering with tools when he was five years old. When he was 10 he constructed a toy automobile that carried him about the streets of Berkeley. Two years later he had invented a model auto racer that could turn out 107 m.p.h.

when guided by a cable around a circular course.

The miniature racer, 19 inches long, was developed into a $100,000 a year business because he decided that $200 was too much to pay for one. He began to build them on a mass production basis in a "backyard factory" and his list price eventually was $28.

The youth's mechanical skill is not the product of any formal schooling. He has been impatient with orthodox education. He chose not to register for his second year at the University of California.

Despite his youth, young Hiller is mature beyond his years. He exhibits no signs of self-consciousness as he sits in his panelled office, responds to inter-communication system calls or handles a Washington long distance call.

He is six feet tall, thin, and energetic. He speaks with reserve until the subject of flying is brought up. Then he becomes enthusiastic. When the conversation is turned to money—he loses interest. A mention of his long, low black roadster will revive his enthusiasm.

When young Hiller decided to build his own helicopter his ambition was to eliminate the tail rotor. He believed it robbed the lifting rotors of too much engine power, estimated at 40 per cent, that otherwise could be used for direct lift. He also decided to eliminate the excessive vibration that he believed existed in the conventional helicopter, increasing the control problem and demanding considerable pilot skill.

As his helicopter experiments developed, he decided to devote his full time to the work. He left Hiller Industries, the die casting concern and the factory continues to operate as one of his father's enterprises.

The helicopter experiments produced a craft with a low slung, tear-drop fuselage whose principal feature is the double rotor mounted on a single, rigid shaft. Powered by a regular airplane engine, the blades revolve simultaneously in opposite directions.

"The entire control of the craft—directional control, its stability control, its descent and ascent control—are operated entirely by the mechanism which operates the speeds and pitches of the rotor blades," Hiller explains.

The pitch of the airfoils may be controlled individually, or in groups of two, or in combination of groups or all integrated into one unit. Such control eliminates the necessity of a tail rotor, required to overcome torque developed by the lifting rotor, and permits use of all engine horsepower for the development of vertical lift.

Specifications and performance figures of the *Hiller-copter* are:

Gross flying weight: 1,400 pounds.
Engine: 90-h.p. Franklin.
Diameter of rotors: 25 feet.
Rotor rpm: 250 at their base.
Maximum distance of direct lift without wind: 38 to 45 feet.
Maximum ceiling at 85 m.p.h. cruising speed, three-quarter throttle: 6,000 feet.
Estimated maximum forward speed: 100 m.p.h.
Cruising range: One hour, 30 minutes.
Span of tricycle landing gear: Nine feet.
Overall length: 12 ft. 6 in.

Height: 8 ft. 4 in.

The outstanding characteristic of his craft, Hiller believes, is the simplicity of its controls which Kaiser learned to operate with sufficient skill to solo 10 feet in the air on his first flight after five minutes verbal instruction.

Manipulation of four control devices is required to fly the *Hiller-copter.* One is the engine throttle. The other three operate the patented rotor speed and pitch control mechanism that is hidden under canvas coverings during public flights.

The first, similar to the rudder pedals of a conventional airplane and moved with the feet, turns the craft to the left or right about its own axis but does not influence its forward or backward motion.

The second, similar to the stick or elevator and aileron control of a conventional airplane, controls the direction of the craft and is manipulated to turn its flight path to the left or right.

The third control, a lever placed low on the pilot's right side, governs the rate of ascent or descent.

The degree of control co-ordination required for efficient operation of a conventional airplane is not required for flying the *Hiller-copter,* its inventor says. Vibration has been eliminated to the point where the pilot may fly the craft "hands off" for periods ranging up to 60 seconds, depending on the gustiness of the wind which sways it from "straight and level" flight.

Kaiser has employed Hiller to develop the craft. The first objective will be construction of a *Hiller-copter* that is reported to interest the Navy—a four-place helicopter that could be used for rescue work, scouting, signaling, and other assignments where terrain conditions prohibit the operation of a conventional airplane.

Details of the arrangement between the young inventor and Kaiser have not been disclosed. Their joint activity is construction of a factory in Berkeley, indicating they intend to build more than one helicopter and, probably, are hopeful of perfecting the craft as a post-war product adaptable to mass production—and mass sales.

Kaiser is interested in any product that he is convinced has a possible post-war use. He has advocated a large scale personal airport construction program and his light metal factories might supply the material needed for construction of the all-metal *Hiller-copter.*

Hiller believes his craft can be mass produced and sold during peace time in the price range of conventional aircraft of comparable horsepower. Its advantages, he believes, will be elimination of high dual instruction costs, hangar rents, and transportation to and from airports.

In the models he would offer in a peacetime market, he would include a device to transfer the engine power from the rotors, which can be locked in a position parallel to the length of the *Hiller-copter,* to the wheels of the landing gear when the craft is on the ground.

"As an automobile," he predicts, "it will do 35 to 40 miles an hour—fast enough for anyone to drive from the neighborhood helicopter landing lot to his home. Its size will permit it to be parked in an ordinary garage."

Mr. Piper Counts the Score

BY WILLIAM T. PIPER

The dean of the lightplane industry reflects on the last 20 years of FLYING and the personal plane.

William T. Piper

FLYING's 20th anniversary is a better time than most to review aviation's past and look at its future.

We have, fortunately, passed the plane-in-every-lunch-box stage of post-war flying and are in a good position to cope with the pessimism that more recently has been sneaking up on the industry. Without undue pessimism or optimism, and assuming that we are willing to profit by the lessons we have been exposed to in the past 20 years, I think the industry is in the best shape in its history. Sales during the next 20 years should make those during the last 20 seem miniature by contrast.

But this will only be possible if we profit by our mistakes and recognize a few fundamentals which we must face before lightplane prosperity can get under way in these air-minded times.

Here are some of the most important facts:

Aviation is not a mass-production industry, either from the manufacturing or the customer viewpoint. It is a craft industry and probably will remain so for at least another 20 years.

Our best potential customers are not the youth of the nation—if they ever were. They are important, but our best potential customers are middle-aged.

We must concentrate on making roomier aircraft with better performance in all categories, and not go overboard on our dreams of airplanes of the future. I would be the last to say that the day will never come when all aviation's dreams will be realized. But in the meantime we must recognize that, for the present, half-flying-half-auto contraptions, machines that can fly at good cruising speeds but will land on a dime, and planes with removable wings or jet power, are still in a highly experimental category.

We must have many more airports.

We must quit over-selling ourselves and our public.

We must improve service all along the line.

Looking at these statements pessimistically at first, let's see what the past 20 years should have taught us.

Take the mass-production problem—and let's not begin by confusing it with the assembly-line problem. We've learned a lot about assembly lines in the past 20 years, some from the automobile industry, most of it the hard way. Even a factory with annual unit production in the three-figure bracket has some semblance of an assembly line and as production increases the line improves. This isn't mass production, it's just horse sense.

But look more closely at anybody's airplane assembly line and you will find that every man working on it is a craftsman. He does a lot of highly technical things with his hands when the fuselage or engine stops in front of him. He and the other workers add up to the many man-hours that go into total labor making an airplane today.

Those man-hours could be reduced amazingly if special machines were designed to do the things men now do with their hands. But the machines will not pay off in aviation unless they can be kept day after day, week after week, doing their stuff on thousands and thousands of planes—mass-production for a mass-market.

We estimate that the point at which mass-production could be begun profitably at Piper would be an annual production of at least 50,000 planes. We do not anticipate such a demand for as far into the future as we can see.

Let's face it—there isn't a mass-market. There can't be now or in the foreseeable future. If the past 20 years have taught us anything, they should have taught us that only a relatively few people can take advantage of the utility or pleasure of the airplane. They add up to a lot of people but not to the millions sometimes envisioned.

We come down to this plain fact—aviation is a craft industry, craftsmen cost money, production will be small compared with mass-market production and prices will remain correspondingly high.

We must concentrate on making roomier aircraft with better performance and not go overboard on our dreams of airplanes of the future.

Again, the plain facts of the past 20 years illustrate the lesson. We know—or we should know—that roadable airplanes, for instance, while feasible, are complicated because they violate one of the fundamentals of aerodynamics and therefore raise costs, operation expenses, and upkeep. A roadable that will not fall apart at the first traffic light must withstand stresses and strains different from those encountered aloft. Construction has to be auto-sturdy and to do that means adding weight. A *heavy* lightplane is an anachronism. Anybody who knows anything about them, especially if he is a potential customer, knows this. Anybody who doesn't will soon find it out and be angry at anyone who misleads him into buying a half-plane. It is bad business any way you figure it, and the sooner we quit compromising and concentrate on building the best possible *airplane,* the better off we'll be.

This applies with equal force to helicopters (even their best friends admit that they are still in a highly experimental and highly expensive stage), and lightplanes powered with jet engines.

Their future is unlimited; their present has a dubious and very limited public appeal.

Today, the lightplane imperative is to build what we best know how to build—airplanes.

If the assumption is true that we are going to concentrate on making and selling airplanes, it is obvious that we are going to have runways to land them on. Airports are an urgent lightplane "must" in aviation today.

Saying it isn't going to get them built. Equally urgent is that we who are interested in the welfare of the lightplane, pilot and manufacturer alike, have to educate the public in the need for ports and their economic benefits to the community.

A steady, consistent, nation-wide campaign is needed to point out that airports are a community asset, that they are comparatively inexpensive, that they are not a nuisance, and that they are a national necessity both for prosperity and for defense.

We cannot point out too often that airports are cheaper than highways. A plane with the necessary range can make a 400-mile trip using only $100,000 worth of airports. An automobile making the same trip will use some $20,000,000 in highways. Airports, too, increase in usefulness as they increase in number. Build two and you have only two places to go. Build 10,000, and the next airport completed increases the choice of destinations to box-car figures for all ports involved.

If the past 20 years have taught us nothing else, they have taught us this need in lightplane aviation, and it's about time we did something about it.

It is also high time we quit kidding ourselves and our public. The airplane has three weaknesses from a utility standpoint. Average Americans measure the utility of a vehicle by the number of steps it saves them. The airplane's utility should be measured otherwise. It is not a pedestrian machine, it is a long-distance machine. Another thing, until average Americans can be educated to use it for long-distance purposes, the airplane's utility is further circumscribed. There is no point in selling it as a useful vehicle when it takes a man to some place he doesn't want to go because of lack of fields.

The airplane's usefulness is further reduced because it is a fair weather, daylight vehicle. In another 20 years it may not be, but I think it the better part of wisdom to go ahead on the assumption that it will. Then all improvements can be declared sales bonuses and in the meantime there will be no disappointed manufacturers and disillusioned customers.

Improved services should, like improved planes, be taken for granted. History of aviation in the past 20 years points the trend in the next 20, for despite all the healthy beefing in the flying world, both services and planes have improved. There is hardly an established airport in the country whose front office does not boast before-and-after pictures proving this statement. If there are not enough hangars to go around today, at least there are more than there were 20 years ago. If services are still sometimes too casual, or inadequate, they are certainly far ahead of what they used to be.

But they must continue to improve if we are to get the most out of our future markets, and improvement is imperative if we expect those markets to increase. A concerted good-will program, deeply rooted at every fixed base, is the only answer. Otherwise, instead of winning more customers we are likely to lose many we now have.

It might be assumed from all this that I'm so pessimistic about the future of lightplanes that I may quit in disgust any day now.

Nothing could be farther from my thoughts or the thoughts of everyone else at Piper Aircraft Corporation. If anything, we are more optimistic than we have been for a decade.

We believe that there are excellent prospects now and for at least the next two decades for improving airplanes and for selling them in large numbers. For example, we sold nearly 8,000 in 1946. We expect to build half again as many more this year. We expect to do an increasing business in three- to four-place planes, but anticipate a continued steady demand for two-placers.

We expect to increase our buying public by cleaner, higher-performance designs, by simplified instrument boards rather than additional complicated safety devices, by better radios, better engines, metal as well as fabric covering, bigger and better airplanes, steadily improving in performance and safety.

We expect to find more and more buyers in the sportsman-recreation field. We do *not* expect to find them among younger people. Generally speaking, they cannot afford a plane or, if they have the price, prefer to spend the money on a house, a

One of Mr. Piper's Cubs, a J-3 flying near Piper headquarters at Lockhaven, Pennsylvania in 1938.

car, or some other equipment more useful to them than today's airplane. If they fly, they'll do it on a rental basis. We expect to find our best customers among older people, who have already paid off on the house, own a car, and have the time and leisure to enjoy flying.

We expect to sell airplanes on the basis that they are not only fun, but get you more for your money than even an automobile. Planes today have a lower first-cost-per-mile value than cars. At 40,000 to 50,000 miles the average automobile is traded in. Nobody knows the life span of elephants or airplanes and we should capitalize on this in our sales talks whenever possible.

Airplanes are safer than ever before, and will become more

so.

All maintenance costs will go down steadily, including hangar rentals, gasoline, and insurance. This will make for easier selling to more people.

More and more people are flying annually—to more and more places. You can go farther and see more than in any other vehicle, for less per mile. Flying also is a major sport—more stimulating, more flexible, more fun than any other form of locomotion.

On this foundation, we can build a lightplane industry. I don't predict that it will be gigantic, but it certainly will be healthy.

COME, BE AN AIRLINE CAPTAIN

Getting to be a captain is tough, but take it from one who has reached his goal—it's worth it.

One of the classic airliners, the Douglas DC-4.

BY HY SHERIDAN

FOR YEARS whenever the airlines could lure more customers into their ticket offices it was necessary to put another DC-3 into service. The DC-3 was the workhorse. It held just so many passengers and it went just so fast. It was as standard as a radio joke.

Thus, if business expanded 20 per cent the crew list expanded 20 per cent. When a *Cub* pilot didn't know what to do with himself, he would go into the airline employment office and allow them to hire him as a co-pilot. The airline would give him an examination, though.

"Don't take this personally, sir, but can you fly?"

"Yep."

"Thank you, oh, thank you, sir. You are hired."

The new pilot would go out and get himself a wife and Duesenberg in good running order, or vice versa, because he knew that he would soon be a captain and rolling in the fat of the land, for those were the days when he got a check instead of merely a list of deductions. But things have changed now.

I am asked on every hand of late: "How can I get a job as an airline captain?" "How should I go about it to get the required training?" "Can a private pilot break in?" "Should I try to get a ground job and so be on the ground floor, so to speak?" "After I get a job as a flight engineer, for instance, how long will it be before I get to be a captain?" "What qualifications do I need?" "What are the jobs of flight engineer or second officer, and first officer, and captain like, anyway?" "What do you do with all that money that they press upon you?" "I hear all kinds of tales," you say, "so what is the low down?"

Quite a number of persons seem to know a good deal about the matter—but what they know mostly isn't so. And thus I intend to write the truth.

As to qualifications for an airline captain today, you need honesty, ambition, intelligence, health, luck, a means of obtaining food for a few years, and an excellent supply of longevity. Of these, longevity is the most important. Before you get to be a captain or even a fist officer you will have worn out lots of longevity.

But isn't the airline business a new industry? Aren't the airlines expanding? Doesn't that mean lots of new jobs and quick promotion? Yes, the airline business is expanding and it will continue to do so for many years, but it does not mean fast promotion for the flight crews.

When traffic expands now, the airlines retire four DC-3s and put on a DC-6, a *Constellation*, or a *Convair*, or the like. That leaves three crews over. Where the DC-3 carried 21 passengers, the DC-6 carries 52 to 58, more than twice as many, and it flies about twice as fast. It can handle four times as much business in a month as the DC-3.

You know what happens. The boys on the bottom of the seniority list just get amputated. In the past three years the larger airlines have furloughed about as many pilots as are left on duty. By agreement with the Air Line Pilots Association the airlines are bound to hire these men back before any new ones are employed. This sounds pretty bad for a fledgling who is alight with hope and hunger, but it is not so bad as it sounds. A lot of those men do not and will not come back.

If a high quality young fellow wants to be an airline captain, he has as good an opportunity to rise in it as in any other profession. The airline pilot's job is a good one, likely to be stable and secure in the future, healthful and satisfying. The job has one tremendous advantage over almost all others. When you step out of that cockpit your worries are over. You never have to take a worry home.

You don't have to lick boots or play office politics. If the boss doesn't like you he can always think of something else, because the Air Line Pilots Association will protect you if you do your work in an honest and expert manner. Those are present facts.

What is the best way to train in order to get a job as an airline pilot? Fred Bailey, Chief Pilot of American Airlines, one of the clearest thinkers I know, said that it seemed reasonable to expect that most of the new pilots would come from the Air Forces. In the Air Force the young hopeful can get wide, varied experience. He can fly big, fast airplanes as much as he wants and, some pilots tell me, more. He can fly long routes in all kinds of weather using the latest radio aids and navigation methods. During this training he has a good job with good pay. In fact the job and the pay are so good that it is a decided comedown to take an airline place as a second officer or even as a first officer.

Our excellent private flying schools also should have plenty to do. Non-airline flying is due for a good, healthy expansion over the next few years. More corporations will have their own planes, not only for the use of executives and salesmen, but for quick transportation of machinery parts, and so on.

How soon will the airlines start hiring outsiders, that is, men who have not been furloughed? Very likely, airlines will be hiring new men before any of FLYING'S present readers could complete their training. The hiring will be at a fairly slow rate as compared with some of the periods in the past but it looks like it will be a steady one.

Last spring the airlines began to call back some of their furloughed men. The call went out according to seniority, and it was a surprise to the bosses how many declined. Most of the men had other good jobs. So the list of furloughed men who are available for return will be exhausted very soon.

What job will you get first? There is a new element in the situation—the flight engineer or second officer—and just how this will work out no one yet knows. You no doubt know that it is now required that a big aircraft on the order of the DC-6 and the *Constellation,* and the larger ones, have a third man in the cockpit. These new jobs have used up a lot of the furloughed pilots.

The third man must be either a mechanic or a pilot. Some of the airlines have put a number of mechanics in these jobs. Some airlines insist on licensed pilots. What policy the airlines will adopt in the future will be determined by experience. Assistant Chief Pilot "Al" Schlanser of American Airlines believes that in the end all airlines will require third officers to be pilots. He points out that a mechanic who is a flight engineer is stuck with the job without hope of promotion. A second officer, on the other hand, who is a pilot can look forward to promotion to first officer and, if he has a telescope, to captain. Your first job, then, may be either as second officer or as first officer, depending upon the current policy of the airline.

How about promotion? How long will you have to work before you get a boost in grade? And in pay? That will depend upon how the airline traffic increases and whether they continue to put on still bigger and still faster airliners. It depends, too, on how they go after new kinds of business.

The regular passenger business of the domestic airlines will expand quite slowly. Already the passenger miles of the airlines is half of the total Pullman passenger miles, and this is an astonishing figure when it is noted that so many more towns are served by Pullman. Some experts believe the airlines now have their full share of the passenger pool and can get more passenger miles only as general travel and living standards rise. Other experts believe the airlines are supplying a new commodity and thus fashioning a new demand rather than taking business away from the other forms of transportation, and careful analyses seem to bear them out.

The so-called coach, or low fare, business has barely begun. This should expand rapidly over the next three or four years and then level off.

The freight business expanded 1,000 per cent in 1946. In 1947, it increased over 100 per cent, in 1948 almost a 100 per cent again. In the latter year almost 70 million ton miles were flown. The freight business is still an orphan and not handled well, but it will be, and it seems likely that it will expand nicely for the next eight or 10 years.

The international and over-ocean business is a big question mark. The potential is huge. If we can achieve some stability in the world it should have a great 10-year expansion period.

Expansion means more crews. The trend to bigger and faster airliners which cut down on the number of crews required, is leveling off fast and tends to be self limiting. In domestic service the airliners cannot get much bigger and be economical to operate. The mechanical complication of the very large craft keeps them out of service too much on account of mechanical troubles. A DC-6, for example, takes four times as much service as a DC-3 when it is on the ground.

Only over the few long-distance, heavily-traveled routes can bigger planes be used at all. Frequency of service is an important thing. People fly to save time, and without convenient departure times, and fairly frequent ones, they will have nothing to buy. It is clear that in a local service very large craft cannot have their seats well filled. Unless the load factor is over 50 per cent it is not likely that the airlines can make money, something they are prejudiced in favor of.

For over-ocean work it is probable that we shall see within the next 10 years on the heavy traffic routes planes that will carry 100 to 150 passengers. Already 75-passenger planes are going into service. But this trend will be restricted to the heavy traffic routes, and most of the traffic expansion will be handled by the planes of today's size. Don't forget that the DC-3 lasted for over 10 years. What international freight will require is anybody's guess.

Except for very long routes, much greater speed could not be used if we had it today. Some people are dreaming of 500 m.p.h. jet transports, but for ordinary airline service we can't use them. We have 300 milers now, and in rough air we have to slow them down because neither the passengers nor the structure can stand the batting. The climbs and most of the descents have to be made at reduced speed, and of course the approaches must be made slowed down. And there is often sharp turbulence in the high clear air that we used to think was always smooth and serene. On the other hand, nothing is impossible. Someone spoke of your reporter as impossible and one of the stewardesses said no, he wasn't impossible, he was just highly improbable.

It does not seem likely, then, that the airliners will increase much in size and speed over the next few years except for special work that would be confined mostly to new business. If there is any effect in slowing down promotion it will be small.

One of the things that has retarded air travel is the weather. Now the weather has been licked so far as certainty of arrival at the destination is concerned; all the airlines have to do is put into service one of the new devices, such as the Zero Reader, so that when a passenger buys New York, say, for his destination we deliver what he paid for, not Philadelphia or Boston. That alone will have a marked influence toward creating new business for the airlines. And it should do much to level off the dizzy seasonal fluctuations that are so costly because so much of the equipment is unused during part of the year. And right here is the meanest part of the otherwise attractive airline job prospect. You should know about this if you are looking forward to being a pilot.

Every fall most of the airlines let out 15 to 25 per cent of their pilots. Perhaps they are not let out—the pay just stops. For one to four winters, then, you can expect to be out of work.

Then in the spring the airlines go out and try to hire them back but, roughly speaking, the best men in the group have obtained good and steady jobs and they don't come back. The plain result is that every year the airlines lower the quality of their flight crews—which is a burden to them and a bitter problem for the Air Line Pilots Association. In order to save a few dollars today, the airlines seem willing to handicap themselves for 30 years. In the long run it is a sad waste, not only on account of the lowering of quality but because the investment they have in each good man who does not come back is considerable. One of the furloughed men said, "That policy is a monument to the intelligence of the managers. You know what a monument is? Well, it represents something that is dead." None of the airlines is wealthy enough to stand this kind of economy.

Even if this is solved, promotion for the second officer and the first officer will be rather slow. On American Airlines which currently has about 750 first officers and captains, about two a month leave the seniority list for all causes, or about 50 a year, including retirements and physical failures. Most of the physical retirements, of course, occur in the older men, and it has been said that the first officer never feels good unless the captain feels bad. When you calculate the promotion rate, to this one-in-fifteen ratio you must add the number of new jobs that we can guess will open up as the result of expansion, new business, and so on.

The airline managers are keenly aware of the probability of slow promotion and the majority of them say, at least privately, that in order to attract the very highest type of man they must pay him well as a second officer and as a first officer. You will have to raise your families in those ratings.

What kind of a job does the airline pilot have? What are its advantages and disadvantages? Well, it is a darn good job, counting out the lay-offs that you have to suffer at first. The job requires a lot of experience but you will have that. It demands the utmost alertness and you must learn how to plan ahead with absolute precision not only for normal operations but for any kind of emergency.

The work is easy but you are paid not so much for the work that you do as for what you know. That is a good deal like the doctor you may have read about who charged a patient $100 for an operation, and the patient complained that the bill was too steep because the operation took only 10 minutes. The first bill read: "To appendectomy, $100." The doctor tore it up and made out another. "To appendectomy, $5. To knowing how, $95."

The work is clean, it is well ventilated, and the view is unsurpassed. Because there is a physical strain, the causes of which are not completely understood but which may include the frequent changes of atmospheric pressure on the system, your hours in flight in a month are limited to 85, the limit having been determined by extensive research by the flight surgeons; the limit of 85 may be cut down a little in the future. A fatigued man cannot be thoroughly alert and efficient, and one place you don't get a chance to learn from your mistakes is in a cockpit. It's against the rules, even if you live.

This limitation means that if you have anything but a shuttle run you will have 10 to 12 days a month off duty. Of course you actually work more than 85 hours a month. Your ground time will about equal your time in the air so that your total hours on duty during a normal month will be around 170. Just the same, the days off are mighty handy.

If for some reason you should lose your job you are out of luck; you have lost your profession. That is the worst thing about the airline pilot's job. But you are not at all likely to lose it; there is lots of security. After you pass your probationary period and you have shown that you have the necessary ability and stability you are not likely to lose your job for anything save physical reasons, and experience has shown that the airline pilot's job is the healthiest in the world.

Failing the physical examination is not such a threat as it used to be because it is becoming more and more well understood that commanding an airliner is done with the head, not the muscles. Airline pilots once had to be athletes, but it is now required that they be intelligent instead. There is another influence working along this line with the larger airliner. There was a time not so long ago if a captain became disabled in the cockpit from some physical cause, it was an emergency indeed. Now that he has two highly trained and capable men with him, the airliner is just as safe as an ocean liner under the same circumstances. The small chance of physical retirement is there, just the same, and you should allow for it and prepare for it.

Here is where a hobby comes in so handy. Airline managers used to insist that when off duty their pilots should rest and nothing else. They learned by experience, as we all do. The flight surgeons noted that the men who had hobbies were in far better shape physically and especially mentally. Having hobbies began to be encouraged. A hobby is known to be a good thing by all of us now, and if it pays it is still more relaxing. This can be, and with many of the pilots is, an offset to the minor chance of losing the job for physical reasons. That you can pursue a paying hobby is one of the greatest advantages of the airline pilot's profession.

Your reporter talked to a senior captain whose hobby was making him lots of money, and asked him about it. "My hobby?" he replied. "Hell, my *hobby* is flying. I love it."

Hedge Hopping Through the Dust...

*How does a duster pilot feel when he's heading for a
towering maple at ninety m.p.h. and the old mill skips a beat?*

BY WILLIAM F. HALLSTEAD, III

SURE, YOU'VE HEARD about crop dusting. You've heard
of DDT and 2,4-D; of toxicology and entomology; defoliation
and weed control. Precise information—scientific.

What about the man who flies the plane ankle-high at 90-
plus? Is he thinking of hexachlorides as he rolls out of a low
turn-around and dives in over the telephone wires? How does
he feel when he's holding her flat on the corn hills heading
straight for a towering maple and the old mill skips a beat?

To get the inside story on a duster pilot's life, its trials and
errors, I spent some time with Jack T. Race of Carbondale, Pa.
Jack, incidentally, was one of the two USAAF flyers assigned to
pilot British General Montgomery's staff during the past war.

He's tall, husky, in his late twenties; he has some 3,000
hours in his logbook—from *Cubs* to *Constellations*. For the
past two summers, Jack dusted for a company near Rockford,
Ill.

Jack came out of the Air Force in 1945 a first lieutenant.
Already a licensed flight instructor, he taught G.I. students
until the program slackened. He wrote letters, as pilots must
these days, and was told by a Midwest dusting company to
report to Illinois for a flight test in May.

The first field Jack dusted was a flat 60-acre pea field. He
wasn't worried about trees, poles or wires but whether or not

he'd spread his DDT in the right proportion—30 pounds to
the acre. That's up to the pilot and he makes his own hopper
opening adjustments by hit-or-miss to regulate the dust flow.
On this first job, Jack hit it right on the nose. His first load ran
out right where it was supposed to; the second load ran out as
he crossed the fence at the end of his last pass. "Almost never
happened again," he marvelled.

A pilot's first dusting job is generally a lulu. In the older
ships, the hopper was closed by pulling the lever back. At the
end of his knee-high swoop, many a novice duster found himself hauling the throttle back and shoving ahead on the already forward hopper handle. The sudden sickening silence
produced a fast correction. On the later planes, everything goes
forward at the end of a pass, except the stick. On his first
passes, Jack found himself ruddering his turns on the inside
and he didn't like it. "But you can't get around fast enough
unless you skid. You've got to watch yourself!" Duster and
sprayer pilots are still being killed in stall-spin accidents.

The hours are another problem. You get up before 3 a.m. if
the weather's right. That means the wind has to be under six to
eight m.p.h. You get out to the loading field in the dark and as
the sun comes up, you start flying. And you keep flying until
the wind picks up, it rains, or thermals lift the dust as fast as

you drop it. Then you start in again around 5 in the evening and dust till you can't see your target field.

These hours and a field man's insistence can lead a pilot into difficulty. The ground crew sometimes forgets a pilot's limitations in its determination to lay dust on schedule. This, plus Jack's conscientiousness, nearly put him in the hospital.

"We gotta have this test plot dusted tonight," the field man pleaded. "We're on a tough schedule and this is the last patch—just this one."

"I don't like dusting this late," Jack told him, "but since it's the last field . . ."

By the time he was in the air, it was dark. "I was unhappy about this trip," Jack said, "but it was an experimental dusting job for the University of Illinois so I went ahead. On the first pass I felt something grab the *Cub* like a big rubber band. It stretched with me as I struggled down the field and finally snapped. I was still flying all right, so I finished the job.

"When I landed, I found I'd ripped loose a transformer a mile and a half away and the town of Flagg Center, Ill., didn't have any power for a couple of hours.

"Know what I got paid for that trip." One hundred pounds at a penny a pound—that's one buck!

"I think right there you have the answer to dusting's high accident rate," Jack went on. "That night I carried a light load, made one dollar. I'm not saying what I usually carried; the CAA authorizes 300 pounds. But you can definitely quote me on this: a duster pilot can't make money without carrying overloads."

Dusting has created an enigma. The dusting company generally pays its pilots 20 per cent of their gross. The canning company pays some five cents per pound distributed; the pilot gets one cent of that. The dusting company gets the rest pays for dust, gas, oil, maintenance, and pays insurance to the tune of, in one case, $36 on every $100 the pilot takes in for his company. And the insurance companies don't get rich either. The more dust pilots carry, the higher their risk the greater the insurance cost, the less their commission. If they carry less dust, the insurance reduction for a good safety record won't go into effect until the next season at which time the duster pilot may be with an airline or the Air Force. Jack regards an average season's take at $2,500. That's for a pilot who's interested in making money. You can make as high as $150 per day; you can make as low as nothing.

Perhaps one of the most unusual assignments handed Jack was the proposition that he fly back and forth over a bean field to circulate the air and prevent an unseasonal frost formation. There were two strings to this one: he had to fly low enough to do some good, and the whole operation was to be done at night.

As pitch darkness fell, Jack fired up his weary Stearman and began the grind at 100 feet, flying between flare-pot markers. He was uncertain about the effectiveness of the 100-foot altitude, so he tried half of it. "I got to worrying so much about the frost that I ended up at 10 feet and flew that way from 11 p.m. to 7:30 in the morning. I stopped only to gas up. That early autumn night I thought I was going to freeze to death. But I made $75."

The following day, the Stearman blurped and had to go back to base for a new carburetor.

Morning flying has its pitfalls, too. Jack began flying at dawn one particularly hot day, carrying heavy loads of rotenone. "My first take-offs from the loading field were fine. There was a low fence at the far end and I cleared it by a good 10 feet. The sun came up and it got hotter. I ordered the same load everytime, and it wasn't long before I just barely got off the ground before I reached that fence. That wasn't all. I hit a post with one wheel and bent the gear. The air had been so thinned by the heat that my lift was cut to the danger point.

"Even though a big load can get you in hot water, my favorite plane for dusting is a *Cub*. I've flown a lot of others, but in a *Cub* you have good forward vision, it's easy to handle in and out of little rough patches, even country roads. One man can push a *Cub* in or out of a hangar or a tie-down. A *Cub* won't do any tricky stalls for you either; it'll stall, but about the same way every time. But I do like at least 75 h.p. A 65 is too critical for dusting."

Asked about the pace of the job, Jack recalled a flight he made during the tomato season.

"Sometimes I sat around for days; sometimes things happened too fast. I was sent on a hurry call once to dust a tomato field where a group of Mexican laborers was working.

"I asked the field man, 'How about those workers—want to move them out?'

" 'No time for that,' he told me. 'Go ahead and dust.' "

"I was using chlorodane and it's hard on the eyes and awful to breathe. I made two or three passes, and those Mexicans were getting mad. Maybe nobody'd told them why I was buzzing over their heads dropping the stuff.

"Suddenly something thudded into the *Cub's* nose, then the fuselage was hit, and the rudder. I finished the load and landed to look over my battle damage—one nicely dented cowling. There was juice running down the side of the ship—tomato juice. They'd bombarded me as I flew cross!

"I didn't blame them too much. I'd had a couple of run-ins with dust myself. I dusted a field one calm morning and the stuff hung in there like ground fog. I had to fly back and forth through it. I couldn't see the end of the field until it loomed up in front of me, and judging distance above the crop was a matter of guess. The effects of rotenone last—my eyes burned for hours afterward. Parathion is worse. There was a barrel of water right on the dust truck for us to wash in if the stuff just got on our skin!"

Jack insists that dusting need not be a death-defying occupation despite dangers of low-flying, weather and the dust itself. The flying becomes almost routine: shallow dive to ground level, clear the wires, hopper open and flatten her out, cruising speed across the field, throttle open—hopper shut, pull her up and bank downwind, swing around the other way, shallow dive . . . It's the difference in a pilot's attitude, he warns, that makes the difference in health and paychecks.

"A man gets money-mad and carries absurd overloads. He sees a little audience of farmers and goes wild. Next thing you know, he's wrapped up somewhere.

"If you use your head, you'll make out. Just remember you've got to fly that airplane. Dusting isn't the most pleasant kind of work, but it's the best experience a pilot can get."

The Rotary Club

BY RICHARD B. WHEEGHMAN

LEARNING TO FLY the helicopter is like learning to play the violin.

Your first encounter is likely to be something of a shock.

But then, hands in your lap, you watch your teacher. He is a virtuoso. At his touch the helicopter performs an obbligato over the grass, a fortissimo climb to pattern. His stops are grand, swooping Shostakovich finales.

But be advised, he is no frail-figured concertmaster. He is hewn from living granite. He is a stone-faced Ozymandius who betrays only a flickering shadow of emotion when your untutored hand threatens crisis during beginning hours.

This is reassuring. For the beginner's playground, in the helicopter, is not the vast, grand panoply of the endless sky, but a small, weedy backyard next to a windsock, a daisy's breadth from that uncompromising obstacle—the ground.

The margin of error is the flick of the wrist. And the flinching reaction of a less stolid instructor would be unsettling.

So you fiddle and fidget. And as your fingers slowly and clumsily learn the fingering of the musical instrument, so do they gradually learn to control the helicopter.

The author, at left, manually and painfully keeping his Brantly B-2 under a semblance of control.

It is an animal process, one of the sinews and nerves and reflexes—not one of the mind.

If you *think* what to do in a helicopter, you will be too late. For the helicopter is a wind devil straight out of Pandora's box—constantly weaving, leaning, swerving, feinting—alive and mercurial. The mind plods behind.

You match wits and reflexes with a coiled spring of kinetic energy, awaiting the tiniest hint of a command—planned or accidental—from the pilot. Receiving this, it bursts the bonds of inertia with an alacrity that astonishes.

Lean slightly on the stick (called a cyclic), and 40 miles an hour is just three seconds and 50 yards across the field with a cavalry charge rush that takes your breath away. Twitch the lever in your left hand (the collective), and you're sitting 30 feet up in the air or, Bump, you're on the ground. Take your left foot off the rudder pedal, and—Zip—you're corkscrewed 90 degrees to the right.

In a word, the helicopter is sensitive.

"Think of yourself as sliding on ice, always slipping around." So warned Maurie Paquette, instructor supreme (6,000 hours-plus in rotorcraft, pilot for Execuplane at Westchester County Airport in White Plains, New York).

"As long as you treat it easy, it'll treat you easy. But if you snap at it, it'll snap back at you."

So you first learn to hover, because if you can't do this, you can't take off or land, or navigate this strange machine a scant three feet above the ground (as you must do to ease yourself around an airport ramp among the parked airplanes).

After a brief, tantalizing introductory flight over the countryside, you are brought back to within a few feet of airport sod and tethered there until you master that basic skill.

So you join the starlings, crows and field mice that inhabit the buffalo grass of the airport and you shadow-box with the ground.

This is complicated by a little neurotic tendency of the helicopter immediately introduced to the beginner. Each control keeps a kind of jealous watch on what all the other controls are doing. So move one, and you must appease the others.

Pull up the collective lever in your left hand (it controls up and down movement), and you must crank on more throttle (motorcycle fashion) with that same wrist. But at the same time, you must shove a little more left rudder (to counter the torque generated by the bigger bite of the rotor blades).

So your ups and downs and left and right turns take on a kind of nervous jockeying and jiggling of controls—with every limb, member, muscle, twitching nerve and facial contortion joining in. As a matter of fact, the beginner throws himself so wholeheartedly into this exercise that he can work up a good sweat just from body English alone. It's something like fencing. After a duel with a helicopter, you walk away rubbing sore muscles you didn't know you owned.

The smug, confident fixed-wing pilot who rummages for analogies from past experiences to ease the transition probably will come up empty-handed. Alas, his carefully nurtured reactions, reflexes, philosophies are next to worthless. He's learned Spanish, and the helicopter comprehends only Chinese.

He'll pull back on the stick in a turn; the helicopter must be pushed forward. He'll use left rudder in a descending left turn; the helicopter demands right rudder; he'll pull back on the stick and flare out for a touchdown; the helicopter must be landed flat.

He'll feel comfortable only with those familiar terms the instructor is likely to toss out, like "rpm, inches of manifold pressure, do a 180, check the mags," etc.

The pilot with military experience, however, may have access to a kind of psychological Rosetta Stone.

The subtle, light, quick control pressures demanded by the helicopter in hovering are closely akin to those required in formation flight. Similarly, they must be made in all three planes simultaneously—up, down, forward, back and sideways. This demands the constant attention of both arms and both feet.

All itches go unscratched. All fascinating, important sights go unmarked. I stumbled across this bit of knowledge one day at 500 feet when I took my right hand off the stick (cyclic) to point out an airplane directly ahead. On this occasion the calm reserve of my Ozymandias cracked briefly. His hand, cobra-like, grabbed for the stick, and we were saved.

(The lesson learned: left on its own, the helicopter may perform a lomcevak. In the abstruse idiom of the aeronautical engineer, that grand cliche: the helicopter is inherently unstable.)

At this point in my learning I also became acquainted with the second of two methods of passing information inside the helicopter. Maurie shouted with a great roar that I heard and understood.

His most common means of communication, however, is by hand signals. These are small casual gestures—pointing, for example, to go that way; palm up, to rise.

At times of great confusion, these may be augmented by some explanatory shouting, for when sitting in the cockpit of the Brantly B-2 helicopter, you share the noise level of the Queen Mary's boiler room.

One common shout during the initial hours is a suggestion that you look at the windsock. In helicopter flying, much more than in fixed-wing flying, the wind is to be followed, watched, spied upon and thoroughly mistrusted.

The wind takes on almost physical presence. It is something you can almost see—like water—flowing from *that* end of the airport, curling and burbling over *those* trees, sliding into a downwash beside *that* hangar. The helicopter is a bubble borne by the wind and subject to all of its vicissitudes.

Air taxi blithely at right angles to the wind and it will kick the tail of your craft out of its way. Turn your back to it, and you invite a vigorous, bodily shove to a head-on confrontation. Swing downwind for a takeoff, and the wind cruelly withdraws support, may indeed drop you to the ground and tip you over.

The helicopter, it turns out, has a secret pact with the wind. Ferreting out the provisions of this pact is the challenge posed to the beginner

For example, although the helicopter apparently flies quite well on the strength of its own thrashing and churning of air, it often needs an assist from the wind. Fly into the wind, and, surprisingly, you find your craft buoyed by invisible lift, much

Hovering near the ground is the prerequisite skill before taking a helicopter to greater heights.

as the airfoil of a fixed-wing airplane would be.

As a matter of fact, this phenomenon determines much of what the helicopter pilot can and cannot do.

To climb efficiently, the helicopter must actually move forward fairly rapidly (40 to 60 mph in the Brantly). Therefore, the typical climbout is not straight up an imaginary elevator shaft, but off across the field in an angled climb much like that of an airplane. In a similar way, the approach for a landing is not straight down, but along an angle.

The beginner also quickly learns that forward speed is a friend in the event of power failure. Without it, the rotor blades quickly lose their speed, and the helicopter then is easy prey to the clutching hand of gravity.

So while the helicopter may actually be able to take off and travel up and down a plumb line, the margin for a safe autorotation (a gliding descent without power, supported by the autogyro-like lift from freely swinging rotor blades) is either slim or non-existent.

And without that same forward speed the helicopter on a hot summer day is a helpless, stationary fan rippling the dandelions. It won't fly. Thereupon the pilot must actually tip it forward on its skids with all of the manifold pressure he can muster and kick the rudders back and forth to jar the vehicle loose from the friction that bonds it.

If the ground is sufficiently level, and the pilot sufficiently skilled, the helicopter then grunts and struggles over the ground like a Dodo bird until enough wind flows under the rotor to allow an inch-by-inch climb to altitude. But don't aim at any hangars.

Then again, the great broiling cascade of wind forced downward by the rotor blades of the helicopter creates odd effects of its own. It will provide the gift of extra buoyancy close to the ground, but may prove embarrassing up at altitude during a hover or vertical descent, especially in still air.

What happens is that the helicopter may get caught in its own downwash. While slowly letting down in a steep vertical descent, with no forward speed, the rotor-craft may suddenly begin dropping faster than its pilot desires. He then adds power and pulls up collective, only to find out that the rotorcraft is exercising another option in the fine print of its pact with the wind, and is descending—no matter how much climb power is applied.

The only recovery, then, is to fly forward and out of the downwash.

The moral: be careful flying into small areas. The helicopter, for all of its sinuous ability to fly in and out of places no other machine can reach, still has its limitations.

Nevertheless, the beginner learning his aeronautical arpeggios will find it all adds up to a great adventure.

Once he starts to get the rhythm of it, he is a zephyr, a thistle seed with a rudder, he's Prometheus unchained.

As in learning the violin, the biggest battle will have been with himself. His psychological plateaus are not plateaus or dips—they're manholes. There will be periods of audacious self-deprecation, of violent slanderous feelings echoed in dank hangars, and grievous grim words hurled at the inventors of the helicopter.

At two hours I knew I'd never fly the helicopter. A little after three hours, I soloed the thing (for low-to-the-ground exercises). But there's a long way to go.

The violin is a most unnatural instrument. My hand goes out to the patient man who masters it, who learns to make it sing.

Similarly, the helicopter is a most unnatural device. And I would extend my hand to the pilot who has mastered it, learned to make it dance and pirouette—if I thought he could let go of the controls to shake hands.

235

AIR TRAFFIC: THE CLEARING SKIES

A STAFF REPORT

The time for readjustment is past. What's needed is a clean sweep.

PROGRESS IN AIR TRAFFIC CONTROL has been grudging. The "system," incredibly cunning and intricate, stands anchored in the bedrock of tradition and investment. And like the Eiffel Tower, though grotesque and impractical, it defies being torn asunder to satisfy logic and expediency. Instead, the pressures are to merely paint and refurbish, to tinker and adjust.

But the hour is too late for this. Planning in air traffic control has groped clumsily behind the ballooning crises of crowding, conflict and delay. Therefore, a wholesale reevaluation is needed, a reformation in the business of shuffling aircraft across the wide blue bowl of the sky.

Whether due to the calamitous accidents in recent months or the vast and increasing inconvenience suffered by both airline and general aviation travelers, or the alarm by people on the ground that an airplane is going to come crashing down on them, the whole subject of air traffic control has been receiving oodles of attention. Much of this has been kindled and fostered by the airline lobby, the Air Transport Association, which feeds its biased grist into the mill at every opportunity. The result is that everyone agrees Something Must Be Done.

Most of the FAA energy is being directed toward refining the present system by introducing more sophisticated and expensive equipment into the ATC centers. The object of such gear is to reduce the workload on the controllers and to permit the system to handle more traffic and reduce delays. At the same time FAA wishes to increase the areas of positive control gradually until a vast blanket of controlled airspace, with a floor at 8,000 or 10,000 feet, covers the entire country. In this area transponders would be important if not required.

In general the airlines support this program. They suggest adding one more fillip, that of building more airports suitable for general aviation aircraft and then enticing or restricting them to such airports. Of course we support the cause of more general aviation airports though we oppose the outright banning of qualified pilots from any airport.

We also urge the airlines to get their own house in order in the matter of scheduling. While Eastern is making a valiant effort in this direction (leave 15 minutes before the hour and avoid the crowd), most of the airline flights are scheduled to arrive and depart during an impossibly short period of the day. While rescheduling is not the entire answer to the problem of delay and congestion, it would certainly help ease some of the strain on present airports and terminal facilities.

The ultimate answer, however, must come from recognition of the fact that we now have and will continue to have a two-level system. One class of aircraft wants and needs a highly sophisticated, no-nonsense, no-delay means of getting from one point to the other with a high degree of efficiency and in absolute safety. The other class of flying is much more casual about its needs both in pilot skills and in the importance of being at a particular place at a precise time. The solution is to realistically divide the airspace horizontally and vertically, providing ample room for each.

This can be done, and you may be sure that regardless of how this proposal is received now, it will be done. (Who can imagine an SST flight holding for 40 minutes over Colts Neck?) The only question is: When are we going to accept the fact that there is enough room in the sky for both the casual flyer a well as the airliner and its general aviation counterpart, the business aircraft?

In general we see certain terminal areas being restricted to aircraft with the equipment necessary to operate safely in a highly sophisticated environment, and flown by pilots with qualifications suitable for such operations. Climb and descent corridors would connect with massive flyways that would lead to other high-density terminals. Aircraft using these flyways would file instrument flight plans at all times and they would be afforded the benefits of POSITIVE positive control instead of maybe-positive control as is the case now. Each aircraft would be required to carry a transponder with altitude readout capability. The area outside the flyways would be uncontrolled.

If this sounds like revolution, be assured it is intended to, for the present archaic system will not respond to anything less.

A raft of failings plagues the present ATC, ranging all the way from the omni navigation network to the handling and coordination of traffic to the airport design itself. Pilots will

also point an accusing finger at the awkwardness of the voice communications system, the haphazard mixing of slow and fast aircraft, an antiquated weather information system fraught with cryptic symbolism and, finally, an all-pervading complexity and inconsistency.

The pilot must match wits with this pernicious system from the moment he begins flight planning until the tires touch the runway at his destination airport.

Well, the time has come for a change. And the ways and means are available—now.

First, the flight planning and weather forecast procedure must be converted into custom computer service tailored to the individual pilot and fed back within the minute.

Next, the business of allowing any aircraft to take off and head for another airport so jammed that it can't possibly accept it without stacking, holding, rerouting and delay, must be vetoed post haste.

Then, the random chugging of uncontrolled aircraft into the approach and departure zones of high-density airport areas must be summarily forbidden.

And finally, the airport itself must be drastically redesigned to eliminate intersecting runways and to provide separate, distinct runways and instrument approaches for fast and slow aircraft.

At the heart of the complexity of the entire system is the omni navigational network. Paradoxically, only a decade and a half ago it was the great emancipator. In one bold stroke it had obsoleted the ancient system of low-frequency radio ranges with their static-plagued aural beams, and had blanketed the sky with 360 electronic roadways apiece instead of the meager four offered by each low-frequency loop and Adcock range.

But this turned out to be a double-edged sword, for while it permitted unheard-of-flexibility in navigation, it tripled the complexity of the airways layout. Thus the fairly limited low-frequency network was transformed into a gigantic spider's nest of Victor airways threaded in every conceivable direction.

And because there was still no effective way of determining intermediate points between omnis except by cross-fixes between two stations, pilots began to see spots, in the form of intersections. These were sprinkled throughout the system and identified by nonsensical names; and they have remained to plague each new generation of pilots by their invisibility in high-density areas and their ability to befuddle attempts at omni indicator interpretation.

Furthermore, the VOR airways have forced the flow into arbitrary corridors. Though this mattered little when traffic was light, it has created tightropes and bottlenecks under the full crush of today's air travel. And the mincing omni-to-omni zigzag steps required to get from here to there have become inefficient and dilatory.

The overall effect is to hinder and confuse the pilot throughout the flight as he is shunted from one fix to another, constantly shifting directions and returning and identifying omnis, outer compass locators and ILS beams and moving up and down a staircase of assigned altitudes.

What's obviously needed is a way to untangle the Gordian knot of the omnirange system, to sweep away the Victor airways and SID cobwebs, and allow the pilot to do what he really wants most, after all: to fly directly from one airport to another without being routed all over the heavens along the way.

And there is such a way. It's by using a beautifully simple concept that labors under the pompous name of "course line computer, pictorial display." It consists of a cockpit device that presents the pilot with a moving navigation chart and a stylus that marks his position. It works off existing vortac stations and offers complete freedom to plot any course anywhere, regardless of airways.

The best organized practical trial of the device is under way at this moment in the skies between New York City and Boston. Eastern and American Airlines are trying out systems by Decca, Collins and Bendix in DC-9s and BAC One-Elevens with an eye toward speeding up and simplifying the routing on highly competitive tracks between the shuttle cities. cities.

The benefits of eliminating the present omni airways system and adapting a pictorial chart environment in its place have not gone unrecognized by the FAA, nor by one of its most prestigious nongovernment advisory bodies, the Radio Technical Commission for Aeronautics.

The Commission last March urged "simplifying the air route structure by eliminating all except major air routes as designated and charted airways. Deletion of much of the present airway network," the Commission said, "would not only simplify airspace utilization complexity but would encourage flexible routing and improve air traffic control."

Decca's Omnitrac System gives an idea of the versatility a course line computer can offer in solving this problem. With ?5 feet of preselected charts, it permits the pilot to plot a course for anything from a transcontinental trip to a local-area jaunt, all in the proper scale. This stylus draws the flight path upon a translucent plastic sheet that can be retained as a flight record or erased and used again. And the computer portion of the plotter will indicate range, bearing and ETA to any point on the chart. Furthermore, the instrument will accept information not just from VOR/DME, but also from other navigation systems such as hyperbolic gear plus self-contained doppler and inertial equipment.

Other general advantages of the pictorial display include these, listed and endorsed by the FAA itself in an Advisory Circular issued in 1965 encouraging development of equipment by manufacturers:

"It permits the establishment of parallel or off-set routes where no vortac facility exists or where a facility cannot be sited. It makes possible the establishment of instrument approach procedures for many airports that do not now have instrument approach capability. It can provide for the establishment of straight-in instrument approaches where only circling approaches now exist. It reduces communications workload by establishment of more diversified arrival/departure routes, and reduces the need for radar vectors."

With credentials such as these, what has delayed adoption of the system? Until recently it was lagging technical development. But now avionics manufacturers are rallying to the concept and new designs are appearing.

For example, besides the Decca, Collins and Bendix models

(Collins and Bendix have off-course computers but do not use a pictorial chart), a new pictorial display system has been proposed by Hughes to the FAA for evaluation, and Narco has been reported working on its own design for some time now.

Another priority item on the agenda of essential improvements in the air traffic control system is the present voice communications system. Even with the newly enlarged 360-channel VHF capacity, the airwaves are crowded in high-density areas, and matters are going from bad to worse. Furthermore, passing information and instructions by voice over radio is at best a clumsy and inaccurate business, as any pilot will testify.

The obvious solution to the problem is adoption of data link. This is a system of ground-to-air communications by means of silent radio relay. Controller instructions in the form of headings and altitudes, airspeeds and radio frequencies are displayed on a cockpit indicator.

For more complex information like en-route weather forecasts or reroute clearances a small printing device actually types out messages onto a reel of paper that can be retained for reference as a permanent record.

Messages and instructions from the ground appear only in the cockpit of the airplane designated. The aircraft data link "listens to" all transmissions, but accepts and displays only those tagged with its private identification signal. Also, data link transmissions are frugal with radio time, because they take only a fraction of a second. This even applies to the paper printing device, which may accept as many as 30 words a second from the ground transmitter, even though it may print them out more slowly (at about one or two words a second).

The next obvious step would be to feed data link instructions directly into the autopilot to allow the airplane to be controlled automatically from the ground.

What is the status of data link equipment? The FAA has tested and evaluated hardware of its own devising with traffic control simulators on the ground at NAFEC (National Aviation Facilities Experimental Center) in Atlantic City, New Jersey, but at this time it appears that no civil avionics manufacturer has a fully developed data link system. (NAFEC did, however, use an existing print-out device built by the Mitre Corporation, Paramus, New Jersey.)

At any rate, according to William L. Sullivan, who conducted NAFEC's data link tests, the technology is easily available, and the new integrated circuitry will permit compact cockpit equipment instead of the cumbersome designs offered in past years.

Another much-needed innovation well within the scope of current computer technology is a custom flight weather forecaster and flight planner.

The foundation of the present weather system, the teletype machine with its reels of cuneiform symbols and fractured abbreviations and irrationally grouped information, must be relegated to the Smithsonian.

In its place at the airport there should be a computer-printer, hooked into a national weather and flight planning information center. The pilot will feed in his destination, airspeed, requested altitude and time of departure, and the machine at the airport will relay the data to the national center.

The national center computer will select present and forecast weather for each segment of the desired flight path and relay it back to the local airport printed on a card the pilot can take with him, along with routing instructions and approved departure time, and NOTAMS for en-route Navaids and the destination airport.

The weather along his proposed route will be illustrated graphically in profile showing cloud heights, icing level, expected turbulence and precipitation and strong surface winds at the destination airport.

But the advantages offered by all this sophisticated equipment will be lost unless we keep aircraft using it in carefully monitored corridors, free from intrusion by uncontrolled flights.

The present system, at least under 24,000 feet, is a joke. Because no matter how beautifully monitored and coaxed and comforted one is on his IFR flight plan, there's nothing in the world to prevent someone else from barreling into him while exercising his uncontrolled VFR prerogatives.

Needless to say, all pilots would also like the benefit of traffic separation, but right now there's no way to get it. Even the so-called VFR radar traffic advisories are served by radar controllers on only an occasional basis when they're not caught in the press of other duties.

So we must stake off two "no trespassing" zones from which uncontrolled aircraft are barred. One of these is an upper-level corridor system or flyway for traffic between really busy terminals; the other is a fairly sizeable security zone around each high-density airport.

We propose the upper-level corridors be given a floor of 8,000 feet. This would permit non-oxygen-equipped general aviation aircraft to take advantage of the positive control offered in the corridors for several thousand feet above the "floor" in good or bad weather.

Initially, the corridors should be limited in number, connecting only high-density terminals such as New York City, Chicago, Washington, Los Angeles, Dallas, Atlanta, Kansas City, Boston, Miami, and San Francisco. They should be sufficiently wide to allow enough parallel routing to accommodate the maximum traffic load that can be fed into terminal airports, without holding or circuitous rerouting.

However, all the airspace between the upper-level corridors, up to the moon, should be open to anyone who has a mind to fly there, without a flight plan—provided above 8,000 feet he is equipped with an altitude-reporting transponder that will allow ground controllers to guard against conflict with flight-planned aircraft in the flyway. Below 8,000 feet a basic transponder will be required. Beyond that, everybody's on his own, except for the high-density airport security zone.

This should be shaped like a top-heavy, machine-tooled mushroom. Its stem should extend out 10 nm from the airport and should rise to 2,000 feet above the ground. On top of that the cylindrical-shaped cap should canti-lever out over the ground to a radius of 20 nm from the airport, and should rise to the moon.

The function of the "cap" would be to allow approaches and departures to the high-density airport without fear of encroachment by uncontrolled traffic. Uncontrolled aircraft would have to fly around the mushroom cap, or under it. Pilots using smaller nearby airports may use the airspace under the cap for entering and exiting, and if the airport is within the "stem" area, they will be given an access corridor.

All other airports will retain the present traffic area of five miles (but nautical, not statute) in radius and 2,000 feet in altitude. Aircraft using these airports and the uncontrolled airspace will be stuck with having to use the better elements of the present system. Eventually some higher density terminal areas will have to be integrated into the flyways system.

But it will do little good to fly about, safely protected in rigidly controlled, sacrosanct sky, knowing each instant to a gnat's eyelash where we are, how fast we're going and precisely when we're going to arrive over the threshold of the destination airport runway, if when we get there we discover that 12 other guys had planned to arrive at that precise instant also.

As asinine as it is, airlines deliberately do this now, routinely and with no apparent intention to change until they're forced to. As many as a dozen flights of different airlines are scheduled in at the exact same time, in the full knowledge that only one can land at the appointed second.

The other 11 must be vectored and slowed down and shunted up and piled in stacks upon which another stack of 11 will be stacked in the next moment. Add to this the chaos on the ramps and in the terminal when thousands upon thousands of people are deplaning and enplaning, each being greeted and wished bon voyage by grandparents and wives and old college buddies. It's hardly civilized. Even a goose has sense enough not to try to lay two eggs at once. The only practical answer is a reservation system for use of runways and other airport facilities.

It's time we stop kidding ourselves about maintaining a democratic first-come, first-served system and get on with a sensible, orderly program. The solution is not impossible. Airlines themselves have had the answer more than 20 years. If an airliner has 100 seats, it's scheduled, through a national reservations system, to have no more than 100 passengers on it at a time.

A typical ILS runway, on an instrument day, can handle up to 46 IFR movements, 23 landings, 23 takeoffs, per hour. This allows a slight bumper for the unusual, such as poor braking action. Therefore, beginning immediately, we must schedule no more than 46 aircraft operations on a runway in any given hour. It's so obvious.

Only by doing this can we insure that when an airplane arrives at the approach point he can turn in, land and disgorge instead of going through the incredibly complex, wasteful and dangerous business of being stacked, vectored willy-nilly, told to speed up, slow down, do this, do that by a fallible controller who's also juggling 20 other aircraft.

The maximum number of aircraft whose flight plans will be accepted will be the maximum number that can be safely handled under the worst conditions. Acceptance of a flight plan will assure a landing time almost regardless of weather. Heavy snow or ice on a runway would be the only conditions that could disrupt the smooth flow of traffic. Zero-zero automatic landings must be standard for airlines and optional for general aviation. Holding for better weather cannot be tolerated because the plane must get on to another airport if it is to do its job.

When a runway is shared by general aviation and airline aircraft, the available landing slots would also have to be shared. The fair ratio is two to one, or 30 airline operations to 16 general aviation types.

Here's how the system will work. The 30 airliners (less, at some hours) will of course have their flight plans prefiled. Flight plans will call for precise navigation and to-the-second arrival at points along the route. Any deviation from the plan will be monitored from the ground and the pilot advised to speed up or slow down.

What about headwinds and tailwinds? Quite simple. The flight plan will not call for the aircraft to fly at optimum speed. Speeds will be perhaps 10 or 15 percent below that so there will be enough speed flexibility remaining to make up for headwinds. Tailwinds will call for a comparable slowdown.

But the airlines will cry that they are being kept from getting best use from their jets. Nonsense. They are not getting anything near efficient use from them now, and at least with programmed flight they would, with rare exceptions, be able to anticipate what conditions would prevail. We have had an executive of one airline acknowledge that he would happily accept rigid, clearly established flight plans on his routes (though they would be slower than presently scheduled) in preference to the haphazard way his flights are now slowed down, vectored and held by ATC.

The general aviation aircraft that will be integrated into this system will benefit too. In general they will be flying at altitudes with other speed-compatible aircraft. There will be no infernal twisting and turning and vectoring to avoid who knows what.

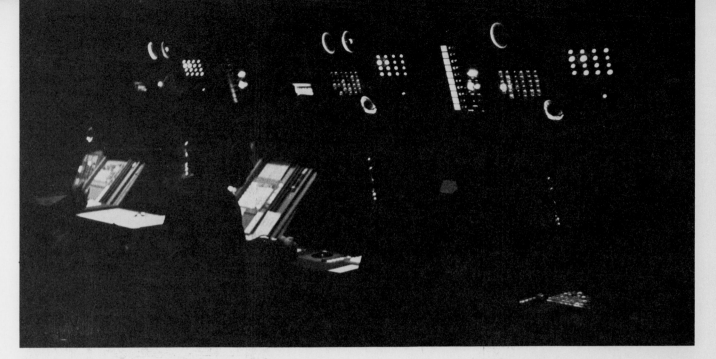

Inside the control cab of an air traffic control center. Each controller has a sector to work.

And though they might not always fly at best speed, there will be no holding and interference from the ground. The flight plan in all cases will describe what to do, and provided it is done there would be no communication with controllers. The only time a controller will speak to the captain will be when he sees a conflict developing; an aircraft slow to clear the runway or an unauthorized aircraft in the approach path.

Hour on hour, day after day, aircraft will arrive at the approach point at the appointed time, turn in and land, taxi out at the proper time and take off. Nonscheduled aircraft will have to request a takeoff and landing slot as part of the preflight planning. When a general aviation pilot in Kansas City, for example, files for Dallas' Love Field, a computer in the national system will search through the reservations for Love, select the next available slot, search through reservations for Kansas City for the most compatible open departure time, figure a speed that will get that particular aircraft to Love's approach point at the proper moment and automatically write all this into the clearance.

The computer will also search through the reservations for alternate airports, such as Addison or Red Bird, north and south of Love, which may have an earlier opening. The pilot may then elect to file for the alternate. If, when he arrives in the vicinity of Dallas the weather is VFR and more than 46 aircraft can be handled at Love, or if a reserved slot has been cancelled while he was en route, he will be cleared to land. Otherwise, he goes to Addison.

But what if the major terminal can take him? How will he fit in with the big boys? He will simply have to maintain a higher airspeed in the approach. The present system of approach speeds ranging from 180 down to 60 knots is almost as ridiculous as having a dozen aircraft arriving at the same moment.

The consequences of a 180-knot plane falling in behind a 60-knot one at the presently legal three-mile spacing are obvious. When a 60-knot plane falls in behind a 180-knot one the gap of course widens. By the time the faster plane crosses the

threshold, assuming a 15-nautical-mile final approach path, the spacing is 10 miles, and two slots in the landing sequence have been lost.

The solution is a standard approach speed for everyone. In the future all planes, on reaching an electronically fixed approach point 15 nm from the airport and perpendicular to the active runway, must be doing 150 knots exactly and this speed will be maintained to the threshold. Procedures for general aviation approaches must be modified to take this into account. Some plane types will be ruled out of making approaches to the main runway because they can't meet this requirement. Some will have to go through approach gear-up, in which event their gear extension speeds must be high and doors beefed up so they can be lowered at the last instant.

The ultimate solution to the mix problem must be more general aviation runways at air carrier airports. It isn't realistic to think we can ever ban general aviation from any airport or totally lure them away with auxiliary fields. On the contrary, their numbers at major terminals will increase.

We envision the day when some airports will have one long runway for carrier jets and several shorter ones for general aviation. For just as surely as we once used cars to take grandmother to the train depot, we're going to see more and more little planes delivering Dad from Podunk to O'Hare to catch a plane to Los Angeles. Plans for this sort of thing should have been made 10 years ago. They *must* be made now.

Our present practice of making a "strip" for little planes on any old unused corner of the airport is unsatisfactory. It solves nothing. If a general aviation runway is to be worth its asphalt, it must be properly spaced from and parallel to the active jet runway. Otherwise it cannot be safely used simultaneously during IFR weather. It must also have an ILS tailored to general aviation needs and ramp connections to the terminal area so Dad can be delivered directly to his LA flight. It need not be long—2,500 feet will do nicely—and the approach paths need not be so flat as those for the big jets. The standard approach speed to it will be 90 knots.

When most general aviation flights are scheduled onto separate runways, the airlines can safely schedule up to 40 operations per runway per hour (leaving some spaces for business jets and other high-performance general aviation aircraft). But for the convenience of the public, all of whom seem to want to arrive in New York and Chicago and St. Louis in a wad, we'll need to increase that. Parallel jet runways are one solution. But for heaven's sake we mustn't repeat the mistakes of the past.

At Dallas, for instance, the city at enormous expense built a parallel jet runway, but there's only 3,000 feet separation instead of the 5,000 feet presently required for simultaneous IFR operations. As a result the airport has a high capacity in VFR weather and the airlines have scheduled flights accordingly. But comes an IFR day with ceilings down to 400 feet and the second runway becomes illegal and traffic stacks up. We've been number 12 in line for takeoff and had to wait 35 minutes, for planes cannot be released on the departure runway until the incoming plane touches down on the approach side. So have 5,000 feet separation between runways, or change the legal minimums, or forget it. Displaced thresholds? No. They require the coordinating skills of a controller and we must keep controllers free to monitor movements for safety.

The final answer will be new, properly designed airports. The present crosshatch of prime runways tangled and knotted with crosswind runways is a farce. The pity is we're now building and planning airports using ideas that were outdated back in the days of the Jenny.

The fact is, we don't fly taildraggers with tailskids and no brakes anymore. Nowadays we can land in crosswinds with ease. The kind of pilot who'll be using our major terminals can handle a 30-knot crosswind, no problem. Design airports with that in mind.

The most pressing need is for longer runways. They should be 15,000 minimum. This will give more room for error in automatic landing systems. But more importantly, it allows us to put more airplanes on a runway each hour. One of the problems now is that it takes from 47 to 58 seconds for a large aircraft to clear the runway after crossing the threshold. And another aircraft cannot be cleared to takeoff or land until the runway is clear. This is why we must allow 78 seconds (46 aircraft per hour) for each operation on a single runway.

A plan such as this will bring howls of protest of course. And the loudest is going to come from the airlines. Their first objection will be that limiting the number of flights scheduled into an airport to those that can be handled under the worst conditions will seriously curtail the overall use of expensive airport facilities. On the contrary, by spreading the traffic over a longer period during the day, it will be possible to achieve more utilization with more safety.

The average number of operations handled per active runway per hour at JFK today is only half the number we propose. What we're saying is, under a rational system of reserved landing and takeoff slots, the capacity and utilization of JFK can be doubled.

The next objection will be that it'll inconvenience the traveling public. That's true to a degree. But the inconvenience of not being able to get a flight to JFK to arrive at 11:45 a.m.

for a business luncheon or connection to Paris is not nearly so great an inconvenience as being able to get the flight not knowing whether it's actually going to land at 11:45 or 12:45 or maybe even at an alternate airport at 1:45.

General aviation is going to scream that this system will discriminate against our basic freedom to use a public facility. On the contrary, the system will increase our opportunity to use tax-paid-for airports. In the first place our proposal will over a period of years affect only 20 to 25 of the nation's high-density airports.

The most thorny question is going to be, who decides how many slots are allotted to general aviation and to the airlines and which slots go to each airline. It's bound to be difficult and the problems are complex. To make the most efficient use of its fleet an airline must schedule with immense care. The plane at Chicago that is due to leave as Flight 86 for Dallas must be there in two hours and 17 minutes so it can be fueled and tidied up to become Flight 53 for Los Angeles. And that same plane must be in Tulsa at 1 a.m. February 21 when it's due for its annual maintenance at the Tulsa overhaul base.

But these things can be worked out; it's not impossible, provided the airlines do not get to bickering amongst themselves. If that's too much to hope for, we suggest that a board similar to the Civil Aeronautics Board be appointed by the President. This body would be empowered to make arbitrary and binding decisions on the allotment of airport use as well as to settle other knotty problems plaguing the national airspace.

It could also assist in the implementation of the flyways system by examining some of the currently inviolate procedures and regulations. Many of these must be scrapped.

For a start, let's drop the requirement that runways be a minimum of 5,000 feet apart for simultaneous IFR operations. A number of major airports with dual runways do not meet this minimum. But, while this might have been desirable separation between air carrier and general aviation runways, and even between air carrier landing runways years ago, it certainly isn't necessary between takeoff and landing runways used by IFR pilots flying sophisticated aircraft.

The fear of course is that an aborted landing will come into conflict with a departure. But in the future, with automated approaches, balked landings will be extremely rare. Even when they do occur, the decision will be made in most instances some distance before reaching the threshold.

If we then make it standard procedure for a 10-degree turn to be made away from the takeoff runway as soon as feasible after a go-around decision (and this can be programmed into the autopilot), and if we make it standard procedure for departing aircraft to turn 10 degrees from the landing runway after takeoff, the danger of conflict will be less than it is now with 5,000 feet separation.

The next rule that must go is the one on three-mile separation of aircraft departing and approaching the airport. Takeoffs from a departure runway should be at the pilot's discretion. In any event they need not be more than 30 seconds apart, for we shall also have a departure speed, which, along with the scatter headings, will make conflicts much less likely than they are now.

On the approach side of the airport we need to make some changes in the runway and the criteria for determining when an aircraft is clear of the active. Our present, so-called high-speed turnoffs are a joke. What we need is a very gradual turnoff beginning about 3,000 feet from the approach end. What this will probably involve will be a huge, abutting ramp so that aircraft landing from either end can begin a gentle swerve after about 3,000 feet of roll and depart the runway at 50 knots or so. Note that this would leave us with an extension beyond the turnoff in case the preceding plane slows down too quickly.

With this arrangement, coupled with an automatic landing system that will insure that touchdowns are made within the first 1,500 feet, we can space approaching aircraft on a timed rather than a mileage basis. And the time can certainly be as short as one minute, including a comfortable bumper. Eventually, as automated systems become more trustworthy and accepted, this will undoubtedly be cut to 30 seconds or less.

Thus, air carrier operations at major airports should easily reach 120 or more per hour in the next two years.

The possible refinements and modifications of what we have come to call the flyway system are endless. But nothing can happen and no true progress can be made until there is acceptance of the basic premise that some flights want, need and must have everything the state of the art offers in guaranteeing safe, swift and on-time travel. These flights must be channeled into airspace where they can be constantly monitored from the ground. They must use equipment that provides absolute navigation accuracy and they must be assured a time for landing before they take off. Holding, slow-downs and delays in the airspace are consummately inefficient and are a clear indication of a breakdown in the system.

Most aircraft in the general aviation fleet will be happy with VFR flying in the remaining uncontrolled airspace. And of course IFR flying in the uncontrolled airspace will be carried on much as it is today. Ultimately there must be a requirement for all aircraft to carry transponders, but this will be a small price to pay for the added safety afforded.

When VFR aircraft are equipped with basic transponder (anticipated cost, about $500) and aircraft flying over 8,000 feet carry transponder with altitude-reporting capability, the need for vastly expensive and complex collision avoidance systems will be sharply reduced. Aircraft in the flyway system will be monitored constantly to assure separation by controllers whose major remaining function will be to do just that. Any intrusion into restricted airspace by a nonqualified aircraft would be dealt with severely. Any wandering off course or deviation from flight plan by an aircraft in the flyway would call for immediate notice from the ground station. Could there be a midair in the flyway? It could happen, but it would be difficult.

How is the establishment of such a system to be paid for? There can no longer be debate about that. President Johnson has decreed that the users shall pay. Department of Transportation Secretary Alan Boyd is busy adding up the bill for the present system and he has been directed to figure out a way to collect for this and for future developments.

It seems certain that those benefiting from the advantages of a new, more sophisticated system should pay most of the tariff. Since airlines and their passengers would be the primary beneficiaries, this could mean an increase of from five to 10 percent in the airline ticket tax plus a fuel tax of four cents per gallon. General aviation aircraft using the system would continue to pay the four cents fuel tax plus an additional surcharge.

Certainly these are stiff penalties to pay. But they are necessary and in many ways they are more equitable than the present system of collecting from airlines and general aviation as though each used all parts of the ATC service equally. Today few quarrel with the notion tht one must pay extra for use of a throughway or limited access turnpike. Similarly there must be a premium charged to those who benefit from on-time takeoffs and departures plus the added safety offered in reserving for each flight a moving block of air from takeoff to landing.

What are the chances for adoption of all or part of this proposal? Not as bad as you might think or as good as we might hope. Much depends on Congress, for it is unlikely that FAA can be persuaded to act without a direct order from Capitol Hill. At present FAA is hung up in the belief that in meeting its basic purpose (to provide for the safe and efficient use of the airspace) it has neither the right nor the authority to deviate from its policy of first-come, first-served for all qualified aircraft.

Unless our Ouija board is malfunctioning, the sense of the people and the temper of Congress is such that great progress can be made now. But regardless of the opportunities for headlines, Congress must remember that progress will not result from hasty and foolish statements to the newspapers (Congressman would ban all light aircraft, Congressman would put two pilots in every cockpit). Congress' job is to restate, reaffirm or redirect the mission of the Department of Transportation and of the FAA if such is needed and leave the details to the experts.

There are those in aviation who howl in protest against any change and there are those who would leave things pretty much as they are. To them "see and be seen" is still the primary truth in safe separation of aircraft and any problem can be solved with another procedure tacked on top of the mountain of procedures we already have. These people resist what they don't originate and they wallow in the outmoded concepts of the past.

There are others who say man is a creature of the clock. He eats and sleeps and works and travels at certain periods of the day. This results in insufferable crowding and delay on the trains and throughways and buses during brief periods of the day while they lie idle most of the time. To such traditionalists air travel is doomed to the same fate.

We suggest it needn't be so. That from the standpoint of safety alone it had better not be so. We believe that the pressures on men and equipment in the present airway system are such that something or someone will surely break, soon. We are convinced that the air traveler in an airliner and the pilot and his passengers in a general aviation airplane will not put up with delay, diversion and danger forever.

Is there a better way to run the airspace? You bet there is.

The Dangerous Life of the Demo Pilot

"Uhhh . . . you're getting a little low there . . . watch your airspeed . . . little more power . . . more power . . . I'VE GOT IT I'VEGOTIT!"

BY JOSEPH E. MESICS

WE ALL LIKE to try new toys before we buy them. This accounts for such time-honored trials as squeezing bread in grocery stores; buy-now, pay-later plans; and, more recently, birth-control pills and airplane demonstration rides. Of all such tryouts, the airplane demo can be the most dangerous both for the demonstrator and his client. . .

The handsome young factory pilot for the new light twin grins amiably as he pushes you firmly into the left seat. "This bird flies itself," he chuckles, not realizing that you've just had three Martinis at lunch in order to get up your 300-hour courage for the ride. You hope it flies itself, since you're in no shape to fly it. As your back arches against the artfully contoured seat, you jam in the throttles and notice the perspiration beads forming on the back of your hand. The demo pilot is murmuring something in an enthusiastic monotone in your right ear, but all you're concerned with is trying to look casual while controlling the airplane.

After half an hour of single-engine stalls and bumps and grinds around the pattern while you fight to look airmanlike, you taxi up to the sales office wet with sweat and secretly relieved that the demo ride is over. "Heh, heh, nice little airplane, but a little squirrelly around the yaw axis," you comment professionally to the demo pilot, who—you sud-

denly notice—is also covered with sweat but still grinning manfully. What he's thinking behind that grin is: "God help me, I hope the spastic nitwit doesn't try to raise the gear right here on the ramp. He's done everything else wrong." But what he says is: "Glad you like her. The way you fly her, it looks like you two were made for each other."

If only demo pilots could simply toss you the keys, the way the car salesmen do, you could investigate an airplane in your own fashion and on your own mission. The only time this happens, however, is when the military straps a single-seater to you and says "go"—after a very complete preflight briefing, perhaps days or weeks long. The civilian demo pilot finds it difficult to be that thorough with his customer preflight. Sales prospects pay his bills, and time to show his product is short. This squeeze can result in rushed, ill-prepared and unsuccessful flight demonstrations.

You can always tell the demo pilots relaxing at the Cumulonimbus Cafe after work: They're the ones who spill a lot. The demo pilot rides with hundreds of pilots each month, and he spills a lot because at times he misjudges a pilot-prospect and has to sit on his hands while his airplane is asked to do impossible things by an unqualified or overenthusiastic aviator. . .

The salesman watches in disbelief as the aileron bangs against the stops and his demonstrator snaps over into a spin entry. His teeth grind together behind his perpetual smile as he watches the prospect pull farther and farther back on the stick when the airplane is high and already slow on final. He chokes back his surprise when the customer abruptly pulls out the mixture knob instead of the carburetor heat. He leans against the unbalanced turn wondering if the guy will ever notice either his off-kilter backbone or the off-center turn and bank. The salesman has told his client to add power, raise the gear, pull the flaps, trim, and now, believe it or not, he's explaining where the horizon is.

It's all part of the lightplane demo pilot's job, so you can't blame him for his nervous tic or his prematurely graying hair. He can help himself and the prospect, however, with thorough preflight and postflight briefings: The demo pilot's credo should be to explain exactly what he plans to do, do it, and then review it.

The only solution to the hazardous nature of airplane-demonstrating lies in proper understanding of aircraft limitations by the demo pilot and proper preflight preparation. Other, more specific, suggestions have been offered to help raise the safety level of airplane demonstrations. Some suggest that only commercially licensed pilots be allowed to fly demos. Others suggest more awareness on the part of the demo pilot of the hazards of demo rides, while still others advocate training in "control-blocking" and other methods of taking over if problems occur.

If both pilots are properly briefed on the demonstration procedures before the flight (including who is responsible for what if an emergency should develop), then the flight should be most satisfactory. No degree of licensing or special ratings will automatically make for smoother demo rides. Even an advanced pilot—whether he's the client or the demonstrator—who is not briefed on procedures can bungle a demonstration badly.

The preflight briefing may not be as easy as it sounds, however, if the prospect shows up half out of breath, thinking of everything but aviation, and immediately wants to leap into the airplane and go flying. Why not hand him a printed scenario of what the flight will include and who will be responsible for various stages, in order to get him in the proper frame of mind for a good demonstration? This briefing establishes the air rules and also allows the client to best describe his aviation capabilities and his personal mission. If he's worth demonstrating to, he will appreciate this preflight briefing as valuable input for his buying decisions. It is also a time during which he develops confidence in the demo pilot's ability.

After the demonstration has been accomplished and you have landed, debrief each other on what has happened. Many of the prospect's questions about the airplane will surface during a well-structured and unhurried debriefing. If there is no debriefing, chances are the prospect's questions will stay unanswered, and the demonstration remains incomplete.

Most pilot-prospects manfully demand the left seat. The demo pilot should offer the left seat only after he has demonstrated the airplane's capabilities himself. In larger aircraft, such as corporate jets, most demo pilots refuse to allow left-seat flying until the pilot has been prepared to participate, even when the trip is paid for by the prospect and his pilot is a professional.

Since a competent demo pilot should be able to fly his airplane equally well from the right seat, however, it isn't unreasonable to expect that good, professionally planned demo flights can be flown from either side. But the prospect should insist on a good left-seat demonstration prior to his flying, so that the aircraft's best performance can be observed. After he buys the airplane, he will normally sit in the left seat; why not show him at the outset the proper left-seat procedures and performance?

Unprofessional demonstrations are endemic to the airplane sales business in spite of attempts to professionalize them. What demo pilot hasn't forgotten the local charts, for example, or left a customer behind by mistake, or forgotten to pull the chocks or close a door properly? Or—horrible thought—closed the door too quickly on the prospect's thumb? Take pity, though, on the demo pilot whose prospect's seat came untracked on a STOL ride just at takeoff. The prospect ended up in the fetal position, still strapped to his seat, the seat lodged firmly in the tail cone of the airplane. It wouldn't have been so bad, except that Mr. Big weighed over 250 pounds, and the pilot had some difficulty adjusting to the abrupt CG change. What a ride. What cursing and swearing. What a job getting the prospect out of the tail cone.

Brakes fade and flaps fail to retract, circuit breakers pop and fire-warning lights flash. Engines won't start because they're overheated, and everybody is inside the demonstrator parboiling while the battery struggles to get a few more turns out of the prop. Maybe one engine on your twin won't start after you've delivered your prospect and his family to O'Hare to make a connecting flight. Try to look extremely cost-conscious as you taxi out of sight with one engine caged while you smartly wave goodbye.

Prospects get sick. Some get angry when you don't give them the expected wild demo ride. Radios quit and the weather gets bumpy. Doors pop open on takeoff and baggage slides out of unsecured baggage doors. You haven't really had a bad day, however, until you fly one through a fence at an airshow demonstration. That—and worse—has happened.

Sometimes, nice things happen to demo pilots, though. The nicest thing that can happen is an aircraft sale. If the demo pilot is fortunate, he might even get some credit for the sale. Some demonstration rides are smoothly executed and thoughtfully prepared. These rides allow the prospect to investigate the airplane's capability without distractions. Even the smoothest ride sometimes backfires, though. One pilot flew a wonderfully smooth demo and was convinced the prospect would be unable to resist signing the check. The client had other ideas. Although the new twin exactly fit his capabilities and his mission, he decided to buy the competition's airplane because it seemed to require more pilot skill. The smooth demo had made the airplane seem too easy to fly for the dashing client.

Invite a demo pilot home for dinner. He needs all the love and affection he can get.

THE RATING

*For the professionally minded pilot, there's always one
more rating peak to reach. Here is one of the toughest climbs.*

BY JOHN W. OLCOTT

"LOOK," the instructor had said to me in the beginning, "I
know only one way to get you ready for this rating. I'm going to
assume you're a corporate pilot preparing to captain a Learjet
every day. If you get the type rating, you will have the same
authority as someone who does, and that privilege is not to be
treated lightly. The fact that you are a magazine writer, not a
full-time Lear pilot, bothers me, and it is sure to bother the
examiner. As far as I'm concerned, you've already got two
strikes against you."

Just before my check ride, the FAA examiner questioned
what I would do with a Learjet type rating and made it clear
that he didn't care whom I worked for or what my objective
was; he'd just as soon flunk me as pass me if I couldn't hack it.
Strike three?

Now the flight test was over. For seven hours, starting at
9:30 in the morning, John Cureton, the FAA's regional Lear
specialist, who works out of Columbia, South Carolina, had
been probing the depth of my knowledge and skill. The last
hour and a half had been an inflight examination of what

couldn't be covered in the detailed oral exam. As I taxied the
Lear 25B to the ramp at the Raleigh-Durham airport, Cureton
was disturbingly silent.

This doubt-filled moment was the culmination of 17 long
days of nearly continuous ground and flight training. They,
too, had been doubt-ridden. From the first Monday morning
when I arrived at the FlightSafety International Gates Learjet
Learning Center in Wichita, there was a noticeable atmosphere of skepticism surrounding my presence. What, after all,
was a magazine writer doing in the company of career aviators
who flew for leading corporations or executive charter
services? My firm neither owned a Learjet nor was an
immediate prospect for buying one. Didn't I know that type
ratings were not for sale at this school? This was a place for
professionals—pretenders and dilettantes wanting to add a
prestigious rating to their tickets need not apply.

The personnel at FlightSafety were courteous, and my
fellow students were friendly. Perhaps my sensitivity
stemmed from my own anxiety that earning a Lear 25 rating

*Seventeen-day scenario: step by step, student John Olcott grinds through the jet-pilot training mill.
First challenge: the abstractions embedded in schematics and manuals; then, the cold hardware
(and sweat) of the simulator. At last, the Learjet itself, gleaming, sleek and hot, and the cockpit setting
of the climactic struggle.*

on my ATP would not go well.

I had long aspired to be checked out in a jet, but I'd had minimal experience in the likes of a Lear. Also, the type rating itself held a certain cryptic quality for me. Required for flying multiengine jets and all aircraft with gross takeoff weights greater than 12,500 pounds, the type rating represents the post-doctorate of aeronautical degrees.

Combined with this desire was a professional purpose for my being there. I wanted to know what such training involved, how high the standards for corporate jet pilots were and how good the instruction was. I wanted to be able to report what I found to prospective jet pilots, to corporate executives contemplating jet purchases and to the aviation community at large. With the inexorable deterioration in airline service, travel by business jet is here to stay, and it will not be too long before smaller turbine aircraft will be offered for general aviation. My objective was to experience what a 6,000-hour pilot with ATP and CFII licenses (much of it in twins) would face in transitioning to jets.

The FlightSafety school in Wichita is operated under a contract with Gates Learjet Corp. that establishes FSI as their official training organization, and each Learjet purchased entitles its owner to train two pilots and two aircraft mechanics at FSI's comfortable facility in the Gates Learjet administration building. Those not buying a Learjet may receive training at substantial fees. There are eight flight instructors on the staff who train students either in FSI's simulator, which faithfully duplicates the 23, 24 and 25 Lear models, or in the customer's aircraft. Four classroom instructors provide a number of ground-school programs ranging from initial and recurrent training for pilots to complete courses for mechanics.

Mornings were devoted to independent study and simulator schedules. Classroom ground school started at noon, under the direction of Earl Drew, classroom instructor for FlightSafety. Drew is a man of studious mien; unhurried and methodical in his attention to each element of the Lear. A retired Air Force F-105 pilot, he also holds a teaching certificate. His approach

was cooly professional—no jokes, no banter, no informality, only the facts. Drew had just 50 class hours, spread over 10 days, to cover each system of both the 24D and the 25B/C. Half of the dozen students were from South America, so Drew had the added burden of teaching in English to several who would have preferred Spanish. One student was from the United Kingdom, another from Canada; four of us were from the U.S. All of these students had considerable experience with jets. Peter Austin, from CSE Aviation, Ltd., in England, is a retired RAF wing commander. Ron Williams is a Lear 24D copilot, with Mesa Petroleum; he was upgrading to captain. Floyd Pepperling is an ex-TWA, ex-U.S. Steel pilot with thousands of hours of turbine time. Pepperling has his Lear 25 type rating and flies as a captain for Thunderbird Airways; he was attending Drew's "Pilot Initial" course to satisfy the ground school requirements of his employer's FAR Part 135.2 jet air-taxi certificate.

The two-inch-thick Learjet study guide prepared by Flight-Safety reflected the magnitude of the material we would have to learn. Fortunately, many of the guide's pages applied to earlier Lear models, but those pages that did describe the 25B—the model I was studying—represented nearly a semester's reading in a normal college syllabus. In addition, there was a Lear 25B/C study guide flight manual, and students were encouraged to obtain the pilot's handbook and operating manuals for the specific aircraft they would be flying.

Detailed knowledge of the aircraft and its systems is one aspect of any type-rating examination that separates this kind of licensing from other flight tests. In addition to being able to fly the selected aircraft to the standards of one's present license—private, commercial or ATP—the applicant for a type rating is tested on his specific, detailed familiarity with a particular aircraft. Thus we had to understand every element of the Lear. It was Drew's task to prepare us for a barrage of questions from the FAA examiner.

After an overall introduction to the 25B, Drew led us through each of the airplane's limitations. The aircraft is

licensed to transport-category standards under FAR Part 25 for a gross takeoff weight of 15,000 pounds, but the student must memorize the roster of other critical weights: ramp weight, 15,500 pounds; maximum landing weight, 13,300 pounds; maximum wing bending weight, 11,400 pounds; maximum fuel in wingtips for landing, 800 pounds. The indicated airspeeds and their appropriate designations also were memory items: V_{MCA} = 102 knots; V_{FE} (full) = 153 knots; V_{FE} (app) = 202 knots; V_{LO} = 200 knots; V_{LE} = 265 knots; V_{MO} = 306 knots; M_{MO} = 0.82 (AFC/ss operating) and 0.78 (AFC/ss inoperative). Charts provide information for the other critical speeds. Engine failure speed, V_1; rotation speed, V_R; takeoff safety speed, V_2 and maneuvering speed, V_A. There are oxygen requirements for various types of operations and various altitudes; there are altitude limitations with and without certain components of the flight control system operational; there are engine limits, takeoff and landing OAT operational limits and en-route operational limits; fuel limits and specifications for approved oils, and, naturally, EGT limits that depend upon duration.

There are endless, seemingly unimportant, minutiae. The windshield is 1.1 inches thick and consists of three laminated layers. The finish is seven layers of polyurethane paint. The number of spars in the wing is eight. The left aileron has the trim tab. Trivia for the casual observer but grist for the professional who is never sated in his craving for more information about his plane.

Drew's explanations extended far beyond the superficial steps of an emergency checklist. We were expected to know each aspect of each system. When the electrical system was mastered, we tackled the environmental and pressurization systems—*"In a Lear 25, do not turn on the Freon cooling during landing roll. You may lose your brakes for three to four seconds."*—then to the all-weather ice protection system—*"Each individual element on the electrically heated horizontal stabilizer stays on for 15 seconds. There are 12 of them. Monitor the ammeter for the appropriate fluctuations."*—and on to the hydraulics—*"Pilot and copilot brake action goes through a common shuttle valve; see it here on this mockup. If both you and your copilot are on the binders at the same time, you can hang that valve up in the center position and get no brakes at all."*—to the fuel system—*"Watch this fuel system mockup. See all the valves that sequence automatically when I hit the start/generator switch?"*—to the General Electric CJ610-6 engine—*"This film is a good review on the theory of jet engines,"*—then he concluded with the warning systems.

Seven days and 35 classroom hours elapsed before Drew had covered all the information needed to comprehend the systems of the Lear 24D and 25B/C, the current GE-powered production aircraft of Gates Learjet. In addition, there were movies on radar, oxygen, aircraft-upset problems, hydro-planing and wake turbulence. We considered normal operations, emergencies and special contingencies. The emphasis was on in-depth understanding—the sort of comprehension that is sought by the true professional, regardless of his field of endeavor. More than learning facts for the sake of aeronautical snobbery or some pedantic urging, we sought information to generate an attitude of complete preparedness. If the system

could behave in a certain fashion, whether by design or by malfunction, we would know about it and would be able to react.

Drew had reserved the remaining 15 hours for weight and balance, performance estimates and a written test. Since I had to attend an important meeting during the last two days of the scheduled ground school session, I made separate arrangements to cover the material I would miss and then take the written exam on my own. How well I did on the oral portion of the FAA type-rating test would determine whether or not I had cut corners on the ground school. Absorbing the material for the oral examination did not throw me, however, since the FlightSafety study guide and the classroom sessions were well prepared and easy to follow. Fortunately, my background in engineering also made me feel comfortable with the bookwork.

The simulator sessions were a different matter, however. Initially, there was a problem setting up the simulator schedule, possibly because of a misunderstanding about my intentions. "But Mr. Olcott, we thought you were here just for the familiarization program. We have you scheduled for four hours next week. We didn't realize you were here for the type-rating program. That requires about 16 hours. That amount of simulator time may be difficult to arrange." Implied, but not stated, in the secretary's apology was the overriding question: why a type rating? "Also, Mr. Olcott, where are you going to find a Lear to fly?"

The problems of both simulator and flight time were resolved, but compromises had to be made. Five consecutive hours in the simulator were scheduled for Saturday, May 3; a four-hour session was arranged for Sunday; two hours for both Monday and Tuesday and four more back-to-back hours Wednesday morning. Wednesday afternoon, I would fly to Washington for the meeting, and Saturday, May 10 would find me in North Carolina, where Raleigh-Durham Aviation would rent me eight hours of flying time in Lear 25B, N427RD, prior to the type rating check ride scheduled for Wednesday, May 14. That compressed schedule was the best that could be arranged. There was neither time nor money available to expand either the number of training hours or the elapsed time allotted to the task. Either I accepted these conditions, or the type rating would escape me.

Was I ignoring all the advice against attempting to squeeze a sophisticated training syllabus into a highly constrained time frame—advice I so sternly had given my own students when they wanted a license or rating in a hurry? Working against a deadline, flying many consecutive hours, is not the way to use your time or the instructor's talents effectively.

My instructor for both the simulator and flight portions of my training was Chuck Cox, a thin, wiry man who had flown PBYs for Shell Oil on the West Coast in the late 1950s and had transitioned to jets shortly after the Lear 23 was introduced, 10 years ago. Cox had been with FlightSafety for three years, and his logbook showed more time in Lears than my total hours. Although he was continually in good spirits and reflected a positive attitude throughout the hectic 12 days that followed, I knew that Cox did not appreciate the compressed schedule to which we were working. He was in a difficult position, since failure would reflect poorly on him and on FlightSafety, even though I was the one whose schedule and skill might prove to be the limiting factors.

I was particularly eager to experience training in a simulator, because I have long been impressed by the excellent conditions for learning they provide. The environment within the simulated cockpit, which was mounted on large hydraulic servoactuators that provide three degrees of limited motion in pitch, roll and yaw, seemed realistic, although I had little with which to compare this analog of the real airplane I would be flying within a few days. The cockpit duplicated the standard Model 24 or 25 front office.

Behind the pilots' seats, in the afterbody of the simulator cab, were located the instructor's chair and control console from which he could create havoc in the form of engine fires, instrument failures, electrical losses and general aeronautical mayhem. In response to either the pilot's control inputs or the instructor's malicious will, the simulator's analog computer generated the signals that produced accurate instrument indications and fairly convincing motion cues. (Incidentally, FlightSafety's Lear simulator was manufactured by Redifon and is representative of the second-generation devices used by the airlines for training until the newer, third-generation digital simulators became available after 1969.)

The first two hours in the simulator progressed satisfactorily. With Gary McVeigh, from Calgary, Canada, as my copilot, I managed not to embarrass myself as we ran through a routine consisting of cockpit checks, takeoff, climb to FL 410, basic airwork, slow flight, approach to stalls, steep turns, descent and a few ILS approaches.

Since the power developed by a jet engine varies with airspeed, it took me awhile to stabilize on specific reference speeds. If I were attempting to set up 250 knots indicated for the steep turn drill, I found I had a tendency to overcontrol with the thrust levers. By memorizing the proper configuration—pitch attitude and percent power—the overcontrolling and the resulting tendency to chase a reference airspeed was overcome. The higher speeds of the Lear did not present much of a problem, mainly because Cox continually stressed the need to think in terms of time, not distance, to a fix. Below 10,000 feet, all jets are restricted to 250 knots, so during those busy times in the terminal area, you can survive by assuming a groundspeed of four miles per minute; three when you are in the initial approach configuration. The secret, naturally, is to plan ahead.

After a quick coffee break, we got back into the box and started on three more hours of simulator work, but this time the tasks involved more abnormalities, more frequent ATC instructions, and more emergency drills. There was less time to plan ahead, and I was being pushed to the limits of my ability to cope with the basic IFR tasks. Airspeed and altitude performance started to suffer and I was becoming saturated. Obviously, it had been a long, five-hour session in the simulator; tomorrow would be different.

It was. It was worse. The interior preflight checks involve 38 different tasks; the checklist only calls out their order, not what to do. I started Sunday's four-hour session by confusing several of the checking procedures. The answers to some of Cox's questions were not as prompt or complete as I wanted.

The airwork was a bit sloppy. Things were not going as easily as I had hoped, and I was trying too hard. It seemed that when I started to get on top of the situation, Cox would throw me another curveball, and I would swing wildly. He had set the pace needed to meet our ambitious schedule, but I was falling behind.

Monday was not much better. The list of errors that Cox had noted after our two-hour session included 14 items, many of them little things that reflected a lack of familiarity with high-performance aircraft, like doing more than 200 knots IAS below 3,000 feet in the airport traffic area. (Hell, I had only been 200 indicated a few times before stepping into the simulator.) Not changing the altitude alerter to the next assigned altitude was another kid-like blunder. Then there was the point about briefing the copilot more often. I wasn't using his talents enough. "The Lear is licensed for two pilots. You are not allowed to fly it alone, and the examiner will look for your ability as the captain to utilize your copilot. Let him fly the airplane while you look over the approach plate; have him hold the plate during the actual approach; have him read off the critical information," Cox would say. But there were more serious signs that I wasn't cutting the mustard, like the time I became so involved in nailing a back-course approach that I didn't realize the needle was centered only because the localizer had failed.

Tuesday's list from Cox had eight items, little things that bugged me and reflected the extent to which I was overreacting; trying too hard. I had thought the simulator would be used more as a teaching tool than as a testing device. I had expected that, when a mistake occurred or there was some confusion, Cox would freeze the simulator's action, explain the situation and let me try again, at whatever pace was necessary, until I mastered the task. Instead, the simulator was used to push me, to force me to cope. Perhaps my errors were just too basic, too amateurish to warrant using the simulator or Cox's time in a more sophisticated manner. Perhaps there wasn't enough time to proceed any more slowly, but if we went any faster, the ego couldn't take it.

Was this the hell I deserved for not realizing when, as a young, full-time flight instructor, I had demanded too much from my students or let them demand too much from themselves? Was this fair retribution for not realizing that a middle-aged businessman who doesn't fly every day can't learn at the pace of a 25-year-old who spends most of his time either inside aircraft or dreaming about them? I was learning more than just how to fly a Lear, but I feared the price I would have to pay.

The first two-hour session Wednesday morning wasn't much better than the previous day's efforts; the list of errors stood at seven. I felt frustrated and disappointed at the apparent futility of my efforts and found myself adopting an attitude of near indifference. Almost immediately, things improved.

It might have been my less resolute attitude, or possibly the change was caused by a less demanding, concerned Cox, but the second and last simulator session, Wednesday, May 7, was the best of the entire program. The error list was down to three relatively general items. That afternoon, after the ground-school session, I left for my meeting in Washington. Cox and I agreed to continue our pursuit of the type rating after my two-day respite in D.C.

On Saturday, May 10, Cox and I rendezvoused at Raleigh-Durham airport. We spent the late afternoon and early evening, through dinnertime, reviewing questions the examiner might ask on the oral examination. I was encouraged by my ability to recall information covered in ground school. It was possible to visualize the schematics Drew had so painstakingly described and then develop the answers to Cox's questions. The in-depth approach was paying off.

Early Sunday morning, after two weeks of books and simulators, I climbed aboard a real Learjet and slid into the left front seat of N427RD.

The scene inside the cockpit was familiar. No, it was more than familiar; it was comfortable, like sitting down at a desk where you had spent many hours working out challenging problems. The interior preflight, starting engines and before-taxi checklists were old hat. Taxiing out for takeoff was a new experience, but one that was easily mastered. The pre-takeoff check completed, I taxied into position and applied full power, just as I had done many times before in the simulator.

The forces of the two GE CJ610-6 jet engines, each producing 2,950 pounds of thrust, created a forward acceleration unlike anything I have ever flown before, and naturally, unlike any forces felt in the simulator. I relied on visual tracking of the runway centerline for the takeoff roll in our Model 25B; all simulator takeoffs were strictly "on instruments." With the exception of the impressive acceleration and

use of natural references, however, the takeoff in the real aircraft handled very much like those I had practiced during my 17 hours in the simulator.

For our introduction to the Model 25 that particular Sunday morning, the aircraft—full of fuel—was about 800 pounds under gross, and the air was smooth, so conditions were near optimum for a complete exposure to all the airwork items of the flight test. We climbed to 16,500 feet, leveled off and tried some normal turns. Then we transitioned directly into steep, 45-degree-bank turns, which must be accomplished at 250 knots within ATP tolerances of $\pm$ five knots on airspeed, $\pm$ 100 feet on altitude and $\pm$ five degrees on both bank angle and final heading. These turns require a light hand on the controls since the instruments connected to the static source, particularly the Lear's altimeter and rate of climb indicator, initially indicate backward if the aircraft's pitch attitude is positioned too abruptly. When a pilot is trying to recover 50 feet of altitude in a steep turn, it is disconcerting to apply a noseup correction and see the altimeter and rate of climb plummet downward. A light touch on the controls, plus careful attention to the artificial horizon and the proper power setting, are the secrets to doing steep turns successfully in a Lear.

Then we did slow flight and approach-to-stall drills; first clean, with gear and flaps up, then with approach flaps and finally with landing flaps and gear down. These maneuvers completed the normal airwork items; now it was time for a simulated pressurization loss and emergency descent. Cox signaled an explosive decompression by a sudden clap of his hands behind my right ear. Almost instinctively, I recalled the

Checkride time, and the author goes through a briefing, while his (temporarily) Learjet awaits.

simulator drill. I snapped on the oxygen mask that was hanging around my neck, reduced the power levers smartly but smoothly to idle, deployed the spoilers, dropped the gear, hit the trim so that the autopilot would be disconnected, were it on and promptly rolled into a 60-degree left bank while I pitched the nose down to about 20 degrees below the horizon. Even though we practiced this drill at less than the 265-knot indicated speed (indicated Mach of 0.82 above 25,000 feet) used for actual emergency descents, the rate of descent, needless to say, approached that of a stone. It was actually about 8,000 fpm. We quickly reached a lower altitude and then proceeded to RDU for some instrument approaches.

When we touched down, after 1:30 under the hood, both Cox and I felt a sense of relief and encouragement. The Lear had been reasonably easy to fly precisely, the knowledge and skill so painfully developed during the ground school and simulator sessions were transferable, and it appeared that I was emerging from the quagmire of doubt that surrounded the simulator experience.

A student's progress is characterized by rapid progress, plateaus, precipitous falls and recoveries. He can go through several such cycles before understanding and skill develop and his performance becomes more stable, more predictable. It was too early to tell where on the learning curve I was with respect to flying the Lear. The next four days, with two flights each day, hopefully would smooth out the hills and valleys of my performance and lead to success.

Fate, however, seemed to see things differently. Cox was preparing a Raleigh-Durham Aviation copilot, Glenn Fulp, for his check ride, scheduled the day after mine. We were to alternate flights, and in that way we both would receive additional exposure to the maneuvers required for the type rating. On Fulp's hop, Sunday afternoon, the pitch trim servo failed during a preflight check and N427RD was grounded until a part could be air-expressed from Wichita. It is a credit to Gates Learjet's customer service department that a replacement unit was on an airliner Sunday afternoon, and the Lear was operational again late Monday, but we had lost a day and we were rapidly approaching another rush situation.

We had to fly eight hours; the examiner said that was the bare minimum time he considered acceptable for a type rating applicant without considerable jet experience. We were able to squeeze in two hops on Monday, one of them at night. Tuesday, we also flew twice, and by Wednesday morning, when the examiner arrived from Columbia, South Carolina, I had exactly 8:00 of Learjet time in my logbook.

John Cureton sat down across a large table from me, carefully positioning the Lear flight manual before him and, with the deliberation of a Supreme Court judge, proceeded with the oral examination. The inevitable question was first.

"Why do you want a type rating in a Lear?"

I explained.

"How do you expect to maintain your currency in a Lear?"

"I plan to be exposed to corporate flight operations, hopefully as a copilot when occasions allow. Also, the magazine assures me that there will be a proficiency program involving refresher training at FlightSafety. We have a commitment to this aspect of general aviation."

"How often do you fly now, and what do you fly regularly?"

"As often as I can, no less than once a week. Our firm has a Cessna 310Q and a Cherokee Six, plus I fly a Beech Travel Air and an Aero Commander 500B occasionally. I also try to maintain my proficiency as a flight instructor."

"Do you think you are ready for this rating?"

"Yes."

"Well, let's see."

Cureton opened the flight manual to page one, where the section on limitations starts, and proceeded to ask his questions.

"What is the Lear 25's gross weight?" "What is its maximum wing bending weight, and what does it mean?" "Describe the pressurization system." "What if smoke is coming out of the pressurization ducts?" "Describe the electrical system. . . ." The questions went on and on as the examiner, literally, went through each section of the manual, page by page.

"Describe what is meant by the departure profile, and define all the terms, such as V speeds, balanced field length and second-segment climb." "Do you know what the required climb gradients must be?" "Let us assume we wish to depart from a 5,000-foot runway at an airport which is located 3,500 feet above sea level and the outside air temperature is 10° C. What is your maximum allowable takeoff weight, and which performance factor will be limiting?" "What is the significance of a brake energy takeoff limit?" "What is the takeoff path distance under the following conditions?" "Show me how you figured your weight and balance."

The questions continued until 12:30 before a suggestion was made that we break for lunch and wind up the oral in the afternoon.

At the restaurant, Cureton was reserved, even a bit distant. An instructor and examiner for many years, he was an old hand at what he was doing, and he had just completed his yearly refresher course in the Learjet at the FAA Academy in Oklahoma City. Without our noticing, he picked up the luncheon check for Cox and me. On his own, he had chosen to fly up from the General Aviation District Office in Columbia, South Carolina, thus sparing us the trip down there in the Lear. He said he frequently flew to RDU for flight tests; there was nothing unusual about that. However, it was obvious that there were going to be no obligations on his part; the slate was completely clean for whatever Cureton wished to write.

After lunch, we finished up a few open items on the oral exam and proceeded to the aircraft. The quizzing continued. "Show me a preflight inspection." "That left angle of attack vane controls which pilot's angle of attack indicator?" "Where are the quick drains?" "Show me the stall warning check, and explain each step." "What happens to the horizontal stabilizer when you do those checks?" It was nearly three o'clock before I closed the cabin door and settled into the Lear's left front seat.

Cureton elected to occupy the jump seat so that he could observe my ability to coordinate cockpit chores with a copilot, who, for our ride, was Cox. Checklist complete, we taxied out to Runway 5 and lined up on the centerline. We were immediately cleared for takeoff.

I depressed the wheel master button on the yoke and held it, thus assuring I would have power steering until I let the button go, and advanced the power levers until Cox called out an exhaust pressure ratio of 2.28. As the airspeed came off the peg at about 40 knots, I had sufficient rudder control to release the wheel master button as 13,620 pounds of Learjet accelerated toward a V1 speed of 127 knots and a guaranteed simulated engine failure.

As Cox called out V1, I took my right hand off the power levers and placed it on the yoke. At the same instant, the combination of deceleration and sudden yaw signaled the loss of the number-two powerplant. Yoke firmly forward and left rudder nearly against the stop, I held the Lear on the runway until Cox called out VR and we rotated to the lift-off attitude. With the airspeed on V2 and a positive rate of climb established, I called for gear up, and we were on our way.

The engine cut on takeoff behind us, I put on the IFR hood and headed eastward toward the Rocky Mount VOR. Both turbines hummed away at 94 percent. Once outside the airport traffic area, we accelerated to 250 indicated and then to 290 as we climbed through 10,000 feet on our way to 15,500.

The first airwork task was to enter a standard holding pattern on the 024-degree radial of Rocky Mount. Planning ahead, I slowed the Lear to the mandatory maximum holding speed of 230 prior to executing a reverse teardrop entry over the VOR.

(Damn, the RMI isn't functioning. Maybe the circuit breaker has popped. Isn't it on my side? I can't remember where it is, and I don't want to hunt for it now. Hell, I have flown holding patterns without RMIs to help me stay oriented; I can do it again.)

The holding pattern entry was lousy. It took me several circuits around the pattern to get situated. What a way to start a flight test. If I had remembered where that elusive RMI breaker was located, or if I had followed Cox's advice to use the copilot more, I bet that between us we could have gotten the RMI on the line.

We broke out of the holding pattern for steep turns. I stabilized on 250 indicated and 84 percent, rolled into a 45-degree bank to the left, added 1.5-percent power and concentrated on maintaining a four-degree, noseup pitch attitude. The task turned out okay. Then we did slow flight and stalls.

After returning for another crack at a holding pattern, Cureton called for a complete engine shutdown, some single-engine holding and a relight. From the holding pattern, Cureton cleared us for an omni approach to Rocky Mount, with a circling approach to Runway 4. While maneuvering during the omni-approach procedure, we again simulated the loss of an engine, which meant the Lear was now constrained

to circle with a maximum of 20 degrees of approach flaps. That automatically placed us under the higher circling minimum of Category D, due to the increased stalling speed with only 20-degree flaps. I executed the procedure the way I had been taught: once we were at circling minimums, I kept the airport on my right and used the copilot, Cox, to concentrate on maintaining visual contact while I concentrated on altitude, heading and airspeed. The procedure worked out well, culminating with the return of the simulated dead engine when we were lined up on final, followed by a touch-and-go. It was not until after the test that Cureton admonished me for not circling to the left, with the airport on my side. He admitted the choice of whether to circle right or left was a judgment call, but, lest I forget, he was the judge.

After the touch-and-go, we executed a missed approach back to the VOR, transitioned to the ILS and shot an approach to minimums. Then, at 200 feet above the runway at Rocky Mount, we declared a missed approach, went to 98 percent, established a maximum rate of climb profile to 14,500 and returned to RDU.

We were light by then, having burned off all but about 1,400 pounds of fuel, and our rate of climb was exhilarating. Even in my somewhat depressed state over the way the flight had gone, climbing from about 350 feet above sea level to 14,500 feet in a little over two minutes had to be the highlight of my flight experience with the Lear.

As we proceeded toward RDU at nearly seven miles per minute, I had just enough time to contemplate how this type rating exercise would end. Cureton's options were all open. There could be outright failure. That would be easy. It wouldn't be like flunking the corporate pilot who did nothing else but fly. There was that other option when an ATP license holder botches up his test: give him the type rating, but limit it solely to commercial privileges. That is like offering the Ph.D. candidate who flunks his general exams the right to obtain a Master's degree. A practical solution but humiliating.

We received vectors for an ILS to Runway 5 at RDU and arrived with an acceptable yet hardly graceful touchdown.

Although the Lear's air conditioning was still pumping out cold air as we pulled up to the ramp, the atmosphere in the cockpit seemed stuffy, almost sullen. I shut down, finished up the checklist items and exited the aircraft. Obviously, it was a time to leave Cureton alone, and as an instructor myself, I knew the anxiety Cox now felt. I waited outside alone as Cureton, emotionless, exited the Lear. But the look on Cox's face as he approached me said it all. I had received an LR-25 type rating on my ATP. Cox's smile went from ear to ear.

Night Mail

BY STEPHAN WILKINSON

THE HANGAR is dark, the night air bitter, the cockpit of the Twin Beech dead and inhospitable. I hold my jacket tight around me and watch Paul Tonnesen carefully run through the checklist, shining his flashlight on the little readout roll attached to his control yoke. "I hope you brought some earplugs," he says—I had—"and long underwear"—I hadn't—"because the heater in this bird breaks down every once in a while."

Too late now. Paul plays some practiced chords on the magneto switches, starter toggles and primer-pump buttons as the left starter wheezes the R-985 over with the asthmatic sound so typical of Twin Beeches—like an old lady sneezing. One cylinder chuffs, then another, and the engine awakens, loping along on three or four, urged on by Tonnesen jiggling the mixture and throttle and poking the primer. The sound booms inside the T hangar while the right engine is persuaded to add its bellow as well. I suspect that starting the engines is probably the hardest part of this job, but I have a lot to learn.

"This job" is carrying the United States mail, for a considerable amount of it—both airmail and first class—travels in the quietest pit of the night in general-aviation airplanes, flown by young pilots no less lonely and alone than were their DH-4-borne predecessors. Despite the fact that airmail stamps invariably show stylized outlines of swept-wing, many-engined jet transports, the letters to which they are affixed are very likely to fly at least partway to their destinations in Twin Beeches, Aztecs, Beech 99s or Navajos. If you're mailing a letter from one major air hub—New York, Chicago, Kansas City, Dallas or Los Angeles, say—to another, the envelope will travel in pressurized comfort in the belly of a Boeing. But if you live in Utica, Knoxville, San Angelo, Casper, Lincoln, Eureka, Fresno or any of almost 250 other cities served by the U. S. Postal Service's network of almost 200 general-aviation mailplanes, the letter will be tossed with thousands of others into the stripped cabin of a Twin Beech, most likely, and humped into place by the pilot, who will then crawl into the cockpit kicking and clawing for toeholds like a spelunker, through the narrow tunnel left atop the dirty orange nylon bags. Though their contents may be marked "Fragile," "Do Not Bend" or "This Side Up," none, it is assumed, are stamped "No Step."

Tonnesen loads his first bags at Topeka, Kansas a little before 10 p.m. The red-white-and-blue mail truck is parked at the far end of the ramp, driven from its usual spot in front of the terminal by the arrival of a local politician in a Convair greeted by a small band and a crowd not much larger. Tonnesen straps the bags down with a cargo net, fires up again and tells the tower he'll be ready to go by the time we reach the runway.

Climbing out over the glowing farmhouses of Kansas on our way to Midcontinent International, north of Kansas City, it occurs to me that some of their lights may have been shining back in the mid-1920s, when the Post Office first tried flying the mail in frail aircraft over many of the very same routes. That attempt, pioneering though it was, was also a disaster for the pilots. Still, those are the days that many fondly recall when "airmail" is mentioned, while the safety record of today's general-aviation airmail seems to be the subject of numerous jokes; when they heard about my intention to fly with the mail, my professional-pilot friends' reactions ranged from mockingly making the sign of the cross over my head to seriously recommending that I not fly in the mailplanes. Reality, however, is somewhat blander: Last year, general-aviation mailplanes flew almost 24,000,000 miles with only 11 accidents and two fatalities—a rate less than half as bad as that of general aviation as a whole.

To the casual, disapproving eye, Tonnesen's Twin Beech looks a mess. "This one's what they call a ten-four mod," he explains. "They chopped four feet off the wings and upped the gross 400 pounds. Tell you the truth, I've always wondered if they could cut 10 feet off and up it a thousand." The paint is worn through to the aluminum in places; the cowlings are oily and dented by decades of being dropped on the hangar floor during maintenance; and the cabin is a denuded, fiberglass-walled refutation of this executive-windowed G model's more glamorous past. The cockpit, however, is efficient and capable. Every instrument on the newly painted, crackle-finished panel works; some of them are new (including an IVSI); none are retired from a B-29; and the avionics are an economical but utilitarian assortment—two King KX 175 navcoms, a King ADF, Narco transponder, Century III autopilot, DME and even an elderly RCA AVQ-47 radar. Paul flies for what many say is the best of the airmail operators—SMB Stage Lines, an Oklahoma outfit that operates 50 aircraft almost exclusively on mail runs—and it is said they cut no corners on the essentials of equipment and maintenance. SMB puts a Twin Beech to work only after completely gutting and rebuilding it. Tonnesen told of one they'd recently bought in Florida, the dealer insisting it was "ready to go right out on the mail run." It took SMB mechanics a week of work before they were reluctantly ready to let it fly. Their reluctance was not unreasonable, for the airplane came back from the first run with a squawk sheet seven pages long.

Tonnesen switches to Midcontinent Approach, and the comforting solitude of our night flight, lit by the drowsy red glow of the old eyebrow lights, is broken by a string of vectors, squawks, altitude changes and traffic call-outs. TWA and Frontier are also on the frequency, but so are Stage 3005, Stage 4005, Semo 4006 and Slick 4002—all of them mail flights, drones converging on the hive. "My wife says that at night, the ghosts and the Twin Beeches come out," Paul laughs. This night, the Twin Beeches—and a pair of turboprop 99s—seem to own the vast concrete plains of Midcontinent, for Kansas City is one of the country's major mail changeover points. The mailplanes bring it in from Topeka, Springfield, Omaha, Grand Island, Wichita and dozens of other points to be put aboard the airliners or other mailplanes, and they leave again for the outlying cities with the bags that have come off the big jets.

Whatever glory might still cling to flying the mail, the Postal Service does a good job of dispelling it at the Midcontinent mail facility: While the bags are being sorted, the pilots drink vending-machine coffee in a room behind a door conspicuously marked "Drivers Lounge." Indeed, the mailplanes are treated as nothing more than trucks with wings, and the pilots do look more like wheelmen than airmen. They are almost all in their twenties (Who else would work from nine at night until four or five in the morning?), and few of the supposed status symbols of their calling are in evidence; no shiny Wellington boots or flight-instructor bomber jackets with pencil holders up and down the arms, and obviously no big sunglasses. The standard uniform seems to be Levi's or ordinary slacks with a work shirt, a quilted hunting jacket or vest, and scuffed Western boots. At one stop, I saw a pilot in his commuter-airline suit of lights, gold stripes on each sleeve, for his mailplane was converted from air taxi to letter carrier in the middle of his shift. He was mercilessly referred to as "The Commander."

"I like this work," Paul says. "I really do. It's good duty because the hours are so regular, and you get used to the late-night routine. My only problem is being able to fall asleep on a normal schedule on Saturday and Sunday nights, when we don't fly."

"Hell, I'm *scared* to fly during the day," another SMB pilot, Ed Carthel, adds. "It's been so long since I've flown in the light, I probably wouldn't know how. You fly at times like this and you know the only other guys out there are mail pilots—or someone lost."

"This is as close as you can get to working for someone else yet being your own boss," Tonnesen explains. Like most mail pilots, he hangars and administers his own airplane at his home base—Topeka—hundreds of miles from his company's office in Muskogee, Oklahoma. "You can go six months without ever seeing a company exec. And you don't have to put up with a copilot, or with somebody in the left seat telling you what to do and how to do it."

"Another nice thing about this job," says Carthel, "is that we've got security. It isn't too glamorous, and it may be the low end of the aviation totem pole in some people's minds, but as long as there's any gas left, we'll be flying. We've got a *job* to do. The airlines aren't giving the Post Office the kind of service it needs, and I think that the Post Office will take more and more of their contracts away and give them to people like us."

The sorting and loading finished, the squadron scrambles at 11:15. R-985s pop and fart up and down the ramp, and the 99s spin up next to their elderly kin with an ease that seems a little contemptuous. A motley string of a half dozen mailplanes trundles to the nearest intersection to be flung into the sky in rapid succession, their lights arcing off like brief Roman candles toward their various destinations. We're up on the mains, the engines howling and the oleos flexing, while Semo Beech is just starting its turn toward Omaha, and Tonnesen has us cleaned up, trimmed, power set and bent toward Wichita in short order.

A broken stratus deck just under us appears filmy and delicate, lit from below by city lights to form other-worldly pools of depthless luminescence. If I lean forward, I can see the river of blue-pink fire pouring out of the right exhaust stack. Tonnesen watches the flame on his side as he slowly leans the mixture, but I have no idea what variations in color or intensity he is seeking. Satisfied, he runs the right mixture back a way as well, and then he turns to me: "How's it look on your side?" he shouts over the roar.

"Uh, fine, Paul, just fine." I peer out with what I hope is a knowing air, then peek at the matched mixture handles and fuel-flow needles and am reasonably content that my ignorance won't torch our valves.

"I cross with an eastbound flight right around here every night," Tonnesen says. "He's out there somewhere—he's already on the frequency—yeah, there he is." A landing light blooms and fades far ahead, ours answers, and soon the other airplane slides soundlessly past, its lights winking rhythmically—wing tips, beacon, wing tips, beacon—making it look like some kind of mystic fish swimming by, gills pumping. Paul recites the other flight's complete route to me, so I ask if he runs into the pilot frequently.

"No, I only met him once, in Wichita while I was holding because of a line of thunderstorms. He wouldn't even get out of the airplane—talked to me through the cockpit window. I said, come on in, I'll buy you a cup of coffee. Oh, no, he said, got to get going. He got as far as Hutchinson and got the shit kicked out of him." Tonnesen evinces a hint of secret pleasure at the poetic justice dispensed to the man who refused to honor the

The Beech 18, workhorse of the night mail fleet.

comradeship of the night sky.

What's the worst weather a mail pilot has to put up with? "In the winter, I like to think it's the summer thunderstorms," Tonnesen says, "and in the summer, I try to convince myself it's the freezing rain. It's the same as any other flying job, actually; I guess the only one you can get where you can avoid bad weather is dusting." Freezing rain is the only weather condition that causes automatic cancellation of mail flights; everyday icing is something they have to contend with. When the Post Office Department started the general-aviation mail flights in January 1967, early operators were even using light singles—until a few disappeared at night and the Post Office decided the minimum equipment requirement would be two engines. Then they began dropping things like Apaches without deicers, and the Post Office declared that not only two engines but full deice capability would be required. "The Twin Beech will hold a lot of ice," Paul opined. "It's probably the best ice machine around. The 99 won't, though."

At Wichita, we trade airplanes, for our final destinations—Fort Scott and Independence, Kansas—are smaller towns with lighter mail loads. The Twin Beech goes westward with Ed Carthel in the cockpit, and we inherit the scruffy, gutted Aztec that Ed brought in. Tomorrow night, the switch will be worked in reverse. Some mail routes are out-and-back affairs or round robins, bringing the pilot home every morning, but many of SMB's routes go outbound one or two nights and back home the next. This night, Tonnesen will stay in the ramshackle roadside motel he frequents, for economy's sake, in Independence. ("We call him The Pilot," the motherly lady in the motel office said proudly the next day when I asked if he was up and about yet.) He'll arrive almost at dawn, for he'll have to drive an hour to hand-deliver a priority package to a Coffeyville, Kansas hospital as well as driving the regular mail into Independence.

Unfortunately, flying the mail is not a land-it-and-leave-it job; the pilot does the loading and unloading himself, and in the smaller cities, he sometimes delivers the mail to the post office as well. "One problem with this job," as one pilot put it, "is that the cargo doesn't get in and out by itself. You have to help it." A Twin Beech carries about a ton of mail, and two full loads a night is normal—one inbound, the other outbound. That means a mail pilot lifts about four tons a night. Plus flying.

Gulfport, Mississippi is hot, heavy with humidity and the threat of distant lightning playing along the horizon. I wait, not without some trepidation, for the pilot who flies the Gulfport-Laurel-Jackson-Memphis-and-return mail route for the Jim Hankins Air Service: That afternoon, when I'd stopped at the Hankins office in Jackson, they'd signed me on as a dollar-a-day mail loader so I could make the trip. "Had a couple of accidents in which right-seat riders got wiped out," the man had said. "Sued us for millions." Great. And I haven't even been paid yet.

If I am waiting for the mantle of age and maturity to reassure me, I shouldn't be waiting for a mail pilot. Jan Gordon, when he appears out of the darkness and sticks his hand out in greeting, turns out to be a smooth-faced young man in his mid-twenties. The Gulfport mail truck arrives just after Gordon does. ("The mail pilot?" the lineman had said in response to my question. "Well, he's due out of here at nine, so he'll show up about 8:58." I couldn't help but remember that SMB's pilots are required to be on duty an hour before their scheduled departure.) We are loaded and off into the low, steamy Gulf Coast clouds soon thereafter, racked into an enthusiastic right turn just after we go on instruments. I grip my seat just a little tighter.

Laurel, Mississippi is Gordon's first stop, and he brings the Beech in with a touch that is reassuringly competent. We sit in the sultry cockpit, the gyros noisily winding down, surrounded by the smell of hydraulic fluid and ozone and Elixir of Elderly Airplane. The mail truck is late, so we talk about mail flying and Twin Beeches and other outfits.

"The trouble with———," Jan says, "is that they think nothing of penciling in a hundred-hour without ever doing one. Their airplanes are in the best shape they'll ever be in when they buy them, and they go progressively downhill." (I find this to be standard procedure among mail pilots. They think their own companies are excellent, admit the competence of a few competitors and are full of scare stories about all the rest.) "And one of the problems of working for ———," he continues in the same vein, "is that you have to call the president of the company before canceling a flight because of weather. He either tries to talk you out of it or says, 'Well, if you can't handle it, I'll have to find me somebody who can.' Hankins leaves the decision entirely up to the pilot." Tonnesen, as well, had told me of one mail operator who had been fined $300,000 for automatically mailing out certificates of six-month-check proficiency to its pilots twice a year.

Unlike Tonnesen, who seems to be a career mail pilot and truly devoted to the job, Jan Gordon doesn't see himself doing it forever. He does enjoy it, though, and points out, "One of the nicest things about the job is the schedule. You're home every night [four or five a.m. being a mail pilot's definition of "night"], you have every weekend off, you hardly ever have to worry about a sudden phone call and the schedule is about as regular as any flying schedule can be." Jan points out that flying the mail also provides a considerable amount of actual flight time per hour on the job: He works eight hours a night and logs five and a half of them. "It's a great time-builder—lots of weather, most of it single-pilot, and the Twin Beech is relatively heavy compared to what most 2,000-hour pilots are flying." The Postal Service requires only a three-axis autopilot in the Twin Beech and smaller airplanes—no copilot. Two-man crews fly the handful of 99s, turboprop Twin Beech conversions, Twin Otters and the three DC-3s that SMB Stage has just put on the line. Federal Express and Executive Jet both operate considerable fleets of Fanjet Falcons, Lears and Westwinds on mail routes as well, and they also require two-man crews.

Jackson is a short flight north, and we park there in the sultry Mississippi night while the mail is sorted among the canvas tubs right on the ramp. Immense roach-like beetles, so big they should have nav lights, bat themselves to death against the ramp floodlights, and inside the open hangar is the real Confederate Air Force—the Governor of Mississippi's starred-and-barred Citation and square-tailed 310. Gordon

Piper Aztecs have often been used for mail carrying.

goes across the ramp to check weather at the FSS and comes back with word that there are thunderstorms between Jackson and Memphis, Tennessee. "They're in a line, they said, but they do appear scattered, and I think we can pick our way through them." I surely hope so.

We are soon cocooned in cloud, the running lights atop the wing tips made even more vivid by the river of rushing fleece in which they glow. The broad, thick wing of the Twin Beech, flat-lit by the lightning flashes, looks reassuring. Lulled by the drone of the big engines and comforted by the solidity of the ship, I regard the now increasingly frequent lightning almost as an amusing diversion. Until a blinding flash fills the cockpit, and we are in it. All I can think of is Ernest Gann's admonition to crank the seat down and the lights up, but I can't even find the seat adjustment. Gordon seems unperturbed, pausing to shout, "This thing'll take on water, more than likely, around the windshield."

The props are ringed with Saint Elmo's fire—a dim, almost imperceptible blueness at the tips rather than the pyrotechnic display that sea stories had led me to expect—and the lightning comes in short, violent snaps, but we don't get into real rain until Memphis Approach is vectoring us onto the ILS. Gordon flies it without an approach plate, but after all, he does it every night, five nights a week; besides, a plate wouldn't be of much use in turbulence in the dimly lighted cockpit anyway.

I call the lights in sight as we break out at about 500 feet and Jan turns on the ancient windshield wipers. They flail about vigorously but ineffectually. One arm is misadjusted, and it slams the cowling on every downstroke, sounding like a broken tire-chain banging around inside a fender. We touch down gently, anticlimactically, and only as we taxi in do I notice that my right sleeve is soaked from the shoulder down by water that has sluiced back along the cockpit sidewall.

Five minutes after shutdown, Gordon is in the employees' lunchroom of the Memphis mail facility, eating spaghetti and meatballs from a cup in his lunchbox and deep in the sports section of a discarded newspaper. We are alone, for other inbounds had canceled.

On the way back to Jackson, we are given superb handling by Memphis Center through what has by now become a nearly solid line of thunderstorms, and, if anything, the trip is easier than the northbound run. Center issues a string of vectors, almost as though they're conning us through a radar approach, and we feel hardly a bump. "That's one nice thing about this kind of flying," Jan says. "There's nobody else up at three in the morning, and Center has time to help you."

The night peters out fitfully, punctuated by distant lightning, as we retrace our steps from Jackson to Laurel, the runway lights growing out of the darkness like an aircraft carrier adrift in a sea of blackness, and finally back to Gulfport, the pile of mailbags diminishing at each stop. Jan parks in his spot on the ramp at four a.m., and though a soft rain has begun, he carefully postflights the airplane, checking for leaks and cracks, seepings and drippings and other marks of the night's combat. He leaves me at a motel half an hour later and drives off with a wave in the little van he uses to deliver the last of the mail to the Gulfport Post Office.

These young mail pilots are looked down upon by a lot of people who couldn't do half the flying job they do, night in and night out, through every kind of weather, with nobody to help them, nobody to make their decisions, nobody even to stroke them with the kind of ego support that most pilots get. Their cargo is unresponsive, their spectators sound asleep. They're driving aging mail trucks with dented wings, and there's nobody around to impress—no stewardess to do a number on, no passengers to remark upon one's gold stripes, no airport lobby or FBO's office to stride through, no fuzzy-cheeked young student in a 150 to watch with awe as the Twin Beech rumbles in, not even a wide-eyed boy at the airport fence.

They and their general-aviation airplanes are doing a remarkable job, though, and as the airlines continue to cut back on flights—especially night flights—and fill their freight compartments with extra baggage that passengers have checked in lieu of baring them for security guards to paw through, the job is becoming increasingly big. It started as a sideline for light air taxis, then was thought to be a convenient thing for commuter lines to do at night with their dormant passenger planes, and has now grown into a specialty operation performed largely by mail-only operators and professional mail pilots. They're flying over 100,000 miles a night, carrying over 1,250 tons of mail a week, and doing a job that would have made even those brave, jodhpur-clad lads in the DH-4s proud.

*Twenty-four years later, Richard Collins rediscovers
a very special airplane he once owned.*

Piper Pacer:
A Nostalgia Report

BY RICHARD L. COLLINS

I'VE HAD A dream. It's about an airplane, and it's been a repetitive dream since 1959. I owned a Piper Pacer that I cherished and sold it in that year, 1959, in the course of raising down-payment money for our first house. The dream has been that I didn't sell the Pacer. It rests in a hangar, like a jilted lover, waiting to reclaim the adoration and attention that has since been bestowed on more modern and fancier flying machines. In the dream, I return to the hangar, walk around and admire the airplane and tell it to be patient, I'll be back.

The dream has come true, at least part of it. I found the Pacer, walked around it, spoke soft and kind words, took pictures, and then we actually flew together again. As we lifted off for the first time in 15 years, I felt a sense of attachment to a

flying machine that I hadn't felt since 1959. Gordon E. Canamore, of Flint, Michigan, might own the body of the Pacer, but its soul is still mine. The Pacer is a classic, and that particular Pacer, N125RC, and I did many memorable things together, things that affected my life profoundly.

The Pacer was born in troubled times for aviation. The Piper Aircraft Corporation (and the rest of the general-aviation industry) took it on the chin when the post-World War II flying boom fizzled after a short period of time. As Piper was working its way back from the bottom, it was still building airplanes that were basically Cubs, and not a wide variety of those. The airplanes that *were* built were built with the same skills used for Cubs, and, in many cases, with materials that had

Above: The cockpit of N125RC, once Richard Collins' home away from home. Collins (right) checks over the Pacer's well-kept logbooks with Piper's present owner, Gordon Canamore.

258   Flying/January 1975

already been purchased when Cub sales virtually stopped. Because an airframe, like beefsteak, tends to cost by the pound, Piper worked at cutting prices by slashing airplane size below that of even the Cub and Super Cruiser.

First came the 1948 Piper Vagabond, a stark, little, two-place-side-by-side with the familiar Cub airfoil but smaller than the Cub by four feet of length and six feet of wing-span. It was powered by a 65-horsepower engine, weighed only 625 pounds and was fairly ignored by the airplane market. So much for the Vagabonds.

Next was the 1949 Piper Clipper, a compact, 115-horsepower, four-place that looked for all the world like a Vagabond with more windows. The Clipper was austere, in keeping with the times, tipped the scales at 850 pounds, and would almost carry its own weight in useful load. The price was right, $2,995, and it started building a new market for Piper.

The Pacer came next, in 1950, and was the ultimate in compact, four-place airplanes. It had more power than the Clipper, 125 horses, flaps to lower the landing speed, 1,800 pounds of gross for a little better useful load and gas tanks in both wings (one of the Clipper's two tanks was in the fuselage, right behind the panel). The Pacer also had more sound-proofing, an attractive interior, a softer landing gear and other amenities that had been left out of the Clipper but that pilots began to want when color started slowly returning to the airplane market's cheeks. The Pacer was also one of the first light airplanes with a vacuum pump to run gyro instruments for IFR capability. (Cessna's 170, the competition, used externally mounted venturi tubes and wind to make its vacuum.) The deluxe 125 Pacer came off the line at $3,795. Empty

Pleasure revisited: taxiing out in a grand old friend.

weight was up to 980 pounds, only 355 pounds more than the two-place Vagabond, and the external dimensions were virtually the same. Piper had taken the Vagabond concept and created a winner. The Pacer was comfortable and capable of cross-country flying, yet it was a small, personal machine that did not overwhelm its owner with size or with cost. The Piper ads said it would fly for bus fare, and it would.

That the Pacer was soon replaced by the Tri-Pacer is common knowledge. The aviators of the nation wanted tricycle-gear airplanes and were led on by advertisements touting the ease of flying them. One ad said that the Tri-Pacer made flying so simple that an eight-year-old girl learned to take off and land in six short lessons. I taught a couple of 20-year-old girls to fly in a Tri-Pacer once, and it took a lot more than six short lessons, but I suppose a Tri-Pacer *was* easier to operate. But the Pacer was a spirited little machine, and the Tri-Pacer had about as much spirit as a grocery cart. Turning the main gear around and adding a nosewheel completely changed the character and complexion of the airplane. I don't suppose most Pacer owners would have traded their airplane for two Tri-Pacers, much less admit that the Tri-Pacer was virtually identical to the Pacer except for the landing gear.

My Pacer was serial 20-615, meaning it was a model PA-20, number 615 of the litter. It bore the registration number N7785K when built and came into our family new. My father picked it up at the Lock Haven, Pennsylvania factory on February 15, 1951 and took it home to Linden Airport, in New Jersey. He used the airplane in his business until April 9, 1955, when he moved on to a Cessna 180 for his primary means of transport and I bought the Pacer for a nepotistic sum. It was well equipped, with a Narco Omnigator, a Bendix LF receiver and VHF transmitter, a Lear ADF-12 with transmitter (making a total of three transmitters), a Safe Flight angle-of-attack indicator, a Javelin autopilot, full panel and other assorted extras. It was, in fact, IFR-equipped, and as soon as I bought the airplane, I set out to get an instrument rating in it. I had been flying four years, had spent a lot of anxious hours chasing scud and was eager to get up in the clouds legally.

It only took a month, and on May 9, 1955, I got the instrument rating. The FAA inspector, a salty one named Al Meyer, insisted on orange glass over all the windows and blue goggles for me, plus low-frequency-range approaches on partial panel. It was a warm and sweaty afternoon in the Pacer, but we pulled through together. When the ride was over, Meyer said that I had done okay and that the airplane was really neat except for its birdcage appearance; he was referring to the wingtip-to-tail antenna for the LF radio.

It wasn't until October 19 that my first actual IFR flight occurred. I was flying northeast, VFR, toward New York and came upon a wall of weather at Morgantown, West Virginia. When I landed and filed IFR, the FAA (then CAA) employee stared down his nose at the jaunty little Pacer resting on the ramp. It sure as hell didn't look like an IFR airplane to him, and he didn't hide his feelings from me—the first of many such experiences.

I was really keyed up as the Pacer punched into the bottom of the overcast; I felt it might all self-destruct; we'd discover that clouds are really solid and that IFR is all a joke. But as it enveloped the airplane, the wetness turned out to be soft, after all, and I relaxed a little and made peace with the environment. The flight was a smooth one; it took only about 25 minutes to get through the weather to clear conditions on the other side. The front had been stationary over the mountains for days and was scheduled to remain so; all the VFR-only airplanes (a vast majority in 1955) would just have to wait for better weather. My Pacer and I went on through IFR. It's of such stuff that buddies are made.

Later that same week, the Pacer and I wound up in some ferocious wind and turbulence over Pennsylvania, and I learned that the airplane handled beautifully in mean winds; better, in fact, than some heavier airplanes I'd flown. The good ride couldn't be attributed to wing loading, for at 12.3, the Pacer's was relatively low. It was probably due to the diminutive overall size of the airplane in relation to weight.

Most of the time, I flew the Pacer solo or with one or two others on board, and it was just as well, because the Pacer was not really a four-place airplane, in the absolute sense of the word. The useful load suffered from all the equipment and was down to around 730 pounds. Subtract 216 pounds for fuel, and that left just over 500 for the cabin, or three and no baggage. The fuel supply (36 gallons) wasn't overly generous, either. At low power settings, cruising around 100 knots, the Pacer had an absolute endurance of almost six hours. That might sound like a lot, but with a headwind, the IFR mission profiles could get pretty short.

My adventures with the Pacer came to a stumbling block in November 1955, when those neighbors I had thought were friendly brought greetings, and I was off to the Army. I started to sell the Pacer before going, but decided to keep it. That was a wise move, for after eight weeks of training in the basics of soldiering, I moved on to the Army Aviation School at Fort Rucker, Alabama. The Pacer went with me, and I found it a home in a T hangar at Dothan Municipal Airport.

I flew the Pacer IFR a lot from Dothan, much to the consternation of the people in the flight service station. One man was particularly bothered by the thought of such an airplane flying around in clouds and rain and would try to talk me out of every flight. Then he'd take the flight plan, get the clearance and would stand outside watching, waiting for the crash. I disappointed him every time.

The Pacer played cupid at Fort Rucker, when I asked a civilian employee of the department of fixed-wing training, Miss Ann Slocomb, if she would like to go for a Sunday-afternoon ride in my flying machine. She had never been flying, and the thought struck her fancy; we were soon darting around together in the Pacer. She didn't like it at first, because of some low-level turbulence, and would actually grab me every time we hit a bump. I never took her flying on a smooth day after that.

The Pacer took me a lot of places in 1956 and 1957, while at Fort Rucker. I often landed on a short sand strip right across the street from the beach at Panama City, Florida. The Pacer's big, fat tires worked well there; airplanes like Cessna 170s, with smaller tires, often had problems. The Piper ad about the Pacer's being as inexpensive as bus fare was right, and when I could get someone else to buy the gas, the Pacer virtually flew for nothing.

Disaster struck the Pacer while I was at Fort Rucker. One warm day in late February, I decided to polish the plane. I was on a ladder doing the top of the wing and felt the fabric tearing when I leaned on it to rub the polish off. I started poking around and found that the fabric all over the airplane was rotten. What a note on an airplane that was only six years old.

Hugh Wheelless, the fixed-base operator, listened to poor trooper Collins describe the plight of his rotten fabric and said that Willie Blanchard would recover it for me, under the supervision of the shop foreman, for $700. I borrowed the money from my father, and Willie, a talented black man of good humor and a true friend, began working hard on the Pacer. He put on extra coats of silver dope that weren't included in the $700 and striped the airplane to perfection. It was painted two-tone green and the registration number was changed to N125RC. I rewarded Willie with several bottles of his favorite potion and continued to do so for some years. When I found the Pacer in 1974, I pored through the logbooks and found that Willie's cover job reflected his talent; it had lasted sixteen and a half years.

My time with the Army ended in 1957, and I went off to Little Rock, Arkansas to go to school and to work as a pilot, flying a Beech Twin Bonanza. The Pacer played a very important role during this time, because the girl I had taken for her first ride in the Pacer was still in Alabama, and the Pacer was my only means of getting to see her. The urge to end such a long-distance courtship became overwhelming, though, so I hopped in the Pacer one day and flew to El Dorado, Arkansas, where old Buddy Paul Rush owned a jewelry store; I made the best possible deal on a ring and took off to transport it to Alabama—in the Pacer, naturally.

It was a dark and rainy night, that trip back with the ring. When I made an IFR position report passing the VOR at Greenwood, Mississippi, the man on the ground recognized my voice and number and asked where I was going in my Pacer on such a black, stormy night. I told him I was going to propose to my true love, which prompted him to suggest that I land at

Greenwood, have some coffee with him and think the whole thing over again. I didn't stop, but he surely planted a seed of doubt in my mind, because I kept the ring in my pocket and my mouth shut the whole weekend. But two weeks later, I got back into the Pacer and flew the ring back to Alabama. Ann would become a full-time crew member.

I flew to our wedding, IFR, through one of southern Alabama's rare snowstorms. The airplane was hidden in the hangar at Montgomery, to keep mischievous friends from tying tin cans to its tailwheel; after the wedding, my father would fly us back to Montgomery in his Cessna, where we'd pick up the Pacer and go on the honeymoon. Those who would have conspired against us went to the airport and asked which plane belonged to my father. Friend Willie Blanchard was there, and he dutifully pointed out a 182—only it belonged to someone else, someone who is probably still trying to find out how and why all that rice got into his airplane.

We moved to New York to start out in the aviation-magazine business in July 1958, and the Pacer went in for more avionics modernization. I wanted more communications capability and bought a "Wright 60" crystal-controlled transceiver. It had 60 channels, a fabulous number then, and was made by a company that was shortly bought by an energetic guy named Edward J. King, who gave his name to the company and developed a few radios of his own. The Wright went over on the right side of the panel, displacing the second VOR I had added earlier. For some reason, the need to talk seemed to overshadow the need for redundant navigational capability.

We moved ourselves to New York in the Pacer, which came to live again at Linden, the airport that had been its home when my father owned the airplane. In October 1958, the engine was exchanged for a remanufactured one at the Lycoming Service hangar. The whole deal cost $1,485.89 back then, when labor was at the rate of $5 per hour.

I only flew that new engine about 120 hours, though, because we had an offspring the next April and planned to buy a house in Princeton, New Jersey. Something had to go. Reluctantly I sold the Pacer to a friend, Rudy Peace, the man who had taught me to fly in 1951. I delivered it to him and didn't look back as I walked away. I guess I felt it had been truer to me than I had been in return. The Pacer had taught me to use an airplane for transportation: It had provided many carefree hours when I was in the Army, had helped me to find Ann, had participated in the wedding festivities and had flown her home to have our first child, Charlotte. I had always lavished loving care on the airplane in return for its favors, but now it was being left in a dealer's hands for sale to anyone who came along.

I kept up with the plane as best I could. Whenever I went to the FAA records branch in Oklahoma City, I'd steal a few minutes with the Pacer's file to see how it was doing. I followed it from Camden, Arkansas, to Medina, New York, to Michigan, where it was owned by a druggist. Then I lost touch and stayed out of touch until I received the following letter from Gordon E. Canamore:

"Hi, Richard Collins. Here's a photo of your old lover. I read your article in *Invitation to Flying* in 1971. Meant to tell you how 125RC is getting along. I'm rather slow at doing things,

and now it's 1974. I bought 125RC in May 1968. She has 3,151 hours total. The engine was topped at 1,027 since remanufacture. In October 1973, I had Ceconite put on her, a new headliner and strobe light on the bottom. I have a Genave Alpha 200 radio in the panel . . ."

I was elated and answered Canamore's letter asking if I could come fly the airplane once more. That led to my reunion with the airplane at Cagney Airport, Clio, Michigan.

The plane was parked in front of its T hangar as I circled to land in my modern-day version, Skyhawk N40RC. After landing, I taxied up, parked the Hawk out of the way and then walked to the Pacer with my son, Richard, Jr. The airplane was beautiful. When Canamore recovered it, he used virtually the same stripe design that Willie Blanchard had put on the airplane in 1957, and even the light green was a close match for its color in my day. The darker green on the airplane now is much brighter than it was, and the number is on the side instead of on the wings and tail, as in 1957. Even from a distance, I could tell that some things had changed: The ADF loop is gone from the bottom, and the artificial horizon that was perched atop the panel is missing, but it is unmistakably my Pacer.

I spent an hour photographing the airplane and chatting with Canamore about it. He had done a lot of work on the airplane himself, and he asked about some things that were puzzling to him. When he stripped the paint off the wheel pants to paint them, for example, he found a thick layer of bright orange paint. What was that? Well, believe it or not, the Government proposed in 1959 that 25 percent of the surface of all airplanes be painted with high-visibility fluorescent paint. I had painted the Pacer's britches to see how it looked and how long it would seem to last. (The answers were terrible and not long.) Canamore asked about the two-way valve that allows the turn and bank to operate either from the vacuum pump or from a venturi, which provides redundant power sources, and about the separator in the vacuum system. He also asked about a sleeve welded onto one leg of the left landing gear. The Pacer shows no logbook record of a scrape in its 24 years, and he was curious about what might have led to the sleeve. I had had it in the shop for a shock cord change once, and the mechanic (who still works at the shop that maintains my airplane) had let it slip on the jack, damaging the gear.

We went back through the log and papers, and I found the records neat and complete. My father had flown the airplane 1,400 hours; I had flown it just slightly over 800 hours. For some reason, I thought I had flown it more than that. Since I sold it in 1959, only 900 more hours had been added to the tach, in increments as few as 35 per year. Canamore flies it about 75 hours a year now, purely for pleasure. He uses it on skis in the winter and on wheels the rest of the year, except for about a month in the spring when it isn't used at all because of soft and wet conditions on the runways.

I started preflighting the airplane and studying the changes. One big change is under the cowling. Canamore actually polishes the metal on the baffles, which makes the engine compartment resemble a highly polished Army kitchen. I thought I used to keep the Pacer in neat shape, but he is surely one up on me there. I noted some baffle patches in the engine compartment that I had fabricated myself. I used to do a lot of work on the Pacer and enjoyed it.

Inside, the Pacer has the good airplane smell it and other fabric-covered airplanes have always had. The panel is still a full one, but the artificial horizon has been moved to the right side. Someone must have objected to the airspeed indicator reading only in knots, and it has been replaced with one that reads in miles per hour. The extra radio I had is gone, and the Genave Alpha 200 is the complete complement now. The front seats have been recovered and are beginning to need it again, but the back seat is original and in excellent shape.

I was the first person other than Canamore to slip into the left seat of N125RC since he had acquired the airplane in May 1968. He was being generous and, at the time, didn't even mention the fact that he had liability insurance only—no hull. He was really letting me fly his airplane.

I reached for the master switch and starter, just as if they were under the seat on all airplanes as they are on the Pacer, and didn't even have to feel around to find what I wanted. The 125 Lycoming needed a little more prime before it would start, but it soon came to life. When it balked, I thought of the day in December 1958, when it wouldn't start at Linden and one of the line crew propped it to start us off on our longest day in the Pacer—from seven in the morning until midnight, 18 hours' total time en route, Linden, New Jersey to Little Rock, Arkansas. I remembered talking to Claud Holbert, who flew Winthrop Rockefeller's Twin Beech from New York to Little Rock that same day. They were about 12 hours en route in the fancier airplane; the southwest wind was very strong.

I remembered, as I taxied, that the Pacer was not too strong on braking, and also remembered its better-than-average visibility for a conventional-gear airplane. The controls seemed to fit my hands and feet perfectly, and as I stopped at a run-up position, I asked Richard, Jr., in the back seat, what he thought of it so far. He grinned approval.

Into position, full throttle, the old familiar sounds and control pressures began flowing back. Run just a little, raise the tail, stay lively with the feet, and lift off at about 60, with the light (half fuel) load and one notch of flaps. The Pacer flew eagerly away from about halfway down the 1,850-foot-long runway, and I enjoyed the light and responsive controls, which seem to have been lost with progress in airplane design.

The Pacer's ventilation is excellent, but I remembered that the heating was poor. I froze my toes in it during the winter, and Canamore said the heater had not improved with age. The noise level in this airplane is higher than in new ones, but not markedly so. Vibration is where there's a great difference. The Pacer has a steady and rhythmic shake that new airplanes lack, mostly due to progress in building engine mounts. The indicated speed was 110 miles per hour, with the engine turning 2,200 rpm. That's what I use to run it, and that's what Canamore selects. It's really not much power, and the reason my Skyhawk is considerably faster than the Pacer is because I

wind it up to 2,500 rpm down low and 2,700 rpm up high. I started to ask Canamore if we could match the Pacer against the Skyhawk in a full-throttle run but decided against it. He would have probably agreed but would much rather not lay the lash to his engine. The outcome of such a race also might have prompted me to make a financially ridiculous swap-offer.

I had often wondered if the Pacer might not seem crude compared to new airplanes. In a way, it did, but in most ways it did not. The Pacer's handling qualities are actually better than those of most new airplanes—both its descendants and the competition—with a good, snappy roll rate, effective rudder control and rather light elevators. In the time I flew the airplane, I remember only one time when I thought control effectiveness was running out. A strong southwest wind was pouring over the mountain south of the Williamsport, Pennsylvania airport one fall afternoon, and, after two passes, I finally retreated to an airport in a calmer spot.

The Pacer's good handling qualities rewarded me with perfect landings in 1974, even though I hadn't flown the plane in 15 years and haven't flown that many conventional-gear airplanes recently. The landing conditions were good, with a light south wind blowing down the runway at Clio, but after shooting four or five landings, I felt ready to fly the airplane again in winds up to 35 knots—which is very close to the limit for modern tricycles.

The Pacer's stubby wings can contribute to high sink rates on approach, and I had to use power on my first approach. The flaps aren't overly large, but the approach can be quite steep, and I recalled having to use a lot of power a few times to arrest an overly enthusiastic sink rate. In fact, I touched (smoothly) with full power one afternoon on the Wolfeboro, New Hampshire airport. The runway was but 1,500 feet long, the approach was made at minimum speed, and the Pacer developed a terrible sinking spell just as the trees at the end of the runway were cleared. Power was the only answer and it worked.

As I flew around over the Michigan countryside, I remembered other days. Looking at the little, unheated pitot tube beneath the wing, I was back to the time when I got a good load of ice on the Pacer, over the mountains of Pennsylvania. The airplane held on for me until I reached a minimum en-route altitude and could descend into warmer air. I remembered an approach to minimums at Richmond. Richmond was the alternate—the destination had gone to zero-zero. When I landed, an airline pilot waiting to take off made a rather uncomplimentary crack about the quality of planes on the ILS course that misty day; I thought bad things about him and went on my way. It *is* a hell of a good airplane—capable, friendly and true. I thanked Canamore for the opportunity to fly my old friend again and for showing it to my son. And I made him promise me one thing: I get the first call if he ever decides to sell the Pacer. Don't forget that, Gordon, *please*.

STEPPING STONES

At first, you were hungry vets selling piston-flight swiftness to eager executives.
Today, you are pros handling turbine
fleets in industry's service. You've come a long way, corporate pilot.

BY GEORGE C. LARSON

YOU CANNOT PIN a year on any of this, because what we are talking about here is an era, and not even about chronological time so much as about an attitude. We are talking about a time when the movers and shakers in this country first began to find their way around in their own airplanes. The brash ones were leading the way, while the more conservative types still took the train.

It is a time when the trains, some of them, are still pumping steam, and the railroad men, filled with the holiness of their profession and the hypnosis of parallel rails, proclaiming the diesel locomotive the very symbol of anarchy—while the railroad's treasurer muses over maintenance costs and fuel efficiency with a slide rule, already composing the speech he will make to the chairman.

The Bell XP-59 Aircomet has flown, although not winningly, and Kelly Johnson's Shooting Star has Lockheed humming. Fathers commute home from New York City with the *World Telegram* under one arm and tell wide-eyed youngsters about airplanes that are "jet-propelled." Still, on Memorial Day, it is the distant thunder of reciprocating pistons that those same young ears thrill to as the wings and flights and squadrons form up to pass in aerial review before dignitaries on battleships. At the candy store, the vendor slips a kid a packet of bubble gum, the kind with flipcards depicting airplanes ("Friend or Foe?") of all nations. The cards compete for attention with the newly formed Little League and with the latest Ford and with V-8 engines. A neighbor drives home in a brand-new Oldsmobile, with something called Hydramatic Drive.

Professional football is still looking for profitability. And television? Too expensive. Better wait a while, until they get the bugs out of it. Truman has surprised everyone by remaining President, and for the first time, dinner is allowed to be interrupted on a Sunday evening, while friends are gathered for the somber but frenetic rat-tat-tat delivery of Walter Winchell, who sees hellish consequences in the military tensions of postwar maneuvering. But not to worry. We have the A-bomb. A for atom. For America.

The bomb would be delivered in a B-29, or perhaps a B-50. Curtiss-Wright knows the secrets of how to squeeze the very last bit of mechanical energy from the heat of combustion of gasoline and air. The oil companies, heroes of the big war, have enough tetraethyl lead in aviation fuel to generate octane ratings that are almost double—145—what they were 20 years ago. The turbo-compound engine reigns supreme. Only those pilots who flew from the Marianas know of the holocaust a gasoline engine can create, of the dulling fear that made each cockpit crew member clench the throttle handles for the pulsation that marked a backfire—a backfire that would light the magnesium in the engine; a backfire that would turn the B-29 itself into a flaming thermite bomb.

The turbo-compound engine is in the Stratocruiser and the Constellation and the DC-7. No engine could possibly surpass its efficiency. Even the last of the exhaust flow has been put to work, to run turbines that are geared down to return torque to the crankshaft. A reciprocating engine is the very spirit of America; in Hollywood, they use the working piston as a symbol of the productive might of this "sleeping giant" that awoke from its torpor and slew the evil weavers of the Axis. We can trust a piston. We can hear it at work. We can understand its timing. It turns a propeller, and the propeller moves air, like a big fan. (The turbojet? Oh, some sort of folderol about Newton and "action has an equal and opposite reaction.")

So it's DC this and DC that and Connies and Convairs. An airport smells of gasoline and blue oil smoke, the same oil that drips from honest round engines and leaves pools on the concrete, which becomes black with the stuff. Starting an engine calls out the fire guard in the event of backfire. The huge pressure carburetors, as big around as a lavatory sink, can go off like Roman candles when the mixture burns too slowly and climbs back up the intake. But we believe in the working piston.

And there are plenty of these kinds of machines to be had. The Western deserts shimmer in seas of aluminum, some of it with only ferry time from the factory on the air-frame. Production had continued to the bitter end, you see, because there was always this doubt about Japan's willingness to surrender. So there is a vast pool of bombers and transports that have gone begging. Back home, under the front-porch flag, there is piloting talent swinging in the hazy summer heat, scanning the want ads and looking for the promised postwar big money. It is a time of heroes, but the public is longer on talk than on material reward. You've got to hustle. There's something out there waiting for the right person.

The oil companies are going like gangbusters. Price wars. Can't drill for enough of the stuff. ("I don't care if you have to drop to nine-nine a gallon, I want 'em *beat*.") The big guys have to be everywhere. If Standard doesn't do it, Texaco will. If Republic doesn't take that, U.S. Steel will. ("I don't care what it costs, I said get me a real *airplane*.")

You've been there, and you know how to swagger. Walking in and selling an executive is kind of different from sneaking through clouds of -109s, but he's a good ol' boy, and by the time you walk out, his eyes are wet with the knowledge you've imparted of what he can do with a Twin Beech or a DC-3. American Airlines has been flying ILS systems with the Sperry Zero Reader into unheard-of ceilings. ("I don't care what it costs . . .")

You can spend $100,000 and know you'll have the ultimate. A DC-3! Good God, that's an airliner! And there you are, in the left front seat. You know how to mix the Man's favorite drink. You've got him buffaloed. He thinks you're a genius, and one night on the way home from Syracuse, he brings his girlfriend along, and you know just when to dim the cabin lights. Later, they come up front for a cigarette, and he starts to tell her all about how everything works and gets some of it wrong, but you just grin and note in some secret corner of your mind what *she* drinks. Got to restock the filter cigarettes, too.

You're not all that well liked in the company. There are people who complain, "Hell, we can't even get a typewriter, and he's getting a budget of twice the president's salary to run an airplane?" But you spend more time with the boss than they do. It's the right kind of time, too. He likes this new airplane. No more waiting in line. No more sleepless nights on a train or in a limo. You fly 500 hours a year, and he counts it all as time you've saved him.

You keep up on your trade. Your competitors are hiring people just like you to set them up in the same kinds of operations. The surplus airplanes are still selling for 10 cents on the dollar, and there's an ocean of spare parts. The airlines are flying the same engines, and each year their mechanical record gets better. That hydraulic system bug got fixed for good. They've come up with a new power-setting schedule that's supposed to prolong engine life umpty-ump hours. The copilot knows what it is.

Oh sure, there's a copilot now. He's even nicely broken in, getting all the ice and stuff as you used to do. Empties the john. At the last stockholders' meeting, some woman made a fuss about the flight department budget, but in an all-male audience, there was just a bunch of polite chuckling all around. No threat. ("I don't care what it costs . . .") One of the members of the board made a stirring speech about the competitive pressures of the business, about the need for industrial "security" that only the company airplane could provide. Why, only last month, the airplane carried a courier with some priceless competitive information.

That Christmas, you get a nice raise. (The copilot had been agitating about how you ought to have an instrument rating, so you fired him and hired an airline pilot who'd been furloughed. He's carried you in some pinches, but he knows his place. What you lack in know-how, you can make up for with flamboyance and office politics.) Another year has gone by, and strangely enough, the airplane has logged just about 500 hours again. These executives seem to have just so much time they can spend away from the home office. If they want to reach farther out, something is going to have to pop, and it will probably be the company wallet. ("I don't care what it costs . . .")

You're beginning to get restless with the DC-3. Learstars and Howards and those glorious converted warplanes are beating your block times. The DC-3 is reliable, which counts for a lot (and it will, in your memory later on, seem more reliable than the new jets because it has so few systems to care for), but you haven't helped the cylinder problems by tightening the cowl, even though it bought you a couple of

One of the big stepping stones in the development of corporate flying was the advent of the Convair 240, meant mainly for commercial transport work but easily adaptable to more private needs.

The Grumman Gulfstream I (left) was for a while the supreme biz plane, until pure jets took the thunder from the turboprops. One such as been the Israeli Aircrafts Industries 1124 Westwind (right).

knots. You remind yourself to check out the new Convair. Think about higher altitudes, about pressurization, about smoother flying conditions and fewer weather problems. This scud-running has a way of catching up with people, and you'd be just as dead at 150 knots, even with the sacred DC-3 under you. The airlines have been fooling with weather radar that could actually see thunderstorms and let you steer around them, the copilot has told you. He has begun to act a bit cocky. On a brief vacation, you check into a flying school out west and get a quick instrument rating. Hell, it wasn't hard; just some brushing up. Once back on the job, you begin to see how handicapped you've been by your fears and your equipment. ("I don't care what it costs . . .")

Basically, you still fly by eyeball and seat of the pants. The decision about whether to land at a short field is made by having a look at it and getting a gut reading on how the airplane is flying today. You've had some close ones. The boss once pulled up with four guests and a pick-up truck loaded with 500 pounds of fresh fruit for the folks back home. You'd gassed up for one passenger and luggage, but he wanted to go *right now!* It was close. Weather was more a challenge to your memory and experience than it was to your knowledge of instruments and electronics navigation. You had dual ADF. Could anyone ask for more? IFR work was manageable at these airspeeds. Of course, if you worked into something hotter, it might be a different story. Of all the systems and machines in the flight department's property records, it was *you* who made the difference between getting there and not.

You are the king of the hill. You had survived much worse in the war, and you had a survivor's instincts and a survivor's confidence. Near-misses only confirmed what you already knew: that you were blessed, that you just *knew*, that's all, you just knew. That's what you told them when they asked.

Once, you were called upon to trek across the water and take the boss and a carefully selected handful of key executives on an international circuit of considerable importance to the company's growth strategy. The copilot, not you, had worried about survival gear, so you had packed a couple of rafts. It had taken weeks to figure out the logistics, let alone the paper-work, for the trip. (The term "Third World" had yet to be

The Howard 500, for a time, prima donna of the corporate fleet, with two deep radials that sang a song of plenitude.

coined, and an American in an American airplane still got white-glove service. Those in charge turned out to be real gentlemen. They had respect for the dollar and knew power when they saw it.) You had done it and brought them back, but it had been a strain. Almost as much of a strain as the paper-work the company was beginning to require of you in the management of the airplane. Couldn't they understand that you didn't have a degree in business, that you were a *pilot?*

That year, there was good news and bad news. The good news was you got the Convair. The bad news was that the week after that, the board chairman had heard of Grumman's plans for the Gulfstream I. It was to be as roomy—well, almost—as the Convair, and it would have jets—well, almost. The whine of the turboprops was becoming a persistent, nagging wail, as bothersome to you as the whistle of the blower on a diesel was to the steamheads on the railroad.

How did these turbines work, anyway? You and your cronies talk over coffee about the new stuff. Nah, it'll never make it. The turbo engine is just too thirsty. Fuel consumption at lower altitudes is incredibly bad, and you have to flight plan like some kind of Einstein. In the Convair, you can drop down without any fuss and even hold for a while and not get too worried. Besides, there will be only brand-new parts for Grummans. You can get anything you need for a Pratt & Whitney at any airport in the country and also find a mechanic who can either fix it or talk to it in the right language. The company executives invite you upstairs, ask your advice. They're willing to go turbine if that's what you want, they volunteer (and you can tell, deep down, that they're really telling you it's what *they* want). You squirm. You are concerned about the costs—and about the check out. Even some of the airline people are down on jets, you say. If a Constellation was good enough for Eisenhower, it was good enough for you.

Big piston jobs seemed like the way to go; to the big airline stuff that was still plentiful. At National Airport, down at the North Terminal in Washington, at the very center of power, you had seen how magnificent it could be if you had the dough and the commitment. The 1254th Wing, 98th Squadron flew all the princes and potentates out of there. Beautiful Connie 649s and 749s (with radar). There were pull-down bunks, like on the trains. Everything was scheduled, controlled. If you were early, you'd throttle back to make your block time. You always shut down the outside props and swung the thing around on momentum to avoid too much blast, but sometimes a little oil ended up on a white jacket. Two stewards: one in cummerbund and tie and the other a master chef as well. The stewards were known internationally; one had given a well-known VIP his Scotch on the "rocks" by pouring the booze over some boiled stones. There was a vibration pick-up in each cylinder to complement the engine analyzer; the pick-ups weren't that reliable and you'd lose a couple on each flight, but they'd chart the valve closing noise and the injectors, too. The airplanes were pressurized giants, with Doppler nav and loran for long-range navigating. Yessir, that was the way to go, with red-hot stacks glowing in the night and the reassuring blue exhaust that told you more about the mixture than the needles did. With the beat under your feet and in your butt, the steady pulse of pistons working in closed cylinders was better than prayer.

But the turbine is refusing to listen to you. On the next visit to the president's chambers, he's not asking you, he's *telling* you to get him into the right turboprop. ("I don't care what it costs . . .") It's a rotten turn of events, but at least the things fly at reasonable speeds and use propellers. Maybe you can get used to it. They find a buyer for the Convair in South America. (Although you don't know it, they had almost found a pilot to go with the turboprop, but the boss, who still likes you, kept you in your seat.)

In 1976, you look back on that era, that attitude, and it all seems dimly quaint. Too much has changed, in your opinion but you are still a survivor.

The two jets in the company hangar have spent more time overseas this year than in the U.S., and the new president is now *telling* you to get in on the ground floor when the intercontinental business jets go into production. This time, what he doesn't know is that the whole thing was really your idea, that you planted it in the right ears and knew it would get back to him. Best to let him believe he thought of it first. Wait a decent interval. Give him your presentation. "No, Sir, the 727 isn't up to what we'll need in terms of fuel capacity, especially westbound, even though it has the three engines you mentioned. I see you've been reading up on this. Ha ha. Let me show you one of our typical mission profiles and we can talk about what the ideal airframe would be . . ." Look at yourself, and listen to what you're saying. You've become a manager. Yes, it's almost surprising, but it's you who run the parts pool and sit on the endless panels to determine how the NBAA will respond to the noise legislation.

Looking back, are things really so different? Last year, the airplanes flew—you guessed it—about 500 hours, but now, they travel at three times the speed of the DC-3 and reach out farther. The weather you work in is no worse, though the forecasting is (all the career met men seem to have disappeared or to have moved to where they no longer know what they're doing). What your money bought you is *comfort,* but more important, it bought you the ultimate in hassle-free mobility.

You can still recall the day the chairman of the board walked up the stairs to the first jet, a brand-new JetStar—yes, you were one of the first in those, too, and you sold *him*—it was a new ball game. He had looked inside the door at the cabin, which was smaller than the Convair's and DC-3's and had nearly knocked your head off. "How do you expect all of us to squeeze into this little thing?" he'd bellowed. Five minutes after takeoff, you knew you'd won when you saw the expression on his face as you walked back to the cabin.

The two jets you fly now are here for him, and he's getting what he wants. You'll solve the systems problems and get the maintenance and parts squared away sooner or later. His next airplane will be just like this one, designed and purchased on paper without so much as a minute of demo flying. It's there because the numbers are right. They say the Grumman G III could run $10 or $11 million.

The other day, you were walking out to the rental car as dusk fell and a DC-3 rumbled overhead on its way down an ILS. There is still nothing quite like that sound. This airport reeks of kerosene now, you suddenly realize. Jet-A is up to where it's almost as much as gasoline, and you can't help but wonder whether you would have been right to stick to your guns back when the question of piston-versus-turbine had been hot. The turboprop certainly didn't seem to last long, but now, with the fuel crisis and everything, some of the rules seem to be curving back around again.

You could, you think to yourself, with very little brushing-up, climb into that DC-3—or one of the few On Marks or Howards or Learstars—and show these kids a thing or two. You had been forced to change and grow with the times, and you succeeded. Those who hadn't been flexible had gone under. The companies that misjudged what really was wanted had simply never understood what was going on.

PART VI

Technique

Consistently over the years the techniques of flying have been matter of vital concern to the Editors of *Flying* Magazine. In most issues, discussions about technique have taken up from a third to one half of the editorial content, being presented in feature articles and in various departments and writers' columns.

As the sampling of articles in this section may suggest, stories about various elements of good flying methodology need not be dull and may well, in fact, be fascinating and even amusing.

There are some selections that have a double notability.

"How to Do an Outside Loop," contains solid advice—for those few at the time who really intended to do an outside loop—from the great stunt pilot, "Tex" Rankin. One wonders how many *Popular Aviation*

readers scored points at cocktail parties describing how *they* did outside loops—on the basis of this article.

More familiar but perhaps far more important is instruction on how to coordinate one's turns as given by Wolfgang Langewiesche, whose classic *Stick and Rudder* was still fresh on the bookshelves. And it is particularly fitting that when William Peter Blatty, author of *The Exorcist,* did a technique piece for *Flying,* it was titled "Ghosts in the Cockpit."

Also represented herewith are three of *Flying's* most popular writers, Frank Kingston Smith, Robert Blodget and the current Editor of the magazine, Richard L. Collins, who shows in "A Day in the Life" that just flying straight and level can be as challenging as attempting an outside loop.

How to Do an Outside Loop

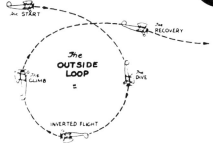

Two Famous Record Flyers, Tex Rankin and Phil Love, Explain Some of the Intricacies and Sensations in Performing the Most Difficult Maneuver in Aviation.

HOW are outside loops made?" is a question frequently in the minds of those interested in aviation. Here's what J. G. "Tex" Rankin, who recently made 19 consecutive loops, for a new record, has to say:

"There are no rules regarding the establishment of a record for outside loops, so when I set out to 'try my luck' the local contest committee, consisting of Captain French, Lieut. Smith, Lieut. Clark, and Vance Breese (who were appointed by the President of the Portland Chapter of the National Aeronautic Association), made up a set of rules covering such a contest.

"The rules in part read that the plane would have to start a dive from a level flying position without first stalling the plane. The motor could be on or off. The pilot would continue in the dive, pushing the stick forward until the plane was on its back, then continue pushing the stick forward until the plane came out level on top in a direction not more than thirty degrees off from the direction in which the dive was started, without stalling or falling off to either side, but continuing a straight flight. There should not be more than two minutes interval between each loop.

"Although I made thirty-four outside loops in a total elapsed time of 50 minutes without stalling or falling off on any of them, I was granted only nineteen by the committee. This was due to the fact that I came out slightly more than thirty degrees off on the top of fifteen of these loops, which was caused by the difficulty I experienced in keeping my feet on the rudder—there being no straps provided for this purpose.

"I found that my best loops were made by starting from level flight with the stabilizer set clear down to make the ship nose-heavy, with the engine wide open, and making a dive of 3000 feet around an arc of 1500 feet, and coming up the other side in the same manner. My net loss on altitude was about 500 feet. I also found that if the ship was pulled into a stall before starting the dive that it usually resulted in a sloppy loop. This also occurred if I did not allow the plane to come up high enough before leveling off. In other words, if I dived 3000 feet and

tried to level off with a new loss of 1500 feet in altitude, it would cause the ship to roll on top of the loop. I experimented on outside loops for several weeks before attempting to establish a record, gaining much useful information from these experimental flights.

"I carried a parachute, but no other safety belt than the regular standard one that was already provided on the ship. I experienced nothing unusual while doing these loops, and I feel that anyone with sufficient flying experience—and particularly with experience in inverted flight—would have no difficulty in making as many outside loops as I did, or even more, providing of course that he be in good average physical condition and use a strong, well-made ship."

The outside loop is probably the most exciting and hazardous stunt known to flying. It is so called because the pilot is on the outer rim of the loop during the maneuver instead of on the inner side as in the ordinary loop.

Lieut. Phil Love, flying partner of Colonel Lindbergh's on the air mail run, has the honor of being the first to do an outside loop with a passenger. The plane used was a tapered wing Waco, powered with a J-6 seven-cylinder motor, but the ship was not quite standard in every respect. An auxiliary tank was built into the landing gear, in order to furnish gravity flow to the carburetor during inverted flight. In normal flight this tank is filled by gravity from the main tank but when flight is inverted a check valve stops the flow back to the main tank, allowing the gasoline to run only to the carburetor. All of the oil breathers are taped up and the main tank is vented through a long pipe to the landing gear so that when inverted the top of the vent is above the main gas tank level.

Shortly after landing from the loops Phil gave a report on his experiences. He said that there was very little experience while actually undergoing the loop itself.

"I have experimented with several different ways of executing them," he continued, "I have tried stalling into the initial dive both with power on and with power off, and with different stabilizer settings, but I find that the most satisfactory way is to set the stabilizer very slightly nose-heavy, and to fly the ship into the dive with power on from cruising speed.

"It takes quite a bit of determination to shove the stick forward until the nose is straight down, and then keep shoving it forward until the ship is horizontal on its back. The execution of the remaining portion of the loop is the most difficult, for after the nose has risen above the horizon there is no reference point to level the ship by, and it is at this point that the pressure tending to throw the pilot out of the cockpit is greatest. In addition to this centrifugal pressure which is aided by gravity it is necessary to apply no little amount of forward pressure on the stick to force the ship up and over the top.

"I find that going up the back side of the loop the ship has a tendency to rotate and care must be exercised to prevent this. The horizon is visible out of both sides of the cockpit, but I have never been able to tell much about the attitude of the ship by looking at it in this position. I find it more satisfactory to make sure that the ship is absolutely level at the bottom of the loop while the horizon is visible and relative to the wings, and then to shove the stick exactly straight forward to prevent

turning with the ailerons, and carrying just enough rudder to counteract 'torque.' If this is done correctly the chances are that the loop will be completed with the wings level, and the ship flying in the same line as when the loop was started.

"The Waco Taper wing does its best loops starting with power on at about 100 miles an hour and about 140 miles an hour at the bottom. The diameter of the loop is about 1000 feet, and very little altitude, if any, is lost on completion of the maneuver."

"After completing several loops in close succession I found that the rush of blood to the head causes an extremely warm feeling about the temples, and if one isn't in good condition he is apt to be a little bit 'woozy,' but this soon passes. Ordinarily upon landing the pilot will find that his eyes are slightly bloodshot, but no after effects are noticed unless too much speed was gained in the dive, and an attempt to change direction too fast was made. It is this changing direction snappily that throws the tremendously excessive strain on both pilot and ship."

During Lieut. Love's flight an ordinary safety belt was used, the only added precaution being a heavy rubber band to hold the clasp from opening accidentally.

Tex Rankin seated in the cockpit of his Great Lakes.

KICK THAT BALL

Do you center that ball in the glass tube with rudder or aileron?

BY WOLFGANG LANGEWIESCHE

THAT little ball in the curved glass tube is the simplest of all aircraft instruments, and perhaps the only one that approaches perfection. It has no mechanical troubles, requires no connections, is without instrument error, needs no corrections, is not affected by temperature or altitude or ice or bugs or little boys who tamper with your cockpit—it always works. And it has another very good quality—it is direct. Many of the more complicated instruments use an indirect sort of approach; they merely hint. The altimeter measures altitude by measuring pressure; the air speed indicator measures air speed by measuring the pressure difference in the pitot tube; the tachometer and manifold pressure gauge merely allow you to guess at your power. But the glass-encased ball always gives you a direct, straight answer.

But you've got to ask it a direct, straight question. Don't call it a bank indicator and ask it about your degree of bank. The ball can stay in the center at any degree of bank and, if the airplane is correctly flown, it *will* stay in the center. In a sufficiently strong, fast airplane you can do a complete barrel roll, still keeping the ball centered. Often in flying the wrong word leads to the wrong idea—and the wrong idea leads to faulty flying. Therefore, don't call it a bank indicator or a ball-bank; call it simply "the ball."

What is the question the ball can answer? Some pilots think the ball shows whether one wing is *lower than it ought to be.* In straight flight, then, it would show whether one wing is low. Some pilots might add that in curving flight it would show whether the bank is too steep or too shallow "for the amount of turn." But this is asking the instrument a question it can't answer—moreover, one that may lead to confusion.

The ball can tell you only one thing: *whether you are slipping or skidding.* When the ball is off, it does not mean that one wing is low, but that the nose is pointing to one side or the other of the actual flight path. And the way to get the ball back to center is not to work the ailerons but, by rudder: *kick* that ball!

This is a basic proposition associated with the art of flying. It should be understood. First, consider what happens when

From left to right, this ball bank says that you are in straight, level flight; making a correct left turn; making a skidding left turn (too much left rudder), and a slipping left turn (not enough left rudder).

the ball is off center during straight flight. Suppose the ball is off to the right. If you look at your wings you will find, true enough, that the right wing is down. But now, just to convince yourself, lift that wing with your ailerons while leaving your rudder pressures unchanged. What happens? The wing comes up but the nose now swerves off to the left and the ball is still off-center to the right.

It's what you should expect. If the airplane was at an angle of bank, to begin with, why didn't it turn? An airplane always wants to turn when banked—that's another of the basic principles of flying. Hence, if you were flying right wing low and did *not* turn, possibly it was because you were holding left rudder. Or, perhaps, it was because the spiralling of the propeller slipstream gave you the same effect as holding left rudder. Thus your right bank was trying to turn the airplane to the right while your left rudder was trying to swerve it to the left. When you level your wings, your rudder is still over and, with no force pulling the plane to the right it swerves to the left.

Now, just *kick* that ball. If the ball is on the right side, a little more pressure on the right pedal than on the left will roll the ball back to center. Here the airplane starts to turn to the right, for its right wing is still low. Now, you may not want to turn, in which case, level your wings. The airplane then will fly straight, the ball centered. But note that this one simple change—different rudder pressure—promptly put the airplane into a normal flight condition. Of course, in actual practical flying you center the ball and level the wings simultaneously. But the way in which you think about it makes a big difference in your flying.

Observe what happens in a turn, say to the left, at 45° of bank. The ball is off center to the right, or on the "high" side. What exactly is wrong? How would you correct it? The rate of turn depends on the angle of bank and the amount of back pressure on the control; the bank is always right for the turn, because it is what causes the turn. Steepening the bank will not return the ball to center. This can be proved if you change rudder pressures and simply steepen the bank without, at the same time, increasing the back pressure so as to hold altitude. You will finally have to give up because the angle of bank, "g"-load and loss of speed will become excessive—and the ball will *still* be on the high side.

The ball stays on the high side simply because the pilot is holding bottom rudder. He is yawing the airplane to the low side, thus making it slide sidewise through the air toward the high side—a good way to spin out of a power turn, incidentally. The remedy here again is to *kick* that ball! Add a touch of right rudder pressure or release some of the left rudder pressure. The ball will immediately go to the center and stay there.

Whenever the ball is off-center it indicates one thing and one thing only—wrong rudder pressure. The rudder was put on the airplane not to make it turn but to allow you to kill slip or skid, i.e., sidewise slithering through the air or, on occasion to allow you to cause skid or slip deliberately. Its main purpose is to counteract various effects which might cause such sidewise slipping flight, primarily engine torque and aileron yaw. Sidewise-slipping flight indicates that you are *not* using rudder when you should, or that you are using rudder when you should not. In any case, the remedy for an off-center ball is rudder pressure—more rudder on the side toward which the ball has rolled.

A smooth pilot never actually will kick his rudder. But it is his foot, not his hand, which should react to the ball. As a simple rule, it may help to think that your foot is kicking the ball *away*.

When you fly entirely by "feel" you use the rudder to maintain lateral balance. When the airplane's attitude feels wrong to the "seat of your pants," you don't change bank, you simply work the rudder until the airplane's seat is squarely and firmly under you. This is why a ball on the instrument board *can* be actually detrimental, and why many instructors don't like one before their students. If you think of the ball as a "bank" or "wing-low" indicator, then in a skid, your "feel" tells you to do one thing, use the rudder, while your instrument tells you to do another thing—use the ailerons. This could result in confusion and rough flying habits. If you accept the ball as an indicator of slips or skids, which call for corrective rudder action, then a glance at the instrument tells you the same thing your senses should tell you, only more accurately.

At this point someone will ask: "If, in instrument flying, you control the ball by ailerons and the needle by rudder, why not in ordinary flying?"

There are several answers to this. First: instrument flying is a special case because you can't see your bank. We don't *actually* control the ball by aileron and the needle by rudder; we only pretend to. In pretending to kick the needle over, we actually kick the ball to the side; then, in getting the ball centered, we bank the airplane. In stopping the needle by rudder at the desired point, we kick the ball off center; and finally, in centering the ball, we keep the bank from increasing. It is merely a round-about way of flying.

Secondly, the needle-by-rudder way of instrument flying is more appropriate to the older airplanes than the newer ones and to slow ones than to fast ones. Many modern airplanes are flown on instruments by controlling the turn indicator hand with the ailerons, blending in rudder as necessary to keep the ball centered. In short, they are flown by a normal technique.

Finally, the present trend in instrument flying is away from all mechanical systems which control specific instruments with certain controls. You visualize the airplane's altitude and flight condition from the combination of all the instruments and then you fly the airplane according to normal technique. Normal flying technique means keeping the ball centered by using the rudder correctly.

GHOSTS
IN THE
COCKPIT

*A unique course on the psychology
of pilot survival helps to exorcise some
of the specters that lurk in the sky.*

BY WILLIAM PETER BLATTY

A DISTURBINGLY LARGE number of pilot errors," according to a California psychologist, "are caused by hallucination, mirage or illusion." And this psychologist should know! Head of the Department of Psychology at the University of Southern California, Dr. Neil Warren also teaches a unique course on the psychology of pilot survival for SC's School of Aviation Safety, the only one of its kind in the world.

For the private and businessman flyer, the school conducts a specially designed eight-week course that deals with subjects not usually included in ground schools operated under FAA requirements, concentrating instead on factors that will contribute to the reduction of flying hazards. Main attractions of the course are a fully operational human centrifuge and Mirage Expert Warren.

Where the centrifuge tests G-tolerances, Warren deals with the apparently ubiquitous specters that haunt the cockpit: autokinesis, inner ear illusion, false horizons and "lean." None of these items is on the side of good.

"Autokinesis," explains Warren, "is an illusion that can be very easily demonstrated if you will stare constantly at a small, fixed light in an otherwise dark room. You will notice, in approximately ten seconds, that the light appears to move." Applied to aviation, when a pilot stares at a fixed light on a dark night, he may believe that the light moves or peels off when actually the light remains stationary.

It wasn't too long ago that a private pilot—let's call him Joy —dramatically demonstrated that autokinesis is more than a parlor trick. Arriving at an airport in New York, he became tense while observing that the weather there was foul. During the 45 minutes that Joy waited for take-off, the fog was boiling so turbulently that visibility fluctuated up and down between zero and one and one-half miles. On the runway, finally, he was delayed another ten minutes for clearance. While he waited, Joy decided he would have to open his cockpit window in order to see the runway lights while taking off. It was a near-fatal maneuver. During the run, he leaned slightly to the left to see through the side window, fixing his gaze on the runway lights. Autokinesis set in. The row of lights appeared to move left and away as he went down the runway, and to compensate for the imagined deviation, he turned his ship to the left. At the end of his roll, 75 *feet to the left of course,* the aircraft crashed and burst into flames. Joy, who survived, recounted later that he thought he was turning *right!*

What sets the stage for autokinesis? According to Dr. Warren, simple eye muscle fatigue.

In addition to autokinesis, there is also some spooky business going on with regard to "leans," says the aviation psychologist. "During instrument flying," he explains, "if attention is not focused on the instruments and if the plane should roll suddenly to the left and then recover very slowly, the pilot will feel that the plane is still banked to the left when actually it is straight and level." This illusion, he continues, results from an inadequacy of the "inner ear" sense: the inner apparatus of the ear has a threshold for tilt and pitch and

movements under this threshold are not perceived.

As dangerous as the lean is the "Graveyard Spin." "Rotary motion which is discontinued," says Warren, "gives the sensation of rotation in the opposite direction. Thus, the pilot who does not have a good visual reference on recovery from a spin to the left feels the false sensation of spinning in the opposite direction and attempts to correct this, causing the plane to go back into the original spin." Probably everyone has experienced the same sensation in childhood, when spinning around very rapidly in an attempt to become dizzy. On cessation of the spin, the world would appear to be moving in the opposite direction. Avoid wanton spinning of this sort like the plague, urges Warren.

The occasional danger that lurks in the skies also takes the shape of "false horizons." In flying with no reference to the actual horizon, cloud banks, as most pilots know, are often misinterpreted as being horizontal, and this false impression can be strong enough to force the pilot to change the attitude of his plane.

Indeed, so many of these illusions are so compelling that pilots flying IFR will frequently prefer to accept the illusion rather than the contradictory—and correct—testimony of their instrument panel! It is for this reason that Warren devotes so much of his course time explaining and describing these illusions: half the battle of psychological air safety is awareness that there are sometimes *fewer* things in heaven and earth than pilots occasionally dream. Trust the instruments!

But for the private or businessman pilot who prefers VFR, the hazards of hallucination are equally great. "When outside visual references are eliminated," explains Warren, "the pilot who has not been trained to fly by instruments attempts to interpret his muscle sensations and 'inner ear' sense of balance in terms of attitude, particularly in reference to the angle of bank."

Cockpit hypnosis? It exists, opines Warren, who insists that the conditions of hypnosis are present in long periods of uneventful, straight and level flying.

But by far the greatest and most fatal error attributed to private pilots by the SC psychologist, is a pole-vault out of the area of illusions and perception into the tricky field of "judgment": taking off VFR in IFR weather. "There," says Warren, "is your greatest killer."

Other words of wisdom from Warren to the private flyer: Reduce autokinesis by not staring at lights. Avoid fascination (hypnosis) by shifting attention from one item to another. Avoid leaning over in the cockpit to pick up something when the aircraft is in a tight turn at night; this reduces chances of experiencing the "Coriolis effect"—leans of the same direction. Avoid violent maneuvers at night, or run the risk of vertigo. Above all, always remember that disorientation and hallucination may occur. When your private mirage contradicts the instruments, trust the instruments!

Your Passenger's First Ride

He'll never forget that first hop if you follow these rules.

BY C. B. COLBY

ONE of the nicest things about taking a passenger up for his very first airplane ride is the fact that he has no idea what to expect. No matter what you do it will be a thrill. Here's how to impress a "first flighter" with what a really hot pilot you are and how wonderful flying really is.

Once he has decided to go up, proceed as follows: do not find out in advance how high he wants to go, where he wants to go or what he wants to do. That would take away much of his pleasure in the hop and it certainly would cut down the element of surprise.

Before taking off with a first-flighter there are certain rituals to be observed. They help impress the passenger with the skill of the pilot and the miracle of flight. If there is a cloud in the sky always say you have to "check with Weather" before take-off. After you have apparently done so inform your passenger you are sure you can "get away with it okay." This will relax him for the hop.

Help him into the plane and be sure to adjust his safety belt. Pull it up until he begins to gag and then release slightly until only his lips are blue and he is nice and snug in his seat. That promotes a feeling of security. Now examine the exterior of the plane where the passenger can watch you do it. This will assure him everything is just peachy. Shake the wings briskly and keep going back to one wing in particular for another shake. If a mechanic happens to be handy point to the wing and mutter something under your breath he can't possibly hear. He will ask you over again and you can mutter it again even more unintelligibly so that he will walk away shaking his head. The passenger will love it.

Waggle the ailerons briskly so that he can see the stick move about the cockpit or cabin. After his knees have stopped stinging go to the rear of the plane and snap the elevators up and down several times. You may miss him completely but it will demonstrate the mental alertness and abdominal agility aviation requires.

Back in the cabin, inspection over and your safety belt fastened, get the engine started. Once it has caught, keep fiddling with the instruments, choke and throttle. Switch from mag to mag and back again with a look of grim concentration. This will impress your first-flighter with the importance of a good engine and a pilot who understands it. If you switch mags slowly enough an interesting skip can be produced. This fascinates a first-flight passenger.

When you taxi out to the end of the runway come as close to other parked and moving aircraft as you can. It will show the marvelous control you have over the aircraft and give him a chance to see the other planes close to. Don't bother to "S" along the taxi strip. He might think you can't steer straight and, besides, it takes too much time.

At the end of the runway, rev up your engine again and switch from mag to mag slowly. It will increase his respect for your knowledge of intricate powerplants. Lean over and check his safety belt once more, and then wipe the palms of your hands with your handkerchief. This is IT.

Your take-off should be made especially interesting . . . not just the routine straight and level affair, but with some yaws and banks to let him know he's really in the air. Of course you might prefer to add a bit of novelty to your take-off with a first-flight passenger by holding her on the ground clear to the end of the runway and then hoisting her right up past the factory windows. That really gives the "first" passenger a bang.

In the air and away from the field you can settle down to giving the passenger a good time. Most first-flighters are disappointed in the smoothness of the ride and the ease with which the pilot flies the aircraft—but not your passenger!

Fly smoothly only until you spot something of interest on the ground (anything will do) then put the ship in about a 50° bank and circle it briskly a few times, yelling to him all the time to look at it. If he refuses to open his eyes, do a fast slip right down to the tree tops so that he can't miss it.

If you have a chance to go cross-country during the ride

make it look rugged. Be sure you have plenty of maps aboard and fold and refold them as you fly, keeping the plane more or less level with your knees. Use a pair of dividers too if you can and check your compass repeatedly. This makes it appear very technical. Peer out of the ship in all directions and do an occasional fast "180" to look over some check point again. That shows caution and attention to detail.

While on the way you can demonstrate some of the more interesting maneuvers such as spins and stalls. Be subtle. Do not tell your passenger that you are going to demonstrate anything. Raise the nose higher and higher as you work the throttle back. The passenger may not notice it at all. If he does belt you in the back and scream "Whasamatter?" just shrug your shoulders and keep that nose coming up until it breaks in a brisk stall. They love it. After you have pulled out of it and the passenger has returned to normal color explain that stalls are harmless unless you fall off into a spin from a stall low down.

A little later, to demonstrate, do a brisk power-on stall but kick it into a spin as the stall breaks to illustrate what you meant by a stall being safe unless you developed a spin from it. To be sure he doesn't miss what is happening shriek: "We're in a spin!" several times before you let the ship come out of it by itself. He will probably remember the maneuver better this way and much longer.

Never let a first flighter touch the controls—not but what he could probably land it himself if he had to—but it makes it look much harder to him if he doesn't find out how ridiculously simple it really is. Keep it looking technical, checking your instruments all the time, looking out at "that wing" frequently, and changing throttle setting suddenly. He'll know you are doing your best to make the flight safe and interesting for him.

When you get back to the field once more, don't just go on in and land. Drag the field first, fish-tailing nicely as you do, then go around again, to let him know it is no easy matter setting a plane down. Go around several times if need be for the desired effect, then make a long straight approach watching your instruments all the while, shaking your head occasionally as though you aren't quite satisfied with the way the landing is going. At the touch-down "stir-the-stick" a little to make it look like a battle to hold it level and let her roll almost to the end of the runway before turning off onto the grass or taxi strip.

Once off the runway, slump slightly in your seat as if it had been almost too much for you, then, with a "carry-on-old-man" smile, taxi back to the hangar, waving to anyone who happens to be hanging around watching, whether they appear glad to see you or not.

Well, you have given your passenger his first ride. You have been loyal. You have been out to sell aviation. You are the kind of pilot we have too many of in this business of ours.

The Game

BY FRANK KINGSTON SMITH

Fellow name of Bob Blodget, who used to be an Editor out in San Carlos, California, before he turned square and began to sell airplanes, invented it. Like most of us, he got miffed all the time by tower operators telling him and other pilots strange to an area to report over such non-Jeppesen points as "the brewery," at Van Nuys, the "hammerhead crane" at Philadelphia, the "packing plant" at Chicago or the "chain bridge" at Washington, D. C.

The game is based on the human weakness of never admitting that one doesn't really understand what someone else is talking about and is designed to get even psychologically by giving a point, called a "Blodget," to a pilot or a tower operator who successfully ploys the other. The rules of the game are simple. First, it can only be played on a really clear day and second, the entire exchange must be conducted with a straight face, for if it ever gets out that there is a game being played, either or both parties may gain only a forfeit, known as a "Halaby."

For example: Pilot calls tower and reports five miles out. Tower responds, "Roger, report the brewery." If the pilot responds with his own "roger" without disclosing that he doesn't know where the brewery is, and fumbles his way into the pattern, then the tower gets a Blodget. If, however, the pilot calls in and the tower says, "What is your position?" and the pilot answers that he is over the clover-leaf intersection, and the tower doesn't ask *which* clover-leaf intersection, then the pilot gets the Blodget.

Five consecutive Blodgets constitute the game and entitles the winner to shout "Schulte!" which is sort of an aeronautical "Eureka!"

The game may be played with all sorts of ground people. Take radar controllers, for instance.

Fellow comes into the Atlanta area in a Cherokee and contacts Traffic Information. Radar says, after making the pilot do all kinds of gyrations so he can be identified, "Roger, radar contact. You have traffic, five miles, twelve o'clock."

The pilot sees the traffic two thousand feet or so above him, but replies, "Negative contact." "Three miles, twelve o'clock," snaps the controller. "Negative contact," answers the pilot, edging under the path of the high flying airplane. "One mile, twelve o'clock," says the controller, in a strained voice. "Negative contact," says the pilot. Then, as the other airplane passes overhead, he shrieks, "Whoops—There he goes!" thereby scoring a Blodget. Of course, it has to be done discreetly, otherwise he might get a Halaby.

If the radar operator catches on before the pilot scores and says, "Radar contact lost," then he gets credited with the Blodget.

I heard of a guy who won a few contests at Washington National but had the score evened one Friday evening when the tower made him do two successive 360s at the middle marker and then made him go around on the next approach, pleading runway congestion.

He pulled ahead for a few seconds the following Friday when the tower told him to report downwind over the Masonic Memorial. "Masonic Memorial," he wailed. "But I'm a Catholic!" He looked like a winner until the tower, without missing a beat, said, "OK, report over the power plant." Even all.

One of the greatest coups of all time was pulled one afternoon at Idlewild Airport, which is, as you may know, the biggest concrete maze ever designed by the mind of man.

These two lads were parked on the itinerant ramp, known as "Taxiway Romeo" and asked for taxi instructions for a VFR takeoff westbound.

"Roger," said ground control, briskly. "Taxi northeast to Taxiway Tango, turn right to the inner perimeter, turn left to Taxiway Baker, turn right to outer perimeter, right to Taxiway Easy, right on runway thirteen left, then cleared across two two right, hold short of runway two five for your run-up and call the tower when ready on one one niner point one. Over."

To the two week-end pilots in that airplane it sounded just like something out of Alice in Wonderland. They knew that they were supposed to read the entire clearance back, but this one floored them.

For several seconds there was a dead silence over the ground control frequency as the two visitors tried to decide whether to ask for a repeat or to try and bluff it all out. The frequency hissed a couple of times as the pilot pressed the microphone button, then after a long delay, the voice of the copilot was heard, not only in the ground controller's receiver, but in every ground vehicle, airplane and office set tuned to that frequency. The copilot knew what to do under the circumstances. "Ah, the heck with it," he said in a voice ringing with decision. "Tell him 'Roger.'

Blodget!

Follow Me Through: Where Would You Put It?

BY ROBERT BLODGET

In the old days, it sometimes seemed that airplane engines stopped as often as they ran. It was not really uncommon for a pilot to have five or six forced landings a day.

Our engines have improved enormously since then. But even though the chances of engines stopping are remote, and quite comfortably so if the captain knows how to manage his fuel, the possibility still exists.

There used to be a lot more attention paid to forced landing practice. Then the FAA found that practice was causing more accidents than genuine emergencies, so they let up quite a bit. New pilots are now getting less preparation for both forced landings and precautionary landings than used to be the case; and it is quite likely that they are not getting enough to be useful.

Flight instructors used to keep asking: "Where would you put it, if the engine quit right now?" And they would keep saying that, frequently closing the throttle at the same time, until the student got the message. The student became accustomed to planning his flight path across the land the same way he would cross a mountain brook. He would get a rock to put his feet on, then keep hold of that rock until he spotted the next.

This often seems wasteful; the short, direct route at the lowest possible altitude is most efficient. But efficiency becomes meaningless if that good old reliable fan should stop and the only place to reach the ground is in the tree tops.

There is no need to become obsessed with this, and no need to lose faith in our reliable engines. But the cycle has swung too far from the old, dangerous days when the instructor pulled power without warning, and neither instructor nor student was quite clear about who was going to restore it, and sometimes each thought the other was going to, and at other times the engine simply refused to cooperate. The present position contains too much complacency.

Fortunately, there is a way to minimize the hazards without losing track of the goal. The techniques of landing do not change; the same airplane needs the same amount of space, the approach needs to be flown at the same speed, the same judgments of height and distance apply alike to airport landings and forced landings.

There are differences: the surface of a forced landing area will be much rougher than the roughest airport, there may be wires to avoid, and there will be no chance for a missed approach. And because the landing area will probably be smaller, there is a greater need for keeping the approach speed low.

The best place to practice forced landings is at your home airport. Find out, from the book, what the total landing distance for your airplane is supposed to be. Pace it off along the runway, and mark it somehow. If your runway has border lights, you can do this simply by finding out how far they are apart, and counting off the distance; then making landings until you can come to a stop opposite the right light.

Next, pretend that there are wires 50 feet high at the runway threshold, and you cross it high enough to clear them. You will perhaps be surprised at how much farther down the runway you will touch down and stop, the first few times.

When you begin to get your approach speed properly nailed down, and perfect the use of your flaps, you will be even more surprised how short a landing you can make.

By now, you have a good mental image of the actual space you would need, if you ever did have to make a forced landing, at least at the elevation of your home airport. At high altitudes, the space you need will be bigger, and on a hot day, it will be bigger still.

Now you can make a game of it, for at least a few minutes out of every hour you fly cross-country. Could I get in there, if the fan stopped? Or there? Are there wires around that field that I can't see from here? Is that surface likely to be firm enough to support us? Are there holes or ditches hidden in that tall grass?

Whenever you can, pick out what looks like a good field from the air, and then walk over it, checking the surface, the obstructions and the length. Before you know it, you will be quite expert at selecting forced landing areas.

Finally, remember that it's easier to repair airplanes than people: the first requirement is safety of the pilot and passengers. If you can get the airplane onto a reasonable piece of ground, at a reasonable speed, rightside up, the chances of serious injury are quite small.

The most likely result of thinking and planning and practicing is that you will develop better habits and take fewer chances, both of which will make the possibility of ever having to execute a forced landing even more remote.

SPINS

In some ways, modern lightplanes are even <u>more</u> vulnerable to this killer than their supposedly dangerous predecessors.

BY ROBERT BLODGET

THERE HAS BEEN a good deal of mystery about spins since the earliest days of man-carrying airplanes. In fact, it was years before anybody even recognized that some kind of rotation was going on before ground impact at the end of what was called a nosedive. Jack Irwin, one of the least known of the Early Birds and pioneer airplane builders, thinks this was because airplanes simply didn't fly high enough for the spin to become developed. He says that he perceived spins by flying scale models from the roof of his grandmother's barn, in Sacramento, California. The effective scale altitude of the barn was high enough for spinning to become well-established.

There are two different kinds of spins. In the "normal" spin, or tailspin, the airplane revolves around a vertical axis, which passes on an angle from nose to tail. In the flat spin, the airplane rotates horizontally, like a Frisbee. The flat spin was supposed to have been eliminated by some remarkable test flying by James Doolittle in the 1920s. Doolittle found that there was no successful recovery technique for flat spins; the modern remedy for such spins is the installation of a parachute in the tail cone of airplanes being tested. Even this sometimes doesn't stop the spin. If it doesn't, the pilot takes to his own parachute.

The conventional spin, in airplanes approved for spins, can be stopped. The exact technique is different for airplanes of different design, but an experienced test pilot, if the spin begins high enough, can usually find the combination fairly quickly. It is not at all a simple maneuver, however; a recent NASA paper on spins says:

"The spin has been defined as a motion in which an airplane in flight at some angle of attack between the stall and 90 degrees descends rapidly toward the earth while rotating about a vertical axis. The spinning motion is very complicated, and involves simultaneous rolling, yawing and pitching while the airplane is at high angles of attack. Since it involves separated

flows in the region beyond the stall, the aerodynamic characteristics of the airplane are nonlinear and time-dependent; and hence, at the present time, the spin is not very amenable to theoretical analysis." (*Summary of Spin Technology as Related to Light General-Aviation Airplanes,* by James S. Bowman, Jr., Langley Research Center, NASA. NASA TN D-6575, December 1971.)

A more familiar description is that the airplane spins because one wing is fully stalled while the other is still producing some lift. This produces a kind of autorotation. A fully developed spin is a stable flight condition, typically with a terminal speed less than the airplane's never-exceed speed: thus, a spinning airplane can lose altitude rapidly without suffering structural failure. In contrast, the spiral dive, or graveyard spiral, allows airspeed to increase well past the structural limit.

There is evidence that the French, and probably the British, understood spin entry and recovery before we did. During World War I, spins were used to escape from an aerial opponent, since the spinning airplane could lose height rapidly and safely while a pursuer in a dive risked structural failure. During the 1920s, pilots teaching themselves instrument flight also relied on spins: When they lost control in cloud, they put the airplane into a spin from which they recovered when they broke out of the bottom of the cloud deck. Spins were used in aerobatic shows during that same period, and are still so used occasionally. In the waning days of World War II, low-time Luftwaffe pilots used spins to descend through low cloud decks to make visual landings at fighter bases on the French coastal plains. Sailplane pilots still use deliberate spins for rapid descents in emergencies.

Since most pilots during the 1920s were either trained in military flying schools or by former military pilots, spin entry and recovery was a standard part of the curriculum. Thus, when the first Civil Air Regulations were adopted in 1926, spin training was required. At the end of World War II, the Civil Aeronautics Administration began promoting general aviation and liberalized a number of rules, one of which dropped the spin requirement for the private pilot. Although the change was well-publicized and supported by a majority of the aviation community, there were some serious flight instructors who thought it was wrong—and some still think so. These instructors continued to teach spins, since there was no law against it, but newcomers did not. In fact, spin demonstrations are no longer required even of flight-instructor candidates; a logbook entry showing spin experience is enough. If proposed revisions to Part 61 of the Federal Aviation Regulations are carried out, even this will be dropped.

The CAA made another change that attracted little attention. It decided that airplanes in what it called the normal category would not have to be spin-tested for certification. (There are two other categories: utility and aerobatic. Aerobatic airplanes must be able to recover from a spin in not more than one and a half turns after a six-turn spin; utility airplanes can meet either normal or aerobatic spin standards, but if they meet only normal standards, they must be placarded against spins. Many general-aviation airplanes qualify as either normal or utility, depending on loading.)

The reasoning that led to these changes is straightforward: If a pilot is going to use his airplane for transportation and not aerobatics, why burden him with learning spins? And if his airplane will never be spun, why require it to meet the old-fashioned spin tests?

There is a serious flaw in this argument. The argument concerns itself only with deliberate spins, and ignores the hidden face of the spin—the accidental, inadvertent or unintentional spin. Actually, even in the days when both pilots and airplanes were required to be competent in spins, very little attention was paid to accidental spins.

Once pilots learned to recover from conventional spins, few crashes followed deliberate spins. Pilots who spin on purpose know what they are doing, know what their plan is (a certain number of turns, or a certain loss of height, or penetration of a cloud deck), and know how to stop the spin and recover. Accidental spins—by definition—occur without the pilot's cooperation. Most often, they happen at low altitudes when there is not enough room for recovery even if the pilot is spin-trained, the airplane spin-worthy, and the pilot recognizes what is happening.

The difference between intentional and accidental spin has to be seen to be believed. Entry into an intentional spin from coordinated flight, whether with wings level or from a turn, requires that the airplane be stalled, and at the moment of full stall, the rudder moved briskly full in the direction in which the spin is desired and the stick or wheel pulled full back. In recovery, the rudder is applied full against the spin, after which the stick or wheel is moved forward.

A small error of control application or timing during entry often results in a sloppy spiral instead of a spin. Some airplanes that can be made to spin will come out of the spin by themselves after a turn or two. Even when spins were

required, pilots who had only been exposed to intentional spin entry were pretty well convinced that the airplane really didn't want to spin at all. They were no better prepared for the suddenness of an accidental spin than if they hadn't had spin training.

The thing that turns an innocuous stall into a sudden spin is lack of stick-and-rudder coordination, so that the airplane is already slipping or skidding when the stall occurs. To help visualize what happens, remember that an airplane always rolls away from a slip or skid at the stall. In a skidding left turn, the airplane is skidding right, so the resulting roll will be to the left, or "out the bottom." In a slipping left turn, the airplane is slipping left, so the roll will be to the right, or "over the top." Things are reversed in right turns.

Warning: Do not attempt the following maneuvers unless the airplane meets the aerobatic-category spin requirements. Do not attempt them solo. Your safety pilot must be fully familiar with spins; an aerobatic instructor is best. Since these are classed as aerobatic maneuvers, be sure you are in an area approved for aerobatics. Clear your area with special care before starting the maneuvers.

When experimenting with an accidental spin over the top, enter a steep turn at somewhat less than normal cruise speed, and unless the airplane has less than 100 hp, use a little less than cruise power. The angle of bank must be over 60 degrees at the beginning; thereafter, do not attempt to hold altitude by shallowing the bank. Instead, keep applying increasing back pressure.

The airspeed will bleed off, and the ball in the turn-and-slip indicator will be out of its center position, or "cage," toward the low wing. Don't try to center it: It is indicating a slip, and that's what you want. Some modern airplanes are quite resistant to spins out of slips, and many won't spin at all out of a left turn. If your airplane is reluctant turning left, try a right turn and add power. Eventually, you will feel the pre-stall buffet. Ignore it; just keep a lot of back pressure.

After buffeting begins, you'll get the stall. The airplane will rear up away from the turn, pitch down, and spin rapidly in the direction opposite that of the turn. It will be momentarily upright on top; and after you have tried this a few times, you'll be able to "catch" it there, and fly it away level. But not the first time. The first time, your reaction will be that you never before had an airplane change direction so fast. If you have never been in any kind of spin, you'll be astonished. Keep the astonishment in mind, because that's why accidental spins are so serious.

Now for the real killer: the spin out the bottom. This time, establish a steep turn to the left with a bank angle of about 50 degrees. Enter with cruise power, from cruising speed. Maintain altitude with pitch. Apply enough rudder into the turn and aileron against the turn to move the ball half out of its cage to the right (skid) side. Center the ball under the right wire or line, and keep it there. As the airplane slows down, increased opposing control pressures will have to be applied.

Once again, time will pass; in fact, you will fly the cross-control turn almost 720 degrees before anything happens. This time, you will have little or no pre-stall warning. The airplane will spin briskly out the bottom—in the direction of the turn—and again you will be astonished at how fast it happened. Once again, keep the astonishment in mind.

The first time, the airplane will probably make close to a full turn in the spin before you or the spin-trained safety pilot can recover. (If he is an aerobatic instructor, he'll recover in half a turn.) As in the spin over the top, after several tries, you'll be able to recover before rotation has gone past a quarter turn. Even then, you will have lost some height.

It might seem as difficult to get an airplane into an accidental spin as it is to persuade it to spin out of coordinated flight; and in fact, spins over the top of a slipping steep turn are rare in the killer spin. Spin entries from skidding turns are, however, quite common; and the cause is the most common error in airmanship. Let's see how this comes about.

All airplanes with propellers that turn clockwise as seen from the cockpit (almost all United States airplanes) have a tendency to turn left. The opposite is true for airplanes with propellers that turn the other way. Usually, this tendency is eliminated in level flight at cruise speed and power by installing the fin slightly off line with the longitudinal axis of the airplane. When more than cruise power is used, the left-turning tendency must be offset with opposite rudder. The effect increases as speed decreases and power increases, and is greatest at stall with full power. In a glide, or descent with less than cruise power, the offset fin tends to make the airplane turn right.

We encounter the left-turning tendency on every takeoff. Until the lift-off point, pilots use enough rudder to keep the path straight over the runway most of the time. After liftoff, most pilots do not use enough right rudder. Few light airplanes are equipped with rudder trim, and holding steady and quite heavy rudder during climb and departure maneuvers is tiring. As a result, very few pilots ever do use enough right rudder; they try to stop the left-turning tendency by unconsciously leaning a little right aileron. The habit is easy to fall into and hard to break. A glance at the turn-and-slip indicator will show that the ball is half out of its cage to the right, just as it was in the split-ball skid exercise.

We don't usually think of skids when an airplane's wings are nearly level—they are here—but it's skidding all the same. In left climbing turns during departure, the skid persists, but the danger increases: the steeper the bank, the higher the stall speed. When the margin of angle of attack above the critical stall angle is less, the steeper the bank angle. Almost always, departure turns are climbing turns, which still further reduces the stall margin. Under these conditions, a sharp-edge gust can produce a stall, and the result will be the beginning of a spin out the bottom. As in the early days, altitude is often so low that spin rotation is not evident even to aeronautically qualified eyewitnesses.

In flight tests, the FAA standards require demonstration of departure stalls. When these are flown with the airplane fully coordinated, they are quite tame and easy to manage. Neither FAA inspectors nor examiners call for the maneuver to be demonstrated in a skid; though this is wise from the standpoint of safety, it doesn't represent the conditions that go with real-life departure stall-spin crashes.

The sequence of events leading up to an approach stall-spin, and a stall-spin that occurs during a go-around, are different in some ways from the departure type, but what does the dirty work is still the skid. In addition, the rudder is usually left of neutral, which compounds the problem. In the approach sequence, the engine is developing low power, or may even be at idle. The nose has been trimmed "up," the amount depending on the airplane and the pilot, but it is not uncommon to find pilots using full nose-up pitch trim. Left rudder is being used to overcome the right-turning effect of the fin.

Turning from base to final, if the pilot sees that he is going to overshoot the proper final-approach path, he will usually increase the bank angle. He may try to keep the angle low while accelerating the rate of turn, which is what a skid is; but that isn't necessary. If he then sees that he is going to undershoot the runway, he will add power. This will cause the nose to pitch up and the airplane to try to increase its turn to the left. Since he already was carrying some left rudder in the approach, his chances of getting in enough right rudder are small. The airplane will spin out the bottom. A gust somewhere during the turn will simply ensure the outcome.

During a go-around, full power will produce considerable pitch-up; and in several light airplanes, a forward push on the yoke of well over 30 pounds is required to maintain attitude with full nose-up trim and full flap. This is hard to hold, and any momentary relaxation of forward pressure may well be enough to start the stall-spin. If flaps are retracted, even partially—and a lot of airplanes cannot do a missed approach with full flaps—this will add more pitch-up. All this without enough right rudder sets the stage: If the pilot then makes a left turn to come around again, it will be a skidding turn, and he may not survive it.

Once again, flight-test demonstration of approach stalls are always flown properly coordinated. In fact, even the one-turn "spin" tests are made out of coordinated flight.

Low-altitude stall-spin crashes have been with us forever. Neither spin training and spin testing of pilots nor insisting that airplanes have good spin and recovery characteristics have been able to eliminate them. It seems most unlikely that reintroduction of spins into the pilot-testing procedure would help, simply because it never did help. Changing the spin requirements for airplanes might allow pilots to become familiar with accidental spin entries, since it is presently unsafe to spin a normal-category airplane. Certainly, demonstrations of accidental entries would help to make pilots aware of the consequences. The rebirth of aerobatic airplanes and training at least provides safe machines for such work.

Better training in use of rudder will help, too; and that is well worth doing if only to show the considerable improvement in rate of climb when the airplane is properly coordinated. Adding rudder trim to all airplanes would be a step in the right direction, and need not be expensive if a bungee system is used. Finally, the FAA should take a much more aggressive stand on the matter of angle-of-attack indicators, which are the only instruments that show the pilot the lift conditions of the airplane's wings.

The necessary equipment is available. The only thing that must be developed now is the pilot's readiness to use it.

LACKING ANYTHING ELSE to do, I leaned against the rolled-back north hangar door at the Big A and watched—mixing resignation and infuriation in equal measure—the grizzled senior mechanic who had just refused the proferred assistance of a junior and untempered associate. While delicately performing some intricate finger exercise on the axle and its fittings, he was filling me in on the caste system among airframe, powerplant and instrument-repair license holders. "Work here for six months and off to the airlines. *They* will take anyone. To be really good at your job is a forgotten value-judgment," he said, which showed he was reading the *Wall Street Journal* on his coffee break. I had half a mind to ask what stocks he was in but decided against it, knowing full well that the brakes on the Bonanza would continue flaccid if he was as hot for his stocks as everyone else seems to be these days. Why did this have to happen the first time I filed my own instrument flight plan?

"Jack Olcott, please," I'd hollered into the phone.

"Jack? I did you proud! I got it, I got it!" Poor Jack. I guess part of being a top-flight instrument instructor is hearing all your students saying the same thing—letting them babble on with all the details and then putting in the clincher: "Remember what I've told you. Instrument flying is for two pilots, with at least one of them a professional, in a well-equipped airplane. And who fly a lot. A lot more than you do."

"Yes, Jack."

"From now on, file flights IFR if VFR conditions hold. It's good practice. Then later, sensible use of IFR."

"Yes, Jack."

"You come out to Morristown one morning and there's a layer of ground fog and all VFR above, then take off IFR. That's the way to use it. To do more than that or for breaking through a high overcast should await a well-equipped twin and a competent copilot."

"Yes, Jack."

This dialogue ran through my head as I bounded up the stairs to the FSS at Teterboro. Always a busy place, always a crowd, the station staff are always helpful.

"Teterboro-Montreal? Looks pretty good all the way. Clear with haze from here to Albany. From there to Burlington, scattered at 3,000, a high deck at 10,000. Montreal is reporting about the same. Forecast is for little change in the next six hours. Moderate turbulence north to Albany. No icing; freezing level will be several thousand feet above you. VFR all the way."

Well, now, this was just what Jack ordered, so I filed for Montreal using as much of the JFK-Montreal preferred routing as possible.

Down the stairs, back to the car, balanced the suitcase under one arm and the Jeppesen books (which always seem to be lead-lined—designed, perhaps, to protect the contents from radiation at FL 600 and up) under the other and headed for the white Bonanza. Just out of Atlantic Aviation that morning from a 50-hour inspection, she should be ready to go.

Gene took the keys from between my teeth and opened the luggage compartment. We stowed the gear and opened the cockpit door. Gene and I work in adjacent offices and have made a number of flights—some good, some bad—in this air-

Ready to Copy?

Fresh new instrument ticket in your pocket, butterflies in your stomach and a cockpit full of qualms . . . are you ready to file IFR for real?

BY ROBERT PETERSON

plane. Gene flew in WWII. Ventura-Vegas and big stuff like that. Across the Atlantic, ferry flights across the Mediterranean. *Real* flying. No longer holds a license, but under his hand, the Bonanza holds heading and altitude, and climbs, descends and turns like the lady she is. During our trips, he had mastered the en-route navigation charts and now handles all the navigation and radio work. The preflight was fine, and I again pointed out that we'd be in Montreal before the Air Canada flight we might have taken and to which Gene now departed to cancel our reservations.

He found me at length in the Atlantic office, asking who signed off the 50-hour inspection when the brakes had no life of their own and were flaccid. I found the word descriptive and used it several times. Flaccid brakes! *Flaccid.*

Blocking out the caustic commentary on the young squirts infesting the honorable preserves of the craftsmen-like A&P ("few of us left"), I watched the day shattering before my eyes. It had started splendidly. Arrived at the dentist at nine a.m., tensed for drilling, and cleaning was the order of the day. Good. Picked up Gene at 10. Teterboro at 11:15. Cleared with the maintenance boss. Plane on the line. FSS. Phone the office. File flight plan. Defective brakes—strike that—flaccid brakes discovered at 11:15. Argument. Inspection of flaccid brakes by Atlantic shop supervisors. Discussion with a foreman. He can get right on it. Shouldn't take 45 minutes. Just before noon, the Bonanza is pulled off the line and positioned for treatment. Noon. Everyone goes to lunch. Due back at 12:30 p.m. Should be ready to go at 1:30 at the latest. Gene and I confer as to the wisest course to follow. Should we drive to Kennedy and go Air Canada, assuming we can get reservations? We compromise by going to eat. When we return, if the plane is ready, we'll go and be ahead of any jet we can get from Kennedy, which is a good hour and a half by car from Teterboro.

The sun feels good when I move away from the hangar door. Better than the day of the flight test last week. I'd arrived early and walked the long walk to the FAA building to find I was 30 minutes ahead of time. Went over in my mind, like beads on a string, the various compulsory reports to ATC, and various other bits of lore that others had led me to believe were subjects on which I would be catechized at length.

I was then invited—no doubt because of my pitiable and strung-up appearance—to join a quartet of FAA inspectors going out to have coffee, which I did, trying as best I could to be animated, much like Marie Antoinette in a tumbrel . . . I even offered, last of the spenders that I am, to spring for the coffee.

Returning to the office, Inspector Claus, to whom I had attached myself like a respectful leech, asked for my documents and assigned me to a room, a table and a flight to Poughkeepsie and its airport. Since I'd been laying out courses for Jack for months, this was no sweat and I phoned in the flight plan. That done, Inspector Claus and I departed, going through the inane procedures men have of holding doors open for each other and Alphonse- and Gaston-ing each other to death.

Then we settled down in the grooviest, fire-engine-red Fiat two-seater in north central New Jersey. Running her up through the gears, he spoke of the mileage he was getting, when any fool could see he was enchanted with the little machine.

Skipping all the folderol, we went and did our thing. Omni approach to Dutchess County Airport, holding at Peekskill Intersection, ILS to White Plains, partial panel. A good hour and a half. I thought I was doing all right—especially northbound to Spring Valley—when I had communication trouble and worked it out according to the Federal Air Regulations.

When he told me to take off the hood and said he couldn't find anything wrong, I generously offered to let him fly us back to Teterboro, freely admitting I was so pooped I'd probably dig a trench down the middle of the active runway.

When I got that slip of paper, I could understand why French generals do what they do after decorating someone. Not that I did it. The test was a satisfying experience and it was conducted by an obviously understanding man, but one with a strong sense of duty. If I were to go for a multi-engine, I would hope to get him again.

Not that all this reminiscence was doing much good leaning against the sunside of the Bonanza. The wheel is back on. The scorned assistant is now assisting in bleeding the system and I'm encouraged to climb in and try the brakes.

Gene and I swing into the wind on Taxiway Bravo and complete the checklist, confident our clearance will be coming through shortly since there is only one aircraft ahead of us. The sky is filling with haze, a usual condition in the neighborhood of Teterboro. The checklist completed, our little group has swollen to four aircraft, with a fifth approaching and filling the frequency with complaints about the delay in his clearance:

"Estimating clearance delay of 15 minutes each aircraft."

"Fifteen minutes? I'll be at the end of this cotton-picking line for an hour and a half!"

"Estimating clearance delay of 15 minutes each aircraft."

"Well [mike button released] you can cancel my [mike button released] clearance. I'm going back to the [mike button released] hangar and walk!"

"We will cancel your flight plan. Good day, sir."

Gene and I didn't think Hartford's trouble was funny for long. Even at 800 rpm, that tach time moves at the rate of $35 an hour. A lot of people were sitting around in Montreal cracking their knuckles, even though we had suspended the meeting until an ETA was handed out by ATC in Montreal.

Number one in line departed. Yes, on inquiry, we were ready to copy:

"ATC clears Bonanza 2024W to the Montreal Airport via Butler Seven departure Sparta then flight plan route maintain 5,000."

With no mention of a departure frequency and the surprising number of stand-bys in earlier clearance recitations, I wondered if there was another tyro at the other end of this conversation?

"Teterboro ground, is there a Newark Departure frequency?"

"Affirmative. Contact Newark Departure on 119.2. Contact tower 119.5 when ready."

One last look at the narrative version of the Butler SID and we were on to the tower, who, perforce to demonstrate virtuosity to the tyro in the cab, ripped off the SID like a stream of fléchettes.

Then the Bonanza gathered itself and headed for the hazy

blue yonder, right up the localizer with the ADF needle aimed at the outer marker. Newark Departure, like a cavalry sergeant, called for a right wheel and 2,000-foot altitude, direct Sparta. The navigating omni was set for that and all was well. The number two was set for the intersecting radials from Solberg that predetermined the climbout. For a guy with a week-old instrument rating, I was in command.

Newark Departure: "Bonanza 2024W, let's see. Radar contact. Direct Paterson. 4,000. Report leaving three."

It's at a time like this that the King crystal-controlled ADF is the ultimate instrument in an ideal panel. You reach over, dial and turn to the indicated course. It's the nuts. We ain't got one. I haven't seen the ground since the nose blocked it from view on rotation and I'm not going to see it now. Our ADF tuner is off the proper readings a hair and getting Paterson in, and climbing, and reporting leaving three is not a make-work project. Gene keeps her in the climb, his view obscured by my arm reaching across his face to the ADF, a blockage he rectifies by stretching his neck like an observant turtle. There's dit-dah-dah-dit, dah-dit, dit-dah-dah-dah. Forget the fine tuning: It may slip away completely.

"Bonanza 2024W at Paterson. Direct Pawling. Climb maintain 7,000."

Yeah, but we're not at Paterson. The needle is glued to the top of the dial. So I'm deaf. Newark talks sharply to some other poor slob who has failed to do something he was told to do and the channel is filled with the wails of a reprimanded first-grader.

There's the ADF needle moving, a fact of which I make Newark aware and emphasize I'm headed for 7,000. Going to seven rather than five is the ball game, as later innings are to disclose.

The run to Pawling is a good 60 miles and 22 minutes, including the climb. Level at seven, we clean and dust the cockpit and examine the world around us. There are 72 billion motes of dust in the air, each and every one of which is reflecting sunlight and you couldn't recognize your own mother if she were 100 yards ahead of the nose. The ground is down there, okay, but constricted, like the view through a reversed telescope.

I urge Gene to identify all omnis and stop mentally subtracting 180 degrees from course lines on the charts; do it with a pencil, for heaven's sake, and then tell me if he can recognize his mother at 200 yards.

This is the time, VFR all right, when the old IFR comes in handy. We have been handed off to various calm voices called New York Radar and New York Center, with whom we have laconic dialogue. Somewhere in the general area of Spring Valley, the communications bug out. Try number two transceiver. No better. Well, this is nothing new. Happened during the test. Back to the last controller.

"Bonanza 2024W. Try again. Controller reads you loud and clear."

So full volume first on number one and then number two. All we get is a reedy, unintelligible squeak. Back to Doubting Thomas, who reluctantly assigns 127.55, probably being held in reserve for CIA Snoopers on Air Force One. Speaking from the depths of the thorax, breathing slowly, enunciating

carefully and with volume on maximum output, we send our signal. Back at full volume comes the majestic voice of 127.55 loud enough to put contusions on the bus bar. Whatever interference lies in wait around Spring Valley, we're past it and communications are splendid hereafter, but one's expectation that they will be befouled continues unabated.

Pawling yields to Cambridge, sign posts on V-487. Glens Falls lies on our left hand and we stare down in its general direction with loathing, since many's the time we've weathered in there on VFR flights to the north country.

The haze abates and darling little cumulus puffs festoon the sky. With the weight and majesty of the FAA and FARs behind me, I penetrate them happily and legally. They, in turn, give 24W little pats on the belly.

Out of Cambridge, as Gene calculates our ground speed and reports 175 knots, I concentrate on determining the new heading to freeze the needle, and find myself peering at the instruments and wondering whether my eyes are giving out. I look out to see solid cloud in every direction, all grey with black mottling. Impressive as finger-painting, but what do I do now?

Re-route? File for an instrument approach to Albany? Keep going? First, a quick look at the gauges: everything in the green. Second, navigation: on course, the needle centered, number two set for Benson, which lies ahead. Check this out.

"Bonanza 24W Boston Center verifies your position on course 15 DME north of Cambridge."

The charts show minimum en-route altitude at 4,000 feet. Check the ceilings.

"Frequency change approved. Advise return."

Glens Falls Radio informs us of the weather conditions at Glens Falls and Burlington. The cloud decks were apparently still there, but more compressed than forecast. Descent, if necessary, to VFR conditions along the route was in the cards.

Press on. Elation and cowardice fought it out. Firmly resolved to hold 7,000 feet, plus or minus 20 feet, and a heading to desensitize the needle, press on we did. Gene, stout fellow, demurred not.

The hours of training were put to use.

Best of all was the absence of the abominable hood. With each passing minute, the tension diminished. In fact, except for an occasional burst of rain and some light turbulence, it was ridiculously easy. The temperature held at 45° F., so ice was no problem.

Concentration, even without tension, builds in its own penalties. Much like prolonged driving on an unlighted road on a black night, the eyes seem to pop out of the head, the self to come out of the body. This happened regularly and was overcome by an exchange of conversation with Gene, a look around the cabin, a shifting in the seat. I'm of the opinion that the controllers could help this. For long stretches, in a radar environment, there is no air-ground chit-chat. A pilot, alone, would welcome the change from the concentration, doubled, no doubt, with all the chores he alone must handle.

The ATC does keep its eye on the sparrow. The poor sparrow, on the longer legs, surrounded on all sides by what looks like the contents of a vacuum-cleaner bag, gets to feeling desolate and forgotten. Some seemly remark like: "East Overshoe Center monitoring Bonanza 24W. Position now 15

DME from Petticoat Junction."

No one needs this kind of solace in the high-density areas where, with three hands and two heads, you barely keep up with the frequency switches.

Boston Center "good dayed" us a little north of Benson Intersection, advising us to dial Burlington Approach 15 miles north at Weybridge Intersection. Gene, who for reason unknown is now reading the transmissions far better than when he had the ground in sight, has the Lebanon 310 radial in place instantly. With infinite regard, we watch the number two needle swing to its proper upright position.

North of Burlington, en route to the St. Jean Omni, surveillance shifts to Montreal Arrival Control. As contact is established, we emerge from the dust bin and a great panorama of sky, clouds, water and earth spreads itself before our eyes, lit in varying intensities and at varying levels by the sun moving downhill toward the horizon.

Gene and I straighten in our seats, lean back and each lights up one of his cigarettes. The relief of hearing from Burlington Approach on schedule at Weybridge and his farewell as he cleared us straight to St. Jean was anticlimatic compared to this view of highlighted fingers of the upper reaches of Lake Champlain.

Into the dust bin again, but heartened by the assurance that there was earth down there and that, as we had carefully noted, the ceiling was well above 1,500 feet. V-487 called for inbound from the Warden Intersection located midway between Burlington and St. Jean.

"Bonanza 2024W your clearance limit is the Mike Radio Beacon. ILS Runway 24L. Report St. Jean."

The seat is no longer a lounge chair, nothing presses against its back. I'm in the *achtung!* attitude favored by the more cowardly of instrument pilots just emerging from the chrysalis stage.

Gene is examining the Montreal area chart, and holds it for me to see. The 324R from St. Jean looks to us a likely routing, which would indicate a left, or westerly, turn at St. Jean. The chart for Runway 24L is examined and the localizer set up is fed into number one. But what and where is the Mike Radio Beacon? Gene finds it—it's the outer marker. Now experienced, his fingering improved by the unbounded confidence he has in me, the ADF springs to life, in a stream of spaced double-dashes. The needle moves down into the upper left quadrant, a location that reason dictates it to be.

"Bonanza 24W your frequency and course are correct for ILS."

Reassuring, since the Montreal charts may be out of date.

"Bonanza 24W St. Jean. Depart St. Jean 350-degree radial. Expect radar vectors to intercept localizer."

"Did he say the 353 radial?"

"That's what I heard."

"Okay, but it beats me. I thought we'd head 325 or 330 degrees."

The frequency, while we'd been on it, had seemed like a private line. Coming alive as a party line, the controller began getting his ducks in a row—or a big arc, rather—swinging in a semicircle north and west from St. Jean and back south and west to the Mike Radio Beacon. Sounded like about six of us

were markers on the string. Satisfied with his inter-fix spacings, Arrival Control now had to get them moving at proper speeds to maintain his planned separations.

"Twenty-four Whiskey understand 160 indicated. Turn left heading 325."

Then he turned us further left to 310, putting us on an inside track on this aerial highway that now began to slope downhill as descent was authorized and maintenance of 160 knots requested. Adding alacrity to the zeal with which we executed these commands was a little pointed commentary by the controller on the slipshod way a Northeast jet was reducing speed—counterpointed, at least in my mind, by repeated references to Northeast's traffic, which was a Bonanza. Didn't that Northeast captain know that it was *me,* father of young, that was his traffic? In fact, resisting the impulse, I was prepared to use the party line to advise Northeast that Gene and I between us were father to 10 children.

"Bonanza 24W five northeast Mike Beacon. Descend to and maintain 2,000. Now intercepting localizer. Contact tower on 119.1."

The result of a gentle turn, the needle moved slowly to center and held in the lovely, smooth air. At 2,000 feet, a big door in the sky opened, the ground appeared, and straight ahead—as welcome as a completed section of the Pan American highway—was good, handsome Canadian concrete laid out on a heading of 24L, all atwinkle with HIRL, ALS, REIL and clear, lighted air.

Enjoined by the tower to feel free to do so, we landed. Changing his tone of voice slightly, he urged us to make best speed, as it seemed our old Bonanza-eating Northeast friend was on a three-mile final.

We were downtown before 5:30 and the meeting convened, freeing us to fore-shorten the next day's discussions enough to catch Air Canada's three p.m. flight to Kennedy. Prudence dictated this, since the weather from Richmond north had turned sour. In such cases, junior birdmen are wise to RON or go on the airlines. We elected the latter, held at Poughkeepsie for an hour, made it to Pawling for a 45-minute hold and then came back to Montreal when Kennedy closed down. So it goes. Some you win, some you lose.

The gold-plated, brass-bound upshot of the trip is that—being led by the hand all the way notwithstanding, and ably assisted by Gene—the flight did move from A to B and then from B to C in cotton-batting and came out all lined up for a landing. No big deal, but still 316 nautical miles.

I was elated. Better, for sure, that it happened so soon. If I'd put it off, I'd probably continue to put it off. But the baptism is behind me, and I can look forward to more instrument flights. Like this one: with someone along who can help with the navigation, where violent air is not anticipated, where the terrain is not ominously inhospitable. Jack says that *real* instrument flying calls for a twin and a copilot, and he's right. Coming into JFK in wild and wooly weather is no place for a tyro in a single-engine Bonanza. But a dozen more flights like this one to Montreal and I'll be confident that I can handle moderate instrument conditions away from the high-density areas. *Then,* when that superbly-equipped twin comes along, I can start, working on JFK, with Jack or another pro along.

What happens when, in a single trip, the elements conspire to challenge all the capabilities—and limitations—of instrument flying.

A DAY IN THE LIFE

BY RICHARD L. COLLINS

Any flight can illustrate many of the fine points about flying, but occasionally a day's flying almost completely frames a subject. The best example of this I've seen came one rainy day in March. The circulation around a low to the south spread a rather complete weather collection over my 300-nm route of flight, putting many IFR theories and procedures to a good test.

The alarm clock rang that morning at 4:30. Heavy rain was falling, but no static was evident as I listened to the transcribed weather broadcast during breakfast. The broadcast, featuring a forecast made the night before, mentioned the low and included the inevitable prediction of embedded thunderstorms.

My call to the FSS cast a pall on the outlook for a successful completion of my planned activities. I had a 10:15 appointment at my destination to test-fly an airplane. Since the forecast called for 300 overcast and one mile visibility at that time and place, why not just cancel and try again the next day? That was my prime thought, but then a better one came to mind: nothing ventured, nothing gained. If the weather should look suitable to start flying in that direction, start. I had caught myself looking for excuses not to fly rather than putting my mind to the task at hand.

The flight service station briefer painted less than a pretty picture of the en-route weather. His weather radar repeater was broken, but the digital radar plot available to him indicated a lot of activity for the first 150 miles. Some of the action was shown as heavy, but there was still no static on the low-frequency radio, so I filed IFR. Temperatures aloft were forecast to be relatively warm. At this point, I planned to go to the airport and look at the radar screen in the National Weather Service office, which is conveniently located near my T hangar. The radar showed a relatively precipitation-clear path in the direction I wanted to go, and I took that as a signal to start.

My Skyhawk was soon in cloud at 6,000 feet. The air was smooth except for an occasional light jiggle and a very infrequent moderate bump. The groundspeed was higher than it should have been, based on the winds forecast, which indicated to me that the low pressure was probably not located where the computer that had forecast the winds thought it

might be. A bad wind forecast is always a sign to watch for other surprises.

Some static was evident in my ADF receiver about 100 miles into the flight, so I started to quiz each controller carefully about any weather return he might have on the scope. There was some weather about 30 miles off course to the south. This had appeared on the NWS radar screen I had looked at before takeoff and was about 60 miles south of my course at that time—a reminder that information obtained before takeoff often becomes obsolete rather quickly.

Two of the flight service stations in the area have weather radar repeaters, and I regularly dropped off the center frequency to talk with the FSS people to compare their observations with those of the center controllers. The latest sequences were also of interest. I had selected an alternate based on legality, and it was barely legal. I had no intention of flying to the destination, shooting and missing an approach and then going to that alternate. If the destination were not above minimums and showing an improving trend as I got closer, I'd stop short, take on a full load of fuel and then proceed. My practice in such a situation always has been to start making the final decision about continuing to the destination when coming up over the last en-route airport with minimums and an ILS approach. As I neared that point, the weather check this day indicated that a continuation was okay.

Two hundred miles into the flight, I had been flying instruments for almost two hours. There was some rain, and at this time I had something new to contemplate. The temperature at 6,000 feet was running lower than forecast: it was only 2° above freezing. So I started asking two questions of each new controller: any weather, any icing reports?

There was a temperature inversion—warmer air aloft—and I thought that 6,000 feet might be the warmest spot. This idea was reinforced by a pilot report about light icing during a climb through 4,000. That added even more interest. If I started collecting ice at 6,000, it might well be colder and icier below. How about climbing? You just don't count on much of an ascent from 6,000 in a Skyhawk as it starts to collect ice. Surface temperatures became my out: 38° at the destination and 42° at the airport I had just passed and was holding open

as a refuge in case of need. The only problem was some heavy precipitation, possibly thundershowers, to the southwest moving northeast. They could make that refuge an uncomfortable alternate, so I called the FSS again and developed plans for a second refuge if any ice started to build and if storms moved over the primary choice.

Then I flew out between layers for a while. No way to have icing problems there. I had been on instruments for two hours and 15 minutes, and it was nice to fly visually for a change. Gauge flying is really no harder than visual once you settle in, but it is somehow confining. When flying out of the clouds that morning, I recalled how nice it used to be to open the top of a Link Trainer after an hour and survey the room, the great outdoors. Same feeling.

I also thought about the differences between VFR and IFR tasks. Some pilots make IFR sound difficult, but it is not. There's no autopilot in my Skyhawk, so the airplane has to be flown by hand from start to finish. That's no problem, though. The airplane is relatively stable, and looking at charts, figuring estimates and writing things down can be accomplished easily. The airplane is relatively slow, which means I have more time to complete the chores on a given flight than does a pilot flying a faster airplane. All the same monkey-motions are required whether it takes two hours or three. There's still not that much to do. The physical tasks are confined mainly to squawking and talking, and neither of those takes any high degree of concentration or intelligence.

In fact, I look for things to do during the en-route phase of flight. I keep the weather carefully checked and recorded—that adds a chore, keeps the brain active and contributes information that can be vital to the safety of the flight. I try to maintain a continuous challenge of the "go" decision. That flow of weather data feeds this thought process, which in turn keeps the mind active. I also keep a log of all the engine instrument readings, including those from my four-cylinder EGT, and find that I'm better about getting this done when flying in cloud than when flying on a clear day. Considering good risk-management technique, it certainly is logical that you have more time to piddle and doodle when flying in cloud, because there is no requirement to look for other traffic. On a fine VFR day, you had better keep alert and look for other aircraft while enjoying the scenery. On an IFR day, you might as well be sitting in the airplane in the hangar—except for the requirement to keep the thing upright, on altitude and heading.

So much for the daydreaming. Back flying in cloud now. It was time to get psyched for the approach. The thermometer said it was freezing, but there was no ice. The pitot heat had been on all morning. I watched for ice on the way in, mainly for information. ATIS gave the ILS to Runway 1R. It was time to study the plate and set both altimeters. Surface wind was light, as was the wind aloft. No wind shear possibilities. I received vectors to the final approach course and set the RNAV on the outer marker site to monitor the vectored interception of the localizer. It would be close, maybe a couple of miles outside the marker and at a pretty good angle. That's okay in a Hawk and especially no problem with RNAV to use in monitoring the vectors. When the RNAV shows a half mile from the localizer course and the controller is talking to someone else, I start around to the inbound heading, in anticipation. It works out just right.

The ATIS had said the weather was 600 overcast and five, and the ATIS was correct. The ILS was thus not a tight one. I parked the Hawk and was at the appointed place at 10:15, right on time. Routine IFR all the way.

Knowing the weather profile through at least 6,000 feet, I elected to go on and fly the airplane I had come to fly. However, when taking off about an hour after I had landed my airplane, I found an entirely different weather profile. We started picking up ice at 5,000 and carried some on top at 8,000. We picked up a lot more ice on the letdown and didn't start shedding it until reaching 4,800. Weather does change rapidly.

After completing that part of the day's mission, I went off to have lunch with some other folks. The next stop was the flight service station, to see what had happened to the 300 nm's worth of air since I last used it. The briefing indicated no real problems for IFR except the continuing forecast of embedded thunderstorms over the route. Local radar showed nothing but light rain for as far as it could see, so there would be no initial problem. That is the "start" signal. Ice seemed the biggest potential troublemaker, but the FSS folk were not concerned about that. No pilot reports about ice. I gave them one from my local flight, and they perked up. It's easy for a pilot to forget the legal and moral responsibility to give pilot reports, but reports should not be neglected. Your fellow aviator might glean the necessary information for a good go/no-go decision from your words of wisdom. Based on my own pilot report, I filed for 4,000 feet—lower than air traffic control usually likes in the area, and the wrong altitude for the direction of flight. But unless they wanted to fool around with a Skyhawk turned Popsicle, they would have to approve it.

The weather had again changed complexion. There was snow at 4,000 feet—a lot of snow. The temperature remained just above freezing, though. The surface temperature at the departure point was still 38°, and it increased to 42° in the first hundred miles of my flight. That made 4,000 even more comfortable. I'd be out in the boondocks, over flat terrain, so if enough ice did develop to outgun the little 150 Lycoming, there probably would be no conflicting traffic below and certainly no escarpments beneath me if I sagged to a lower altitude while flying toward warmer air. Still flying in solid cloud, I was to get almost my whole six months' worth of required gauge flying this day.

The controller called to say that my radar contact had been lost. This fellow reported it very routinely, but some controllers say those words with an ominous tone. What difference does it make to the pilot? I certainly couldn't have cared less at this time. I had carefully checked for a weather return on radar, and there was none. It was comfortable flying, and my only interest was in not having a conflict with another IFR airplane. Presumably, the controller was taking care of that little item by not clearing anybody to fly in the same area at the same altitude. I would be willing to fly to the end of the earth out of radar contact as long as I knew there was no conflict with another airplane flying in the clouds and no thunderstorm at 12 o'clock.

After reporting at a specified distance from a vortac, I was back in radar contact. I kept quizzing controllers about the weather as I moved along, as well as about cloud layers. Another pilot volunteered that there were clearly defined layers farther down the line, and I looked forward to new scenery.

Airline pilots were reporting vigorous turbulence at higher altitudes (as they had done during the morning trip), and I derived some satisfaction from that, since the Skyhawk twitched only occasionally. On most days, the Skyhawk wallows around in the bowels of cumulus while the airliners whine serenely along on top. No way my Hawk can climb to FL 370 for smooth rides on days when it's rough down low; no way those jets can descend to maintain 4,000 when it's rough up high. Each airplane has its moments.

When I flew over the last ILS approach before home plate (111 nm away), my decision to continue again took some thought. The FSS at home is remoted through the vortac at the point I was approaching, so I could talk to them in person. A familiar and friendly voice gave me the latest weather: 200 obscured, two miles, moderate rain and fog, wind 070 at 20 with gusts to 35. Nothing like that on any forecast I had written on my piece of paper, and it did not sound like an appetizing environment for a Skyhawk. I was between layers at this point, though, and the upcoming problem was defined in the windshield: higher clouds, lower clouds, and a very dark gray line in between. It was not the type of situation you'd stumble into. My FSS friend said the first of the precipitation was about 40 miles from home. I was still at least an hour out. I decided to continue: the airport I was passing became my newly designated safe haven in case of need.

That 200-and-two report was actually for my alternate airport. I had filed to my home-base airport across the river, five miles from the municipal airport, which I had used as the alternate because it was legal. Now it wasn't. Someone once said that in such a situation you should call and change the alternate on your flight plan. Why, though? I knew where I would go, and that's what counted. I had plenty of fuel to fly on home, fiddle around there and then return to an airport that had all the appearances of breaking wide open (from an IFR standpoint), since its ceiling was still 600 feet, but the visibility was 15 miles and radar information indicated nothing moving in.

I elected to fly the north airway. It adds only five miles to the distance flown and avoids an area of rough terrain, an area that would probably be obscured. Why not stay over or within gliding distance of a nice wide valley? Engines don't quit; they seldom ever go putt-putt-putt, but it's wise to keep flatter terrain beneath you if there's a choice. Five miles is not much to invest. I wondered why I hadn't taken the same course on the flight up, earlier in the day.

As I got closer, it appeared that the weather would be okay for the RNAV approach to home base, minimums 458 feet agl. The wind might be a problem, though. The runway is north/south, and 70 degrees at 35 knots would add more spice to the conclusion of the flight than might be necessary. An air carrier and a light twin reported various and sundry combinations of turbulence and wind shear, and the question was whether or not I wanted to risk having to fly extra time in such conditions. If I made the RNAV and missed it or was unable to land because of the crosswind, I'd have to fly a missed approach and then the ILS at the big airport for a landing on Runway 4.

The surface wind began to abate as I got closer; the highest value reported was 21 knots as I began the RNAV approach.

I laughed, too, at a question I had asked the controller before starting the approach: "Would you call the tower cab and ask them if they can see the Lakewood House?" That's an apartment building near my airport, and if the tower can see it from their perch three miles away, conditions probably will be pretty good for my approach. Bet that question would cause consternation if I arrived at some point other than the runway and the NTSB played the tapes. They'd just know I had developed some clandestine apartment-house approach to cheat on minimums, based on the tower's observation of that building.

The weather was improving, the crosswind was bumpy but manageable, and I soon had the Hawk gassed and rolled back into the T hangar to rest and await the next flight.

The day had been interesting; a good review of many things. I came away with a slight headache, which suggests the IFR effort is a little more demanding than when the weather is clear, but I think that is related more to an increased demand on the thought processes than any actual increase in physical or mental workload caused by the simple fact that the airplane is in cloud. In fact, I've come away from VFR flights, during which thunderstorms had to be circumnavigated, with much more fatigue than I had felt after spending almost six hours flying inside clouds on this day.

IFR makes an airplane work, too. I keep a little tally sheet on the Skyhawk's performance in reaching appointments. As the airplane nears its thousandth hour, which will have been flown in a bit less than two years, it has done well. Six trips have been canceled: two because of thunderstorm activity, one because of surface winds, one because of an inoperative beacon on the ground, and one was as impromptu en-route cancellation for a mechanical problem (lead poisoning from 100-octane fuel). Five flights have been delayed substantially (two hours or more) because of weather, but all were completed on the planned date. Four hundred and twenty-seven cross-country legs have been flown more or less on time; 94 of the 427 would have been difficult or impossible to fly VFR.

So, the use of IFR *does* make it possible to keep a 10:15 Monday morning appointment 300 miles away and to be reasonably confident that you'll be there on time and will get back home that evening. That is what transportation is all about.

PART VII

A Good Idea at the Time

A lot can happen in 50 years, among them excellent intentions gone astray leading future readers wondering, "Now why did they ever print *that*?" Excellent intentions can also be fulfilled, which is borne out by the growth of the *Flying* readership over the years and the generally fine reputation held by the magazine.

Some articles in this section have an unintentional quaintness to them, which is in part due to the periods during which they were written. For instance, it may seem hilarious to us today to think of people trying to learn to fly by sitting at their radios manipulating a broom handle and applying body English in pretending to turn their "ships." But the 1920s were like that, full of "innocent" ideas about the uses of commercial radio. Similarly, "What Men Flyers Think of Women Pilots" bespeaks a male chauvinism that was taken for granted, especially in the aviation world, which did not make it easy in pre-women's liberation times for women even to get started as pilots, much less make careers out of flying. In 1938, an article by J. Edgar Hoover about the use of airplanes by the FBI must have seemed downright brilliant, given the awe in which the FBI was held at the time.

The stories in this section have been included for amusement's or curiosity's sake except some that have a special *quality* about them. Robert Peterson's "This Is a Very Peculiar Assignment," an "inside look" at the inner workings of *Flying*, is funny and an exaggeration, but at times hits tellingly close to home. After all, a magazine should not take itself too seriously.

The two pieces by Gordon Baxter are from his very popular series, "Bax Seat." In "Hawk Lips" and "Garbage," Bax has summarized many of the feelings that we have about flying, the exhilaration and wonder of it, the homey feelings one gets when being with fellow pilots, the exuberance at thriving little airports and also the feelings of sadness that accompany the losses that pilots sometimes suffer, be they friends or beloved airplanes.

We think this section is aptly named. These stories, and others we would have liked to include, *were* good ideas at the time of their writing—and many of them still are.

Left bank (note the body English).

A Chair-A Stick-A Radio

BY WILLIS PARKER

WE never will know just how many of the thousands of radio fans in the middle west and Rocky Mountain regions placed kitchen chairs before their receiving sets, seated themselves more or less comfortably with their left hands firmly grasping broomsticks between their knees and eagerly followed a course of ten lessons on how to fly as presented over Broadcasting Station KOA, Denver, by Chief Pilot Cloyd P. Clevenger of the Alexander Aircraft Company.

The fans were assured that, if they supplied themselves with kitchen chairs and broomsticks, they could get more out of the lessons than otherwise. However, the chair and broomstick idea were more for the purpose of creating atmosphere than anything else, for it is obvious that learning to fly by radio or by a correspondence course is out of the question. The entire purpose of the radio lessons was to create air-mindedness and arouse in the listeners a greater desire to learn to fly. Each lesson was fifteen minutes long and they were given Friday evenings from 8 to 8:15 o'clock.

To arouse interest and hold the attention of the unseen audience and create aviation enthusiasm, the lessons had to be entertaining and chuck full of "atmosphere." To do this, the lessons were not presented directly to the audience, but the listeners received the information secondhand through Clevenger's efforts to teach an alleged gawky rube how to fly. The alleged rube was none other than Gene Lindberg, one of the most popular newspaper humorists in the Rocky Mountain region. Gene's comments upon the pilot's instructions, his remarks concerning the various parts of the plane, his soliloquies while "in the air" provided enough clean amusement to prevent radio fans from turning to other stations where jazz orchestras might be performing. There was also a fortunate coincidence in names, for Gene's name is exactly the same, as far as pronunciation is concerned, as that of the famous trans-Atlantic flyer. In spelling it lacks the letter "H."

Through the liberal use of accessories, the audience was given the representation of the roars of airplane motors, which were produced by turning electric fans into the microphones. For ordinary cruising the fan was turned only part way on. For racing the motor while warming up or at the takeoff, a full blast of the fan did the work.

The lessons were in dialogue form and began with the repartee that resulted upon Gene's arrival at the field. The first was a simple explanation of the controls and the various parts of the ship. Gene asks some pointed though humorous questions regarding the qualifications for good pilots, which Clevenger answers and in doing so indirectly assures every listener that almost any person can learn to fly.

In a subsequent lesson Gene asks "When a fellow like me puts up his hard cash for flying lessons, he wonders when the investment will pay dividends. Do you think I'll make my first five hundred thousand by 1930."

This gave Clevenger an opportunity to explain that there is a demand for salesmen with flying ability as well as a demand for good pilots. This should have aroused interest among the radio fans who were seeking financial opportunity and should have created in them a desire to get into the game.

As an illustration of the humorous touches, note this bit of dialogue:

"GENE—Clev, this air feels awful thin to me, today. Are you sure there aren't any holes in it?"

"CLEV—Nope, no holes. An Eaglerock just came in with a fresh load of air from sea-level. I patched all the holes myself."

Illustrating how some of the points may be presented through the medium of dialogue, mixed with a little humor, note the following which took place, supposedly, when the duet were preparing to go up and were starting the engine:

"CLEV—Don't say 'ON;' say 'CONTACT.'

(Sound of motor.)

"CLEV—Hold the stick back, now, and give it the gun. There are blocks in front of the wheels."

"GENE—Then I can't shoot anywhere?"

"CLEV—Nope."

"GENE—Why give it the gun, then? * * *"

"CLEV—You're green, all right! Always remember this. The minute your propeller speeds up, it naturally tries to pull the ship ahead. It can't go ahead with blocks under the wheels, so it tries to stand on its nose. With the stick held back the propeller blast presses on top of the flippers and holds the tail down."

(Blocks are removed and the ship taxis off.)

"GENE—Hey, Clev. This thing wobbles all over the field! What have you got in the radiator, alcohol?"

(After much additional repartee they start down the field with Clevenger issuing instructions, among which are:

"CLEV—Use the ailerons to keep your wings absolutely level."

"GENE—Aha! Go straight and always be on the level! You can't cheat in these things!"

"CLEV—If you get the impression your right wing is low, move the stick to the left enough to bring it up immediately. Now, for the elevators. When you're on the ground, your nose is way above the horizon."

"GENE—I can't help my nose. I was born with it."

"CLEV—Cut out the horse play. Hold the stick forward now, at the start of the takeoff, until the nose comes down almost to the horizon, then return the stick to an approximate neutral."

(They are in the air.)

"GENE—Wow! Something's gone fooey. Wat the————?"

"CLEV—Gene, I'm ashamed of you. I thought I taught you not to skid."

"GENE—So that's what I was doing? You ought to have tire chains on these babies."

"CLEV—There's no mud or soft sand in the air, Gene. It's all in your head. You're in too big a hurry to get your nose around the horizon. Take it easy. You're putting too much inside rudder on your turns. Now! Straighten her out and fly south."

"GENE—South, South! Lord, man, which way is South? Think I'm a duck?"

In one lesson Gene is taught the principles of stalled flight, which Clevenger explains is necessary to know how to prevent getting into accidental tail spins. He explains that Gene probably will go into a spin and Gene promptly has a heavy feeling in his stomach and declares he feels pale. However, the tail spin is made, corrected, and everything ends with a good landing.

In the lesson on looping, Gene has some thrills that are humorous yet dangerous. Clevenger speaks at length on the need of watching the horizon, keeping the wings level and pulling the nose up to the horizon when the loop is completed, to which Gene remarks that "You sure keep your nose on the horizon a lot in this flying business." In reference to keeping the plane straight in a side drift wind, Gene remarks, "Funny things, these planes. Steer 'em straight and they go so crooked you've got to steer crooked to go straight."

The lessons were based upon Clevenger's book, "Modern Flight." All through the dialogue, where appropriate, mention was made of what was said in the book.

Just how effective the radio advertisement was, will never be known, but there were hundreds of copies of the book sold shortly afterwards and an increased visitation to the aviation school. Undoubtedly scores of listeners were led to investigate especially after hearing Gene's soliloquy on his solo flight, parts of which follow:

"Holy Cats! Here I am, 500 feet up in the sky and no Clev in the front seat. * * * Maybe I shouldn't have done this—maybe I'm not good enough yet—maybe. Aw, rats! Clev knows his stuff. He wouldn't have let me solo if he didn't think I could do it. * * * Wonder what Clev's doing now? I'll bank and get a good slant at him. There he is, the little runt, no bigger'n a bowling pin. * * * He ain't making any motions. Just watching. Why shouldn't he? This is a darn keen bank. No skid, no slip. No wind on the cheeks. * * * Gee, this thing works just as well now as when Clev was with me. * * * Talk about having your hands full—a fellow's got his feet full, too. * * * There's Clev, waving his helmet. Guess I'd better drop back down. * * * No wind on my tail this trip. Down she goes in the glide. * * * Easy there. Round her out nice and gentle. Ooops! That left wing's a bit low. There, level as a lake. Now we're slowing. No you don't, you son-of-a-gun, I won't let you settle yet. Three feet off. Sitting pretty. Two feet. Little more stick. Ahh! A three-pointer. I knew I could do it. I'M AN AVIATOR NOW."

Pre-War Stuff

BY BERTRAM W. WILLIAMS

FOR some reason the aviator—the "intrepid aviator" as the press termed him—was always taken more seriously in Europe in the pre-war days than he was over here in the land where human flight was first proved possible. In England, France and Germany he was a public hero, little short of a demigod. Here in the States, people looked upon him as a kind of glorified circus clown with more nerve than brain.

Now, ever since I was a boy, I had wanted to be a hero. Not the kind where you "did your duty" and were rewarded with the hand of the heroine, but the sort who got pointed out in public places, at whom people stared and whispered, "That's him!" A popular idol who received a salary for doing daring deeds—that was my real rôle, I felt sure.

True, the novelty of flying had worn off a good deal with the public. They had come to accept aviation as an actual fact. But the daring soul who piloted a plane outside of an aerodrome was still a person ranking above ordinary mortals in popular esteem.

Six months after first seeing an aeroplane in actual flight I was in London, and the second day of my arrival there found me hovering outside the great flying field of Hendon where there were several schools of aviation. Somewhat timidly I entered the gate—paying a shilling for that privilege—and stared around. It wasn't so impressive as I had imagined. A few planes were droning monotonously round and round less than a hundred feet high, and occasionally a small group of men in greasy overalls would come out of one of the big hangars to watch them. When a machine landed, it was instantly dragged inside its shed to be inspected for damages. I found out later that very few pupils ever did land without breaking a wire or

shaking loose a bolt or something. The planes of those days were not very strongly constructed, neither were the landing fields particularly smooth. Also, if being an "aviator" gave one a certain amount of prestige with the man in the street, it meant absolutely nothing to the real pilots and mechanics in the aerodrome. A student flyer was looked upon as a nuisance. Of course they had to let him take a machine out now and then; but woe betide him if he cracked it up. Parachute packs and safety belts were unknown. If a pupil and his bus parted company in the air—which really did happen sometimes—a small crowd of anxious-eyed men would dart from the hangars to first inspect the wreck. Afterwards, if one of them happened to think of it, they might stroll over to see what was left of the human element. A thing of scorn was he who crashed a plane. This was a somewhat drastic attitude, perhaps, but it weeded out both the timid and the reckless. For in spite of low wages and the comparative cheapness of materials, aeroplanes and aerial motors were proportionately far more expensive than they are now.

Of course, on the voyage to England I had read a great deal about flying and had assimilated a vocabulary of new and strange words, most of them quite Frenchified. In my first half-hour at Hendon I was thus enabled to pick out the different types, although I didn't know any advantage one had over another. They all seemed to make a great deal of noise and to hurtle through the air much faster than was pleasant.

I liked the look of the Caudron biplanes; they had pretty blue wings and a little canoe-shaped structure where the pilot sat amid a maze of struts and wires. The Farman "box-kites" did not seem so good to me. They were dual instruction machines with the pupil squatting in an open seat below the instructor and a great big engine perched overhead ready to give both a nasty poke in the back in the event of a bad landing.

Then there were the monoplanes. There were the long-bodied Deperdussins with their square-cut brown wings, and the trim little Blériots, the same type in which their inventor had first flown the English Channel. They attracted the eye, but I remembered having heard that monoplanes were much more difficult to fly than biplanes.

I was standing in doubt when a young fellow in a leather helmet came over to me. People in England don't as a rule address perfect strangers unless they are drowning or their house is on fire. So I was quite flattered when this chap showed an interest in me. I told him I was thinking of learning the flying game and had half decided to enroll in the school that specialized in Caudrons.

He laughed scornfully. "A biplane! What does a biplane pilot amount to?"

I couldn't tell him.

"Now if one can fly a monoplane," he went on, "one can fly anything. The army's crazy to get monoplane pilots."

"But isn't the other kind easier?" I asked timidly.

"Aw, any fool can take one of those box-kites around. Better join our school. I'll introduce you to the instructor.

The latter gentleman proved to be a rather supercilious young man wearing razor-edged trousers. He was sparse with his conversation, yet somehow he managed to convey the impression that he was doing me a favor in accepting my check.

The standard fee in England at that time was seventy-five pounds, $375 in our money, according to the then prevailing rate of exchange, but in reality more like $1,000. In Germany the fee was higher and there were clauses about responsibility for breakages. All the European schools, however, guaranteed that each pupil be given instruction till he passed the tests laid down by the Federation Internationale Aéronautique of France which entitled him to an aviator's brevet or certificate. These tests would appear almost childish nowadays, but on the flimsy machines then in vogue they were far from simple of accomplishment. Only two flights (solo) were required at an altitude of not less than fifty meters, each flight to consist of five figures of eight 'round a given course, when a landing had to be made with motor cut off within ten meters of a designated spot. The landing was less easy than the flight, as a pupil would usually misjudge his distance or forget to switch off his engine as soon as he touched the ground.

Once I had signed his long and vaguely-worded contract my new teacher seemed to lose all interest in me. He answered my questions with curt brevity, and after recommending me to buy myself a suit of cotton overalls, excused himself.

Much to my disgust I found that overalls and ordinary caps worn reversed (the latter, not the former) were considered all that was necessary for a student flyer to wear. To have worn a leather suit and helmet such as I had wished to get would only have caused ribald remarks. Even goggles were taboo. You might get a gob of castor oil in each eye during flight, but what of it? It was the custom of the country. Most of the pupils were British army officers who during leave of absence were paying for their own tuition. They were a cliquey lot with a great disdain for mere civilians. Even the French mechanics and instructors appeared to share that prejudice—a shadow of the war to come.

Actual flying instruction in those days consisted of visiting the aerodrome before sunrise, where if there was no wind, a battered plane with a 25 h.p. Anzani engine was wheeled out of its shed. The ignition was retarded on the latter so that there was no possibility of the machine leaving the ground. When your turn came, you climbed into the oil-spattered cockpit, a mechanic swung the propeller, and you were left to your own devices. Fortunately the field was big and there was plenty of room to go rolling about in it. Rolling about, that was all one could do. A Blériot monoplane had caster wheels which swung in all directions, making it impossible to steer a straight course unless top speed was attained.

Round and round one would go like a dog chasing its tail. There were two of these ancient crates to play with, but they were both so decrepit that the least rough spot on the field would upset their internal economy. Bolts were continually working loose, wires snapping, or their fan-shaped air-cooled engines going wrong. I didn't know much about motors in those days, but I am now firmly convinced that neither of those Anzanis gave more than five or six brake horse power. As we were never permitted to take a machine out when there was the faintest breath of air stirring, our wings did not sprout very fast. The one aim of our instructor was to get the planes back into the hangar undamaged. He would then fill the tank of his car with the school gasoline and motor airily homewards.

Although there was seldom anything but jealousy among the pupils themselves, a common grievance drew us together. After all, we had paid our money to learn flying, not to ride on a freak-wheeled contrivance that splashed oil all over us. The result was that I composed a letter which we all signed and sent to the great Blériot himself at his Paris home.

Instead of sending the usual promises and regrets, that distinguished pioneer came over to England immediately to personally investigate the manner in which his London school was being conducted. No one had ever seen our original instructor—he of the creased pants—fly; but he certainly "got the air" when his employer visited Hendon. A young Frenchman was appointed in his place, a man who did not think it beneath his dignity to give a pupil the benefit of his own experience. Also, and what was even better, a new Penguin was placed at our disposal. This was an ordinary Blériot plane having a more powerful engine, but with wings cut down so that they could not lift the machine off the ground more than a few inches. One was thus able to get up enough speed to steer the machine much the same as though it were in the air. The confidence thus gained was invaluable. A few runs across the field and I was ready for the next stage, a 28 h.p. engined machine that actually flew. By tugging violently at the *cloche,* one could lift the plane ten or twelve feet in the air and skim proudly across the drome. The control-lever or *cloche,* it must be explained, differed from the usual joystick in having a small fixed wheel on top in the center of which was set the magneto switch. Usually a pupil finding himself leaving dear old Mother Earth would hurriedly cut off the ignition and plop heavily downwards. It was more than funny to see some embryo pilot go bounding across the field like a scared kangaroo.

I soon got quite at home on this machine. Too much so, for I became firmly convinced that my ceiling was about ten feet! Jules was beginning to give me sour looks, which changed to harsh-sounding words in French when one day I crashed the "28" into a fence. However, I somehow convinced him that it was only my high spirits and that I was past the diaper stage of aeronautics, anyhow, and ready for the special "brevet bus."

This machine was a real he-pilot affair. Its 35 h.p. motor was quite new, certainly not more than five years old; and the wings were fairly well lined up with no more than a four-inch list on one. It was wheeled ceremoniously from its hangar one morning, and I clambered in, looking, I hope, a whole lot more confident than I felt. After being warned by the instructor that she was "veree power-ful," I switched on and was off. We didn't know anything about chocking the wheels at that period. There was plenty of room, and as long as one could dodge the cattle and sheep which dotted the aerodrome, one would eventually leave the ground.

Contrary to my expectation, I found the increased power made the machine quite easy to handle. After a few straights at my usual ceiling, I set my teeth, pulled firmly back on the stick, and found myself flying slowly round the aerodrome at the stupendous height of 200 feet. Circles and figures of eight followed without difficulty, though I was careful never to bank my slow-moving bus. I never experienced the least nervousness in landing. Of course, some of my descents were rather bumpy and I never gained enough confidence to come down with the motor fully cut off. In those days a volplane was considered quite a feat and not encouraged among pupils. Several oldtimers commented on the remarkably steep angle of my landings; but it was not till many months later that I learned this was the result of keeping the stick rigidly fixed in the position I had pushed it into to descend. No one had ever told me to return it to neutral once the bus had started down. Imagine what would happen on a horizontal plane, if when you rounded a corner in your automobile, you locked the wheel!

My first brevet test came off all right, though I landed much further away than was prescribed in the rules. In the early stages of solo flight one is apt to concentrate so intensely on the handling of the machine that one has no room for any vagrant thoughts. The arithmetical problem of counting five— the number of "eights" required in each test—was too much for me. On my second brevet flight I lost count entirely about the third round. After that I didn't know whether I was on my seventh or merely the fourth till I noticed a little knot of men in the field below excitedly waving to me. I was badly worried,

thinking the under-carriage had dropped off or that something equally disastrous had happened. Then it struck me that they were merely encouraging me to keep on going. With an effort I dismissed them from my thoughts. A little later I took another peek over the side. The fools were now dancing about in their excitement. Did they think I couldn't fly? I'd show 'em. When I actually did descend I was still a little doubtful whether my quota was complete; yet they told me they had tallied up to thirteen and then given up the count, and a fellow-pupil even hinted that I was afraid to come down!

Yes, learning to fly in the early days was a catch-as-catch-can proposition. The machines were underpowered rattletraps that had been cracked up and patched together a dozen times. The aerodromes were surrounded by tall pylons waiting to be bumped into, and the air was often crowded with a dozen planes flying erratically in all three dimensions of space. Still air? Not so you could notice it. The cross-currents caused by the backwash of all this aerial traffic were fierce and very unsympathetic with our feeble crates. To enhance the pleasures of landing and taking off, live stock grazed unconcernedly all over the field.

Hendon was a popular resort at that time. Celebrated French and German flyers would cross over from their respective countries and land there. Weird-looking machines were tested by the pilots and criticised by the groundlings. Races were held every Sunday which the highest in the land came to see. An aviator was still a person of consequence no matter what his former station in life had been. When Pegoud first looped the loop in a Bleriot monoplane, London went crazy. People were excited, not because it was a spectacular stunt, but because it was taken as a proof that an aeroplane was really a safer contrivance than a motor car—in competent hands.

Some of England's most aristocratic daughters paid fabulous sums to be taken up and "looped." When Gustav Hamel, the foremost British pilot, was reported missing in a cross-channel flight, the Admiralty sent a fleet of destroyers to try and locate him. Winston Churchill, secretary for the navy, was a constant flyer. As for the Anglo-French *entente*, whether it was a result or a cause of this mutual aerial activity, there is no doubt that the adulation which was showered in England on all the famous French pilots of that day had a lot to do with cementing it.

I was up 16,000 feet in a navy plane not long since, and on the trip down the pilot put his machine through many strange evolutions. Yet somehow I did not get the thrill I had in that ancient monoplane with which I made my first flight fifteen years ago.

What Men Flyers Think of Women Pilots

BY MARGERY BROWN

A WELL-KNOWN air-mail pilot, questioned as to what he thought about women going in for aviation, retorted that it wouldn't be printable! Assured that his opinion would be carefully censored before publication, he consented to cite a few of his objections, which, (carefully censored) follow:

Women make poor pilots because (1) they are too emotional, (2) they blow up in a crisis, (3) airplanes are mere playthings to them, (4) their flying is not constructive, since they cannot use it as a business, (5) they usually enter aviation for publicity purposes only.

Another pilot of equal experience, who is also a field manager, replied that ten percent of the women have enough intelligence to learn to fly, and the other ninety percent haven't!

This is a hard indictment. But I have survived severe blows where men flyers' opinions of women flyers are concerned. To be quite honest, I have yet to find a pilot who whole-heartedly endorses the idea of women flying. If one assures you that he approves, beware of taking him at his word; he's only being polite; dig a little deeper and you'll find a sharp resentment or downright opposition.

Women, so a third pilot informed me, don't take flying seriously enough, and they haven't the proper qualifications. They are far too inconsistent.

"If a woman had to make a forced landing," he explained, "she would pick out a field, and immediately change her mind; pick out another, get nearly there; change her mind again, start back to the first—and crack up midway between them!"

He added that due to temperamental tendencies, women make poor automobile drivers, and they will make equally poor pilots for exactly the same reasons.

These arguments are sound. Men flyers aren't nursing a grievance without reason. But it so happens that I am a member of the sex that is under criticism, and having derived incalculable benefit from flying, I am compelled to defend a woman's right to fly.

Yet I see the man's viewpoint, and I agree with him. On the whole, I think he is right. Women are not well fitted by temperament for flying under present conditions.

But, what of it?

Let them change their temperament.

What are some of the objections? (1) Women are too emotional, (2) they blow up in a crisis. Here are two reasons cited. The idea is, that in order to be a good pilot, one must be calm, unemotional, and have control of his mental reactions.

So, the barrier to women becoming good pilots is not the fact of sex at all; it is their apparent lack of the essential mental qualities. That ought to be quite clear.

But mental qualities are primarily thoughts, and thoughts are not the exclusive property of the male sex. There is nothing to prevent women from learning to think the thoughts men think about flying. A woman is a free moral agent where thinking itself is concerned. She can have wobbly, inconsistent, indecisive thoughts; or else she can have cool, calm, collected thoughts. The way she thinks determines the way she flies.

If one is content to be a fair-weather flyer, and is satisfied with occasional "field hopping," there is very little risk, that is certain. But who wants to circle around and around a field indefinitely? If one is at all venturesome, forced landings or other difficulties are a possibility, let it be admitted. The question is: how are you, a woman, going to react to trouble?

If you let your equanimity be disturbed when a hat turns out all wrong and makes you look a fright, ten-to-one you'll let that same equanimity be disturbed when a few cylinders begin cutting out! For the truth is, that if you want to be composed and to think clearly when you fly, you will have to cultivate clear thinking and composure under all circumstances. Then it will become a habit with you, and won't desert you in a crisis. You can't very well let yourself boil over with rage because a man (husband or otherwise) breaks an engagement with you, and then go into the air and take it calmly when a motor suddenly breaks its engagement with you.

They say that if one does not get excited, or lose his head, accidents can often be averted; or at least, personal injury can often be averted. But if a woman loses her head half a dozen times a day on the ground, she is bound to lose it frequently in the air.

The time probably will come when flying will require no more ability than automobile driving requires now, but that day is still quite distant. At present flying demands discipline, mental and physical. It is well that women should understand this fact, before they rush into aviation. Women need discipline, even more than men need it. Then, why should men think these moral benefits accrue to themselves alone?

If one were flying in the right spirit and really trying to overcome her faults, by degrees her nature would be transformed, and she would become a different person.

Nothing would do the average woman more good than to spend a few months working in "the shop," which is operated in connection with most flying fields. Here planes are mended, and motors repaired and overhauled.

These men are occupied, they haven't time for the amenities and pleasantries most women expect. If one borrows tools and neglects to return them, her femininity doesn't soften the sharpness of the reprimand. She is told bluntly that if she can't return tools to their proper places, she had better give up working in the shop. When she starts gossiping agreeably with an harrassed mechanic, he curtly says that he is busy, and turns his back.

If her vanity survives it, this is the best possible cure for what ails a woman! She learns to do exactly what the rest of them are doing—pay attention to her job. When she learns to pay strict attention to what she is doing, she will have overcome one of her most obvious faults.

The air mail pilot mentioned at the beginning of this article, who claimed that woman's flying cannot be constructive, certainly was mistaken. It should be remembered that the word "constructive" isn't limited to mere money making. Anything is constructive that teaches a woman to mind her own business.

The contention that most women enter aviation for publicity purposes only would have borne more weight a year ago than today. Newspapers aren't giving women flyers nearly as much publicity as they did, and this is something to be thankful for, because the public has grown weary of persistent harping on the personal note. That sort of thing is dying out.

His claim that women are too emotional is a sensible objection. This defect is responsible for about three-fourths of their misery, and the sooner they get over it, the better. It is a quality that does nobody any good, and it can be eliminated. The main reason most women are emotional, is because they have nothing better to do. Let them get interested in aviation, and they will quickly become less emotional—or else be forced to give up flying.

I was exceedingly amused one day at a remark made by a pilot who was selling tickets for passenger flights. I said to him, "Aren't women more willing to fly than men?"

"Yes," he replied earnestly, "on Sundays they are; but not on week days."

I laughed heartily at the time, but in thinking it over I discerned what is probably the psychological reason. On Sundays, women are dressed up in their best clothes, are stimulated, looking for excitement, for a thrill, for "something to happen"; hence they are receptive to the suggestion that they fly.

Now, a man would never fly because he was dressed up in his best clothes, and emotionally stimulated. This incident illustrates the fundamental difference between a man and a woman. It is one of the basic reasons for men flyers' objections to women entering aviation. They think women are apt to be too flighty and harum-scarum about flying.

If you are a woman, and are coming to the flying field seeking stimulation, excitement and flattery, you had better stay away until flying is a little bit safer. Where women are concerned, at least, sex and flying are a very poor combination. Sex means vanity, and it is almost impossible to fly safely, if you are thinking of how you look or the impression you are creating. One needs to concentrate all of his attention when flying, and a self-conscious person cannot do that.

If you have visions of yourself cutting capers in the air, and thrilling your admirers, perhaps you had better postpone flying for the time being. If you are planning attractive costumes, and wondering what color most becomes you in a flying coat—again, perhaps you had better postpone flying for a while.

But, if you are thinking that flying will develop character; will teach you to be orderly, well-balanced; will give you an increasingly wider outlook; discipline you, and destroy vanity and pride; enable you to control yourself more and more under all conditions; to think less of yourself and your personal problems, and more of the sublimity and everlasting peace that dwell serene in the heavens . . . if you seek these latter qualities, and think on them exclusively, why—FLY!

It will not matter to you what men flyers think of women pilots.

Roscoe Turner, with his familiar self-designed uniform and pet lion.

LANDING AN AIRPLANE BY PARACHUTE

BY CAPTAIN ROSCOE TURNER

NOW that I have demonstrated that an airplane may be landed by a parachute I believe that the time is near when every up-to-date plane will be equipped with a mechanically released, built-in parachute. By this means a crippled-in-the-air-plane may be landed safely saving the entire load of passengers, pay load and the airplane from wrecking.

It requires no stretch of imagination to visualize every airplane of the near future with a built-in parachute snugly stored in the center wing section in such a way that no evidence is in sight of the latest safety device. Neither will the lines of the plane have to be changed. It has been fully demonstrated that parachutes can be and are being used to land individual passengers safely. It is only a step further to apply the same safety device to a group and to the airplane.

It was only after many experiments with parachutes and weights and much study as to air resistance and other conditions entering into the project that I landed the first airplane by aid of a mechanically released built-in parachute. Had I not been confident of success I would never have tried it before 15,000 spectators at Santa Ana, California. Now that the initial test has proven so successful I look for rapid development along this line of safety first.

In this particular instance the result was accomplished by replacing the second rib on each half of the wing with a heavy ply-wood to take all the compression strain of the two ribs in general use. A steel rib was placed at the center section of the two half wings and the whole compartment thus created was floored with heavy ply-wood. This gave greatly increased strength to the wing and at the same time afforded a solid compartment for the pack.

At each of the four corners of the compartment a heavy coil spring was placed, also two at the center. Each spring had a recoil force of 2,000 pounds and was placed in a solid steel container bolted through the spars.

By this arrangement the releasement shock was equally distributed and a force of 12,000 pounds thrown against the pack at the release of the springs.

On top of the six springs rested the basket which was the parachute carrier. The basket was of heavy fabric webbing strips around a flexible steel frame. To load the parachute into the compartment, it was folded on top of its connecting cables in the center of the carrier which was then placed on top of the springs. Over the whole was placed a cover of ply-wood shaped to conform with the wing curve. In the cover were openings to fit over spring-locking devices. Thus in place, jack-screws were brought into use and the whole forced down into the compartment until the coil-springs were sent down to be caught by the triggers which held them until released by the pilot.

The trigger release was mounted on ball bearings to eliminate as far as possible all friction and afford a quick release. Connected with the springs were six synchronized levers attached to a cable running down the inside of the right center strut to the pilot's cockpit. The cable was attached to a lever on the instrument board. To release the parachute the lever was moved from left to right and the six springs were released simultaneously. At the same moment, the trigger released the fastenings on the cover of the compartment. The force threw the parachute well clear of the plane thus avoiding possibility of entanglement with the rudder.

The parachute may be of any standard make and of size in proportion to the weight of the plane and load. The one used in my test at Santa Ana, California, was a Russell Valve type, 60 feet in diameter with an effective lift of 1,600 square feet. It was made of the heaviest parachute silk and had 60 silk shroud lines, each of 400 pounds tensile strength. Each shroud line measured 60 feet from lip of the parachute to the Dee rings, of which two were used, this precaution to help stop oscillation of the parachute.

The airplane was attached to the shroud lines from the parachute by four flexible cables running from the bottom longerons at the two main stanchions up through the wing. Between these cables and the shroud lines of the parachute a connecting link was so arranged that it could be released by the pilot on landing—this to provide against the plane being dragged in case of strong wind.

Two days before the exhibition test a preliminary one was made from a straight flying position with the motor turned on. A load of papers of the weight of the parachute was used. From that test I saw that there was danger that the huge parachute, much heavier than the individual pack, might envelope both the tail and pilot's cockpit and render escape impossible. To reduce this danger to a minimum I had the extra steel cables added to get more distance between the plane and the shroud lines. The extra length of cables was used for the first time in the exhibition at Santa Ana.

At this test a three-place biplane was used. So far as the spectators could see it was the regulation biplane as there were no visible evidences of the parachute. I circled over the airport gaining altitude, also making observations as to wind currents and possible drift.

After a few maneuvers I selected the most desirable position and pulled the release lever at the same time side-slipping the plane a little. Hurled by a 12,000 pound force the great parachute came out of the compartment thrown well clear of the tail. It took form almost instantly bringing the plane into a combination loop and wing-over at right angles to the arc of the parachute-pulling shroud-lines, and the mammoth billows of silk settled down slowly earth-ward with plane and pilot.

E. H. Dimity, head of the Aircraft Safety Equipment Corporation, says, "All manufacturers of airplanes, for all purposes, are using every effort to provide the greatest possible safety. The greatest drawback to air travel today is fear on part of the public. Once this is overcome, they will all take to the air as they have during past few years to automobiles. I believe that the day is near at hand when every passenger plane and postplanes as well will be equipped with a built-in parachute, either the same or along the lines of the one used by Captain Turner in his experiment at Santa Ana. This is the day of progress in aircraft."

Dillinger: Captured in 1934 by G-men, Dillinger (without hat) was flown to Chicago.

Kelly: "Machine-Gun" Kelly, one of the kidnappers of Charles F. Urschel, is shown in the plane's door after being flown to trial.

G-MEN IN THE AIR

AVIATION has played an important role in the work of the Federal Bureau of Investigation. Looking back over some of the more recent cases we have handled, I note that at certain strategic points in their development, the FBI has had to utilize air transportation to make our plans more effective.

The speed and the fine cooperation of aviation personnel has helped us immeasurably in so many important instances. Probably the most recent aviation "incident" with which the FBI has been linked, is the transcontinental flight we made to return John Henry Seadlund to a spot near Spooner, Wisconsin, where we found the body of Mr. Charles S. Ross, a prominent Chicagoan, who was kidnaped by Seadlund and James Atwood Gray. Seadlund had later killed Gray, whose body had been hidden with that of Ross.

Of necessity a meticulous cloak of secrecy must be connected with the various flights we have had to make on important cases, since criminals like Seadlund are constantly on the alert in reading about the investigative activities of Special Agents of the FBI. If he had known what we were doing toward apprehending him, he could easily have avoided capture. By sticking to the only logical course, then, we were able to capture him, return him—by air to the scene of his crime, and close the entire case within a few days. As is the custom of the Federal Bureau of Investigation, when time is of the essence, an airliner was chartered, the prisoner and several Special Agents were loaded aboard, and from that time on, that airplane disappeared from in front of the public eye. After all, we did not know but what Seadlund had desperate accomplices that might have tried to repeat the horrible slaughter of June,

1933, in which gangsters machine-gunned to death three police officers, and a Special Agent of the FBI, in an effort to free a bank robber.

In extreme emergencies we use chartered airliners. The airlines always have—and for it I am deeply grateful—extended the utmost in cooperation to our men. Airline officials, pilots, dispatchers and radio men have covered the movements of their airplanes that we charter with the greatest efficiency.

When aviation must help us in solving a case, the men of the FBI seek the closest available airplane. It is understood that the pilots and other persons connected with the flight say nothing to anyone concerning our movements. Refueling stops attract little or no attention, largely because no one realizes the mission that airplane is on. When pilots purchase maps of the area we must fly to, they say nothing about their reason for doing so. Lest our movements be checked by short wave radio, airline radio men make no mention of us, and refer to our chartered airliner merely as "Flight So-and-so."

The aviation industry has contributed to the personnel of the FBI, since there are 23 of our men who were formerly actively engaged in the aviation industry. One was once a parachute jumper. Several of these hold pilot's certificates today. Not only do we have pilots in the FBI, there are Special Agents on whom I can depend for aeronautical engineering work. It was an FBI man, working in the Zeppelin plant at Akron, who broke the sabotage case in connection with the U.S.S. Akron.

The extensive laboratories of the FBI in Washington sometimes perform examinations that are not always in

BY J. EDGAR HOOVER

The world's most famous crime-breaker wrote this article exclusively for us. The airplane is an important FBI tool.

connection with crime detection. The FBI laboratory, for instance, has handled special identification work that has come up in connection with puzzling airplane accidents. FBI laboratory men helped determine the cause of one airline crash by deciphering the contents of the dead pilot's log book which had been soaked in oil. Infra-red rays brought out the pilot's notations in the book.

Getting back to aviation, though, I might point out that even our use of the airplane is not always infallible. That again is an important reason why we must cloak all our flights with great secrecy. And once a suspect knows what we plan to do, there is no reason why he can't make use of aircraft himself. Alvin Karpis did, for instance.

Though aviation has been a boon to us in many ways, there is no way we can prevent criminals from taking advantage of flying. The airlines could no more keep track of their passengers' occupations than could the railroads. Hence, while we have access to the most modern transportation facilities, so has the criminal. The modern criminal can conduct his operations on the Atlantic and Pacific coasts—*all within a few days*. He uses every facility in his efforts to avoid the law. Consequently, he will use the airplane in much the same manner as we do.

In the following I will attempt to show how we used aircraft to help in "cleaning up" the Barker-Karpis gang. It was on January 17, 1934, that Mr. George Bremer was kidnaped at St. Paul, Minn. Following negotiations that resulted in his release after payment of $200,000 ransom, the FBI quietly launched a nation-wide hunt for the kidnap gang. During the work that followed (for more than 2½ years) we used airplanes extensively. Rapid transportation to various parts of the country was often essential. Through the course of our investigation it developed that the kidnaping had been perpetrated by members of the Barker-Karpis gang. We finally caught "Doc" Barker in Chicago on January 8, 1935. He was one of the leaders in the gang, and chief participant in the crime. Because of his known criminal associations and the necessity of moving him from Chicago to St. Paul for trial, "Doc" Barker was transported to the latter city by airplane. Clues found in Barker's Chicago apartment, meanwhile, pointed to other members of the gang in Florida. In reading letters from the

301

gang to Barker, we found that they were living near a lake where "Old Joe," an alligator that had apparently become a tradition because of his ability to avoid hunters, was known to hold forth. Our special kidnap squad, operating out of Chicago, boarded a plane for Florida.

After a bit of inquiry in the vicinity of Ocala, Florida (the letters bore this postmark) it was found that "Old Joe" lived in Lake Weir. Persons answering the description of Ma Barker and Fred Barker were found to be living in a house facing that lake.

FBI Agents surrounded the house. An order was called to the occupants to surrender. The answer was a hail of machine-gun bullets. The Barkers' fire was returned, bullet for bullet. Firing continued for about an hour. Soon more ammunition was needed, as were more FBI men. A hurried call and airplanes were flying reserves in from Charlotte, North Carolina, Birmingham, Alabama, and Jacksonville, Florida. When the firing finally ceased, Kate and Fred Barker were found dead.

That did not end the Bremer kidnap case, however. The investigation continued and it was not long before we captured Volney Davis, another member of the kidnap gang, in Chicago. He too was flown from Chicago to St. Paul for trial. Then we set out after two more of the gang—Myrtle Eaton and William Weaver—who were believed to have headed for Florida. Our agents knew they were rapidly moving from place to place. Finally a "hot clue" was received and airplanes again were needed. The special kidnap squad flew to Jacksonville. On the morning of September 1, 1935, the couple was located and captured. They too were flown to St. Paul for trial.

Though we had captured several important members of the gang, the FBI had done no more than break the strength of the kidnap band. Alvin Karpis was still at large. He had escaped a trap set for him by the police of Atlantic City, New Jersey. He was traced to Ohio where he was identified as having taken part in certain bank and train robberies. Karpis, with John Brock and Fred Hunter, held up a train at Garrettsville, Ohio, on November 7, 1935, and escaped with $34,000 in cash and several thousand dollars worth of securities. The trio made elaborate plans for a getaway. Fast automobiles were too slow for Karpis. Early on the morning of November 8, 1935, the three bandits boarded an airplane at Port Clinton, Ohio. They had chartered the ship and its pilot, who flew them to Hot Springs, Arkansas. Soon, however, we received word that Karpis was in New Orleans. It was there, on May 1, 1936, that we arrested him and Fred Hunter. Karpis was subsequently flown to St. Paul.

Seven days after Karpis' capture, the airplane again came to our assistance in this same widespread case. Harry Campbell, another member of the gang, had been caught by FBI Agents at Toledo, Ohio. He was speedily flown to St. Paul, tried and sentenced.

An unusual case in which airplanes were involved was that in connection with the kidnaping of Charles F. Urschel of Oklahoma City in July, 1933. Indirectly, it was the airplane that ultimately led us to the solution of this case. Mr. Urschell was taken to a rural community by his kidnapers and held captive. An acute-minded person, he noticed that every day at a certain

hour he heard an airplane fly over the house in which he was being held. Shortly after the airplane passed over, he made it a point to ask his guard what time it was. One day the airplane failed to fly over. Mr. Urschel made particular note of that.

Following his release after the payment of $200,000 to his captors, Mr. Urschel carefully related the events that took place while he was held prisoner. He told about the airplane flying overhead each day. He told of the day the regular plane did not fly over. Working on that single clue, we checked airline weather reports throughout the southwestern states. Soon it was found that an airplane flying its regular route was forced from its Fort Worth-Amarillo (Texas) course by weather conditions. On a map, this alternate course showed the airplane had avoided the territory in which Paradise, Texas, is located.

An investigation was started immediately in the vicinity of that town. Sure enough, our search resulted in the location of the kidnapers' hideout on the farm of Cassey Earl Coleman, a relative of Kathryn Kelly (wife of "Machine-Gun" Kelly). The kidnapers, though, had left. They were trailed through St. Paul, Chicago, to Memphis, Tennessee, where, on the morning of September 26, 1933, local police and FBI Agents closed in on them. The couple was immediately placed aboard an airplane and rushed to Oklahoma City where they were tried and sentenced to life imprisonment.

I have attempted to show the important part the airplane plays in the work of the Federal Bureau of Investigation. It can be seen where, in the few cases mentioned in this article, the absence of aircraft might quite logically have written an entirely different ending to our law enforcement efforts. The past century had witnessed crooks traveling about a community or limited area in which their activities were restricted.

But all that has been changed tody. There now are speedy automobiles—and airplanes. These two modern developments alone have widened both the field of acquaintanceship, and of activity for the criminal. To a degree, his old-time specialization has vanished. A law violator is no longer a safe cracker and nothing else. He is versatile and the whole nation constitutes his field of operation. He knows where he may secure assistance in his efforts, and also where he can hide out. He can operate as well on the east coast, as he can on the west coast. All he needs is a list of names and telephone numbers. With the aid of the airplane those distances are quickly covered. The thief who "pulls" a bank robbery in the east today can be spending his stolen money tomorrow in Los Angeles. A murderer in the middlewest can be sitting in a Broadway theater that same night.

So it is that crime detection must always be just a little speedier than crime. We must catch our man as soon after the crime as possible. Here, then, is the all-important role of the airplane. Speed always will remain the all-important factor in our operations. For that reason, we depend a great deal on the cooperation of local police authorities as well as that of the crime-conscious general public.

That is why there always is a line open—day and night—in the FBI headquarters in Washington. That is for *you* to telephone us from any section of the nation in case of a kidnaping, or other emergency.

ROGER!

BY ROARK BRADFORD

LITTLE did the skipper realize that the rugged spearhead of the sky fleet would wing to victory that day in the plane flown by Lieutenant (j.g.) Wilco, NA, USNR. As victories go, it was not such a howling holocaust. No enemy was damaged. No problems of training or of logistics were solved. Yet, it was a major victory—a victory in semantics. For on that fateful day, the big bomber was flown 1,400 miles over the trackless sea, with the full crew talking all the time and nobody knows until this day just what. . . . But that is getting ahead of the tale.

Briefing was brief.

Lieutenant Wilco listened attentively while Acey spoke. Acey was officially nicknamed "Acey" because he was the A.C.I. officer, although everybody called him "Jack" because that was his name.

"Baker—2," said Acey, crisply.

"Roger," responded Wilco.

"Bellenger," continued Acey, handing Wilco a piece of paper. "Your sweep."

"Vectored," admitted Wilco. He walked to the strip and climbed into the big bomber. It was ready to go.

"Taxi," said the tower.

"Roger," said the pilot. Then, a moment later, "Taxiing."

"Over," said the tower.

Squared away before the strip, Wilco (his real name was Wilcox, but a French grandmother dropped the "x" because she could not pronounce it) clamped the intercom to his ears and pushed a button. "Radio!" he snapped into the mike.

"Roger," snapped AMM2c.

"Engineer?" asked the pilot.

"Roger."

"Navigator?"

"Roger."

"Co-pilot?"

"Soigned," said the co-pilot. He was a native of Bayou LaFouche, Louisiana. He could fly hell out of an airplane but he was slow at learning languages.

"Roger," said the pilot

For six hours the big bomber bounced along in turbulent tropical fronts. Then the pilot touched the intercom button once more. "Peterman!" he called.

The plane captain came forward and touched the pilot's shoulder. "Peterman," he repeated.

"Chow," said the pilot.

"Wilco," said the plane captain.

"Mister Wilco," said the pilot, who was a stickler for military courtesy.

"Roger," said Peterman.

Chow was served in paper plates. It was chicken a la king, baloney sausage, fruit salad and carrots. It all came out of cans. Except for the added woody flavor from the plates, it tasted as though it had all come out of the same can.

"Pilot to navigator," said Wilco.

"Navigator to pilot," said the navigator.

"Over," said the pilot.

"Over," said the navigator. And then added desperately, "I mean 'Roger,' Sir."

"Roger over," replied the pilot.

"C'est la guerre," said the co-pilot.

"Roger," said the pilot.

Ten hours later, the pilot touched the button once more. "Roger," he commanded, briskly.

The navigator's voice came in, cold and clear. "Wilco," he said.

"Over," said the pilot.

"ETA over," said the navigator.

"PLE over," said the plane captain.

"Alnav," chirped the radio man.

"Over," said the pilot.

"ComFourFleet," said the navigator.

"Repeat repeat repeat over," shouted the pilot.

"Error error error," whined the navigator. 'Stand by for correction. Correction coming at stroke of next gong. GONG! Com4Flt."

"Roger," acknowledged the pilot.

"ComSinkSank," shrieked the radio man, now hysterical.

"Overallcognizance," the pilot sneered.

"Over Roger Wilco," moaned Peterman, sorrowfully.

"Cojones," observed the co-pilot, who could habla la español also.

Presently the pilot set the big bomber down and taxied it to a halt. The skipper came up in a jeep.

"How was the hop?" he asked.

"Wonderful," said Lieutenant Wilco. "Everybody talked all the time and nobody knew what the hell anybody was saying."

"Wonderful," said the skipper. "Wonderful. That's the spirit that it takes to win."

"Roger," said Wilco.

Sunny Rest is a family lodge. Father, mother and children
vacation together; here they have fun touring grounds in Model T.

Pennsylvania's
Nudist
Airpark

BY DONALD JOHNSON

*For years pilots have kidded wistfully
about being forced down in a nudist camp.
Now they don't have to be forced down.*

JUST two miles north of Palmerton, Pa., there is a new airport which is probably unique in the world. It has excellent runways, clear approaches in all directions and a large clear airmarker. On the ground the visiting airman will find a cordial welcome, meals, lodging if desired, a wide variety of sports activities, a particularly inviting swimming pool and probably also the surprise of his life—for he will have landed right on the grounds of one of America's largest nudist parks.

In the summer of 1948 a pilot flying in the vicinity of the Lehighton Water Gap had the unhappy experience of hearing his engine cough and quit. He quickly spotted a level hilltop which looked ideal for a forced landing. Everything went fine until the pilot, flustered, flared out too high and dropped in hard. He climbed out to find his landing gear spread and his prop splintered. Just as he was getting his composure back, the pilot saw a tall athletic man coming towards him wearing nothing but a deep suntan and a pair of sandals. Then he knew that it had happened at last, someone *had* made a forced landing in a nudist camp and he was it.

Within a few minutes the plane was surrounded with men, women and children all equally unclad and apparently concerned only in finding out if the pilot were safe and the ship something less than a total wreck. One of the nudists, who is an airline captain, checked over the engine and told the still startled pilot that the next time he wanted to practice spins he should put the carburetor heat on first.

By the time two days had passed and the ship was once more signed off as airworthy, the pilot had paid his dues as a member of the club and had already turned a lovely shade of blister pink. After he had taken off, the mechanic who had repaired the plane got the manager aside and said that he would like to join too. That did it. A long huddle took place during which all concerned debated the pros and cons of doing what had never been done before, opening an airport as part of a nudist park. After much discussion they decided to go ahead.

Sunny Rest Airport, el. 950 feet, is two miles due north of Palmerton, Pa. NW-SE 2,000 feet, no obstructions. E-W 1,600 feet 2½ percent downgrade to west, no obstructions. Airmarker on dining hall. No landing or tie-down fee. Meals, room, sports, approx. $6 per day. 80 octane. Open May–Sept. 15.

For several years I had known that Reed and Zelda Suplee ran a large nudist camp somewhere over in Pennsylvania. That, I figured, was their business; we were good friends and saw each other often despite the fact that I had never accepted their invitation to visit the camp grounds. When Reed told me that he planned to go ahead with the airport idea, I wished him luck and told him that if I could ever help him out in any way to let me know. That was my mistake. Right then and there I became chairman of the Sunny Rest Airport Development Committee and chief technical adviser. For a while, all I had to do was look at drawings and answer routine questions as to what is and is not necessary for the operation of a small private landing strip—which is all that was contemplated at that time.

Being human, I began to wonder what the proposed strip was going to be like. It occurred to me that there is many and many a landing strip, but never a strip as stripped as this one. Then I got an official notice in the mail that the field would be formally dedicated and opened to traffic on May 13, 1950, and that so and so would arrive at Sunny Rest for the first time and would make the first official landing. "So Reed has finally gotten his strip in," I mused happily. "Well, I hope he enjoys it. And as for the first landing—." I looked again. Yep, I had been elected.

That landing I will always remember. There have been times in the past when I have wondered exactly where I was, but when I saw three or four nude people waving toward my plane from a hilltop, I knew that there was no doubt this time. Reed had told me to fly north through the Lehighton Gap and to look for the field just two miles due north. I followed his directions easily; on a clear day the Gap is visible a good 40 miles away. I came through, began a let-down, heard my wife let out a startled gasp and knew that, as usual, she had been the first one to spot the field.

I dragged it a couple of times, of course, before coming in. The strip looked rather short, but otherwise okay. There were no markings to indicate just where to sit down, but I had a pretty clear idea by the way the people on the ground lined up; I'll bet that no other pilot ever had an approach set-up like that. I warned the back seat passengers to tighten up their seat belts, turned onto the final and set up a graveyard glide. The ship touched down nicely onto a field which was rough enough to make a crocodile swallow his uppers. I hit the brakes as soon as I could, bumped to a stop and cut the mixture. The prop had barely stopped turning when the plane was literally surrounded by about 70 or so people who were all congratulating me on having just landed and survived the experience. They were packed around the plane and the only pair of pants in sight were those covering my landing gear.

The funny thing about landing in a nudist park is that if you know what's coming, it doesn't seem to bother a bit. We all trooped down to the swimming pool where the canteen yielded up some cold soda and a chance to sit down and rest. In the water a number of people of all ages were swimming and paddling as people do everywhere except for the fact that there wasn't a bathing suit in sight. In the shallow end a bunch of youngsters, boys and girls, were playing together and splashing madly, apparently unaware that there wasn't as much as a stitch amongst them.

Over on the tennis court a game was in progress between a man wearing a T shirt and a girl who had donned a sunshade. The volley ball court was busy, some more kids were playing shuffleboard, while on the broad lawn a number of people were just loafing, reading, or listening to portable radios. Just then the prettiest redhead I'd ever seen came up to tell me that we had better wash up for dinner. Each one of our party had a nudist escort, so I shortly had the pleasure of sitting down to a turkey dinner with a charming companion who wore a happy smile and a rose in her hair.

About three in the afternoon I began to look for the rest of my gang to talk about starting back. I couldn't find them anywhere. Reed invited me to take a swim in the pool, which invitation I got out of by explaining that the fellow and his wife with me were bigwigs and that I had better not on their account. I walked up to the pool again, still looking, and heard my name called. Mr. and Mrs. Bigwig were waving at me from the center and yelling, "Come on in, the water's fine." I know when I'm licked. I accepted the loan of a cabin and got into uniform. Then I took a deep breath, flung open the door, and stepped bravely out and just about knocked down the redhead who was walking by. She looked me squarely in the eye, said "welcome to our midst," and that was that.

Before we left I gave Reed the address of Max Karant [AOPA Assistant General Manager] and suggested that AOPA be notified of the new field. I also recommended some improvements if any traffic at all came in.

The Sunday after the AOPA newsletter came out with an announcement of the nudist landing strip, I flew up to see if there would be any business. There was. Two of us were busy all day parking planes and directing traffic; generally we had three in the pattern at the same time. We drew everything from a 40-h.p. *Cub* to a *Mallard* and ran out of space to park the planes. The AOPA has a long and powerful arm. By mid-afternoon the pool was full of airborne swimmers while a visitors vs. regulars volley ball match drew a large gallery.

That day proved that the airport was going to do business and that improvements were definitely in order. I talked it over at length with Reed with the result that he decided to go all out. Work began immediately. I laid out plans for grading and lengthening the existing runway, spotted the locations for corner markers, indicated where some trees should come out, and found an excellent site for a six-foot high air-marker. Personally I am highly air-marker conscious and try to get one put in every time I have a chance; many a student pilot has saved his skin by finding out where he was and where he was not because of an air-marker. That afternoon Reed and I discovered that by purchasing two more farmers' fields, a 2,000-foot clear approach runway could be constructed.

Within a week everything was in full swing. Graders and rollers were hard at work. The rough spots disappeared and the rocky soil was packed down into a hard, smooth surface. The new 2,000-foot runway was cut and cleared in two days. Corner markers built to CAA specifications were put in place and painted international orange. The parking area was rolled smooth and tie-down stakes sunk into place. The Pennsylvania State Aeronautical Inspector arrived and officially licensed the field. The CAA arrived from Allentown to check on the traffic pattern. Meanwhile the planes continued to come in so that Reed had to hire a full-time line boy. The gas company arrived to check on the location for the fuel pumps.

Just five weeks from the day of dedication a rock-upholstered, rough-as-a-cob narrow strip had been converted to a sizable airport able to handle almost anything likely to want to land there. Apart from the fact that the short runway has a 2½ per cent grade which makes caution advisable in landing on it towards the west, the field is at this moment an exceptionally good one without a single dangerous obstruction to interfere with any of the runways.

So far, the experiment has worked out astonishingly well. A large number of pilots and airport managers have flown in, including some whose names are known wherever planes fly. Many of them have become members of the group together with their wives and families. Arthur Godfrey has made frequent reference to the field on his broadcasts. Many newspapers and columnists ran stories on the field, a few doubting that it really exists. Definitely it does. It will be on the next New York sectional chart and in the next Airman's Guide. The pictures accompanying this article are the first to be published anywhere.

Just a few evenings ago Reed Suplee, the owner of the new field, his wife Zelda, and I stood on the edge of the now completed airport watching the sun sink behind the Pennsylvania hills. The view from the field was a glorious one. My plane was tied down for the night; the setting sun was reflected in a bronze glow off the metal wing. Darkness had just begun to sharpen the outlines of the surrounding mountains when a small plane came into view, flying back and forth in a zig zag pattern up the valley.

We waited until it had worked its way within view of Sunny Rest when it swung once more and made a bee-line for the field. The little *Champ* made a quick pattern, landed and taxied up. Two thoroughly frightened young chaps were aboard; they were lost but good. They had two gallons of gas left, no night lights and were more than 40 miles off course. Had they missed Sunny Rest, they would have found no other field for many miles in the direction in which they were going, and few likely places for a forced landing.

After they recovered from their surprise at finding out where they were, we tied the plane down and sent them off with Zelda towards a hot supper and a telephone to reassure anxious parents who were waiting. Then Reed turned to me.

"I know just how much this airport has cost me, Don," he said, "and I feel as of right now that I have gotten my investment back in full." I knew what he meant. As we walked back through the camp together, I wondered myself what will be the next chapter in the story of the world's first nudist airport.

The Flying Saucers I've Seen

Strange objects in the air are nothing new to this airline pilot who has seen more than his share in 13,000 hours of flying.

BY H. A. SHANKLIN

ONE SUNNY AFTERNOON I was flying a DC-3 airliner along the Wichita-Kansas City airway and had just turned north toward Topeka, the next scheduled stop, when I spotted the silver disks. The first flying saucer sped across the nose of the ship going from the right lower part of the windshield to the left-upper corner.

I stiffened in the seat. My hands gripped the control wheel. My eyes, I know, were bulging. At last I had seen one. A flying saucer!

I turned my head slightly to see if the co-pilot had noticed. But he was adjusting the prop controls, so I turned back and waited.

In a moment another disk sailed across in front of us. What was this, I thought, an invasion? I wished desperately for a camera, but in lieu of that, I would study them carefully. There would be no vague descriptions, doubts as to their size and shape and approximate speeds. Coolly I would look them over so as to give the military correct information about these space ships.

Breathing as though I were at 20,000 feet with no oxygen, I stared straight ahead and waited. Two more came into view following the same course as the others. All seemed to be traveling at the same speeds. And I noticed that just before they went out of sight they seemed to pause and then shoot out into space with a terrific burst of speed.

One thing I was certain about now: these mysterious ships belonged to some type of outer space military, for the last saucer had a wing man just to his right and rear, a formation common to our air force. And comparing the space ships with the background terrain, I judged they were about the size of one of our four-engine airliners. Their shapes were the same as reported by other observers: flattened saucers with highly polished metal. Their manner of propulsion puzzled me, for I could see no flames or any whirling parts.

After the two flying in formation had disappeared, I knew it would be best to tell the co-pilot so he could help verify the weird scene. After that I would ring for the hostess so she could corroborate also.

When the next disk came into view, I leaned over to speak with the co-pilot. When I did, the disk disapeared. Somewhat confounded, I straightened up again. The disk reappeared. I sat there moving from side to side, I know the co-pilot thought that I had blown a cylinder, but I was discovering the origin of my flying saucers.

They were caused by the sun reflecting on the farm ponds or tanks that dotted the land below us. I looked down at the little lakes and saw they were silver disks. Their movement across the windshield was caused by the speed of the plane passing over or by them.

I also discovered why the disks seemed to pause near the edge of the windshield and then fade with a burst of speed. It was because I unconsciously moved my head slightly to follow their line of flight. My head movement caused the pause and the speed.

If we had passed over a cloud deck or I had varied our course a few degrees to either side, I never would have discovered that my flying saucers were reflected images of farm ponds below. I would have gone to my grave swearing by all that was holy that I had seen flying saucers. No argument could have torn that idea from my mind. And the strange part of it is that, although I have over 13,000 hours, this was the first time any such combination of sun's rays on water had fooled me.

There is no doubt that I was fooled, for the co-pilot told me later that my face had turned white and my eyes had a glassy look.

Of course, reflection of ground objects in an airplane's windshield is a common occurrence, but usually the airman can recognize the objects. But, once in 13,000 hours, I ran into a combination that duped me completely.

Another windshield reflection so deceived a co-pilot of mine that I honestly believe it caused the lad, who was in his mid-twenties, to have premature gray hair.

We had taken off out of Lubbock, Tex., toward Dallas on a gloomy black night. A high overcast blotted out all sky light. Our cruising altitude put us on top of scattered clouds. I was flying and the co-pilot was staring dreamily through the windshield, his head, I presume, filled with visions of buxom blondes.

About 40 minutes out I noticed through the wide breaks in the clouds a grass and brush fire just to the north of our course, on my side. There were two lines of fire, and I realized the ranchers were backfiring to halt the main blaze.

That's all I thought of the fire until we had just passed it. Suddenly the co-pilot reared in his seat and cried, "Look out! Break left!"

Startled, I banked sharply to the left. "Where's the other ship?" I yelled. It's foolish trying to avoid something you can't see. You might zig when you're supposed to zag.

The co-pilot had whirled in his seat and was looking down and back. "It passed right under us," he said, in a strained and quavering voice.

"What passed?" I asked.

"A round saucer with flames shooting out of its sides," he said, his own eyes about as big as saucers.

"I'm glad it was round," I told him. "I'd sure hate to meet a square saucer."

What had happened was that his windshield had not been reflecting anything because of the dark landscape and the scattered clouds. Then, all at once, it picked up the two wavering lines of brush fire. To this day he firmly believes we were nearly rammed by a jet propelled saucer spitting flames from its sides.

Scanning the skies to an airman is a means of survival. A pilot is like a motorist forever approaching an intersection. Always you look to the left and the right, up and down, sideways. And you don't fly very long before you realize that your eyes, viewing things against the immense backdrop of sky, are unreliable.

Light rays reflected from wisps of clouds, dust layers aloft, and layers of warm air with high moisture content, make you see things that aren't there. Your eyes record the illusions honestly; your imagination gives them substance. For example the disks over the Sangre de Cristo mountains of northern New Mexico have been giving my imagination a workout for the past ten years. On the route from Dallas to Denver our course angles us near this range in the vicinity of southern Colorado. One afternoon, there were, lenticular, lens-shaped clouds over these mountains and the air contained much moisture content in the lower levels, when I saw silvery disks fluttering southward at high rates of speed.

I quit breathing. These were space ships sent to watch our atomic installations in New Mexico. I picked up the mike with the thought of warning our company radio operator about the visitors; but just then, from the corner of my eye, I saw a silvery flash from the base of a high lenticular cloud. Then another. I put down the mike. Now for the past years, when the moisture content of the air seems to be right, I've watched the disks fluttering over the Sangre de Cristo mountains. Always heading southward, always going at terrific speeds.

I have tried to put port holes around them, jet exhausts and identification marks. Sometimes I succeed in installing accessories on the flying disks, especially when my side window is dusty, or I am very tired.

One evening in this same vicinity, my heart was nearly jarred from its mounts by an awesome sight high in the sky . . . a red disk traveling eastward at a steady clip. The co-pilot and I rubbed our eyes and muttered to ourselves. It was many minutes before we realized it was a translucent balloon catching the rays of the sun which had settled behind the mountains. We knew the jet stream was in that area so that explained its speed. But to show how unreliable are eyes when looking at an object which has no background, we could make ourselves believe it traveled thousands of miles an hour, while a minute later we were convinced it was not moving. Yet it must have been moving at a constant rate of speed.

You can look in the sky and see things that aren't there, but what the ordinary person does not realize are the many material things actually in the sky.

The air above the States is filled with our jet planes on routine and training flights. And because these jets use too much fuel below 20,000 feet, they make their home miles above the ground.

I don't say there are no flying saucers, either Earth or Mars made; I've just never seen one. But I do know there are many things in the sky that look like flying saucers and are not.

The Blowfly: Gone But Not Forgotten

BY PAUL SWAN

IT HAS BEEN almost a half-century since the first Gormley-Bulsh Blowfly rolled out of a small, white frame factory in Aldershot, England. Built at the request of the Italian Army, it was intended for use as a defensive weapon against German observation balloon attacks.

The Gormley-Bulsh people were well-known automobile manufacturers, but had no experience in the new-born aviation industry. This led to one of the most unusual features of the Blowfly—its four-speed transmission (one speed in reverse). The company's chief engineer, Sir Ian Hickey (later to become Mr. Ian Hickey), had proceeded along familiar automotive design concepts; his inclusion of the transmission in the airframe design was not discovered until the prototype had already been completed. As the Italians were most anxious to combat the threat of the German balloons, Gormley-Bulsh decided to go ahead with Hickey's original design. These transmissions used straight-cut, nonsynchromesh gears, necessitating additional flight instruction in double-clutching.

Vee belts ran from the transmission pulley to the propeller assembly, mounted above the radiator, to provide the final drive. Another of Hickey's mechanical innovations was the use of burled walnut exhaust valves. These beautiful, hand-turned valves gave the Blowfly a distinctive and not unpleasant castanet-like chatter at idle. The presence of the transmission enabled the designers to employ the standard automotive cranking system in starting the engine. After the engine had been started, the clutch was engaged and the plane could then be taxied in low gear, via a steerable tailwheel. Lubrication was by a rather primitive splash system, which caused the engine to smoke badly whenever it was operated at any speed over 100 rpm.

When compared with other aircraft of the day, the Blowfly's performance was not spectacular. Its top speed of 47 mph (in top gear) was found to be inadequate, especially when coupled with the somewhat severe glide ratio of 3:1. The operational ceiling of 420 feet, however, provided to be of some defensive value: German antiaircraft gunners were reluctant to use explosive shells at this low altitude for fear of causing casualties on the ground.

The theoretical range of 200 miles was never reached under actual flight conditions, as the valves tended to char well before this distance could be achieved. Armament consisted of two Savage single-shot .22 caliber rifles—manually loaded and fired through the propeller. The engine turned at about 120 rpm, eliminating the need for the pilot to worry about hitting his own prop. A three-pronged grappling iron was supplied, but there is no recorded instance of its use.

Instrumentation consisted of a tachometer that was redlined at 200 rpm, and a barometer/altimeter of the ever-popular "Witch-and-Children-in-the-Cottage" model. Peak altitude was indicated when the witch came all the way out of the cottage. An oil-temperature gauge was installed in the prototype, but as the oil rarely remained in the engine long enough to become very hot, it was felt that this instrument served no real purpose; it was not included on the production model.

The three-cylinder, water-cooled, in-line engine was the same the company had been using in its fantastically unsuccessful two-passenger drophead coupe, the Bolide. The engine was rated at 26 bhp, but again, this figure was never actually reached in the production engines because of various material limitations, including the fact that the valves had a tendency to splinter at high revs. Many components of the Bolide automobile were used in the Blowfly: the muffler, radiator, windshield, horn and hood ornament.

An interesting aside to the history of aircraft development was recorded on October 12, 1915, when a Blowfly was used in the first successful test of a phonograph in an airplane. The failure of the Royal Signal Corps to pursue this line of research further was instrumental in assuring the radio's early domination of aircraft communications.

By the end of the war, a total of seven Blowflys had been built at the Aldershot plant, two of these actually being delivered to the Italians. Only one of these planes was ever involved in actual combat. On November 9, 1917, Lt. Giuseppe Imbroglio, a member of the four-man *Dolce Far Niente* squadron, was flying a patrol mission, his first in a Blowfly. He reported that while flying over the Tolmino-Caporetto Sector north of the Bainsizza Plateau at an altitude of 350 feet, he was attacked by a German observation balloon. What ensued was to be the longest recorded dogfight between a captive balloon and an airplane. After a number of furious onslaughts, the Blowfly was brought down when it was hit by a map case thrown by the German observer. Lieutenant Imbroglio was able, by skillful downshifting, to bring his ship to a relatively safe landing, but records captured during the latter stages of the war indicated that the German observer was badly shaken in the encounter and was sent back to Berlin for R&R.

Even though Imbroglio's was the only Blowfly involved in combat, the Germans claimed the destruction of 27 of the planes! It was later determined that this error was due to the plane's smoky lube system: What the Germans had seen was the *same* Blowfly 27 different times, trailing a smoke cloud as it cruised toward its home base.

It is unfortunate that no Blowfly has survived. The last known example saw some use in 1923 after having been converted to a crop duster, but as the oil smoke tended to damage the crops, it was shortly taken out of service. The engineering team responsible for the design and production of the Blowfly remained neutral during World War II, at the urging of the British Government, so there was no chance for development there. The last Blowfly was last seen in late 1956 filled with helium, being used to promote the opening of a supermarket in Grimsby. An ironical use of a machine designed to be the scourge of the balloon! On this same occasion, a sudden squall caused the mooring lines to the floating Blowfly to part, and the sole survivor of the breed was seen drifting out over the North Sea, where, presumably, it finally fell.

One of my most cherished possessions is a polished mahogany gearshift knob embossed with the famous Gormley-Bulsh emblem. A silent *memento mori* of the gone but not forgotten Blowfly.

"This Is a Very Peculiar Assignment..."

You're telling us—but then we get some real beauties, especially where new products are concerned.

BY ROBERT PETERSON

LTR, MARCH 7, *to R. B. Parke, Editor* [*extract*]: "Thanks for the assignment to evaluate the two products. The Morse code computer is on hand, since it came with the letter, but where is the celestial computer? For background: how do you compute stars? Can it be done only at night? Hate to waste that time."

Telecon March 10, with RBP [*verbatim extract*]: "Parke here. What are you calling about? The celestial computer? My stars, didn't that go? Hold on. [*Raising voice*] Marianne, where is that . . . oh. Gone, huh? [*Into phone*] It's gone to you. What? How did it go? [*Raising voice*] Marianne, how . . . oh. [*Into phone*] Mail. Have you got the story done yet? No? Excuses, excuses."

Ltr, March 20, to RBP [*complete text*]: "It is important to me to get the guidlines on this job firmly in my mind, because if I don't I might go bananas. I wouldn't write to you otherwise, sir. I do know how hard you work and the crushing responsibility of getting out that Grand Magazine at a profit, whilst wrestling with that onerous budget, which we all know you truly hate. Well now, what was I saying? Oh, yes, my frame of reference, the guidelines:

1/I am a GA (see glossary) pilot, not a GD (not in glossary, NIG) cub scout.

2/I have passed the test attesting to my being a private pilot, SEL.

3/I'm a reasonably proficient, careful and avoid single-engine-night-flight and-flight-over-oceans SEL. In fact, even if you asked me, I would not fly solo to Tasmania. Except for a lot of money.

4/Most of my higher math is shrouded in the mists of time.

5/People who horse around with stars should have NASA backing and a console to punch and an IBM to do the dirty work.

Does all that sound like sniveling?"

Telecon March 23, with RBP [*verbatim extract*]: "Parke here. Have your letter and I agree. THIS GREAT MAGAZINE represents the interests of everyone—duffers like you, as well as the others, the smart ones, who make up all the rest of the readership. Stop sniffling. [*Rounder tones*] Your postulates are acceptable to us, and we find them suitable as a basis for your work. [*Sharper*] And stop slipping in all that drivel about the budget. I KNOW WHAT YOU ARE DOING AND YOU CAN'T GET AWAY WITH IT. [*SLAM!*]"

Ltr, April 4, to RBP [*complete text*]: "This is a very peculiar assignment, and I'm wondering whether you get many products coming in like these two. Not that that has any bearing. Cotton twine in the bottom of the cap, with the ends hanging out about an inch on either side. Pour in the wax and let cool. These make good kindling when your wood is too wet to burn; the fire is hotter than a match and lasts longer.

Those two items—matches and candles—are the only two parts of the kit that the user will have to manufacture. The remaining items are readily available at the grocery or drug store and the local Boy Scout supplier:

1/*Signal mirrors*: Probably the simplest signaling device ever conceived, flashes from a mirror have been spotted from 12 miles away. The instructions are printed on the back and you can even hang one from a pole and allow the wind to blow it about when you're otherwise occupied.

2/*Boy Scout compass*: Leave the crash site only if you know the direction of a road, house or water source for sure. The plane is the best guide for searchers, so try to stay close by it. In any case, should you leave the site, make a big arrow on the ground indicating the direction you've taken, and use the compass to avoid circling and becoming lost.

3/*Fishing line*: One roll of 40-pound monofilament line is sufficient. It can be used for fishing, snaring small game or catching birds. (All U.S. birds are good to eat, as are all mammals).

4/*Fish hooks*: Wrap an assortment of sizes 1, 5 and 8 hooks in a piece of emery cloth. Sharpen each before use to prevent the big ones from getting away.

5/*Iodine*: Even in remote areas it is a good idea to purify your drinking water before consumption. To make most natural water safe for drinking, add four drops per quart, shake, and allow to sit for 15 minutes.

6/*Salt tablets*: If you have a good water supply, begin taking salt tablets on the second day. In cool weather, one per day is sufficient. However, if the days are hot, or if working about the camp makes you perspire, an additional tablet in the evening is in order. If your water supply is limited, simply crush several and use the powder moderately to season whatever food you may forage.

7/*Condoms*: The dry type can be used as waterproof bandages, canteens to transport water between the supply site and the camp or to hold leftover food between meals. Remember to tie the top closed with a piece of string; a knotted condom will not untie. To make a canteen, insert it into a sock and tie the top of the sock to a belt loop.

8/*Insect repellent*: Stock a box of the towel type. Sealed in a foil envelope, they take up less room, weigh less, and won't shatter on impact. They are usually flammable and can be used to start a balky fire.

9/*Multivitamin tablets*: Don't use the chewable type as they tend to induce thirst. They are an important item since most of us would not be capable of foraging up a completely balanced diet.

10/*Knife*: Don't use one of those Swiss Army Specials with every gadget under the sun in one compact, six-pound bundle. Buy a surplus military knife with the sharpening stone included in the sheath. Regardless of the tales you may have heard about these, they are probably the best all-around knife for survival.

That's it. Now wrap the entire kit in two or three of the plastic shrouds that the dry cleaners drape over your clothes. They will serve as rainwear or windbreakers.

The kit plus your imagination and the insurance company's airplane (you probably can't use it anymore) should serve the average pilot well. Anything you may add will increase your comfort and well-being. It's best to avoid perishables, though, because you have to watch those carefully and they only become a burden. Perhaps for energy, some instant coffee, some rock candy for morale and comfort, or even these pages for review or kindling a fire, would be helpful.

Remember to give jobs to everyone to keep them busy and to bolster sagging spirits. Plan ahead for a long stay and you'll be more comfortable, more secure and the time will pass more quickly.

Store your survival kit in the map compartment and forget about it. It automatically becomes the most useless thing in the plane until you need it. And as for ever needing it, heed the words of Aristotle: "A likely impossibility is always preferable to an unlikely possibility." There is wisdom in being prepared for something that probably won't happen anyway.

Or is it a track? Vector? Wander error correction?

"You know, I think they put that whole celestial computer together just to help me to throw off the wander error, and I'm a failure every time. Where was I? Oh, yes. I'm coming along just great on this assignment and I have only good news for you and it's all lousy. My wife is opposed to flight, you know. I can see her lose weight on takeoff, which is, of course, helpful on density-altitude days. To keep her calm, I urged her to be the frequency switcher, which she enjoyed. Too much. Got cut off right in the middle of a transmission. Solved that by

secretly installing a dummy nav head on the panel. Later, it occurred to me to get her to identify each new frequency, and that added to the zest. She used either the sectionals—which frequently threw her a curve—or the radio charts. She learned to check frequency errors against the Jepps if the identifier didn't match the sectional. To this day, she has not linked the fact that the dit dah in dit dah/dah dit dit dit/dit means the A in ABE. To my wife, Morse code is what the kid in *How to Succeed* laid out, but her fingernail follows the dah dits inside the little frequency box on the charts and 'That's it' is the cry of triumph.

"After this insight got established in my notes, I did a little checking. I found that there are no routine transmissions between pilot and ground or vice versa in which Morse code is used, except for identifiers. I can't find any reason why most pilots in the U.S. need to know the Morse code. Convenient, yes, but not mandatory. Just checking omnis in flight will teach the bulk of the code in a short time. There is enough miscellaneous jazz that we feel obliged to carry around in the backs of our minds without cramming in the coded alphabet.

"So what is this product I'm to evaluate? It is bugled as a New Aid to Navigation. Yet. Another. A Morse Computer, Model E Pat. Pend. It's a pistolero. A rectangular white sleeve over a red insert. The whole thing about the size of a postcard. The sleeve slips to and fro over the insert, thus exposing letters when properly manipulated. Of interest to the solo pilot, the thing can be worked with one hand if the other end is gripped in your teeth. Little hard to read that way or move accurately. The most likely process is to use the Morse code, which—get this—is printed on the back. The way I figure, by the time you work out the directions for using this little beauty, you could probably learn half the Morse code by rote. Better yet, buy a roll of toilet paper and print, at random, sheet by sheet, the 26 symbols over and over and learn a little each day.

"I'll neat up this evaluation in the next week and submit it in the approved style: dry, with little witty elegances. If you want an honest opinion, I'm dedicated to those vortacs that have a female voice cawing the identifier name over and over. Lends a homey touch to flying, like that controller in Richmond. I'd bite my tongue rather than get heavily jocular or arch with her, but it's a temptation.

"Is it okay to put down this product for good and sufficient reasons? Now for the bad news. I know now what a celestial computer is. It's a big, multilayer, expensive, SOB-type (NIG) device with tabs. What it does I know not. Will get back to you on that."

Telecon, April 6, with RBP [verbatim extract]: "Parke here . . . hello? Hello. Hello. HELLO. . . . [*elevated voice*] Marianne, he's not on the phone. He what? Fainted? Well, get him. [*Later*] Hello. What the hell is the matter with you? You sick? Surprised? What are you surprised about? Oh. Well, I do call people when it's important. Now what did I call you about? Right. You can always say exactly what you want in this magazine, any time. Our writers have editorial integrity. Just be sure to clear it with me. And don't try slipping in stuff like that way to learn Morse code. Slipping in stuff like that is a dirty trick, and no writer with editorial integrity would do that. When will the article be ready? Well, hurry up."

Telecon, April 10, with Marianne [verbatim extract]: "You talked so fast I'm not sure I understand it all, but let me go over my notes and read it back so I can tell Mr. Parke when he returns from Palm Springs. Number one, you said you finally got through the instructions on the celestial computer and found out that it is an aid—you laughed at that point. I try to be precise—all right, I'll change it to pitiful laugh—to celestial navigation and that it appears that before it can be an aid-to, you have to have a sextant, so that the computer will aid in solving the problem that starts after you have used the sextant, right? That you had a friend who would let you borrow his sextant, which he uses to grow ivy in—but it works—for only $20 a day, and that if you don't hear from Mr. Parke by tomorrow you'll rent it because you know Mr. Parke is anxious to get the story because he keeps yelling at you about where is the story. Before I go on to number two, I think I better tell you he isn't going to like this, no sirree.

"Now then, number two: You said you needed to learn how to use the sextant because without using the sextant to get fixed on Pollux, Antares and Dubhe, you couldn't use the computer to get the motion of the body—isn't a sextant something like a telescope?—the motion of the observer—well, never mind—Coriolis effect, Q Factor of Polaris and I didn't get the rest; and then you said that you understood they had a course at the museum on 'How to Use a Sextant and Celestial Navigation'; and that you were a member of the museum and the cost of the course was nominal; at least it was compared to the rental of the sextant for the six weeks the course ran; and that you had signed up for it because if you can't use a sextant what good is a computer that helps, and something about for want of a nail . . . a *horse* was lost? Six times six 36, times 20 is uh, 720. SEVEN HUNDRED AND TWENTY DOLLARS! He'll die, he'll just die. If Mr. Ziff ever hears about this . . . where can I get a shredder? Can I rent one for 20 minutes for these notes? Forget we ever spoke."

Ltr, May 1, to RBP [complete text]: "I'm punchy. Up every night at the museum taking this course that would spook Einstein. Everything is a welter of maybes. Do you know how fast the earth rotates? By the time even a sharp navigator works out a fix, it's gone. In the old days of a DC-6, the fix was somewhere nearby. But now, going west to east in a jet with a strong tailwind, if you are a slow plotter or a slow fixer—a plodder about fixes—you can wind up with a virtual impossibility: The next fix is right on top of the last one and you're belting along with a groundspeed of 800 knots. Or else I'm going to flunk this course. There is a message in what I just wrote, but I'm too strung out to get it. It'll come to me.

"For homework this week, we had to plot fixes. There are two other students, both Australians, who are going to do charter work over the ice cap. One of them lives on a mountaintop up toward Albany, and he hasn't had any trouble. The other guy and I live around the city, and we can't even find the stars because of the air pollution. I've got a great idea, though. You need a bubble canopy to shoot—that's what we nav freaks call it, shooting a fix; and don't get any funny ideas; I'm clean. But this friend of mine has a two-place Mustang that's ideal. I figure a fix a night for a week would give me

enough experience to get back to the computer which is an aid-to. The Mustang—you know, they have fuel lines the size of your forearm and you can imagine the consumption even at cruise—is cheap, my friend says, at double the price, so $200 per hour seems reasonable. Tonight is the first outing, and it will definitely be no lark. Boy, what I have to do, sir, for you on an assignment. Not that I would ever fail to do my utmost, you know that, but you must admit I've devoted a lot of time to this project and who was it said that time is money? Don't bother writing me just to answer that. I know you know that it was Benj. Franklin. Just wanted to keep you posted and hope you are enjoying Papeete."

Telecon, May 5, with Marianne [verbatim]: "This is Marianne. You're where? The Eye, Ear, Nose and Throat Hospital? Why, pray tell? I don't believe it. Stuck in your eye? They got it out? No permanent damage? To your eye, not the sextant, silly. Black and blue? Have you tried Visine . . . sorry, it must be painful. You don't mind if I cable Mr. Parke? After your last letter, he'll be happy, I mean relieved. He'll want to know what caused it. Low canopy and high turbulence; okay. Take your time getting out of the hospital. Ciao. [*Dimly*] Alice, ya know wha hoppen to that nut . . . [*Disconnect*]."

Ltr, May 10, to RBP [extract]: ". . . but no permanent damage to the eye or to the eardrum from the emergency descent when the canopy was ruptured. Most of the shock was transmitted to the bridge of the nose. Plastic surgery will fix that, I'm told. The time here at the EEN and T (see glossary) has let me kick the celestial-navigation habit. Cold turkey. Horrible experience. CN is addictive, and I'm lucky I was only part-way into it.

'You remember that last letter where I said there was a message but I couldn't get it loud and clear? It's all clear as a bell now, and the CN computer is the other end of the spectrum from the Morse code teaching aid. For the average GA pilot, both are equally useless, for opposite reasons. Do you know anyone other than an occasional airline navigator who needs celestial navigation? For him, this computer will be hot stuff. But for how long? As long as it takes for his airline to switch to inertial navigation, which is accurate to the distance between two Pacific waves. How many bizjet jocks are going to bore big holes in the flight-deck roof, and for what? Just guessing, but I'll bet there aren't three who can do celestial navigation. So if they can't use CN, why should I or 100,000 other pilots sweat about this, especially with no autopilot to hold her steady while we shift the sextant around looking for Pollux. So I can do it. I get a fix. There isn't enough light in the cockpit to enter the data into the computer and read the results from the messes of little figures peeping out of little windows and all around the circumference. Neither one of these computers fills a need for a great bulk of your readers, and it would be a shame to waste the space in the magazine to try to discuss them. Use the space to tell them about the Wankel engine.

"You won't mind, will you, if I keep the computer for a while? There are templates that fit on its back that my wife can use for patterns and designs. She found out about the dummy head and does needlepoint now."

Garbage

BY GORDON BAXTER

WE WERE INBOUND to Montgomery County Airport, at Conroe, Texas. I had told my wife about the long lines of surplus airliners parked there, old Convairs and DC-3s, parked nose to tail in elephant rows of dull aluminum. In her 10 years with Braniff, Diane had started back in the old Convair days and had bid for them right up to the end. "One girl, one cabin, my airplane."

Then we were over the rows of pines on approach, and I was searching for the rows of Dumbos we had planned to visit, when out of the corner of my eye I saw something that stirred in me a sinking kind of horror. It looked like a giant air crash. The bodies of the airliners were broken and spilled, the silver wings lay torn off. The neat line of transports had been reduced to an aluminum garbage pit. People were cutting them up. Some junk dealer had bought them, stripped the engines and instruments and was cutting them up for scrap aluminum. Tossed around, the carcasses lay in pools of black, dried oil blood and bright shards.

We parked the Mooney and walked slow and silent among the dead sky queens. A few DC-3s reared up in their final agony, wings grounded and trampled, empty windows gaping at the sky. All that was left of the Convairs were the midsections of the cabins: "Mohawk" in black peeling letters, and ice dimples in the skin from winter storm battles over the Hudson Valley. One lay on its back; wind tinkled dangled aluminum strips in the ghostly cabin. "Somehow, I never thought of a Convair like this," Diane said in a tiny, controlled voice.

As we moved to another gutted cabin, her face was a study. She reached in and touched the bare floor stringers that once held her flying feet. "Stout built, weren't they . . . they seemed so much bigger then . . ." *On the ramp under the lights, all snorting and smoking with power and a very young South Park High School girl standing in the doorway. So serious in that old black motorman's uniform, and all those important places to go.* She reached up into the wreck and touched the boarding-stair lever. It moved strongly, as always. The last girl's hand to touch it.

A little apart and at the far end of the carnage, stood a lone DC-3, not yet savaged except for the engines. It stood proudly, still burnished aluminum. The hulk called to us with a very special dignity.

We entered and I made my way up the slanted deck of the stripped-out cabin, looking for the name plate. I used to do this in gentler times when Trans Texas Airlines flew DC-3s. The game was to read the manufactured date and then ask the copilot how old he was. The Gooney Bird was nearly always older than the kid flying right seat. And this one looked even older, something about it, worn smooth as a plate.

I found the stamped numbers, rubbed away the grime and looked again in the faint light: 1936, serial number five. Good grief, this was the fifth DC-3 out of over 10,000 made, and it had flown here. Donald Douglas himself had once stood where I was standing, fresh back from receiving his Collier Trophy from President Roosevelt for this revolution of the skies. C.R. Smith, president of American Airlines, had taken delivery of this one, named her for some proud American city and painted "Flagship" on her. The airlines were being born. She had seen ribbon cutting, flashbulb popping, and had taxied up to the terminal crowds with movie stars aboard and little American flags whipping in the window brackets.

And God only knows where she had been after that, wandered all over the world, through three wars. Thunder in distant mountains, lightning slashing her old vee windshield, and now to earth, in this oil patch in Texas, waiting for the cutter's torch, to become beer cans. Flagship Sixpack, and one last flight, out the speeding window and into the roadside weeds.

We came down. Diane carefully latched shut the cabinet doors swinging slack in the galley. We walked under the nose, and I looked back and said, "Lady, I wish there was something I could do . . . I really do . . ." And the winds blew through the myriad empty holes and slots, and the DC-3 whispered, sighed and murmured something to us, waiting, staring at the sky she had opened to us all.

Going back to our plane, past the Convair corpses again, Diane recognized something, a battered rectangular aluminum box, and lifted it by its carrying cut-out. "This is a Convair garbage container. It had to last us eight hours, 40 people per flight. I'm keeping it. A good stew was recognized by how well she managed her garbage. Sorted it. At every stop, I got off looking for a place to dump my sack of dry stuff; the wet stuff went in here. This thing is Convairs to me."

We left this graveyard, thinking how strange that we outlive these marvelous modern machines, and so quickly, too.

PILOTS TAKE themselves too seriously. Doesn't anything funny ever happen to pilots? I know that some of my own I-learned-about-flying experiences have been more hysterical than terrifying—but then, maybe I haven't learned much.

One particularly hazy day, for example, I had the impression the tower thought I was on a right downwind instead of a left. I had a passenger with me who had a small dog on his lap.

Instead of keeping my mouth shut, I keyed the mike and said, "Tower, this is 75 Pop."

"Go ahead, 75 Pop."

The exact moment I pushed the button to answer, the dog took over, yapping out his whole life history.

"Say, again," came from the tower. "You're garbled."

I got as far as pushing the button down the second time and saying, "This is . . ." when the dog took charge again. For his size, he was the loudest blabbermouth dog I ever sat next to.

A long minute of contemplative silence came from the other end. Finally, with a swift maneuver, I outwitted the dog and made my speech; but by this time, the problem that caused the original call had taken care of itself.

"Seven-Five Pop is turning final," I said, lamely.

"Cleared to land, Bowser," said the tower. "You fly pretty good for a dog, but you're nowhere till you learn to talk."

Believe it or not, I learned something from that dog experience: Keep your communications to a minimum when things are busy.

As frustrating as the dog thing was, it wasn't as bad as the time I had with an alarm clock. It was a big, old-fashioned wind-up clock I was taking to a friend, and I stuck it under the seat.

"I've Got Two Bees..."

BY GUERNSEY LE PELLEY

It was a lazy, soggy instrument day, and after takeoff, it was all routine. Departure had identified me, gave me my altitude and heading and went to talk to somebody else.

I drove along for a few minutes, then departure called and told me to report Duxbury Intersection. I pushed the button and said I would.

A few seconds later, departure told me again, and again I said I would. After a third peevish request and some knob-jiggling, I realized my transmitter was out on the number-one radio. Departure control was insistent, as usual. "Seven-Five Pop, do you read?"

I got the second radio tuned in, but when I pushed the mike button, the alarm clock went off. Not only did it scare me speechless, but it led to a confusing scramble trying to find the clock.

An audible sigh came from departure. "Is that you, 75 Pop? Good morning. Now that you're awake, do you mind reporting Duxbury?"

When I reported Duxbury, the alarm was still ringing in my red ears.

If there is anything to learn from that experience, I suppose it is to make alarm clocks part of your checklist and put them where you can reach them.

Ridiculous things can happen when you least expect them. It was a beautiful, smooth CAVU day and I leveled off at 8,500, cranked the trim, settled back and opened a stick of chewing gum. It was all very peaceful, but while part of the gum was sticking out of my mouth, a bee landed on it.

I exploded the gum as far as the windshield. This must have put the bee in a bad mood, because he did an Immelmann and came at me out of the sun. As soon as he got me in his sights, he was joined by another bee.

I made a rather haphazard attack with a folded low-level chart, but the situation deteriorated when the bees made a flank attack up my trouser leg.

By this time, I imagined I was sitting on a whole nest of bees and began looking for an airport. In answer to my screaming into the mike, a pedantic voice told me wind direction and velocity, barometric pressure, runway, and then, to report downwind. I was hoping for a straight-in approach, so I began to shout about bees.

Of course, the tower said, "Repeat."

I suppose I sounded something like: "Blah blah blah, Comanche, two bees. . . ."

"Comanche Bravo Bravo, go ahead."

"Negative Bravo Bravo. Bees. I've got two bees."

"You've got to what?"

"Seven-Five Pop has got two bees!"

The tower somehow got the idea that I wanted to use the facilities, and cleared me straight in. I went literally buzzing up to the wire fence beside the terminal, leaped madly out on the wing and took off my pants. Not until there was a burst of applause from a Girl Scout troop did I realize how totally I had been routed by the emergency.

Now bees are part of my checklist, just like birds.

One night, coming into a large airport, I was handed off from approach control to the tower, but I had no more than dialed the frequency when the tower broke in. "Seven-Five

Pop, get the hell out of there."

I squeegied around a bit, wondering what horrible violation I had committed and who was going to come after me.

"Seven-Five Pop," said the tower, "Do you read? Make a 180. There's a C-47 right behind you."

I flapped myself frantically around to the right, but I couldn't find any C-47.

"Seven-Five Pop, what's the matter with you? Make a 180 to the left. Turn left."

"Roger," I said. "Left." While I went whirling back around in the other direction I heard a voice from the C-47 say some bad things about student pilots.

The tower came in again. "Yeah, I'm trying to get that nitwit out of your way. Seven-Five Pop, will you move off in the direction you're pointed?"

I went off in a silly direction.

"Seven-Five Pop, what are your intentions?"

I'd had enough of all this. "My intentions," I bellowed, "are to land. If you don't clear me pretty soon, I'll be in Canada."

"Who are you?" asked the tower.

"I am Comanche November 8575 Papa, on what might waggishly be called a downwind leg."

"Oh, sorry," said the tower. "I've got a 75 Pop down here on the ground blocking a runway . . . radio must be off . . . never mind, cleared to land."

Everybody could learn something from that fiasco: Whenever there is the possibility of doubt, use the full number.

Humor can be the panacea for panic. On one of my earliest instrument flights, I had taken off after dark in the pouring rain with a fellow pilot, hoping to get an easy few minutes of actual instruments and then come out on top. No such luck, of course. It all stayed very black and boiling. To make matters worse, instructions came to my sweating ears to hold at an intersection. So squeezing tight on the yoke in only the first stages of panic, I took off on a vector given me.

Then came the voice of doom. "Seven-Five Pop, you are moving 45 degrees off course."

I had set my gyro before takeoff. I was right on. I flapped around a bit, checked the ADF, and continued.

"Seven-Five Pop, you are still way off."

"I'm right on the number," I squeaked.

"No sir. You are nearly 50 degrees off course."

My companion, who had a few more hours than I did, picked up the extra mike. "Departure, we're holding steady on course."

"Then your gyro must be off. Check your magnetic compass. Take an omni radial to get your position relative to the intersection . . ."

The magnetic compass was not only very dark, it was dancing. But it was dancing with all the wrong numbers. I reached my shaky fingers to twiddle the omni, but in this dark hour, the omni bearing indicator flicked up its flag and quit.

But the rain didn't quit. The clouds didn't quit. And my sweat didn't quit. I began my attempt to tell departure what new thing had befallen me, but now departure couldn't hear me.

After what seemed like the worst three minutes in aviation history, we figured out that I could only receive on the top radio and only transmit on the bottom radio. I was still blessed with one working omni.

I babbled my sad story to departure.

"Understand," said departure, as if he couldn't care less. "What's your airspeed?"

I was about to say 150 knots. But when I saw the needle pointing to only 80, I couldn't say anything. I spread the news to my partner by pointing my finger, woodpecker fashion.

"Must be busted," he decided, as if this had become the natural sequence of events.

In a voice that somehow had gone falsetto, I notified departure that there must be mud or ice in the thingamajig.

The voice from departure came back laden with forced tolerance. "Okay, 75 Pop. Tell you what: This will be quicker if you gimme a list of the parts you got left. Start with the propeller. . . ."

I relaxed. Maybe I even grinned. Anyway, my friend did. It may have been the world's worst actual instrument flight, but I cherish it in my memory. We even continued to our destination with a flashlight on the magnetic compass and a lot of helpful hints from center.

I learned a lot of things from that one, but the biggest was that laughing can help you learn. Especially if it's at yourself.

HAWK LIPS

BY GORDON BAXTER

"ALAN," I SAID, "let's take the kids and go see what's happening at the airport." We collected little Chris, Eric, Matthew and Gordon IV and ended up in a motherly old Cessna 172. Man, it looked like a barn loft full of owls. When I looked back over my shoulder, I couldn't see anything but big, round eyes everywhere.

"You little punks really want to fly with Grandpa?" I asked.

"Yeah, yeah, let's go! Take off!"

So we went up and bumped around for a while and it got very quiet; they were strapped down so deep they couldn't see out. What difference will it make, I thought, if we have a few kids floating around in the cabin? We unstrapped them and let them stand up. "Now don't play with the door handle. That first step is a long one."

"Hey, just like little model cars down there. Hey, look at that little-bitty train! Gollee, ain't it purty?" We got to see eight parachutists come out of two jump planes, and we got to see the glider cut loose from the towplane, and we watched all this high color floating and gliding down through the student traffic.

After we landed, I parked the car out near the runway overrun, where they launch and land the glider. It was a wonderful place to park the kids, who were pretending they were sky divers and were practicing bailing out of the Buick. They put on sunglasses and undid the shoulder harnesses and shouted commands. Periodically, all four of them would fly out the windows.

It got to be my turn to fly the glider, and Alan made all the kids come over and kiss Grandpa good-bye and get a last look at the old man. The towplane hauled me up to 2,000 feet and I cut loose.

I had noticed this hawk circling over our corner of the field. A hawk is always hunting a thermal, especially a thermal that might be over mice, and I think he figured the grandkids were mice and was planning on carrying one off with him. Anyway, I shouldered into the thermal with the hawk and circled at about 40 mph in zero sink. He had his wings spread all the way out, glorious and free, and his pinions out on the ends were trembling, like the glider's wing tips. He was right out there, looking me over like he had pulled up alongside and was going to write me a ticket.

Well, the hawk and I started to play. He could turn faster than I could, but I could ease in on him. It was so beautiful to look out there and see that big bird so close. If I got too near, he would duck under and come up on the other side. I dropped the nose a little and picked up speed, began to whistle at about 60—which, for a glider, is moving along—and I looked out there and the hawk was still with me. He had sort of hunched up his wings, and the slipstream was just tearing at his feathers. He looked at me with hooded eyes, and that hawk had a savage little grin on his mouth, the wind blowing his lips back over his teeth, and his eyes sparkling. I could see glints of red and gold as he flashed in the sun. I swear he was enjoying himself.

Then it dawned on me that we were gliding farther and farther from the airport. I remembered a story about a sailplane pilot on the West Coast who had followed a seagull, riding a wave of lift, and when they got way out to sea, the gull sort of laughed, turned around toward land and started flapping his wings. Ain't no way you can flap your wings in a glider. So I let the hawk win our game and eased on back to the landing slot. I was a little low, and I'm sorry about getting bark on the wing of the glider, but they ought to do something about that tree at the end of the runway. Actually, it's more like a bush.

We were driving back to the hangar when I got to thinking what it would sound like if a transient pilot who had never seen the grass airport before were to call in, "Ahhh, this is Piper 34 X-ray, give me your airport advisory. . . ."

I can hear our line man come on and say, "Well, we got three students in the pattern on touch and go, we have two jump planes at 10,000 feet, there are eight jumpers free-falling, a glider has just been released at 2,000 feet over the active runway, radio-controlled models are aloft over the ramp, six Japanese kites at 1,500 on the downwind leg, we have a hawk on short final for a mouse, a '73 Buick on the active with four kids just bailing out of it, the active runway is 12 and look out for the tree. Actually, it's more of a small bush."

The guy would think he had called up the zoo, and he'd go off and land somewhere else and have his radio worked on. It was a beautiful day at the little grass airport. As long as I live, I'll remember looking out at that hawk right beside me, feathers rippling, tears streaming out of his eyes and a big grin on his face. That's kind of hard for a hawk to do, because a hawk has real stiff lips.

PART VIII

Space

Flying is essentially an aviation magazine and has generally not dwelt on the challenges and accomplishments of space exploration and travel. However, the magazine found itself unable to resist the space fever that overtook the nation at the end of 1968 when men first reached the moon, if only to orbit it. In more terrestrial affairs, that year had been awful, as strife, violence, assassinations and confusion seemed to erupt in a weird concentration. The flight of Apollo 8 during Christmas week was a badly needed antidote to all that, and the excitement continued even after Apollo 11's successful moon landing and return in July 1969.

The Editors therefore wanted the Apollo story told in their pages and sought out the best writer they could get for the job. They got him in Ernest K. Gann, and the story Gann wrote (it was published in September 1970) was a vivid and moving portrait of the Apollo program, including a thrilling you-are-there description of a moon landing.

It more than made up for past neglect. It was a classic.

The Magnificent

Apollos

BY ERNEST K. GANN

IT IS A GLORIOUS morning with the sun risen mightily out of the sea and a fresh breeze riffling the surface of the water. The broad beach seems to stretch beyond the hazy infinities of north and south, and it is deserted except for a stunning blonde. She is nearly naked and she is running swiftly along the hard sand, her back arched, her fine legs extended, and her taffy hair thrashing behind her uplifted head. She is smiling, and her eyes are wide with the joy and wonder of her personal exploration. One small hand is extended above her head clutching the string of a scarlet kite, which is nearly as big as herself. She glances back and up frequently to admire its flight.

If you are only mildly interested in aviation, the chances are space flight has also captured your attention; and if you are a pilot, it is natural that your interest in astronauts might be more than ordinary. If you are mechanically inclined, then the complexities of spacecraft certainly merit your inspection; and if electronics is your forte, the exotic array involved in any Apollo flight will certainly intrigue you. If you are mathematically endowed, or chemically, or astronomically, or photographically, or you name it—there is something in space for everyone.

Let us suppose our name is Schirra, Cunningham, Eisele, Borman, Lovell or Anders. Or would you prefer McDivitt,

Scott or Schweikart? We are the contemporaries and comrades of men like Stafford, Young and Cernan. We sense the special mystique of Neil Armstrong, the quiet intensity of Mike Collins and the warmth of Buzz Aldrin. Our training for this single Apollo has endured for three years in the company of such men as Conrad, Gordon and Bean. We have been backup crew, capsule communicators, friend and confidant to Lovell, the dashing Swigert and Haise, sharing alike in their laughs and frustrations.

Let us suit up with such Americans, all of whom have looked down upon all men and now look down upon no man. Let us join these very first universal citizens and penetrate beyond the aerothermodynamic border, where the atmosphere is so rarefied there is no longer any heat-generating effect on the skins of fast-moving craft; let us soar into the land of apolunes, apogees, pericynthions and perigees, for there in the unlimited antipodes of outer space will be found inspiration, sympathy and a strange affection for our fellow men suddenly so far removed. As a dividend, we will inevitably rediscover that very breath of life—pure adventure.

When we decide to go for a space flight, the very first thing we discover is that tongue and mind must take on new disciplines of expression. Such mouthfuls as triple modular redundancy, integrated thermal micrometeorized garment, unsymmetrical dimethylhydrazine, height-adjust maneuver and coelliptic-sequence initiation are not practical in a world that recognizes the nanosecond—which happens to be one billionth of a second. So to save spittle and jaw fractures, we employ acronyms that are quick and clear. Translunar injection becomes TLI, Trans-earth insertion becomes TEI, lunar module becomes LM and command module CM. It's as simple as that. And lo, we have a new vocabulary.

It is now T minus one hour 10 minutes 51 seconds, which is that time before the world's most intimidating sky rocket lights off under our butts and transports us with considerable dispatch toward what was, until very recently, the unknown.

We are 214,369 nautical miles from our destination, and our flight plan looks like the telephone book of a major city. Not so long ago, there was only one very lonely and brave little man— a Shepard, a Grissom, a Glenn, a Carpenter, a Schirra or a Cooper—perched atop such a rambunctious device. Now there are three of us strapped in our personally contoured couches and it is some comfort to know that several thousand people scattered around the world are concerning themselves directly with our immediate and future welfare, every last one of them wholly dedicated to disproving Murphy's Law. One of the most refreshing factors about the whole space program is this determination, which begins with fabrication of Apollo components and carries right through to launch, flight and recovery. Almost every one from the redoubtable Thomas O. Paine, NASA's chieftain, to the man who tightens the last nut on any one of the 50 engines presently a part of our spacecraft has rediscovered pride in his work, if indeed such people ever lost it.

About an hour ago, a quiet, strangely tender man named Joe Schmitt helped us suit up in what has to be the world's most uncomfortable garment. He has assisted all the astronauts since John Glenn and Mercury flights, and is as proud of the suits as if he had personally tailored them. This mod of mod garments is actually a balloon shaped like a man, although when pressurized, as it was before we left the dressing room, it obliges him to walk as if he had wet his pants. It is made of Beta fabric, Nomex, nylon and neoprene-coated nylon, and is strictly a piece of engineering rather than tailoring. It is festooned with intake and exhaust valves, radiation meters, pressure meters and a communication umbilical. Joe Schmitt helped us into these rigs as solicitously as a *novillero* might help a matador into a suit of lights, and when he was done, we picked up our portable life-support systems and waddled to the paddy wagon, which brought us to the launch pad. Now Joe has departed, being one of those last few fellow earth inhabitants to see us properly tucked in. When the hatch was secured, that was that.

We are configuring the stabilization and control system, aligning it with the guidance platform. We are checking each propulsion system toward the final countdown, moving our heads inside the plastic bubbles Joe fixed around our neck ports, and communicating with each other and that already distant-seeming world called launch control, which is located in a building some three miles away. All the voices are familiar to us. We have done this so many times over the past few years that we know instinctively what each voice will announce in the interminable litany of countdown.

It is T minus 49 minutes 10 seconds. We are waiting for Umbilical Swing-Arm Nine to retract. All is still GO, so let us pray a little. Let us commune first of all with our Maker because we are in good health and it is a fine thing to be alive on this launch morning. Then let us gently remind the Almighty of our enduring friendship and of the fact that this is not the safest job in the world. Next let us try not to dwell overlong on the fact that there were three eras in aviation, the first of which was barnstorming via wood, fabric and wire. And it was not a very safe time either. And then there was the adolescent era of Trimotor Fords through the various Stratocruisers, Connies and DCs up to -4s. And a lot of nasty things happened then, too. There followed that very brief transition through -6s and -7s to jets. In the air, we have matriculated, but let us admit that we are still barnstorming in space, and whereas there were once upon a time such nuisances as "gremlins" in our relatively uncomplicated aircraft, we now have graduated to the "glitch," which is spaceese for something inexplicably gone wrong.

It is now T minus 34 minutes. We are running power-transfer tests. The last thing in the world we want is an abort. Will everything operate on our big bird's own power? If only somehow we could harness our own concentration at this time, it seems we might dispense with ordinary energy. We are keyed up, as Schirra quips, "for the Space Olympics." And yet

frighten the boat-tailed grackles, the yellow-throats, rufous-sided towhees, scrub jays, doves and lesser yellowlegs—a few of the other birds native to the Cape—if we could see what transpires directly below us moments after ignition it would as certainly scare the hell out of us. Fortunately, there is little temptation to view the outside world. The command module's five windows are covered by the forward heat shield, and this effectively spares us the sight of the Saturn's pyrotechnics.

T minus 20 minutes and counting. We are still GO, having completed all the vital checks of our carefully stowed lunar module and found an affirmative to every electronic question. The LM is our precious outbound cargo, which is now stowed beneath us in its special adapter. The LM holds the certain distinction of being the most expensive and ugliest flying machine in the world. It looks like a nightmare insect crept out from under a rock.

In startling contrast is our beautiful Saturn, which is really a projectile aimed up an invisible cannon barrel. It has been sighted most carefully for our "launch window," a relatively minute hole in the sky. The precise location of this target changes constantly and is dependent upon several interrelated factors important to the accomplishment of our mission. We are self-centered enough to prefer a daylight launch from Cape Kennedy, because if we splash down prematurely, finding us at night bobbing around somewhere in the Atlantic Ocean might ruin someone's whole evening. Next, our actual launch azimuth is restricted between 72 and 108 degrees east of north to minimize the odd chance we might land on the Champs Elyseé after an abort, or stew in the Sahara until someone came for us. It just could happen.

And then there is the all-important time of descent to the moon's surface, which also affects our launch window. We want low sun angles, ranging from five to 13 degrees, which will create visible shadows and thus aid our landing. We also like radar coverage of our lunar activities by the big Mother of all Antennae, at Goldstone, California. We must calculate backward to make certain its position will coincide with our descent and ascent stages, and predicting forward, we prefer to set our flight plan so that we have a daylight splashdown in the prime Pacific recovery area.

The first stage is at the bottom of the stack and contains the five monster F-1 engines plus well over a million pounds of RP-1 fuel and more than three million pounds of liquid oxygen. It will take us to 200,000 feet, which in our league is still hedgehopping. The second stage has five smaller J-2 engines, which still manage to generate a total thrust of one million pounds. If all goes well, stage two should run us up to 606,000 feet. Then finally the third stage, known as our S-IVB, will burn for about 2.75 minutes and slip us into a 115-mile earth orbit. Eventually, it will burn an additional 5.2 minutes and send us on the way to the moon.

The whole stack is as tall as a 36-story building and weighs 13 times as much as the Statue of Liberty. Also on the statistical snob roll is the machine that brought our Saturn

the first game is played lying down. We have trained so interminably in such clever and malicious simulators that the unreal is difficult to tell from the real. We are dressed up for the party and what happens if there is no place to go? Worse, if there is a serious glitch, we might have to beat a very hasty retreat down the long escape wire that extends from this 40-story tower to the ground. No one has, as yet, found nerve to try it even when they are not, as we are now, reclining upon a stack of volatile chemicals, which, when properly united, match the power of 37 Bonneville Dams at full thrust. While the very sudden release of such brute strength will doubtless launch vehicle to the pad. It is the granddaddy of all tractors, and since each lug on its tracks weighs a ton and its maximum speed is one mile per hour, the man who drives it is inclined toward the ultraconservative.

If you have wondered how such a tall and unwieldy-looking structure can maintain any sort of equilibrium after liftoff, be advised that after the restraining claws have opened, the four outside F-1 engines in the cluster are gimbaled and computer controlled. Thus they automatically balance our great bird as you might a long stick held vertically in the palm of your hand.

Once the tower is cleared, the computer commands a pitch and roll maneuver to seek the correct flight azimuth, and the preprogrammed flight path is assumed.

All the while, big brother is watching the function *inside* the engines and fuel-supply tanks via special telemetry—the while being a grand total of two and one-half minutes, at which point in time retrorockets fire automatically to sever the first stage and the entire marvelous assembly is dumped into the Atlantic Ocean some 350 miles down range from Cape Kennedy.

Important to our morale is our confidence in NASA's Manned Space Flight Network, which is also poised for liftoff. There are 14 land stations, five instrumented ships and eight aircraft in addition to the principal facilities at Madrid, Canberra, Australia and Goldstone, California. The technicians in these stations are gravely concerned with our impending severance of earthly ties, and their vital information on our progress will be swallowed by Maryland's Goddard Communication Center, then spun off to Cape Kennedy's launch control and in "real time"—which is as close as you can come to immediately—it will be handed over and digested by the voracious family of computers at Mission Control, Houston. It is somewhat like several thousand people flying the same airplane at the same time. Ironically, if all goes well, the actual crew aboard will serve primarily as monitors from the time of liftoff to orbit. The action is about to become so hot and fast that computers will do a better job than man. They can sense the precise moment to abort, a decision that might be physically impossible for us to make if our spacecraft should suddenly go berserk.

Computers send steering signals, as many as 25 times a second to gimbal the great first-stage engines. If the computers sense a malfunction worthy of an abort—presto! The escape tower separates from the big bird, hauling our command module after it. The canards deploy automatically, turning our module's blunt end forward. The tower itself is jettisoned on a previously programmed cue, our descent parachutes, which were originally intended to blossom over the Pacific recovery area, then lower us into the Atlantic, hopefully gently, hopefully this side of Africa, hopefully at a location convenient for swift rescue. Atlantic recoveries sound very well organized during chalk talks and appear to be a sure thing on paper, but we are here because we are realists. We have flown enough extraordinary aircraft to know how matters have a tendency to snowball once an initial glitch shows its

ugly head, and we know that if we are going to be so unfortunate as to experience an abort soon after liftoff, there is going to be a lot of hoping involved in the pickup.

T minus three minutes 51 seconds . . . and counting. The third-stage start tanks are chilled down and we have completed checking our launch digital computer. Our eyes are alert but have a tendency to linger on the flight director attitude indicator (FDAI), an instrument that may be liberally compared to an aircraft's artificial horizon. However, it is a far more sophisticated attitude display showing yaw as well as pitch and roll. We have surveyed the balance of the display console so many times in the past few years the action has become like watching the face of a bosom friend. We will recognize the most minute change of expression whether benign or challenging and until such change occurs there is simply the waiting.

Now a brief ceremony. Our commander has long cherished a special and most significant key. He holds it up for us to see, and his next gesture, faked so many times in simulations, is now for real. He inserts the key in the guard assembly that protects the sequence control pyro-arm switches from accidental closing. In a sense, he is unlocking the door to our immediate future, for these rather ordinary little switches activate the many powerful pyrotechnic systems involved in our flight . . . including those that fire off our escape tower.

This is undoubtedly it.

We breathe deeply and with deliberate slowness, as if to prove via the telemetric probes fixed to our body that we really are the sort who never flap under any circumstances. We breathe of this sterile air circulating through our helmets, which has a strangely mechanical odor, as if too many pumps and pipes were involved in its creation—but then all of NASA and even our homes and cars in Houston are air-conditioned, so who knows any longer what real air does smell like? Back in Houston? Who lives there? *Me?* Then what am I doing here, listening to the faint whine of the inverters, watching the seconds snap down on the digital timer?

Ah, there is a familiar voice again calling out numbers, a voice that fails to recognize there will soon arrive one of the very great moments of our lives.

T minus 45 seconds. We switch over to our own four batteries. Another faint thumping; another umbilical service arm has retracted and locked.

Just now, some of us believe that any man who would deny the existence of a few butterflies fluttering through his intestines is very probably a liar.

. . . 15 . . . 14 . . . 13 . . . 12 . . . guidance release . . . ignition sequence has started . . .

Our eyes halt on certain status lights that are presently inactive and that we prefer remain so. There is the launch vehicle rate light, which will illuminate if our bird's roll, pitch or yaw rates exceed limits. There is the red LV guide light to indicate loss of attitude reference in the guidance unit. There are the yellow LV engines lights, which go on if an engine is developing less than required thrust. Then the cue lights for ignition, cutoff, engine below thrust, and physical stage separation.

. . . eight . . . seven . . . six . . . five . . . we have had ignition command to the first stage.

This is not a simulation, of course. We are, if anything, over-rehearsed. We can reach out with our eyes closed and activate almost any of the multitude of gray switches overhead—overhead being where the action is; our feet are where the instrument panel of an aircraft would be. We are lying with our legs in the air, heavy and restless on our couches. We are not utterly dependent on automation if we must abort. There is a relatively large pressure switch staring right back at us. Reach out—one punch, and *sayonara*.

Just below that switch, the digital timer is snapping out the remaining seconds . . . four . . . three . . . two . . . one. A light glistens simultaneously when that familiar voice calls to us. ". . . zero. You have ignition."

"You are GO. . ."

"We have liftoff."

Journalists and TV pundits from all over the world have momentarily laid aside their habitual masks of indifference and pounded their imaginations in vain attempts to adequately describe an Apollo liftoff. It is difficult because the sound itself had never been heard by man in the past, and there is nothing with which to compare it. The very air shudders so that the senses are stunned and do not recover until it is all over. Many of the most hardened journalists cheer, while others simply stand and weep, repeating over and over again involuntary mumblings that sound very much like "Oh God, oh God," or "Jesus Christ . . . Jesus . . . Jesus . . ." And once a very young lady named Kate wrote to her friends afterward, saying from her poet's heart, "There was a strong thunder shaking the

world . . . It burned a hole through the cloud and then it was out of sight."

In the command module, we experience nothing of the sort. Besides the instant sense of exhilaration at now being committed, there is only a solemn devotion to the information on the display console. With rather secondary attention, we feel the thrust pressure our backs, although it will never exceed more than 4.5 G. The overall vibration is impressive yet easily tolerable. There is very little depreciation in our ability to read instruments.

We are only vaguely aware of the sound created by the five first-stage engines, which so appalls the earthbound spectator. At first, the noise is like that of a tremendous, distant waterfall; then, as our velocity increases, the rumbling diminishes.

Only one manifestation continues to remind us this is not a simulation. We are on the receiving end of the world's biggest pogo stick, and the vertical oscillation makes the G load seem inconsequential. Everything in our little capsulized world including our prone bodies has joined in the dancing.

"Apollo, Houston. GO at 30 seconds."

"Roll complete and we are pitching."

"Roger that. Stand by for Mode One Bravo."

We are at an altitude of 5,200 feet. Our speed is approximately 850 miles per hour. We are of a new generation. We think and speak in nautical miles for distance and feet per second for velocity. At one minute 10 seconds, we are at an altitude of four nautical miles, or if you are still thinking in terms of rate of climb, how does 24,288 feet suit you?

We are on trajectory and very suddenly GO for staging. At 17 miles altitude, the inboard F-1 engine shuts down. Then its four companions.

"Apollo, Houston. We confirm inboard is out. You're

looking good."

The four companion F-1 engines stop. We jettison the first stage. There is a shudder as it leaves us, as if the rest of our Saturn felt relieved. After a momentary lull, the second-stage engines take over.

"We have S-2 ignition."

"Roger, Apollo. Trajectory is good. Thrust is good."

Now the five J-2 engines give us another boost toward the limits of the atmosphere. We are already 46 miles high, 72 miles downrange. The vibration is easing. We jettison the launch escape tower. It takes our forward boost protective cover with it and we are instantly assaulted by something we have not had time to think about. Daylight! All five of our windows are now clear, even though there is no recognizable object to be seen. And we are now very, very busy. Our altitude is 75 miles, our velocity up to 7,500 miles per hour and we are 222 miles downrange. Check cabin pressure. Normal. 6.1 pounds.

"Apollo, Houston. Stand by for S-IVB to COI capability."

"S-four-B to COI. Roger."

"You are still looking good, gimbals are good . . . trim is good."

There is something about the voice of our capsule communicator when he confirms such news that makes us love him.

We are still burning the five engines of the second stage and will continue to do so for another three minutes. Then, if all is serene, we will jettison the second stage, light off the S-IVB and go for earth orbit.

"Apollo, Houston, Your level sense arm time eight plus 38 nominal. Stage two cutoff time nine plus 48."

We are at eight minutes 17 seconds and showing a velocity of 20,672 feet per second, which is about 80 percent of what we need for minimum orbit. Eight minutes! Has it been *that* long since we shared a planet with four billion other people? It seems incredible that up here we are the only three of the species. If we face the truth, we must admit we are only frail little men trying to be brave. We are at the shores of a great sea, dipping our toes into the unknown.

The second-stage engines shut down with a sigh.

"Apollo, Houston. We have S-IV ignition."

"Roger. Thrust looks good. Trajectory guidance CMC are all GO."

"Thank you. As advertised." Such remarks at such vital moments come straight out of our stiff-upper-lip basket, which is carried on every Apollo flight.

At 10 minutes 30 seconds, we are 1,080 miles downrange. We have passed through 100 miles altitude. "Okay fer high," said the farmer as he watched the Jenny fly backward across his field in a high wind, "but she ain't so impressive fer far."

Patience. In a very few moments, we are going to redirect some of this energy.

"Apollo, Houston. You're GO at 11 and one-half and predicted cutoff time is 12 plus 34."

"Understand—12 plus 34."

At 12 minutes and 34 seconds after our departure from mother earth, then, our final-stage engine will shut down automatically. We will presumably then be on a circular flight path around the earth, which is a rather dignified way of describing what we will actually be doing. The truth is once that S-IVB engine cuts, we are not going to be doing much flying. We are going to be *falling*, which is quite an art in itself.

If the earth were square or even had square corners, we would have to be out considerably more than this 115 miles to avoid a collision with the first corner. Fortunately, the earth is round, so while we are falling toward it, the earth is falling away from us. If we maintain a certain speed, we will never catch up with it; that speed, known as orbital velocity, is determined by earth's gravity. It happens to be 17,440 miles per hour (25,580 feet per second), which is what we are doing right now. If we reduce this speed, we will be drawn back to our home planet; if we increase our speed, we will leave the influence of earth's gravity and escape into space. We will do just that when the time comes to shove off for the moon.

Meanwhile, back in our ultra-modern little home, let us make a complete systems check and determine that everything is working right before we do anything rash. It's going to be a long day. We are only 17 minutes out of Cape Kennedy and have already lost radio contact with Houston Control. But regard the view! There is Africa, scorched a light umber by a million suns. There, to the north, Cleopatra held court and a little earthwhile later Rommel and his Panzers knew defeat. We have these few coasting moments to reflect upon our unique vantage and how from here you can see both ways forever. And suddenly, profoundly, we understand how every last man who has been in space returns to earth more humble than he left.

We are one hour 50 minutes into the mission. We have sent Houston our gyro torquing angles and realigned our inertial measurement unit in accordance with computer program 52. Since we have passed through the umbra, the earth has been lost in darkness. Below, in the night, a billion people and incomprehensible numbers of fishes and animals slumber. The people and the animals trust in the coming of the sunrise, which we already behold splashing across the miniature Himalayas.

As soon as we complete our TLI (translunar injection) checklist, we will play musical couches, with our commander shifting to the center position and the command pilot taking over the left couch. The left side for the man doing the job—another hangover from aviation, along with the word "roger."

We are soon to be the only three citizens of a neighborhood that happens to include several billion square miles, so we had better discover what kind of human beings we are. How do *we*

In a sea of flame and smoke, shaking the coast of Florida, an Apollo vehicle tears itself loose from the earth: Destination Moon.

come to be here rather than, say, a learned physicist, a celebrated astronomer or a renowned geologist? The reason is simple enough. In time, such scientists will take their rightful place in space, but right now, Apollo has room only for pilots. And very good ones.

There should be—although there is not—a sign over Building Four at Houston Manned Spacecraft Center where each of us has an office, and it should say immodestly, "Through these portals pass the most versatile pilots in the world." At least we used to be. Actually, since we shifted titles and became "astronauts," our ordinary flying has been very ordinary. We are encouraged to fly one of NASA's T-33s whenever possible, but when you are boning up on an Apollo mission, your life and time ceases to be your own. There is a hard-eyed, sergeant-major-jawed man named Slayton whom we know as "Deke." He is the equivalent of our chief pilot and very much the right man for the job, since, in addition to his combat air force career during World War II, he is a graduate aeronautical engineer. He is very approachable but also very no-nonsense, and more than any other single individual chooses the crews for Apollo missions.

Since the first day of our selection, we have been constantly aware of all our fellow astronauts and their almost pathetic yearning for a couch in space. Since we have been the prime crew, they have watched and helped us unselfishly for more than a year, but when they ask, "How are you feeling?" there is sometimes a temptation to question their absolute sincerity. None of these generous men would wish anything *serious* might happen to us, of *course* not! Yet if just a teeny-weenie, inconsequential, very temporary something would suddenly oblige us to move over and leave an open spot for them— "Well, fella . . . it sure is a damn shame, and I cannot tell you how sorry I am." No, through those crocodile tears, he certainly cannot.

Our pay is approximately $1,800 per month, so it is not high finance that lures us out to such altitudes.

Out, not up. One of the first things we learned is that there is no up or down in space.

We applied for the job and got it because we were qualified to relatively the same degree as those selected before and after us. We were less than six feet tall and under 34 years of age. We had a bachelor's degree in engineering, or in a physical or biological science. We had acquired at least 1,000 hours jet time. These requirements remain the standard except for scientist-astronaut applicants, who do not need the flight time, but had better be extraordinary in their own field.

We have joined and become as brothers to such men as Schirra, who deserves to be known as the "grand old man" of space although he has survived just 47 years of earth-life minus his time in orbit. He did not just fall into the job, nor has nepotism yet helped anyone toward a place in space. Schirra graduated from Annapolis as a bachelor of science, learned to fly at Pensacola, and during the Korean conflict flew 90 combat missions. His distinguished career in space is already legend.

We are also brother to an energetic, ski-happy man named Cunningham, who is a physics graduate from UCLA. He was a Marine aviator and is still in the reserve. He has the snap-trap mind of an eloquent scientist and laces it with a sense of humor. And there is Cooper, who has a bachelor of science degree as well as experimental test-flying experience in the Air Force. There is Eisele, who is no less than a *master* of science in astronautics with Air Force research-pilot training as sauce. There is Gordon, another bachelor of science degree who also served as the first project test pilot for the F-4H Phantom II. He is a man of firsts, having won the Bendix race in 1961 with a handy Los Angeles-New York time of two hours 47 minutes. There is Haise, with a degree in aeronautical engineering; and Cernan with his masters in the same art. There is Bean, another B.S. in aeronautical engineering; and the jaunty Aldrin, whose manner is more like that of an ex-fighter pilot—which he is—than that of a doctor of science in astronautics, which he also is. There is Swigert, who holds a B.S. degree but also happens to be one of the few astronauts who is a real-life bachelor. At this writing. Among other honors, he was co-recipient of the Octave Chanute award in 1966 for his part in demonstrating how the Rogallo wing could be employed to return space vehicles.

There are presently some 50 astronauts, and it should be obvious that if you have the slightest urge to join the league, listing a bachelor of science degree on your application—plus a generous dosage of smart pills, plus luck, plus military flying experience, and health—will be an assist.

There is no typical astronaut, although one does stand out as non-typical. His name is Armstrong, and there is something strange about him, a sort of innate greatness such as strikes you the instant Lindbergh shakes your hand and smiles. It is not the event having made the man, but the man making the event. It is as if Armstrong knew all his life he would be the first man on the moon, accepted his assignment forthrightly, and then tried to forget it might separate him even a little from anyone else. His time is NASA's, which devours his talents as well. He finds it extremely difficult to get away for a day of his favorite sport, which happens to be soaring. Armstrong is a brilliant research test pilot, and was deeply involved in flying the X-15 at some 4,000 miles per hour, to say nothing of the X-1 rocket research airplane, the F-100, -101, -104, F5D, B-47 and paraglider. In a way, it is reassuring that his rather mischievous grin still seems to come direct from Wapakoneta, Ohio.

NASA *has* made one considerable and wise concession to public support, and that is a wide-open television policy. As a result, "Mission Control—Houston" has been viewed by a large proportion of the world's population. Whether even a miniscule number of such viewers understand what those men in white shirts sitting in front of their rows of flickering lights are doing is another matter. While they may seem at ease or even dozing on Government money, be assured they are most solidly preoccupied with the constant barrage of information revealed to them by every means of electronic communication so far devised. And each man is part of a team that operates in eight-hour shifts or 28 or 38 if an emergency arises.

The booster engineer occupies the far left console down front. He's responsible for the three stages of main engines in the Saturn launch vehicle and is particularly active at liftoff time. Next to him is the retro officer, a trajectory specialist who maintains a continuous and detailed knowledge of the very best ways to bring us back home safely. Also down front is fido, also a trajectory specialist, who insures that the ground computer knows what is going on away out there in relation to the spacecraft's position and velocity. (Fido is the flight-dynamics officer, and yes, the "I" was inserted to make a pronounceable acronym.) Next to him is the guidance controller, who monitors the all-important behavior of the command module, lunar module and abort guidance computers.

In the next row is the surgeon, who monitors our health via constant telemetry; and adjacent to him is our direct link with life on earth. He is variously known as capcom, capsule communicator or spacecraft communicator, and he is always a fellow astronaut. He has worked with us consistently through endless simulations, understands our problems out here and knows us well enough to identify our individual voices by inflection alone, and even can detect confidence or concern in the tone we unconsciously employ to read off a set of numbers. In addition to straight business, he is, in moments of easy going, our friend. Listen to these history makers of Apollo 13, who at the time of this exchange happened to be 102,342 nm out from earth. They had not as yet made further space history by surviving a nearly disastrous explosion in their service module.

Command module: *Hello there, Houston . . . 13. We'd like to hear what the news is.*

Capcom: *Well, let's see . . . the Braves got five runs in the ninth inning . . . the Cubs were rained out. They had earthquakes in Manila, and the Beatles have announced they will no longer perform as a group. Rumors they will use their money to start their own space program are false. Uh oh, have you guys completed your income tax?*

Command module: *How do I apply for an extension?*

Capcom: *You're out of the country.*

In the same line of consoles are the controllers responsible for our environmental systems in the command and service modules. Two more controllers monitor the similar systems in the lunar module. Often it is a waiting game, so most of these men consume too much coffee during their shifts and do more than their bit to support the tobacco industry.

On the next level are consoles manned by the communications-systems engineer, the procedures officer, the flight director and his assistant, the flight-activities officer and the network controller. Finally, flying consoles in the topmost echelon are the public-affairs officer, the operations director, mission director and a Department of Defense representative who would work with our NASA recovery coordinator if the Army, Navy or Marines were required to retrieve us from jungle or ice cap.

We are now two hours 49 minutes and 39 seconds into the mission. Just over five minutes ago, we lit off our S-IVB engine to kick us out of close earth orbit and place us in one with an apogee altitude of 320,000 miles. That is a fair piece out, and it would take us nine days to get there if it were not for our target, the moon.

As we shut down the S-IVB third stage and thus temporarily retire from the boosted-projectile business, we begin our coast through space along what is known as a "free return trajectory." In essence, this means that if we are able to start our service propulsion engine, which is just behind us in the service module, we will not be lost in space forever. Our so-called "free return" trajectory has been most carefully selected so that when we arrive in the vicinity of the moon, our spacecraft will be swung around the moon in response to its gravity at the precise velocity and direction for a coast back to earth. We are a boomerang . . . almost.

There are some catches involved. If our present velocity is even slighty below the requirement for free return, the moon will deflect our trajectory too much and that would really be farewell sweet friends. The same dismal future applies if we are going too fast for free return. Thus, the precision required to assure us of a free return is within a 20-mile limit for safe capture. This is just too much to ask of even our presently developed technique, nor will the old airplane adage of safe arrival via superstition and fear satisfy anyone who treasures life as we do. So we cheat a little and make comparatively minute mid-course corrections using our service engine if it is working and falling back on our reaction engines if it isn't.

We will soon become used to the flat pop-popping of our reaction engines. Thanks to them, we perform "flying" maneuvers in space—and do a lot of other things.

The most difficult, initially bewildering and frustrating affair any pilot can have is with something that he is inclined to regard as a flying machine yet that is not. Take away wings and air and what do you have? Mix in a few confusers such as weightlessness, a stall speed you had better not think about (an L/D ratio of 0.35), an extremely tricky horizon if any exists at all, cruising speeds almost beyond your natural comprehension, and a something that maneuvers vaguely like an aircraft around all three axes and you wonder what you are doing out here. Such semifamiliar controls as you may find for your seeking hands are appropriately emasculated; one, which might be described as taking the place of what was long known as a joy stick in aircraft, being now dubbed the rotational controller. The other handle, which manages acceleration along any of the three axes, is called the translation controller. Opinions vary as to whether or not their short length of about six inches is significant.

Since space is a vacuum, it would be nothing more than wistful of us to put any sort of elevators, rudder or ailerons on our craft. Regardless of our speed, we could flop them about as much as we pleased without the slightest effect. So we swipe something from Newton's Second Law of Dynamics and install 16 reaction control engines on the service module, 16 on the lunar module and 12 on the command module. They all burn a mixture of nitrogen tetroxide and hydrazine, which ignite on contact and provide an instantaneous burn. The burn results in thrust of modest strength, which is what we are after.

These small hypergolic engines are made by Rocketdyne and are true gems of workmanship—and priced appropriately. On the service module, they are placed around the circumference in "quads," which are so arranged that any of the three axial movements can be accomplished or any combination thereof. It is the service module's reaction engines that will keep us rolling slowly around and around in the "barbecue" mode so that the sun will not overheat one side or lack of it overcool the other. These same little engines will keep us awake with their automatic pop-popping as we coast toward the moon.

Activating our reaction engines via either translation or rotational control to achieve pitch, roll and yaw, we can make minor course corrections to our trajectory, adjust the position of our command module so that the heat shield is forward during earth reentry, and maneuver for photography, star observation or rendezvous with another spacecraft.

We are three hours and nine minutes into the mission and all is GO for the next act. We are about to bring out the star of the show, whom we might as well call "Goldilocks," since there is something peculiarly feminine about her as compared to the rest of our now considerably smaller spacecraft. Goldilocks is expensive, wears gold pants and is formally known as our lunar module, or LM.

It is something like bringing a rabbit out of a hat. We must first separate from the third stage, then turn around and dock with the LM, which is tucked neatly inside. We will perform this maneuver while traveling in excess of 15,000 miles per hour.

"Okay, Houston . . . do we have a GO for pyro arm?"

"Affirmative. You're GO for pyro arm and recommend you secure the cabin pressurization."

Explosive bolts separate us from the S-IVB and at the same time jettison the protective panels around the LM.

"That's quite a bang, Houston. We've separated and we've pitched around about 60 degrees now."

"Roger. We see you pitching."

After this minor fourth of July, we fire the reaction engines until we are turned through 180 degrees and behold the third stage and the LM through our docking windows.

"Okay, Houston. "We've got the SLA panels . . . one of them is out front now."

"Ah so?"

"Got the S-IVB. Guess we're about 70 feet away."

"How does it look?"

"Pretty. Everything is all shiny. But there's some debris hanging about."

"Like what?"

"Noodles . . . and around our windows something that looks like billions of swirling snowflakes."

The alignment sight in our docking window resembles the range finder of a camera. We use short bursts of our reaction engines to roll and pitch our command module until the probe is precisely lined up with the LM drogue. Take it very easy. Our relative velocity is less than one mile per hour, but Goldilocks does not like rough stuff. Easy. A kiss and then we slip in, our probe seeking a socket at the bottom of the drogue.

She is captured. We activate the nitrogen pressure system, which pulls us even tighter together, and are married almost instantaneously. Twelve contact latches on the docking ring automatically form a pressure-tight seal between us and Goldilocks. What a romance. Now to spirit her away.

We remove the hatch and enter the connecting tunnel to check its pressure integrity. We notice a slight burning smell, but it has been reported on other flights, so we proceed with our inspection and discover two latches have failed to catch properly. We secure them and connect the umbilical that can bring Goldilocks to life. We power her up and find her voltage is normal at 13.

All is GO as we return to our command module, secure the hatch once again and prepare for final separation.

"Okay, Houston. We're ready to pull out Goldilocks."

"Let her come."

We flick a switch and the instant result is more fireworks. They separate the tie-down straps that have held the LM in place, and spring actuators at the attach points of its landing gear eject the LM along with us at a velocity of less than a foot a second. We help matters slightly with a short burn from our appropriate reaction engines and pull away at one-half mile per hour.

"Looks like we're in the clear, Houston."

"Apollo, Houston. That's all we're waiting for. Stand by for some action."

As if we haven't had enough recently.

We peer out our windows at the S-IVB. Now the third stage is not a very pretty thing, but it did give us a good boost and we rather regret parting with its company. But Houston controls it now. They will divert it from our own trajectory to eliminate any possibility of collision. And what happens to discarded S-IVBs when no one needs them any longer? The one used for Apollo 13 was deliberately impacted on the moon, thereby giving the scientists a Richter thrill. The S-IVB for Apollo 12 was supposed to take off and go into orbit around the sun. Obviously there was a glitch: At this writing, it is still in an orbit of 447,300 miles by 39,700 miles—around the earth.

We are, at long last, on our way to the moon and cannot avoid reflecting on the emotions of Borman, Lovell and Anders, who in Apollo 8 were the first human beings to embark on this fantastic voyage. *Everything* out here was theory then, tested to the ultimate, but still to be performed in the real. One of that intrepid crew, appalled at the sight of his native planet diminishing in size so rapidly, shook his head and sighed, "Sweet Jesus! I wonder if we'll ever get home again."

We have removed our uncomfortable pressure suits and helmets and are now in our long johns, which NASA has rather stuffily labeled "constant-wear garments." It is a time of relative inactivity, which hopefully will endure for some 12 hours before we will be called upon to make mid-course correction.

Here in space we have been provided with those same fundamentals man has needed since he first prowled the earth. We have food to survive and shelter from the hostile elements.

We have created a little world of our own, which is dominated by the phenomenon of weightlessness. No training on earth can truly simulate weightlessness over prolonged periods, and it does take some considerable readjustment in your physical reactions. You can achieve at least some appreciation of this spooky condition by skin-diving or just swimming as effortlessly as possible under water. But in true weightlessness there is *no* real effort required; with a feather push on anything handy, you can sail right across the command module and you will not stop until you come up against

something fixed. Even that something may simply divert you to one side, and off you go like a cue ball. As Walt Cunningham predicted, "Once you get the hang of it, weightlessness becomes great fun."

Weightlessness also has some drawbacks, and man being a beast of habit would prefer being able to turn it off and on. Eating becomes a problem. Our food will not stay on a plate nor would a plate stay put. Hence our breakfast, lunch and dinner come in plastic bags that have a spigot at one corner—somewhat like an old goatskin wine-bag. We add water to the dehydrated ingredients through the spigot, knead the contents to a fairly viscous consistency, then cut across the top of the bag with scissors, dip a spoon in whatever it is and argue that it is at least better than combat rations. Earlier Apollo crews were obliged to suck their food directly from the bags, but improvements in the viscosity of gravies now allows us to use a spoon. The menus have been chosen by us individually and are as widely varied as our tastes. Breakfast can be peaches, cornflakes, scrambled eggs and coffee; lunch can be turkey and gravy, chocolate pudding and an orange drink; and dinner, salmon salad, chicken stew and fruitcake. That is what it says on the package and that is what it tastes like because the biologists, physicians and dieticians at NASA have spent a long time slaving over their hot stoves to keep us gastronomically content. Yet even though one of them is as comely a young lady as any astronaut would care to remember while he is squeezing away at his chicken and rice, it is still a long way from being mother's cooking. And because a gas bubble will not rise while we are weightless, no one has yet found a way for us to burp.

There is also a series of psychological blocks that tend to discourage any ravenous appetites in space. Once the novelty of chasing a bite of cheese and crackers across your field of vision wears off, eating just isn't much fun. The all-important and integrated physical motions of selection, seizure and real animal chewing are largely absent. Furthermore, we have had little exercise to create appetite and finally, if we eat we must eventually defecate, and that, in a weightless condition, is definitely *not* fun. There are cleverly designed plastic bags available that we paste to our buttocks. After use, we insert a germicide in the inner bag, seal it in the outer bag and stow the combination in the waste-disposal compartment, which has a split-membrane trap to keep previously used bags from floating back out when the door is open. As if this rather indelicate performance was not enough, consider that the human stool drops away from the body because of gravity—a something we do not have. Thus it becomes a force we miss very much on certain occasions, and perhaps the lack of it combined with a formidable variety of other factors explains why every astronaut has lost considerable weight on Apollo flights. These losses, which range from seven to nine pounds, are considered most serious by the directors at NASA, particularly when they contemplate future Apollo flights of much longer duration, as well as prolonged residence in space stations. It is somehow reassuring that in spite of scientists' and computers' best efforts, man stubbornly insists on being just man.

We are now 30 hours and 55 minutes into our Apollo mission, and let us say "Good afternoon, God," toward whom, if we retain the usual conception, we are fast approaching.

We are now more than 120,000 nm out from earth, and since its gravitational force still has its claws on us, our velocity has fallen off to a mere 3,100 miles per hour. Fifteen minutes ago, we performed a mid-course correction by firing our service engine 2.2 seconds. As a result, we picked up an additional 15 feet per second in velocity and shifted the plane of our trajectory one tenth of a degree. A "tweak," as it is called, but enough to insert us into a hybrid trajectory, which is a fancy term for a free-return trajectory slightly altered to suit our mission needs.

Among our less demanding chores, we dumped our waste water, did our housekeeping around the module, took some earth-weather photographs and did a TV show for earth—which was a nuisance for us, although it helped our consciences by allowing us to share all this with the people who are paying the bill. And God, who we trust is out here with us, knows NASA needs all the public-relations help it can get. Finally, we changed the lithium hydroxide canisters, which remove carbon dioxide and odors from the air in our command module.

Our first night aboard was not exactly a slumber party. The continuous whirring sound of the inverters and cabin fans was soothing, but the popping of our reaction engines was not. We slipped into our sleeping bags to keep our arms and legs from floating around, but it was not very long before the habits of our lifetime took over and we began yearning for that exquisite feeling of our tired bodies pressing against a bed. On the long haul, weightlessness has its disadvantages.

As on earth, one thing still leads to another, and when we awakened, there was another quite normal yearning. So we reached for the hose that leads to our waste-water stowage tank, rolled the rubber condom over the head of our penis and relieved ourselves. And we sighed. For certainly when we make the next waste-water dump it will be the ultimate testimony that man finds a way to pollute everything—even unto the purity of space.

In a weightless situation, the act of spitting is asking for trouble, so we started our day by brushing our teeth with ingestable toothpaste—a concoction deliberately designed by NASA experts for swallowing. We made our good-mornings to Houston, who responded in a cheerful voice with a sort of morning newspaper recounting of the most recent earth-side revolutions, assassinations, riots and sundry natural disasters. There are times when we wonder if we really do want a free return.

We are still barbecuing our hull to keep the temperature even all around, and we still have a look at mother earth every 10 minutes or so. It is lying on its side with the Arctic polar cap

still in the same relative position to the left of the sphere, and the umbra line cuts horizontally across the middle of the North and South Atlantic Oceans. Africa is in the night, and some fairly large chunks of the North and South American continents are visible beneath the usual swirls of cloud. The earth is markedly smaller after our sleep, so the facts must agree with what the computers say. And they say a lot.

Near a garish, Muzak-river motel, which survives only because of the many pilgrims visiting the Manned Spacecraft Center, there is an inconspicuous, fortress-like building. The only visible windows are vertical slits across the front, which apparently have been cut as a grudging concession to those few reactionaries within who might still care about the natural light of day. In front of the building, like a tombstone to individualism, stands a graven granite image—I B M.

This dreary symbol more than any other removes the gee-whiz from space. Because of computers, space flight is possible. It therefore behooves us to accept their dictates with graceful resignation.

For example, we must now make another realignment of our guidance and navigation system. The heart of it is our inertial measurement unit (IMU), and it is basically a gyroscopically stabilized platform set within gimbals that can sense any movement of the spacecraft about it. Before launch, we aligned it with earth-centered coordinates. Now, since the stars do not appear to move in translunar space, they become our inertial reference.

If we take a celestial fix at sea or in the air, we normally sight on at least three stars, and compare their azimuths and altitudes above the horizon as viewed through our sextant with what the good books say they should be at our assumed position. The result is a small triangle formed by the intersecting lines of position, and we may reasonably assume we are in the center of that triangle. In space, the Doppler system actually handles our position, course and speed simultaneously, and it is extremely unlikely that we should ever have to navigate in space solely by the stars. However, it is important that we keep our vehicle in the correct attitude; it is critically important that we be properly aligned for any trajectory burns.

Our computer has the inertial positions of 37 stars stored in its memory. We call for some brief reaction-engine firing to change our attitude so the most convenient star group comes into the field of our one-power telescope. It is like walking into a dark hall and suddenly discovering familiar friends. Greetings Nunki, and Delta Capricorni, and Peacock and Formalhaut. What a pity that we must ignore your lovely names and report you to Colossus as star 37, star 221, star 42 and 45. And we will only use two of you.

When we are settled in position, we move over to the 28-power sextant, bring Nunki precisely to the center of the reticle, then punch a button on the control panel. Mark! The mark tells the computer to note the attitude of the IMU in relation to Nunki's programmed position. Two such marks provide realignment reference and verify our spacecraft attitude, so we do the same with Peacock.

The on-board computer can also calculate the relationship of the measurements to position and speed. This ability to keep abreast of our own position would be essential if we should have a communications failure.

Very well, Captain Hornblower, who is sailing the ship? A computer? Extraordinary. Never heard of the chap.

It seems that *our* computer is nothing if not autocratic. For example, primary circuits in our computer and data adapter are triplicated. Their signals are sent to the appropriate circuits, where voting takes place and the majority rules. If one signal does not agree with the other two, it is ignored. You there in the minority group! Stand down!

Among innumerable other favors, computers calculate future positions and velocities along our predicted track through space. They can give us a two-minute warning that the earth is about to appear in our window, thereby saving us a vigil if we want to take a weather photo or navigational fix. They simulate the performance of our spacecraft flight maneuvers and tell remote earth sites where to pick us up on their radars. They tell us how to orient for our celestial-navigation fixes and process radar data to determine our trajectory. They are absolutely essential in our moon landing-site determination.

Altogether, Colossus swallows one of the longest and most complex programs ever written for computers, including every facet of the mission from launch to re-entry. Some of the sub-monsters in the system add or subtract in 0.4 microseconds, but they become inexplicably sluggish and burn up as much as 2.8 microseconds when asked to multiply.

Censors constantly monitor our flight performance and environmental data, and the results are telemetered to earth in the form of electrical signals. Colossus absorbs all and knows all. You ask him. He speak.

Through their computers and associated spies, the controllers back in Houston can monitor much more than we can though we are actually on the scene. And Houston is even now paging us.

"Apollo, Houston."

"Go ahead, Houston."

"Just info on your PTC. Looking good. Pitch and yaw excursions are very low."

It is nice to know our revolving mode, or "barbecuing," for thermal control is keeping our vehicle's skin at tolerable temperatures, because in a sense, its skin is our very own dear epidermis.

We ask the flight-dynamics officer if he has any new comments on our trajectory.

"He's happy. Says you probably will not need any more midcourse corrections. But remember, fidos never guarantee anything."

"So true."

"Can you see the moon? Is it looking any bigger?"

"No . . . not really. Maybe a little, but we've still a long way to go."

"Understand. They're estimating lunar pericynthian as now 62 miles."

We are secretly pleased with the word pericynthian; it specifies our closest distance to the moon during initial orbit, but we also are pleased to remind ourselves that with such a melodic word being bandied about, not even Colossus can erase all romance from space.

We are 36 hours 52 minutes into the mission. Earth, where we were conceived in a society not yet emerged from savagery, is 137,250 nm behind us. Our speed is slightly more than 3,000 miles per hour, a velocity that cannot be thought of in habitual terms; for here in space we are only drifting, a lazy boiler-plate balloon. We are explorers of a microscopic path toward the stars, and we will never reach them or their planets until we can greatly increase our speed. At this rate, it would take us 12 years just to reach Pluto, the outermost planet of our galaxy. Our families, professionally attuned though they may be, would unquestionably have some very firm views about such a prolonged business trip. Beyond Pluto, to the mysterious depths of the universe, traveling time is measured in light-years.

On this mission, we will be off earth slightly more than eight days. The sun will not rise and set for us as for the rest of humanity. All the basics in the heavens—the sun, moon, earth, Venus, Mars, Mercury, Jupiter, Saturn, Uranus, Neptune and Pluto—appear to have halted. "Time, then, does not pass—or does it?" Are we going to be any older when we splash down in the earth's ocean; and if so, how much older? What units of measurement apply to us here? Einstein theorized that if we could go faster than the speed of light, we might go back and witness such events as the Battle of Waterloo or the Crusades. Impossible? Now that we are here, having gone even this pitifully small distance toward another world, we are determined to drop the word impossible from our vocabulary.

As for our age on return, surely Colossus will be able to come up with some kind of an answer. (Worshipful Master, what happened to those eight days?)

It is much easier to stare at the "ceiling" of our marvelous craft and remember how not so very long ago the instrument panel of an airplane looked complicated to us. What we observe now will soon be as primitive as the panel of a DC-3, yet when we first saw our command module's display console we thought alas, the day has come when we have met too much for our competence. And then in time, like all other instrument displays, it became quite simple.

While waiting for drowsiness, our eyes rove the console, pausing sometimes on certain instruments that have already been important to our journey and others that soon will be. On the top left (commander's) side is a familiar face to all aircraft pilots—an altimeter. But this one is small, in keeping with its actual importance. During the launch, it reported our altitude to 60,000 feet and we barely had time to glance at it before it was out of business. But it will regain status during our re-entry, because our altitude in feet above the earth's surface as we make that final descent will undoubtedly fascinate us.

Just below the altimeter is another display that is extremely important at re-entry time. It is unique to spacecraft and is called the "entry monitor." Fundamentally, it is a sort of combination DME, airspeed, and turn and bank instrument. It captures and holds our most devout attention just before and during our plunge into the earth's atmosphere. By turning a selector switch, we observe a scroll that indicates our all-important velocities during entry and our distance to point of planned splashdown. A small gravity sensing light marked "0.05 G" illuminates when we receive our first fiery kiss from the earth's atmosphere. This subsystem display also incorporates a roll-attitude indicator, which relays our corresponding axis attitude.

Directly in our commander's line of sight is the light, which can warn us an abort has been requested by earth control; to the left is the master alarm, which not only glows red for any out-of-tolerance condition but sounds off a no-nonsense tone in our earphones.

A considerable portion of the commander's panel is devoted to launching information, the obvious philosophy being that if the first phase from liftoff to orbit is not successful, nearly everything else on our seven-foot console would suddenly become meaningless. Thus, the whole right side of his panel bristles with launch-vehicle lights to indicate the proper ignition and burning of our engines, staging and separation lights, and switches to cover manual deployment of our launch escape system should the automatic elements fail.

The commander also has meters to inform him of the pressures in the fuel and oxidizer tanks. Just below are thumb wheels with which he can set the position of the pitch and yaw gimbals in our service propulsion engine (SPS).

The center panel faces the command-module pilot's middle couch and is surmounted by a jukebox array of 37 warning lights covering possible trouble from over- or under-pressures, or electrical overloads. It was this display that lit up like a theater marquee when Apollo 12 was struck by lightning, and confirmed serious trouble immediately after the explosion that nearly finished Apollo 13.

Below the warning lights are a host of switches and meters governing a variety of equipment from extension of the docking probe to the interior environment of the command module. The numerous valve-position indicators are possibly the most logical and somehow delightful informants ever to ascend from the surface of the earth. If you observe striped lines—"barber pole"—in the miniature window, the valve is closed. If the window is plain gray, as if some miniature Martian within had pulled down the shade, the valve is open.

The command-module pilot's display also contains the number-two flight attitude director; just below it stands Colossus's all-important nephew, DSKY, who is referred to as Disky in polite space society. He is just another computer, but we find it very difficult to consider him anything less than exceptional when we push the appropriate "noun" or "verb" keys followed by a few numbers, then press "proceed" and lo, there is our question answered just as if he knew it all the time.

DSKY is mercilessly quick to inform us when we have made an error in our button punching. OPR ERR, he proclaims. He is also pleased to tell us when the temperature of our CM stable platform is out of tolerance, our inertial system is not in the mode to provide attitude reference, or the gimbal angle exceeds 70 degrees. DSKY's wisdom is not entirely his own. When his "uplink acty" light glows, we know he is receiving advice from big uncle on earth.

Much of the CM pilot's console is concerned with our environmental control. There are switches and meters devoted to the temperature, oxygen flow, pressure-suit-circuit atmosphere, potable- and waste-water quantities, and all of the redundant equipment specifically designed to keep us continuously breathing, faithfully tax-paying U.S. citizens. Even pressurized aircraft are obliged to maintain a hospitable environment for only a few hours, and if something does go wrong, they can make a quick descent. We try not to remind ourselves that if we have a serious environmental glitch, it's a long way home.

Out here, when is tomorrow? Is it when the oblong-framed glare of the sun ghosts diagonally across the right side of our display console? Is it another night when our slow rolling through space dips the same crowded area in shadow? Whatever happened to our innocent and reassuring world, that long-ago existence when as youths we crawled into a sleeping bag for the first time and watched the forest wave at the stars? Looking up, our view is now solely of man's ingenuity rather than natural beauty. Now the sun glare, too white and unnatural by our earthborn standards, is once more passing across the rows of switches and families of gauges. Each has a message concerning such matters as the remaining strength in our service propulsion engine or the status of our communications department: transponder . . . S-band normal . . . S-band auxiliary . . . voice, data, TV, VHF-AM . . . ranging . . . tape recorder . . . AC inverter . . . bus one . . . bus two. It is all there, so many informants trying in hopelessly inadequate mechanical jargon to tell us what marvels we can achieve with a flick of our finger.

All of this is the stuff of heroes? Of course, we secretly warm to the appellation. As astronauts, we are welcome everywhere in the world, honored and offered new opportunities that are sometimes very difficult to resist—at least financially. If there had never been a space program, it is doubtful that our names would have become household words. Probably, most of us would have completed rather distinguished military flying careers, retired with the customary honors and emoluments and that would be that. Yet the moment we accomplished liftoff, we did become a part of history. Our names will be listed 500 years hence along with such explorers as Cook, Vancouver, Magellan, Shackleton, Amundsen, Marco Polo and Eric the Red. Though our immediate descendants may presently classify us along with all other adults as unworthy of more than momentary attention, our grand-offspring unto the tenth generation and perhaps beyond will meticulously trace their blood lines to "those primitive and most daring adventurers in space."

We will sleep now, clinging to our belief that we are far more than button-pushers. We must remind ourselves continuously that computers cannot *really* think. Like so many people, they are programmed, and if an event occurs that is not included in their repertoire, they are suddenly struck dumb.

We *can* think, and improvise if the need arises, which of course is really why we are here.

We have been falling toward the moon at ever-increasing speed and are now cruising at 2,549 miles per hour. When our "free return" course takes us closest to the moon, we will be doing 5,690 miles per hour. Very obviously, something must be done about this speed or our arrival on the moon will create far more spectacular effects than we intend.

We have meticulously plowed through a lengthy check list to assure we have all the criteria necessary for LOI. Every system and department reads GO—Service propulsion system fuel tanks nominal and no leakage, with the same reports from our reaction control engine tanks.

We are behind the moon in relation to earth, on the side that never has and never will be viewed from the earth—unless someone makes some drastic changes in our planetary system. Could there once upon a time, ten million years before earth human history began, have been an atmosphere about the moon? Is it possible the inhabitants might have determined it was of less importance than their angers with each other, and their reward was this stunning desolation? Now here, with the ultimate wilderness so closely exhibited before us, we can understand why no earthling who has beheld the back side of the moon can ever be quite the same again.

We have set up thrust program 40 for our service engine, which will accomplish a very important fix of our present situation. If we do nothing but sit back and watch the forbidding scenery roll by, we will soon be on our "free return" way back toward earth. Our speed is too great for the strength of the moon's gravity to capture us. Therefore, we must reduce it, which is simple enough if done exactly right. Our service engine is programmed to burn five minutes 56.5 seconds in retro-fire, which will put on the speed brakes and bring us down to 4,070 miles per hour. The moon's gravitational field can then reach out and grab us as if we were a brass ring on a merry-go-round and hold us in an elliptical orbit—still not satisfactory, but better than making a mere passing acquaintance. After two revolutions, we will refine our orbit to as near our special mission requirements as flight dynamics, Colossus, fumbling man, fate and benign God can make it.

The bulk of the moon blocks all communication with earth until we come swinging around the horn again.

We roll to burn attitude, make a final sextant star check and then recheck all the criteria that might say a no-go for the burn. We are in our couches, restraining straps secured, and a little too casual with each other on the interphone. We may be mature men and combat veterans, but there are still some mothers back on earth who are not entirely convinced they raised their boys to hide behind the moon.

Well, Mother, it is 77 hours 24 minutes and 52 seconds into this mission, the SPS thrust switch is direct on and the delta V thrust is on Bank B for a starter and . . .

Two seconds . . . three seconds . . . shoot the whole bowl of sherbert by also switching on Bank A.

The burn lasts for over five minutes. We are not flying but orbiting, which is a handy word for a nice balance between speed and gravitational force. At the end of our burn, our orbit is 168.3 miles by 57. We are on the orbital downwind leg.

We make a few short firings of our reaction engines, maneuvering into communications attitude, and have no more aligned roll, pitch and yaw than advertised as we emerge from behind the moon and there is Houston on the phone.

"Apollo, Houston. How read?"

"Greetings . . . we're rolling around the horn with Goldilocks on our arm."

"Roger. Understand. Suggest keeping your hands on your wallets."

"Are you ready for a burn status report?"

"Affirmative. Go ahead."

We give Houston our fuel residual readings and actual burn time.

"Apollo, Houston. Fine business. Now if you'll give us POO and accept, we can give you a clock update.

"POO and accept it is." We clear DSKY to zeros so his brain is clear and open to accept new data from earth.

And now after our successful and most precise LOI what marvels does our flight plan next demand? There is inscribed in capital letters right on page three dash 56—EAT PERIOD. It is more than a simple physical recommendation. We have 30 minutes to masticate a pack of frankfurters and applesauce, wash the meal down with an orange drink, top it off with an honest chew on a chocolate cube, and remind ourselves that although we may be embarked upon a most extraordinary voyage we are, alas, only the most ordinary of human beings.

Our second revolution around the moon is a busy one. We maneuver for landmark tracking and optically check our exact course over the surface. We are passing over Mare Tranquillitatis; the "highlands" are ahead, with Fra Mauro, where Apollo 13 was supposed to have landed and Mare Cognitum beyond. In a series of observations, we establish our track and position angles in relation to the center of the moon, feed the information into our computer and run through the complete checklist for Descent Orbit Insertion (DOI).

Actually, we are creeping up on our grand objective. It would be much more convenient if we could simply land our whole spacecraft on the moon's surface, but doing it in pieces saves tons of fuel that would otherwise handicap us from the moment of launching. Soon we will part company with Goldilocks, who will deposit two of us on the surface. When our lunar tasks are done, Goldilocks will divide herself like an amoeba, and a relatively small part of her will ascend for a reunion with the spacecraft that now shelters us.

Eventually, we will abandon all that is left of Goldilocks and start home. Employing this ruthless program of casting away what is no longer needed results in acceleration and deceleration of much smaller masses in space, with the consequent economy of power demand. Do not pick up the whole piano, just one key at a time, please.

Our second burn, of only 23.1 seconds, established us in the orbital equivalent of base leg. We slowed to 3,643 miles per hour and established ourselves in a 57-by-seven-mile orbit. Later, when everything had settled down, we instinctively curled up our toes during the seven-mile perilune. We were in "bail-out" attitude and set for an escape burn on two seconds' warning; but now, after our sixth revolution, we are becoming accustomed to hedge-hopping over the craters.

All the while, we have been keeping out of mischief by realigning the IMU, reassuring ourselves via the old-fashioned eye-ball method that the landing site we are looking right at is the correct one, photographing Censorinus, changing lithium hydroxide canisters and making preparations to enter Goldilocks. We have done our housekeeping, eaten lunch or dinner, (no meal seems to fit into the time of day out here), and have just maneuvered to "rest attitude"—an orbiting angle selected for maintenance or communications with earth and most efficient thermal control as we pass alternately from full sun to shadow around the moon. We put up the window shades and bid each other good-night. The flight plan says we must rest again for over eight hours.

Men of the American space program have now logged more than seven months in the far-out-there. The total endeavor has not been easy and promises to become more difficult. We have had one tragic episode in the deaths of Grissom, Chaffee and White, plus a very near miss for Armstrong and Scott during Gemini 8 when a thruster jammed in open position. We are almost as shy as the Russians in recording our failures, but at least no one attempted to disguise the cliff-hanging features of Apollo 13.

Aviation did not progress without considerable human sacrifice, and there is no reason to suppose our ventures into deep space will be any easier on the pioneers. To pretend space travel is safe would be as ridiculous as to claim flying is entirely without hazard. Like flying, takeoff and landing present the most likely areas for sudden trouble, but at least in space we do not worry about hijackings or midair collisions. The 2,000-odd pieces of international junk now in various space orbits are so insignificant in relation to the area involved that their presence is not even considered in Apollo flight planning. Of course, not so long ago we felt the same way about the situation in our skies.

There is one major loophole in every Apollo flight: Every conceivable contingency is covered except the one that would undoubtedly increase the telemetry readings of any astronaut's heartbeat and breathing. And it is fundamental enough. Fortunately, the ascent engine's hypergolic principle is so simple the chances of a glitch are almost nil, *but if* it should fail to start, there is no possible rescue for the men on the moon; when the inevitable time comes for the command-module pilot to commence his return to earth, he will certainly be the loneliest man in all creation.

No chopper pilot could long resist the charms of our Goldilocks. He is bound to feel comfortable with her in spite of her outward resemblance to a tarantula. Both chopper pilot and LM are devoted to relatively vertical arrival upon and departure from whatever is handy. In Goldilock's case, it is the moon.

Because of the vacuum environment in which Goldilocks operates, there is not even the grudging concession to aerodynamic design found in most helicopters. Her strictly utilitarian figure was dictated by component location, by the fact that she would only make one landing—which had better be near-perfect—in her entire career, and by the accommodation of two astronauts who would be vitally concerned that she also makes her one takeoff with absolute perfection.

Goldilocks's lower descent stage represents almost two-thirds of her total poundage. Like many women, the reason is simple enough, and not dietary: The lower stage must support the upper ascent stage, provide storage for communications and scientific equipment, serve as a foundation for a landing gear and ultimately serve as a launching platform for the ascent stage.

The ascent stage is where two of us will soon take our places, airplane style—with the commander on the left and the LM pilot on the right. It shares the same environmental characteristics found in any other manned spacecraft, and is accelerated, decelerated and maneuvered by employing the same thrust principles.

About there, the resemblance ends. Fixed-wing pilots are inclined to stare at the contraption and mutter the age-old aviation paternoster, "I said it to Wilbur and I'll tell it to Orville . . . the damn thing will never fly."

Let us now, while we are resting before the big show, consider some of the peculiarities to be found in any close association with Goldilocks.

To leave our orbit, we must again slow the LM for descent while simultaneously aiming for our landing area. Our descent engine, which is capable of nearly 10,000 pounds of thrust, is unique among rocket engines in that it can be throttled down to a mere 1,000 pounds of thrust. Thus, all the way down to the surface, we can control our *rate* of descent. The engine is also gimbaled to a maximum of six degrees, so that it can be controlled to compensate for the LM's center-of-gravity changes during descent. It has a 20-time restart capability, a life of 910 seconds with the fuel we have aboard, (Neil

Armstrong came the closest to using it for "long range" cruising), and it weighs a mere 360 pounds.

Both the descent and ascent engines use a fuel known commercially as Aerozine 50 and an oxidizer of nitrogen tetroxide. Starting either engine brings us right back to the matter of weightlessness. We must somehow settle the fuel to the bottom of the tanks, so we perform a maneuver called "ullage"—a word with rather pleasant connotations, since it originates in the solemn exercise of filling a wine cask. In spacecraft, we accomplish much the same business by a downward firing of our small reaction control thrusters.

We will arm and fire the descent engine, but from start until we are almost ready to touch down, its operation is better left to Goldilocks's guidance computer, which automatically governs throttle and gimbal angle. For almost any job, her guidance, navigation and control subsystem (GN & CS) gives astronauts the equivalent ego-beating suffered by airplane pilots when they sit back and monitor a good autopilot: "Impossible! The goddamned thing flies better than I do!"

But *all* is not left to Collossus or his nephews. We can and often do control the LM manually. Flying Goldilocks is quite different from flying the CM, if only because we stand at the controls, which are at least third cousins of those found in aircraft, and first cousins to those in helicopters.

There are dual controls in the LM, so that either commander or LM pilot can take over. The "attitude controller" is of pistol-grip design with a trigger-type communication switch for communication with Houston or the CM. As we move the controller, so responds Goldilocks, with a yaw to the right or left in answer to a clockwise or counter-clockwise turn of the handle. If we tip the handle left or right, aileron style, we achieve a roll left or right. If we move forward or back, elevator style, we have pitch up or down.

Our ailerons, elevators and rudder here are the 16 small reaction control engines arranged about the ascent half of the LM in four quads. When they push in response to our commands, something happens. Right now. Which brings us to a basic philosophy that might cover the operation of *all* rocket engines. Like twisting the tail of a tiger, do not start unless and until you are prepared for something interesting to happen.

World War I rotary aircraft engines shared one characteristic with those in Saturn V: Both types, well over half a century apart, operate either at full throttle or stop. The magnitude of their power does not change, only the time used in minutes and seconds according to the mission requirement. However, the rocket descent engine in the LM is throttleable, and is managed by a control that resembles nothing hitherto observed either aloft or alow with the possible exception of a plumber's helper.

It is located about where a single-engine aircraft's throttle might be if you flew standing up, but it does considerably more than vary the descent engine's thrust from 92 percent to 10 percent. It is called a "translational controller." With it, we meter power to the descent engine and maneuver about the other two axes with the smaller reaction engines. Or, by moving a select lever at the side of the controller, we can maneuver about all three axes using only the reaction engines.

We can still be Goldilocks's direct master if we deem it necessary, but we will probably leave most of her behavior to our on-board computer and guidance devices. Once we leave the CM, we are going to have more than enough to keep us busy. And let us remember that although we want to descend upon the moon, we are also most interested in ascending from same.

One of the handier things about flying an airplane from Ottumwa to Elgin, or London to Paris or Honolulu to San Francisco is the way your point of departure and destination stay put. So the guidance problem is only two-dimensional, involving longitude and latitude. In space, we must consider a third dimension, which cannot be plotted in the customary fashion. Objects in space are moving targets; the CM is going to be moving when we eventually must rendezvous with it, and our landing field on the moon is going to be moving—both at not-inconsiderable velocities. As a consequence, we must create some sort of a reference frame with three-dimensional coordinates and thereby specify a location where our target is not now, but will be when we arrive.

We are now in our twelfth revolution of the moon, and on the fourteenth, we will go for the surface. We have donned our pressure garments, helmet and gloves. We have crawled through the tunnel from the CM, activated Goldilocks in all departments, and here is Houston on the phone.

"Goldilocks, Houston. Have you finished landing-gear deploy?"

"Roger. Down and locked. We're ready for undocking."

"Okay, you're GO for P-47 in one minute and 30 seconds."

"Houston, Goldilocks here. We mark GO at one and 30. We're going to dial the verb 77 and stand by on the enters as we're released."

"Roger. Copy all that."

We feel a slight jar as the command module releases us. As soon as positive separation is confirmed, we execute a yaw left 60 degrees. Our pitch angle is up 90 degrees.

"Goldilocks, Houston. You're looking great. If you'll give us POO and data, we'll start your uplink."

They want our computer cleared and in the accept mode, so they can send us guidance for the next event on the program. We advise Houston our DSKY reads all balls—all zeros— and that Colossus's nephew is theirs. And welcome.

We maneuver Goldilocks around for a better look at our CM. Beautiful, but now which one is the orphan?

For the next 10 minutes, we keep the line hot with Houston repeating numbers to feed into the DSKY. They are all pertinent to our FDAI, our coming power descent, and abort procedure if something goes sour.

"Goldilocks, Houston. If you're ready, here are your gyro torque angles."

"Roger. Standing by."

"X is minus zero zero zero four five, Y is minus zero zero zero three five and Z is minus zero zero zero nine two."

We read back the numbers and afterward remove our helmet and gloves. Our command module is in formation with us, standing out against the black sky like a chaperone. Rather reluctantly, we give him a GO for greater separation. He performs a four-second burn on the CM service engine and dwindles into the distance.

Goldilocks is alone, and here comes the back side of the moon. Soon we will come swinging around the horn, and after a program-52 observation through our alignment telescope, we should be open for business on the moon.

Now we must make a final check on our inertial measurement unit (IMU). In the telescope, we observe intersecting crosshairs and a star friend known as Aldebaran. There is also a hairlike object across the viewing lens that looks like a dead spirochete, but which is officially known as an Archimedian spiral. There are two of them, and we will need them on the moon. At the moment, we are satisfied with fixing the vertical crosshair over Aldebaran. Mark! Next the horizontal line and another mark. The intersection of the planes defines the direction of the star.

We select old comrade Nunki for our second star and repeat the act. Upon that invisible, three-dimensional navigation grid we have now oriented our IMU. We have a place to call home in space.

Meanwhile, we're radar-locked on the command module. He's only 4.7 miles away—and hello there, anyone, if you want to call us try 296.8 mHz.

And apparently, our CM pilot already misses us.

"Goldilocks, Apollo . . . How about some exterior lights for me?"

"Okay, coming on. You lonesome?"

"I'll admit I have more room than I really need. Do you have my lights?"

"Sure do. You did a pretty burn, Apollo."

"Thanks. Your tracking light's on."

"Roger. We know it. It's our candle in the window. Just for you."

"Who else?"

Poor fellow. Around and round the moon he goes, all alone for all of tomorrow and a half.

And now, mother, here it is. In two minutes and 10 seconds, your boy starts down for the moon.

It would be a lot less complicated if we could just roll over and split-S down to the surface, but there are too many constantly changing variables in a lunar descent. As we burn fuel, our weight becomes less, and the descent engine must change thrust with it to maintain a desired state of equilibrium. Our center of gravity also changes all the way down, so the engine must be constantly tilted to keep thrust in direct line.

Our orbiting velocity is now 6,577 fps (approximately 4,000 mph). We are coasting at an altitude of 45,000 feet with our cabin windows facing down upon the lunar surface. In another half minute, when we receive a GO, we will be 250 miles down range from our landing target. When we retro-fire the descent engine, we will "fall" down in an arc at a gradually decreasing rate of descent. About 100 miles out and at 35,000 feet, we will rotate Goldilocks so that our windows will be up. Our reaction engines will gradually tip us over until we are nearly upright in relation to the surface of the moon. It is somewhat like lining up for an ILS approach.

At 10,000 feet, we are tilted 23 degrees heads-up and reach the end of the braking phase.

The next phase is like latching onto a glide slope, with our landing radar calling the plays. We tilt to 49 degrees upright and between 9,680 and 3,000 feet crank around farther to 57 degrees.

At 500 feet, we will be descending 27 fps, tilted 80 degrees, and thrusting downward with the descent engine at 2,800 pounds. At 200 feet, we will be vertically upright and our rate of descent cut to three fps.

During our descent, the LM pilot will read from our guidance computer, saying range to landing site, horizontal velocity and the angle at which our commander should look to spot our landing site. There is a scale etched on his window to help him establish this most important search, which is known as LPD angle. The scale is our ominous reminder that no man has ever landed here before. For a certainty, *never.*

We are number one in the pattern with no runways, approach lights or any other designators in sight. We may be operating an expensive collection of most sophisticated machinery, but it is still barnstorming, and for another certainty, during the final phase of our descent things are going to happen very fast.

"Goldilocks, Houston. You're GO for PDI in 35 seconds."

"Roger. Average G descent engine is on and the velocity light. Ten, nine, eight, seven, six, five . . . we have ignition. . . ."

"Throttle up to 26, Goldilocks. Looking good."

"Good . . . good."

Suddenly we realize we are experiencing the sensation of gravity gain. It is not much, but it is there and it is most pleasant. We are two minutes and 30 seconds into the descent. We are rolling around to windows up and already out of 35,000 feet.

"Apollo, Houston. RCS looks good. Electrics look good. Here's a five-minute hack coming up."

Five minutes already? It seems incredible that we have been so long in descent. Yet the moon glare seems greatly intensified, and man oh man, right there, right out the front porch window is the big girl herself. What sheer class!

"Goldilocks, Houston. Throttle down at six plus 22."

"Copy that. We can just now see the horizon. And we're getting considerable RCS activity. We're out of 19,000 feet."

"Goldilocks, Houston. Monitor descent fuel two."

"Roger, Looks good."

We are descending at 160 feet per second, which is slightly faster than programmed and coming up on eight minutes in descent. We just passed through 12,000 feet, according to our tape meter. From now until we reach the surface, we'll be talking so fast to each other that Houston will be lucky to squeeze a word in edgewise. Our commander admits he cannot resist looking out the window for our landing target, which is a smooth area near a high-ridged crater.

"I'm cheating, but think I have it." His voice does not betray any excitement. Much.

"Goldilocks, Houston. You're looking good at nine minutes."

"Roger. We are passing through 3,500, coming down 99 feet per second. Got 15 percent fuel."

"Goldilocks, Houston. You're GO for landing."

"I'd like to see our LPD to the right a little."

"Roger. Try 40 degrees. You're at 2,000 feet. We have 94 seconds descent time left."

"Okay. Let's move forward a little bit."

"38 degrees . . . 36 degrees . . . 1,200 feet . . . looking good and coming down at 30. Got 14 percent fuel."

"Keep talking."

"32 degrees . . . 800 feet . . . 35 degrees."

"Look at that crater. Right where it's supposed to be. Beautiful."

"Two fifty-seven feet coming down at five. Hey, Houston, you sure this isn't a simulation?"

When you are just a wee bit scared, ever notice how your mouth turns out limp humor?

"Ten percent fuel . . . 190 feet. Come on down, boss. You're doing fine. 180 feet and nine percent. We should be kicking up some dust soon. Seventy feet looking good. Forty-two feet coming down in three. You do nice work. Coming down in two. Okay . . . start the clock. Forty feet still coming down in two. Watch the dust. Stay in there, boss. You've got plenty of gas."

Houston calls out 30 seconds of descent propellant remaining, which is not bad eyeballing from 240,000 miles away.

"I can't see anything but dust. This is strictly IFR."

"Eighteen feet coming down in two. You have it made. Contact light."

Two blue lights come on, indicating the probes that extend from the four legs of our landing gear have touched the surface.

"Okay. Ignition off. Cycle the main shutoff valve."

"Done. You get those RCS switches?"

"Closed."

We are on the moon.

But how long? Before we even have a chance to celebrate, we must check our ascent situation. We quickly check fuel, water and helium. All of our ascent expendables are at required quantities.

"Houston, Goldilocks has landed . . . a little on the rough

side account of the dust."

"Congratulations. At least there was no one there to give you a wave-off."

For the next 20 minutes, we are busy performing our stay/no stay check list. We work a program-68 landing confirmation, update and align our guidance system and load it for ascent target. We align our IMU and do a gyro calibration. Finally, we install the LM's window shades and doff our helmets and gloves. We have a GO from Houston on everything.

We are on the moon to stay awhile.

Our first outside chore, after accustoming ourselves to feeling like Batman, is to lay out ALSEP (apollo lunar surface experiments package), a product of the Bendix conglomerate. The unit will record moonquakes and meteoroid impacts, measure neutral particles at the moon's surface, record the energy and direction of charged particles and measure heat-flow characteristics of the moon's crust to a depth of 10 feet. From a previous ALSEP package left by Apollo 12 and an earlier type unit left behind by Apollo 11, we have so far learned that the moon is very elderly and very sedate. Somehow, this does not come as a complete surprise to many Americans. When pressed for more remarkable revelations, the rather thin-lipped response is that the information acquired is "so enormous it is still being evaluated." Amen

Next, we are flight-planned to perform certain lunar field geological activities that involve drilling the surface for samples; collecting surface material (dust and rocks); and studying as well as describing various patterns, shapes and distribution of the local scene. Much of this information is aimed at future moon landing-site selection and sundry problems involving the movement of men and roving vehicles. We are pathfinders, marking the way for reasons that may possibly not be realized in our lifetimes. Only another era will reveal whether a colony on the moon itself or squadrons of orbiting space stations will provide the domination of one self-appointed group of earthlings over the seas of tomorrow.

During our EVA (extra vehicular-activity), we will erect the solar-wind equipment to measure the constant stream of high-speed, electrically charged particles given off by the sun; take some high-resolution photographs; and set up a laser reflector that will bounce additional beams back to a something on earth called a "high-peak Q-switched gigawatt system." It is designed to live up to its imposing title by measuring the distance between earth and the moon with an accuracy of six inches. It will also determine the fluctuation of earth's rotation rate around its axis, the extent of its wobble, and so precisely fix the latitude and longitude of various earth stations that the theory of continental drift, which some scientists claim may lead to the eventual disintegration of these land masses, may at last be confirmed. (More cynical observers may wish to point out that we have already developed techniques necessary to disintegrate continents without any assistance from drift.)

With these thoughts competing for attention with the

customary ceremonial planting and saluting of the American flag, let us now climb back into the protective womb of Goldilocks, where certain of NASA's triumphs become more readily apparent. For example, all aircraft manufacturers would do well to permanently jettison their wretched cockpit lighting systems and make a direct steal from Goldilocks. Most of the LM's instrument displays and all panel placards are integrally lighted by white electroluminescent lamps with overlays. The numeric displays show green or white illuminated digits on a nonilluminated pointer traveling over an illuminated background. The effect is guaranteed to make any aircraft pilot who has endured our standard gaslight-era cockpit illumination want to sting his hands in applause.

When our command service module passes overhead, we are going to light the fire under Goldilocks and start the most important and complex chase of our lives. If all goes as it should, we will rise vertically for a short distance, then pitch over for a more horizontal climb to orbit and final rendezvous with the CSM. Right now, we are most seriously dedicated to working a program 57 for the alignment of our guidance computer. Our rendezvous target is moving very fast, and he is also very small. We are 20 minutes from ascent-engine ignition. It is time to once more don helmet and gloves.

"Goldilocks, Houston. Mark three minutes from liftoff."

"Roger. Checklist is complete. Standing by for TIG minus two . . . mark one minute . . . master arm is on."

"Twenty seconds . . ."

Program 12 starts on the computer as our commander pushes the proceed button. DSKY says continue.

". . . four . . . three . . . two . . . one . . ." Our commander pushes the engine-start button and we have liftoff. The ascent engine is so quiet it is hard to believe its true power.

"Goldilocks, Houston. Ignition guidance looks good."

"Pitchover looks good and we're 316 feet above the lunar surface. We're on our way."

At one minute in ascent, we yaw right 20 degrees and our velocity increases to 264 feet per second. We are wobbling slightly, and while we are still worrying about it, we pass through 9,000 feet. Not bad for two minutes.

Except for this strange gyrating, it is a smooth ride. We ask Houston about the wobble. While they are still searching for

an explanation, everything steadies out and we're on a beautiful trip in a fast elevator.

Seven minutes from liftoff, we're at 57,000 feet and whizzing along in moon orbit at a comfortable 3,409 mph, which should convince even her critics that Goldilocks is quite a flying machine.

"Engine arm off. Okay?"

"Okay."

"Houston, Goldilocks. We've had shutdown."

"Copy that. But you took a four-second overburn, which puts you off speed 32 feet per second. Nothing serious, though."

"We were preoccupied with a master alarm light flashing. We thought that was serious. It's off now."

"Everything looks fine. All three data sources show GO."

Fair enough. If our memories are correct, it seems that every Apollo flight crew spent at least a few anxious moments fretting over inexplicable warning lights, and our older flying brothers still speak of a period during World War II when it was standard for all engine-fire warning lights to flash on just at takeoff.

Now we must concern ourselves utterly with closing the range of 198 miles between Apollo and ourselves until we are only a few feet apart.

The techniques for this rendezvous were developed during the Gemini missions and the earlier Apollos. We are confident they will work because such pioneers as Borman, Schirra, Armstrong, Stafford, Cunningham, Lovell, Conrad, Gordon, Scott and Aldrin damn well went out into space and *proved* they would work.

We are inserted in a lower orbit than our command module, so we are traveling around the moon faster. Since we have the inside track in this race, we deliberately waited until the CSM was ahead of us about 300 miles. Now we are in hot pursuit and will do a sequence of catch-up maneuvers so that ultimately our speed and altitude will match Apollo's.

Rendezvous with our CSM is one of the most critical and demanding phases of the entire mission. It is pure orbital geometry, which begins 58 minutes after liftoff from the moon and is called concentric sequence initiation, which fortunately has a handy abbreviation—CSI. During it, we will add 34 mph to our speed, raise our altitude to 52 miles, and sneak

up on the CSM until we are only 33 miles behind. Half an orbit later, we perform a second phase maneuver called constant differential height (CDH), and if you flunked geometry be content to believe the altitude difference between the CSM and ourselves will remain constant until the next act.

This is a very busy game of catch-up, with our commander operating the radar and DSKY frothing at the mouth with thrusting programs. In the midst of our rather profound calculations, we are having a problem as simple as the first cabin airplane: Our windows persist in fogging over, which is an annoyance we can fix with our heaters, but no amount of antenna switching will improve our vile communications with the CSM. We have apparently come upon one of our spacecraft equipment items manufactured by the lowest bidder. We should soon have visual contact with the CSM, and the temptation is to holler out the window for more reliable results. "Hey! Wait for us!" And here is Houston with their own communication problems.

"Apollo, Houston. We're going to do a handover now from Honeysuckle to Madrid."

"Okay. If that's the way the world turns, I can't stop it."

Earth, 240,000 miles distant, can talk to Apollo, but we who are looking him right in the eye cannot.

"Houston, Goldilocks. You have any ideas on the ground why our communication with the Apollo is so bad?"

"Are you on the starboard antenna?"

"Absolutely. Lower antenna. Lower right." To hell with it. We are in the moon's shadow—our nighttime. We have our high-intensity flashing lights turned on. Let's get this important terminal-phase burn initiated and talk to our CM pilot face to face. We are presently 74 nm behind and closing at about 100 mph.

Our burn is short and sweet of result. Apollo's transponder is belting out a strong signal for our radar, and there are his lights. In a few minutes, we will be out in the sunlight and can move in for docking.

"Hey, Apollo. I have you now at 1.2 miles. Do you copy?"

"Loud and clear, for a change. And I concur on distance. How can you look so good when you're so ugly?"

Apollo's and our own optical alignment sights are now providing us with range and closing-rate cues, and we constantly exchange updated data with each other. There is a tee-cross target on our LM and a corresponding illuminated image in the CM's sight. We are the passive vehicle now and he is maneuvering on us. At least he is one up on us in docking experience, having been at the controls when Goldilocks was originally pulled from the lunar adaptor. Our calculations "mirror" those made by the other crew, which is our best guarantee we won't scratch any fenders.

Apollo is firing his reaction engine very conservatively—moving only a little each time.

"Hey, Goldilocks. I have you at half a mile."

"Yes. Okay. And forward down."

"Now eight feet a second at 1,200 feet."

"Roger. Concur."

"Five feet a second."

"You're looking good. Keep coming."

We are slightly below and ahead of the CM. A further touch of the CM's reaction engines and we slip into line. Now a few feet at a time, with Goldilocks in the role of a coy maiden. Then, at last, a faint metallic clinking as Apollo's probe captures us. Almost instantly, we are locked on and home.

"Houston, Apollo. Goldilocks is mine."

"Roger. We copy. But not for long."

We have crawled through the docking tunnel, brought our moon loot with us, and secured the hatch. We have removed our pressure garments and now the time has come to part with Goldilocks. We are tempted to break out in a chorus of *Auld Lang Syne* and blow her a farewell kiss.

Goldilocks's final departure is triggered from Houston, and our noses are pressed against the window as she speeds away for impact on the moon. She was—at least while we were in her company—a most honest and faithful girl. As they discovered on Apollo 13, she could also be an angel of rescue. Thank you, Grumman Aircraft.

On our forty-seventh revolution of the moon, passing around the back side, we perform a two-minute 15-second burn of our service engine to put us on course for home. TEI, trans earth injection, has begun.

Just before the burn, we were orbiting at 2,000 mph. We are frankly astonished at how relatively little additional velocity is needed to take us away from the moon and point us toward home. Now our DSKY tells us we are already 940 miles out and the familiar features on the moon are reducing in size very rapidly. There is the Sea of Crises and the Sea of Fertility below it. And beyond, just coming up over the moon horizon, the Sea of Serenity.

"Apollo, Houston. You're really moving out fast."

"Well the moon is a nice place to visit . . . but we wouldn't want——"

"We know. Never mind. We know."

Because of the lesser gravitational influence of the moon, our speed after the burn is decreasing much slower than after our departure from earth. Our altitude is 1,865 miles and our velocity over 3,500 mph.

We dump our urine tank, cycle the fans and chlorinate our water. Houston reports that our burn was so accurate it varied only one-tenth of a foot per second from the predicted velocity gain, so we will probably need only very small mid-course corrections.

The moon is the size of a basketball held at arm's length. We start our barbecuing for passive thermal control. Now, for some reason, home-bound, the faint "thunk-thunk" of our reaction engines as we commence revolving seems almost soothing.

Our 160-page flight plan now calls for re-entry preparations. We are 240 hours into the mission, 32,696 miles from earth, and picking up speed fast because of gravity—beautiful,

beautiful earth gravity. The diameter of the earth is now about 15 times that of the moon. Down there somewhere, the Navy awaits us below an 1,800-foot ceiling with waves of eight feet and a southeast wind of 15 knots. It seems strange to be aware of weather again.

We have taken optical sights on Alpheratz, Mirfak and Venus, using the earth's horizon as our marking platform, but the stars were so dim and there was so much stray light in the sextant only our sight on Venus could be called reliable. We also hàd some mysterious high-gain antenna troubles, but they appear to have solved themselves and we are back in communication with Houston.

"Houston, Apollo. Does it look like we are going to have an update to our entry pad after that last course-correction burn?"

"Apollo, Houston. They want a little more tracking on you. It looks good now. Some of the times may change a second or two. As soon as we get a good track, we'll send it on up."

"Okay. Not pushing. Just curious."

Yes indeed. Re-entering the earth's atmoshere is a rather delicate performance, with timing being as important as one trapeze artist catching another, and some little law-of-survival policeman keeps reminding us we are performing without a net. Our re-entry corridor is a mere two degrees wide, because if we penetrate at too shallow an angle we will bounce back out of the atmosphere like a stone skipping across the water. If that should happen, it might be some time before anyone saw us again. If the angle is too steep, then our deceleration would be so great we would disintegrate or at the last pull G forces beyond our tolerance. So we have to hit at just the right angle, which is approximately six degrees—halfway between too shallow and too steep. Our small reaction engine burns thus far have established us in this precise trajectory, but we are in no shape to relax until we say hello to the Navy after splashdown.

Our other problem will be solved by Colossus' nephew. Just penetrating the earth's atmosphere and landing anywhere at random might prove to be politically embarrassing or extremely uncomfortable, or both. Our landing area, called our "footprint," is about 2,500 miles long and 300 wide, which is still a lot of ocean to be sailing in a vessel that can hardly be regarded as a cup winner. So we must transform our command module into a real flying machine by controlling its aerodynamic lift capability with our small reaction engines. If you will bear in mind that *anything* will "fly" if it is going 25,000-plus mph, then it will be apparent that by rolling our CM about an axis parallel to our direction of motion, we can effectively control our lateral and horizontal movements.

We are fortunate in having a remarkable Apollo history to reassure our splashing down on target. All flights have been within three miles of the predicted impact point, with the record held by Apollo 13 at 800 yards.

We are still 20,000 miles out from earth, but our velocity has picked up to 8,000 mph, so our leisure moments are definitely over for this voyage. We peg splashdown at two hours 17 minutes 40 seconds. Houston has told us that not only will an aircraft carrier be waiting to greet us but three helicopters, two E-1s and two C-130s with pararescue units aboard. We suddenly realized we were not as yet earth-oriented when Houston gave us a plain, old-fashioned altimeter setting. Like 29.96.

We are making final guidance alignments and running down the checklists for re-entry.

"Houston, Apollo. Are we GO for pyro arm and for command-module RCS prep?"

"Stand by one. . . . .Affirmative. You are GO for both."

We have finished program 61, which called for us to pitch over manually until we had an earth horizon against the window marking of 31.7 degrees. Now we are setting up program 62, which will separate us from the service module 18 minutes before we re-enter the earth's atmosphere. There will be more fireworks; this time, the detonators will activate guillotines to cut the connecting wires and tubing and the three tension ties between the service and command modules. (While we have never counted the number of explosive devices in a Saturn V stack, there are certainly enough to fight a small war.)

Our velocity is building up fast now, showing better than 19,000 fps or 13,000 mph. Sailors used to make "all shipshape and Bristol fashion" before making port. We are doing the same thing, with all our gear stowed and all lines taut—including those to our sense of anticipation.

We perform a yaw maneuver to 45 degrees out of plane. It is time to dispense with our service module. Just below our DSKY are two guarded switches marked CM/SM SEP. Our commander now lifts the guards and pushes up on the spring-loaded toggles. Instantly, we hear a distant report and feel a slight shudder. Our service module is gone. We are all that is left of the once mighty Saturn V.

"Apollo, Houston. We'd like POO and accept for your final-state vector."

We clear the brain of Colossus' nephew. "Okay. You've got it."

We initialize our entry monitor and set it on entry. Our homing pennant is really streaming now. Only 3,688 nm out and bowling along with a bone in our teeth at better than 17,000 mph.

Less than six minutes to go for entry with a distance to our "window" of 3,223 nm. Under full sail at 22,000 mph.

It is still night on earth as it is in heaven.

Two minutes from re-entry. Our eyes alternate between the flashing green numbers on the DSKY and the monitor. We wonder why we have never before noticed how the digits flashed by our computer lean, as if they were going someplace in a hurry. We are waiting for the little amber .05-G light, which will announce the first tissue of the earth's atmosphere. It should light up at about 400,000 feet altitude.

"Apollo, Houston. Coming up on blackout. We'll see you in about 18 minutes. And welcome home."

"Pleased to almost be there."

Our velocity at entry will be 24,692 mph. One of the inconvenient aspects of such speed is that the earth-air cannot get out of our way fast enough. Our plug-shaped entry end compresses the air, which consequently generates so much heat that ionization takes place. The result is total lack of radio communication. Now, with only a few seconds left to think about it, we wonder what we would say to anyone even if we could.

There is our .05-G light. We have slipped through 400,000 feet and are watching the scribed line on the entry-monitor paper. Our guidance control is set to program 64, so our reaction engines should start thrusting as required to keep us in the descent corridor. Not too shallow and not too steep, but it will not be a consistent straight-line descent.

One minute and 18 seconds into re-entry. We are slowing 700 fps and there is the devil's own breath outside our windows. We are feeling a rapid increase in the G load—up to four now. It will peak out at a maximum of 6.5, if all goes as programmed. A hell of a fine conflagration outside.

Our backs are to the earth's surface, our feet aimed at that element we can again refer to as the sky. The scribe line shows we are "lofting," having descended as low as 200,000 feet and now rising up to 290,000 feet, a wavelike, roller-coaster pattern that we expected.

The ghostly green leaning numbers are whirring along the bottom of our entry monitor. We are 600 miles down range of target.

We glance at each other. Automatic guidance ends just after seven minutes in descent. It seems to be a long, long seven minutes. We are three men in a tub again. Inert, watchful men. You do not move much at better than six G.

Everything reads right on the entry monitor. We keep telling ourselves how very right it is and take comfort in the presence of each other. How did John Glenn, Scott Carpenter, Wally Schirra and Gordon Cooper feel during these strangely unreal minutes? They were each alone.

Coming up on eight minutes six seconds. Our velocity has greatly decreased as the atmosphere thickens. Our altimeter reads 55,000 feet and the fire outside is almost gone. We should have drogue-chute deployment momentarily. Our commander's hand strays toward the three float-bag switches. Since he won't hit them until splashdown, we assume he just wants to make sure they are still there. His hand goes to the drogue-deploy backup switch just in case, but it proves unnecessary.

We have automatic drogue-chute deployment as advertised at 23,800 feet. .

Ten thousand five hundred feet. That same protective hand reaches for the main deploy switch, but automation is still on

the job. A low-altitude baroswitch closes, dumps our drogue, and one second later we are swinging beneath our three main chutes.

We dump our remaining reaction-engine fuel, which is a possible hazard to the swimmers who will first open the hatch to our old, familiar world.

We have about four minutes to wait. We can see the chutes above us. They are incredibly beautiful.

"Houston, this is Recovery. Tally-ho, I have a visual. Apollo is bearing 120 from me at 7,000 and looking good."

We are now about to experience one of the factors in the Apollo program that is still geared to pure luck. With scientists coming out of the very concrete at NASA, no one is able to predict whether we will make a soft or a hard landing. Both varieties have been the lot of previous Apollo crews. It is also a matter of sheer chance whether we will float apex up—which does not make for bad yachting—or apex down—which, since we are returned to the land of up and down, can be miserable for the several minutes it takes our automatically inflated flotation bags to right us.

Our luck throughout this mission has been so sweet and gentle and the glitches so minimal we could hardly expect it to last. Expectations confirmed. We swing in our chute risers at just the wrong time at just the right angle and slam against the face of the only really big wave in a thousand miles.

As soon as we push our eyeballs back in place, we hit the main release switch and get rid of our chutes. Thereupon, we immediately and quite involuntarily roll over into a "stable II" position, which is apex down. And while we are hanging from our straps in this thoroughly undignified attitude with the blood flushing our faces, we try to divert ourselves by supposing what we might say to the President of the United States when he makes his customary telephone call of congratulations. We would like to tell him and also Director Tom Paine of NASA that after beholding the heavens from the windows of our magnificent Apollo it seems strangely appropriate we should suddenly be staring straight down at the dark and mysterious depths of earth's greatest ocean.

Would such an observation make sense to them or would they fear we had lost our reason in space? Better not. Because then we would have to confess that what seems truly incongruous when viewed through the windows of our marvelous flying machine is how the entire Apollo program, which alone revived our morbid international prestige, contributed most importantly to our national security and created a bonanza of benefits for our society—still costs less than the women of America spend each year on cosmetics. We might at least ask them if they also thought it odd.

PART IX

Epilogue

Flying Magazine's most ambitious issue in all of its years thus far was published in September of 1977. This was a 432-page celebration of the magazine's half-century anniversary and the fiftieth anniversary of Lindbergh's flight. Included among the writers were William F. Buckley, Jr., Richard Bach, Ernest K. Gann, Edward H. Sims, Martin Caidin, Robert J. Serling and many other distinguished non-staff contributors. The anniversary issue turned out to be a gala celebration of flight itself, with a strong emphasis on the historical development of the world's aviation industry.

The fiftieth anniversary year was therefore in many ways one of looking back, of rejuvenating memories, often on a very personal level for members of the staff.

For instance, Editor Robert B. Parke had an opportunity to recall a dramatic part of his flying past in two ways. He met an old German fighter pilot who had been among those battling our bombers and fighters in the skies over Hitler's Fortress Europe. Parke had flown some of those bombers—he captained B-17s—and had very likely been one of this German pilot's targets. Now, more than 30 years later, they were sipping drinks together and calmly rehashing their attempts to kill each other. And, as part of the same story, Bob also went to Harlingen, Texas, where a B-17 was made available for him to fly. We've included that story here to convey something of the nostalgic atmosphere that was part of the fiftieth anniversary project.

Our anniversary issue was but one event among many commemorations of the Lindbergh flight during the year. A special Lindbergh fund was established; a replica of *The Spirit of St. Louis* flew over Long Island imitating Lindy's departure on his Paris journey; millions of words were written and uttered honoring the great flier. Lindbergh, ever retiring, might well have been embarrassed by it all—had he lived to witness it. But he had died in August of 1974.

He was buried in simple dignity, dressed by his request in work clothes as one last gesture of the spirit in which he lived. Renowned as a pilot and a strong influence on the growth of aviation, Lindbergh had become much more than that. He was widely respected as an author, philospher, scientist, naturalist and as a man increasingly devoted to the preservation of our natural environment. A man whose name was synonymous with Progress, he viewed true progress soberly as part of a larger natural context. He sought a synthesis between nature and machine. In his later years, he believed fervently in the oneness of all things.

Flying marked his passing, also with simplicity, with a moving memorial written by pilot and author Len Morgan, "Lindbergh Is Gone."

An era is gone. Generated by one man's "outstanding feat of the ages," an ever-quickening transformation swept over the world as aviation made all places swiftly accessible. Now, even as "the Lindbergh era" has slipped away, a new era has come upon us—one hopes for the good.

Will aviation still be at the core of things? We expect so, and *Flying* will be there to tell the world about it. Even as our fiftieth anniversary issue was being put to bed, the next regular issue was well into preparation. *Flying's* second half-century was under way.

B-17

What can you say to an old friend when the memories are still painful?

BY ROBERT B. PARKE

I MET A MAN the other day who had tried to kill me. I was only going to drop some bombs on his hometown. I was flying a B-17 at the time, and he was in a Focke-Wulf 190. When I spotted him, he was barreling out of the sun at 11 o'clock and a bit high, in a lovely sweeping turn with the four wing guns winking as pink as Christmas-tree lights. I called bandits and his bearing, and almost immediately the top turret and ball were pounding away futilely, and I could see the tracers from .50s on other aircraft in the formation trying to get an angle. He came straight at us and suddenly half-rolled overhead, diving inverted in a vast arc, then vanished in some scattered cloud below. It was over in a few seconds. He had missed, and so had we.

Now he and I are having lunch. The splendid restaurant is in a largely rebuilt part of a bustling city of the Ruhr. The schnapps is deeply chilled, and the beer is rich and dark. He is managing director of a large German aviation firm, and we are talking about his airplanes, his war, our memories. It's as though we had attended the same university but had studied different specialties, in another life.

I bemoan how painfully slow the B-17 was, making me feel sluggish and helpless as the fighters danced around us, knifing in while we plodded dutifully on. He reflects on how throat-tightening it was to spot those awful waves of heavies coming

on relentlessly and knowing that sooner or later he must find an opening and strike, lining up carefully, gripping the stick trigger, darting in close, all the while feeling naked and exposed to the avalanche of bullets from a dozen guns in each airplane.

There is fractional relief as we leave the fighters behind and enter the dirty cloud of antiaircraft smoke. First there is the faint smell of gunpowder, and then a hard, flat exhalation, exposing ahead, as if on a transparent screen, a Rorschach-test splotch of black smoke in the clean, thin air. Then there are two more bursts just ahead and we sail through the already-dissipating puffs. Suddenly it's all around us, some bursts close enough to jolt the airplane. Miraculously, two full minutes go by and nobody in our formation has been hit. The tempo increases yet still no one is touched. It's as though the invisible projectiles are carefully being launched from five or six miles below to explode into clumps of shrapnel that will come as close as possible to the airplanes without hitting them.

Then, up ahead, two airplanes are gone—only brief flashes and debris where they had been. The disembodied tail section of one airplane careens into another and takes it down too. Now a half-dozen B-17s are in trouble, engines smoking. The formation is suddenly a shambles, and the colonel in the lead plane is calling on the command frequency to fill the holes and tighten up the group. The lead bombardier calls Initial Point, and the remaining airplanes strain to close the formation for the bomb run. A few aircraft with feathered engines or severe structural damage dump their bombs and flee to an almost certain and uneven battle with the fighters.

It has been four and a half minutes since we entered the target area. Now, steady on the run, doors open, the bombs finally are released and we start an agonizingly slow turn toward home. Two more planes are hit on the way out, and they slowly tumble into awkward spins.

The fighters watch us from above, circling lazily, well clear of the antiaircraft fire. As we break out, they form gracefully in trail and start sailing in on us, but something distracts them. They quickly break off and climb sharply. A handful of -51s come charging up, clearing their guns. We cruise on by, leaving the fighters in an angry tangle behind. It's three hours back to base.

It should be easy going now. We've taken some hits in the tail section, and one crewman has a superficial flak wound in the leg. The copilot is flying, and the formation has loosened up. We're all beginning to relax when the tail gunner shouts, "Bandits three o'clock!" into the intercom. Sure enough, two specks on the horizon gradually become airplanes bearing down on us. The waist gunner lets go a few rounds, though the airplanes are still far out of range. Then, to our relief, the two fighters swing parallel to us and flip up on their wingtips, showing us the elegant twin booms of the P-38. They sidle up alongside, being careful not to let their guns bear on us lest some anxious gunner mistake their intentions. Dropping half flaps, they cruise along until their fuel runs low, and then politely rock their wings and go slipping off, rolling, until they are out of sight.

The Channel slides safely into view, the letdown starts. Oxygen masks slip off, and the crew begins to move about. The formation lumbers across the airport, peeling off in steep 180-degree approaches. Finally, ponderously, the wheels touch down. As we taxi in, the crewmen are congratulating each other, and someone lets fly a celebratory shot with the flare pistol, for this is our last mission. The next day, we climb on the tailgate of a halftrack and ride out.

Having spent almost 1,000 hours of my life in a B-17, I never cared to fly one again.

Until one autumn day in Harlingen, Texas 30 years later. The Confederate Air Force, that group of dedicated war buffs, has resurrected a tattered B-17, turretless and with a hysterec-tomy-like scar down its flank to strengthen the area where a gaping cargo door has been installed. It's rumored that the weary bomber spent its postwar days carrying pipe for an oil-drilling operation in South America. It squats wearily on the ramp beside a B-29, a couple of -51s and a P-38. Close by are other World War II aircraft, preserved and revered by the ubiquitous colonels of the CAF.

I find myself by the hatch beneath the cockpit area and reach for the handhold, surprised that I can still swing up onto the catwalk beneath the pilot's seat. I hoist myself into the left seat while J. K. West, who regularly flies the -17 in the CAF air shows, waits in the right. As I settle in place and look out over the familiar humps of engine on the broad wing, a young kid appears from somewhere to assume the stance of a crew chief by the number-one engine. He smiles encouragingly and holds up his hand, circling one finger.

Start one. I fumble through a half-remembered procedure, and the propeller turns ponderously as the engine, coughing and farting smoke, catches, rumbling up to 1,200 rpm. Two, three, four. Brakes off, tailwheel unlocked, and I taxi out to take off. Checklist complete, line up and ease on power up to 45 inches. There it is—that lovely, reassuring sound of four Wright R-1820s beating in unison. Lift off at 90, go to 110, climb out to 2,500 feet and just fly around for a while.

It's a great airplane—heavy on the controls and needing lots of rudder but maneuverable as a Cub and reeking of strength and reliability. The kid is still there when we taxi in, giving hand signals for parking and shutting down. As we climb out and walk to the officers' club, he follows at a respectful distance. I suppose he'll get his chance.

For me, it has been a stirring experience—like meeting a dear old friend. But we didn't have all that much to say to each other.

My German acquaintance is now talking about the years after the war, when building and flying airplanes was not allowed. He got some more schooling, built and flew hang gliders and, to keep his hand in, secretly constructed his own airplane in his basement. He'd roll it out, attach the wings and fly from a field in back of his house on moonlit evenings. He joined his present firm, did some test-piloting and aircraft designing, and patented some of the features he'd developed on that airplane he built in his basement. Now that we're leaving and shaking hands and making arrangements to meet again, he speaks for both of us: "It is much better this way."

He was the hero of an era, and, as the years passed,
he became an example to us all.

Lindbergh Is Gone

BY LEN MORGAN

THE FLIGHT ITSELF was a spectacular success and worth all the tumultuous applause, yet it was but one of scores of notable flights completed in the 1920s, many of which called for equal risk and daring. Charles Lindbergh was not the first man or even the fiftieth to fly the Atlantic. Had another pilot flown the identical route the same year, his feat would doubtless have amounted to little else than another record for the books. After the furor had died away, the public would have sat back and waited. What new thrill will these brave aviators dish up for us next?

The times and the man: These are the keys to it. It was a different life in 1927, another world. Wrenched from isolationism by the Great War and suddenly aware of their inevitable involvement in future world affairs, Americans in their victory were unsure of the world and of themselves. The great adventure lay behind; it was history. There was already reason to wonder if World War I had been the war to end them all, and great questions lay ahead. With the casting aside of old ideas, all values were questioned and it was a time ripe for something or someone new, dynamic, positive.

The immeasurable impact of Lindbergh has never been better explained than by Frederick Lewis Allen in *Only Yesterday:*

"For years, the American people had been spiritually starved. They had seen their early ideals and illusions and hopes one by one worn away by the corrosive influence of events and ideas—by the disappointing aftermath of the war, by scientific doctrines and psychological theories which undermined their religion and ridiculed their sentimental notions, by the spectacle of graft in politics and crime on the city streets, and finally by their recent newspaper diet of smut and murder. Romance, chivalry and self-dedication had been debunked; the heroes of history had been revealed as people with queer complexes . . .

"Something that people needed, if they were to live at peace with themselves and with the world, was missing from their lives. And all at once Lindbergh provided it. Romance, chivalry, self-dedication—here they were, embodied in a modern Galahad for a generation which had forsworn Galahads."

But you had to be around then to feel it. In the years following, he was aviation's best friend. He remained very much on the scene, ignoring a thousand chances to otherwise cash in on his popularity. He was everywhere, flying new equipment, laying out airline routes, making goodwill tours, writing, speaking, lending his fame and talents to the promotion of commercial and private flying. There was never a better ambassador. Here was no wild daredevil but a reserved, soft spoken, intelligent young man who made good sense, even when talking about future trans-oceanic air travel. Lindbergh's flight captured the world's attention, and Lindbergh had won the world's heart. The effect on aviation was astonishing.

Then, at the height of it all, the horror of the kidnap murder of his baby son, leading to an investigation and trial that may well have been the most widely publicized of all crimes involving private citizens. We can only wonder how the rest of his story would have read but for this. When it was finally over, the Lindberghs packed up and moved to England. Never again was he to be owned by the press or the people.

Since then, we have known little more than the bare facts of this great man's life. The work in France on a mechanical heart; the tours of Axis war plants; the prewar speeches that set well with few of his countrymen; the resigned Air Corps commission; the war years, a contribution to which was obviously minimized in his own *Wartime Journals*; advisory work with Pan Am; his book in 1953 and then the movie, which was very nearly as good; speaking out against building the Concorde and our own SST. Aviation, he said in recent years, was of real value only to the extent that it improved the quality of human life. The man who had showed what the airplane could do spent his life showing what it should do.

It was reported that in later years, Lindbergh was able to leave his office, walk through Grand Central Terminal and catch a train home to Connecticut without even being recognized.

His fellow airmen understood and cheered his success at achieving anonymity. Charles Lindbergh had hundreds of thousands of fellow airmen, few of whom ever saw him in the flesh. To the armies of boys who followed him into the air , 5, 10 or 15 years after that hop to le Bourget, he was always "one of us." And he was. Five years ago, he was awarded honorary membership in the Air Line Pilots Association, the second such membership he ever accepted. The event was informal and free of publicity, at his request. In his brief remarks, he said, "I have always regarded my profession as that of a pilot, and while it goes back many years, as a transport pilot." The occasion honored the country's airline pilots, not him. Harold Gray, chairman of the board of Pan American, had this to say that memorable evening: "I don't give a damn about the fact that Charles Lindbergh flew the Atlantic solo. It was not the feat, gentlemen. It was, and is, the man!"

A month or two back, papers carried word that he was in a New York hospital and had forbidden staff members to say anything about his illness. Who among us did not grin at that and admire the hero who was still very much his own man? Then the final report on the news and splashed across half the front page; who among us did not sit back and think about the man?

At his request, he was flown to his Hawaiian retreat to finish his life. Hana is a remote place, far from tourist trails, a place of incredible peace and beauty, a fitting place for an eagle to live and die. He was buried in ordinary work clothes in a eucalyptus coffin built by local ranch hands; a quiet service in a tiny church with a handful present. All of it was over and done within hours. And so completely in character.

That Lindbergh is gone comes as a shock, much more of one than we would have expected. The Lone Eagle. Lucky Lindy. Every boy's hero for a generation. A nation's adored son, held high for all to see as an example of American youth. And, in later years, the strong, quiet, lonely symbol whose presence was often felt, if seldom seen, in his profession and far beyond it. Gone. It hardly seems possible.

You had to be around when he made his name and a witness to what followed to understand it. The men who decide what goes on newspaper front pages were there; they remember. You have to understand how it was to believe that a single flight, or an individual exploit of any kind, could electrify the world and transform an unknown 25-year-old into an international idol.

Three small things since then sum it up for me, a lifelong admirer of Charles Lindbergh. This, in a letter from a lady who watched the legend unfold from the beginning to the end: "So the Lone Eagle died as he lived, with dignity." This from a retired airline pilot: "My whole career sort of revolved around him as a central figure. I never met him, but I will surely miss him."

And this: In the long lobby of Mexico City's terminal are displayed six bronze busts—five of Mexican air pioneers, the other of Lindbergh. The American figure is flanked by two white floral displays. Yesterday, as the crowds rushed past to catch their planes, it stood there almost ignored. One small, graying man in the uniform of an Air France captain was standing before it, hat in hand, reading the inscription.

———————◄•●•►———————

INDEX